Glycopeptide Antibiotics

DRUGS AND THE PHARMACEUTICAL SCIENCES

A Series of Textbooks and Monographs

edited by

James Swarbrick

Applied Analytical Industries, Inc.
Wilmington, North Carolina

Glycopeptide Antibiotics

edited by

Ramakrishnan Nagarajan

Lilly Research Laboratories
Eli Lilly and Company
Indianapolis, Indiana

CRC Press

Taylor & Francis Group

Boca Raton London New York

CRC Press is an imprint of the
Taylor & Francis Group, an informa business

Reprinted 2010 by CRC Press

CRC Press
6000 Broken Sound Parkway, NW
Suite 300, Boca Raton, FL 33487
270 Madison Avenue
New York, NY 10016
2 Park Square, Milton Park
Abingdon, Oxon OX14 4RN, UK

Library of Congress Cataloging-in-Publication Data

Glycopeptide antibiotics / edited by Ramakrishnan Nagarajan.
    p.   cm. — (Drugs and the pharmaceutical sciences ; 63)
  Includes bibliographical references and index.
  ISBN 0-8247-9193-2 (alk. paper)
  1. Glycopeptide antibiotics. I. Nagarajan, Ramakrishnan.
  II. Series: Drugs and the pharmaceutical sciences ; v. 63.
  [DNLM: 1. Antibiotics, Glycopeptide. 2. Vancomycin. W1 DR893B
  v. 63 1994 / QV 350 G568 1994]
  RM409.G58 1994
  615'.329—dc20
  DNLM/DLC
  for Library of Congress
                                           94-585
                                           CIP

The publisher offers discounts on this book when ordered in bulk quantities. For more information, write to Special Sales/Professional Marketing at the address below.

MARCEL DEKKER, INC.
270 Madison Avenue, New York, New York 10016

Current printing (last digit):
10 9 8 7 6 5 4 3 2

To my parents and teachers,
who stimulated my interest in
the fascinating world of science

Preface

The introduction of penicillin in the 1940s to combat bacterial infections in humans marked the dawn of a new era in anti-infective therapy. Penicillin's high efficacy and low toxicity profile helped save millions of lives. However, by the early 1950s, the emergence of penicillin-resistant strains of *Staphylococcus aureus* resulted in a need for an alternative antistaphylococcal drug. The other two drugs of choice at that time, erythromycin and tetracycline, also developed resistance. Vancomycin filled this need, and following its introduction in 1958, it was widely used. In 1960, new drugs—first methicillin and then the cephalosporins and lincomycin—displaced vancomycin's use. However, in the past 10 years, owing to several reasons discussed in Chapter 9, vancomycin has made a dramatic resurgence and is now the drug of choice to treat serious Gram-positive infections. A second glycopeptide antibiotic, teicoplanin, was recently launched in Italy and France and is under clinical trial in Canada and the United States. Avoparcin is used commercially as a feed additive for growth promotion in domestic animals.

In addition to renewed clinical interest in glycopeptide antibiotics, there has been a marked rise in general scientific interest in these compounds. This has resulted in a large volume of literature in many areas of research. This book, the first authoritative one on the subject, attempts to cover the present state of knowledge in one volume.

Chapter 1 classifies the presently known glycopetide antibiotics on the basis of the differences in the chemical structure of the core heptapeptide aglycone. The authors Yao and Crandall briefly describe the producing organisms and give detailed descriptions of all the known screening assays used to discover new glycopeptide antibiotics.

Chapter 2 (Sitrin and Folena-Wasserman) describes the purification of glycopeptide antibiotics from fermentation broths, and the separation techniques used. These span from the older classical ion exchange to the modern affinity and HPLC methods and from microversion isolation of antibiotics in fermentation broths to large-scale separations. Detailed descriptions are given for the more important members of this class of compounds.

Chapter 3 describes the advances made towards the total synthesis of vancomycin, vancomycin aglycones, and other similar chemical structures. Evans and DeVries describe new methods involving the synthesis of chiral α-amino acids, peptide macrocyclization, and phenol-oxidative coupling. There are interesting sections on the atropisomerism and a brief summary of the fascinating story of the structural elucidation of vancomycin itself.

Sztaricskai and Pelyvás-Ferenczik discuss all relevant chemical aspects of the carbohydrate components of glycopeptide antibiotics in Chapter 4. In section III.B, there is an exhaustive survey of all the synthetic routes to the 3-amino-2,3,6-trideoxyhexoses, the amino sugars unique to this group of antibiotics.

A large group of chemically diverse analogs of the vancomycin-type antibiotics are available as naturally ocurring metabolites, chemical degradation products, or semisynthetic derivatives. Chapter 5 explains in a logical fashion the structure–activity relationships based on the chemical structure, antibacterial activity, and mode of action of this group of antibiotics. This study, in turn, has helped pinpoint the most fruitful area of chemical modification for the preparation of highly active, semisynthetic vancomycin and A 82846 glycopeptide derivatives.

Nicas and Allen discuss the mechanism of action of vancomycin-type antibiotics in Chapter 6. The development of resistance; the mechanism, type, and genetics of resistance; and finally the clinical implications and the need for an ongoing search for new antibiotics effective against resistant strains are discussed by the authors.

Inman and Winely describe the analytical quantitation and characterization of vancomycin and related glycopeptides in Chapter 7. The first two sections deal mainly with analytical methodologies for vancomycin and the third section reviews analytical procedures for other related glycopeptides.

An overview of the discovery, chemistry, antimicrobial activity, pharmaco-kinetics, and toxicology of teicoplanin, the only other glycopeptide antibiotic approved for human use, is covered in Chapter 8, by Goldstein, Rosina, and Parenti.

Chapter 9 provides an overview of the clinically useful information regarding the pharmacokinetics, clinical uses, and safety profile of vacomycin. Zeckel and Woodworth suggest that new agents more potent than vancomycin, with activity against vacomycin- and teicoplanin-resistant enterococci and staphylococci, and with an improved safety profile, are needed.

This book will be of interest to pharmaceutical researchers and academic research groups involved in (1) the screening and purification of new members of this class, (2) structural elucidation and modification efforts to obtain more potent semisynthetic glycopeptides, (3) total synthesis of this group of compounds and deoxy amino sugars, (4) understanding the mechanism of the mode of action and resistance, and (5) developing analytical methods for further study. Hospitals, clinics, and research institutes active in infectious disease research will find this book a useful addition to their library. Individual scientists involved in any aspect of glycopeptide research will find this book a ready reference. We hope that scientists who use this book will be encouraged to enter into many other aspects of glycopeptide research that have not yet been explored.

I would like to thank all the contributors of this book for the eight excellent chapters in their area of expertise. I am indebted to Ms. Louise W. Crandall and Drs. Margaret Faul and R. Nagaraja Rao, who read sections of the manuscript, for their suggestions and criticism. Finally, I sincerely appreciate the help of Mrs. Amelia A. Schabel in preparing the index, and Ms. Louise W. Crandall in proof-reading the entire manuscript. I am responsible for any omission or errors in the final draft.

Ramakrishnan Nagarajan

Contents

Contributors

Norris E. Allen, Ph.D.   Research Scientist, Infectious Disease Research, Lilly Research Laboratories, Eli Lilly and Company, Indianapolis, Indiana

Louise W. Crandall, M.A.   Assistant Senior Biochemist, Natural Products Research, Lilly Research Laboratories, Eli Lilly and Company, Indianapolis, Indiana

Keith M. DeVries, Ph.D.*   Department of Chemistry, Harvard University, Cambridge, Massachusetts

David A. Evans, Ph.D.   Abbott and James Lawrence Professor of Chemistry, Department of Chemistry, Harvard University, Cambridge, Massachusetts

Gail Folena-Wasserman, B.S., M.S., Ph.D.   Director of Development, Med-Immune, Inc., Gaithersburg, Maryland

Beth P. Goldstein, Ph.D.   Head, Department of Medical Microbiology, Lepetit Research Center, Marion Merrell Dow Research Institute, Gerenzano, Italy

Eugene L. Inman, Ph.D.   Head, Animal Science Chemical Research, Lilly Research Laboratories, Eli Lilly and Company, Indianapolis, Indiana

Ramakrishnan Nagarajan, Ph.D.   Senior Research Scientist, Lilly Research Laboratories, Eli Lilly and Company, Indianapolis, Indiana

Thalia I. Nicas, Ph.D.   Research Scientist, Infectious Disease Research, Lilly Research Laboratories, Eli Lilly and Company, Indianapolis, Indiana

Francesco Parenti, Ph.D.   President, Marion Merrell Dow Europe, Marion Merrell Dow Europe AG, Thalwil, Switzerland

---

Present affiliation: Process Research and Development, Pfizer, Inc., Groton, Connecticut

István Pelyvás-Ferenczik, Ph.D.   Deputy Head of the Research Group for Antibiotics, Hungarian Academy of Sciences, Lajos Kossuth University, Debrecen, Hungary

Rita Rosina, Ph.D.   Regulatory Compliance Europe, Marion Merrell Dow Europe AG, Thalwil, Switzerland

Robert D. Sitrin, B.S., M.S., Ph.D.   Senior Director, Bioprocess R&D, Merck Research Laboratories, West Point, Pennsylvania

Ferenc Sztaricskai, Ph.D., D.Sc.   Head of the Research Group for Antibiotics, Hungarian Academy of Sciences, Lajos, Kossuth University, Debrecen, Hungary

C. L. Winely, Ph.D.   Research Scientist, Lilly Research Laboratories, Eli Lilly and Company, Indianapolis, Indiana

James R. Woodworth, Ph.D.   Research Scientist, Lilly Research Laboratories, Eli Lilly and Company, Indianapolis, Indiana

Raymond C. Yao, Ph.D.   Senior Research Scientist, Natural Products Research, Lilly Research Laboratories, Eli Lilly and Company, Indianapolis, Indiana

Michael L. Zeckel, M.D.   Senior Clinical Research Physician, Lilly Research Laboratories, Eli Lilly and Company, Indianapolis, Indiana

Glycopeptide Antibiotics

1
Glycopeptides
Classification, Occurrence, and Discovery

RAYMOND C. YAO and LOUISE W. CRANDALL

Lilly Research Laboratories, Eli Lilly and Company, Indianapolis, Indiana

I. INTRODUCTION

A. Historical Aspect

Vancomycin was first discovered at Eli Lilly and Company in a natural products screening program in the early 1950s directed at new antibiotic-producing microorganisms (1). This antibiotic, active primarily against Gram-positive bacteria, was produced by a new species of actinomycete isolated from an Indonesian soil; subsequently, two other vancomycin-producing strains were isolated from samples of Indian soil (2). The culture, originally designated *Streptomyces orientalis*, was later renamed *Nocardia orientalis* based on the tendency of the vegetative hyphae to break into small, squarish units often referred to as "fragments" (3). With the advent of chemotaxonomic techniques that include the analysis of cell wall amino acids and sugars, whole-cell sugars, fatty acids (including mycolic acids), menaquinones, and phospholipids, this nocardioform, which lacks mycolic acids, was eventually reclassified as *Amycolatopsis orientalis* (4).

The discovery of vancomycin was followed a year later by the isolation of ristocetin (5). These two antibiotics were recognized as belonging to a chemical class of antibiotics called glycopeptides, but their complete structures were not determined until some years later. The glycopeptides are complex molecules, characterized by a multiring peptide core containing six peptide linkages, an unusual triphenyl ether moiety, and sugars attached at various sites. Over 30 antibiotics designated as belonging to the glycopeptide class have been reported. As is expected with microbial secondary metabolites, most of these antibiotics have first been isolated as families of closely related factors.

B. Medical and Agricultural Uses

Even though a number of glycopeptides have been discovered since the 1950s, vancomycin was the only member of this class introduced for clinical use until the recent approval of teicoplanin. Vancomycin is effective at low concentrations against the majority of Gram-positive bacteria, but toxicity problems encountered in the early years of its application precluded its widespread use despite its selective action against bacterial cell wall synthesis. The introduction of semisynthetic penicillins, later followed by cephalosporins and lincomycin, overshadowed vancomycin. In the past 10 years, however, as the result of the emergence of multiple-resistant and methicillin-resistant staphylococcal (MRS) infections (6), vancomycin has become the drug of choice as well as the drug of last resort. Vancomycin is a valuable antibiotic for the treatment of endocarditis and MRS infections and is the first line of treatment for pseudomembranous colitis.

Teicoplanin has been introduced in Germany, Italy, and France and is under investigation for clinical use in several other countries, including the United States (7). Like vancomycin, it is particularly effective in the treatment of severe infections caused by antibiotic-resistant, Gram-positive bacteria.

Ristocetin appeared to be a promising antibacterial agent for use in human medicine, but it was withdrawn following a high incidence of thrombocytopenia (8). Further investigation of its hemostatic effects indicated that ristocetin causes platelet aggregation by interacting with a plasma factor missing from patients with one form of von Willebrand's disease. This property has been exploited, and ristocetin is used in the differential diagnosis of this disorder.

Several glycopeptide antibiotics have shown growth-promoting activity in farm animals; however, only actaplanin and avoparcin have been tested extensively. Avoparcin has been marketed in Europe since 1976 as a feed additive. It increases growth rate and food conversion efficiency in both ruminant and nonruminant animals and is commonly included in milk replacer compounds for dairy calves (9,10). Avoparcin is also used in the dairy industry for improvement of milk production of lactating dairy cows. Actaplanin is also effective for these indications (11) but has not reached the marketplace.

II. CLASSIFICATION

Lancini (12) subdivided this class into four groups based on chemical structure. The first three groups are distinguished by variations in the amino acid residues at the N-terminal end of the peptide core; the fourth group can be designated the lipoglycopeptide group since fatty acid moieties are present. In structure 1, vancomycin, the aliphatic amino acids N-methylleucine (amino acid 1) and asparagine (amino acid 3) are indicated by the dotted lines.

In Table 1 the glycopeptide antibiotics are listed by structural type, with the

Amino acid 3   Amino acid 1

1 VANCOMYCIN

characteristic skeletal fragment shown. The reference for the structure determination is included.

Group I, or the vancomycin (1) type, has aliphatic amino acids at positions 1 and 3. Antibiotics of the vancomycin complex were the only representatives in this group until the last decade, when there was a dramatic increase in reports of new glycopeptides. The same antibiotic, designated variously as A82846 B and chloroorienticin A, was isolated independently by investigators at Eli Lilly and Co. and Shionogi and Co., respectively. Factor A of A82846 (Eli Lilly and Co.) was isolated as eremomycin (Institute of New Antibiotics [former USSR]) and MM45289 (Beecham Group). UK-72051, isolated at Pfizer Central Research (UK) was shown to be the same as factor A of the orienticin complex from Shionogi and Co.

Group II, illustrated by avoparcin (2), has two aromatic amino acids residues at positions 1 and 3. We have listed synmonicin with this group; however, it represents a distinct type with an aromatic amino acid at the N-terminal site and an aliphatic amino acid (methionine) at the third position (13).

The structure of ristocetin (3) is shown as an example of group III. Ristomycin, whose isolation was reported by investigators at the Institute of New

2 *Beta*- AVOPARCIN

Table 1 Classification of Glycopeptide Antibiotics

Antibiotic	Reference	Antibiotic	Reference
Group I vancomycin type			

Antibiotic	Reference	Antibiotic	Reference
A42867	14	MM45289	21
A51568	15	MM47761	22
A82846	16	OA7653	23, 24
A83850	17	Orienticin	25
Chloroorienticin	18	UK72051	26
Decaplanin	19	Vancomycin	27
Eremomycin	20		

Table 1 Continued

Antibiotic	Reference	Antibiotic	Reference
Group II avoparcin type			

Actinoidin	28	Galacardin	31
Avoparcin	29	Helvecardin	32
Chloropolysporin	30	MM47766	33
		Synmonicin	13
Group III ristocetin type			

A35512	34	Actaplanin	38
A41030	35	Ristocetin	39
A47934	36	Ristomycin	40
A80407	37	UK68597	41
		UK69542	42
Group IV teicoplanin type			

A40926	43	Kibdelin	46
A84575	44	MM55266	47
Ardacin	45	Parvodicin	48

Antibiotics (former USSR) in 1963, has been shown to be the same as ristocetin. The antibiotics of groups III differ from those of group II only in having an ether linkage joining the aromatic amino acids at positions 1 and 3.

Group IV, or the teicoplanin (**4**) type, could be considered a subgroup of III since the arrangement of the amino acids in the peptide core (aglycone) is the same. The antibiotics in this group have a fatty acid residue attached to the amino sugar.

3   RISTOCETIN A

III.   OCCURRENCE

All the antibiotics belonging to the glycopeptide class isolated thus far have originated from fermentations of actinomycetes. In Table 2, the antibiotic complexes are listed in alphabetical order along with the structure type, producing organism, and isolation reference. The name for the organism listed in Table 2 is the most recent taxonomic designation, which in many cases does not agree with the initial classification. It is interesting to note that these metabolites are elaborated by a diverse group of actinomycetes ranging from the more prevalent *Streptomyces* species to the relatively rare genera of *Streptosporangium* and *Saccharomonospora*. The less common *Actinoplanes* and *Amycolatopsis* account for almost half of the producing organisms.

4 Teicoplanin A$_2$-1

IV. DISCOVERY AND SCREENING

After the discovery of actaplanin in 1974 and teicoplanin in 1978, no new glycopeptides were reported until 1984. It was not that the microorganisms had nothing more to offer, nor that the interest on this class of metabolites had dwindled, but rather that the assay methods available were both insensitive and nonspecific. For decades, agar difussion assays of different types were the methods of choice for the random detection of new antimicrobial agents. In the past 10 years, however, the application of mechanism-based and target-directed approaches in the search of natural products has revolutionalized the discovery process. Such approaches have greatly enhanced the sensitivity of the assays and at the same time provided specificity in the detection of a desired biological activity. The success of such approaches is evident in the recent discovery of new β-lactam antibiotics, such as the formacidins (76) and the cephabacins (77), as well as the threefold increase in new glycopeptides reported since 1980.

Table 2   Glycopeptide Antibiotics: Producing Organisms

Antibiotic	Type	Producing organism	Reference
A477	ND[a]	*Actinoplanes* sp. NRRL 3884	49
A35512	III	*Streptomyces candidus*[b] NRRL 8156	50
A40926	IV	*Actinomadura* sp. ATCC 39727	51
A41030	III	*Streptomyces virginiae* NRRL 15156	52
A42867	I	*Nocardia* sp. ATTC 53492	14
A47934	III	*Streptomyces toyocaensis* NRRL 15009	53
A80407	III	*Kibdelosporangium philippinensis* NRRL 18198 or NRRL 18199	54
A82846	I	*Amycolatopsis orientalis* NRRL 18100	55
A83850	I	*Amycolatopsis albus* NRRL 18522	—[c]
A84575	IV	*Streptosporangium carneum* NRRL 18437, 18505	56
AB-65	ND	*Saccharomonospora viride* T-80 FERM-P 2389	57
Actaplanin	III	*Actinoplanes missouriensis* ATCC 23342	58
Actinoidin	II	*Proactinomyces actinoides*	59
Ardacin	IV	*Kibdelosporangium aridum* ATCC 39323	60
Avoparcin	II	*Streptomyces candidus* NRRL 3218	61
Azureomycin	ND	*Pseudonocardia azurea* NRRL 11412	62
Chloroorienticin	I	*Amycolatopsis orientalis* PA-45052	18
Chloropolysporin	II	*Micropolyspora* sp. FERM BP-538	63
Decaplanin	I	*Kibdelosporangium deccaensis* DSM 4763	19
N-demethylvancomycin	I	*Amycolatopsis orientalis* NRRL 15232	64
Eremomycin	I	*Actinomycetes* sp. INA 238	65
Galacardin	II	*Actinomycetes* strain SANK 64289 FERM P-10940	31
Helvecardin	II	*Pseudonocardia compacta* subsp. *helvetica*	66
Izupeptin	ND	*Nocardia* AM-5289 FERM P-8656	67
Kibdelin	IV	*Kibdelosporangium aridum* ATCC 39922	46
LL-AM374	ND	*Streptomyces eburosporeus* NRRL 3582	68
Mannopeptin	ND	*Streptomyces platenis* FS-351	69
MM45289 MM47756	I	*Amycolatopsis orientalis* NCIB 12531	21
MM47761 MM49721	I	*Amycolatopsis orientalis* NCIB 12608	22
MM47766 MM55260	II	*Amycolatopsis orientalis* NCIB 40011	33
MM55266	IV	*Amycolatopsis* sp. NCIB 40089	47
MM55270	ND	*Amycolatopsis* sp. NCIB 40086	70
OA-7653	I	*Streptomyces hygroscopicus* ATCC 31613	71

Table 2 Continued

Antibiotic	Type	Producing organism	Reference
Orienticin	I	*Nocardia orientalis* FERM BP-1230	25
Parvodicin	IV	*Actinomadura parvosata* ATCC 53463	48
Ristocetin	III	*Amycolatopsis orientalis* subsp. *lurida* NRRL 2430	5
Ristomycin	III	*Proactinomyces fructiferi*	72
Synmonicin	II	*Synnemomyces mamnoorii* ATCC 53296	73
Teicoplanin	IV	*Actinoplanes teichomyceticus* ATCC 31121	74
UK-68597	III	*Actinoplanes* ATCC 53533	75
UK-69542	III	*Saccharothix aerocolonigenes*	42
UK-72051	I	*Amycolatopsis orientalis*	26
Vancomycin	I	*Amycolatopsis orientalis* NRRL 2450	2

[a]Not determined.
[b]This organism is probably *Amycolatopsis* sp. (Labeda, D.P., personal communication, Northern Regional Research Laboratories, Peoria, IL.)
[c]Personal communication, Frederick Mertz, Lilly Research Laboratories.

A. Molecular Basis of Vancomycin Action

Reynolds (78) and Jordan (79) were the first to demonstrate that the inhibition of Gram-positive bacteria by vancomycin resulted in the accumulation of the nucleotide-linked cell wall intermediate UDP-*N*-acetylmuramylpentapeptide. Their observation was soon confirmed by Wallace and Strominger (80) using ristocetin, suggesting that cell wall biosynthesis was the primary target of this class of antibiotics. It later became clear that vancomycin binds specifically to the cell wall precursor UDP-*N*-acetylmuramylpentapeptide (81,82). Extensive studies employing nuclear magnetic resonance methods principally by Williams and coworkers (83,84) concluded that these antibiotics form noncovalent complexes with natural and synthetic peptide analogs terminating in acyl-D-alanyl-D-alanine (acyl-D-Ala-D-Ala; Fig. 1). As vancomycin is a fairly large molecule (molecular weight = 1949), it is unlikely to cross cell membranes. The most likely target is the lipid intermediate containing the disaccharide-pentapeptide wall subunit and the nascent peptidoglycan as it is extruded through the membrane before cross-linking (85). Thus the tight and precise binding of the antibiotic to acyl-D-Ala-D-Ala at the growing points of the new peptidoglycan chain prevents the transglycosylation and, subsequently, the transpeptidation enzyme reactions in which the new chains are linked to existing chains in the cell wall. Details of the mode of action of glycopeptides are reviewed by Nicas and Allen in Chapter 6.

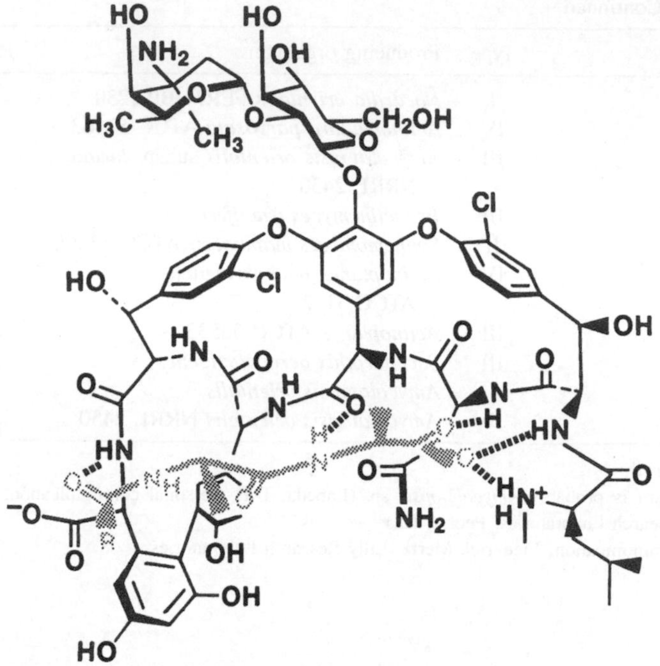

Fig. 1 Binding of vancomycin to cell wall precursor acyl-D-alanyl-D-alanine. (Courtesy of R. Nagarajan and the American Society of Microbiology.)

B. Need for New Agents

The resurgence of nosocomial infections due to Gram-positive organisms as well as the increasing incidence of MRS infections found in neutropenic cancer patients has resulted in the use of glycopeptides as the drug of choice. Because of their unique mechanism of action on cell wall biosynthesis, the emergence in clinical practice of strains resistant to this class of antibiotics has been relatively low. In the past few years, however, numerous reports of resistance found in isolates of *Staphylococcus epidermidis* or *Staphylococcus haemolyticus* and, more significantly, *Enterococcus faecium* have demonstrated a need for new glycopeptides.

C. Peptide Reversal and Antagonism

In a concerted effort to search for new glycopeptide antibiotics, investigators at Smith, Kline and French Laboratories reported a comprehensive approach to discover new members of this class (86,87). This included the use of special culture selection techniques to isolate producing organisms and optimized growth conditions conducive to the elaboration of such metabolites, as well as a mechanistic approach in screening. Since the binding of vancomycin to the cell wall target sites is noncovalent in nature, the inhibition of bacterial growth or peptidoglycan synthesis could be reversed by adding a peptide terminating in D-Ala-D-Ala to the growth medium. This peptide would compete effectively with the natural wall peptides at the growth points for the available glycopeptides (88).

In this assay a tripeptide diacetyl-L-Lys-D-Ala-D-Ala was used specifically to antagonize the antibacterial activity of glycopeptides. Fermentation broth samples were added to paper disks with and without the tripeptide (100 μg/disk). A reduction of 5 mm or greater in zone size in the presence of tripeptide was indicative of the presence of glycopeptide antibiotics. This tripeptide reversal assay was found to be entirely class specific. The sensitivity of this screen to known glycopeptide antibiotics ranged from 3.1 μg/ml for vancomycin, avoparcin, and actinoidin to 100 μg/ml for A-477.

The detection and identification of glycopeptides were further enhanced by the application of affinity chromatography with an activated agarose support coupled to ligands terminating in D-Ala-D-Ala. This provided an efficient one-step separation from fermentation broths with greater than 80% recoveries (86). New glycopeptides can be determined quickly using reversed-phase high-performance liquid chromatography (HPLC) and fast-atom bombardment (FAB) mass spectrometry. The use of tripeptide antagonism combined with optimized culture selection and growth conditions yielded novel glycopeptides at a rate of 1 per 320 cultures screened. This mechanistic approach proved to be extremely successful, yielding a number of new entities in this chemical class, including kibdelin, ardacin, and parvodicin.

D. Selective Bioaffinity Adsorption Detection

The interest in glycopeptide antibiotics in the early 1980s was not confined to the group at Smith, Kline and French Laboratories. Investigators at the Merrell Dow Lepetit Research Center reported the use of a selective bioaffinity resin, again based on the interaction of this class of metabolites with D-Ala-D-Ala (89). Broth samples (3 ml) were passed through 0.5 ml columns packed with the affinity resin. The columns were then washed exhaustively at neutral pH. The presumptive activity was eluted at alkaline pH. After neutralization, samples were tested for antimicrobial activity against susceptible organisms. With this highly specific assay, a total of 72 strains were identified as producers of glycopeptide antibiotics, 49 of which were fully characterized. Among the known antibiotics, ristocetin was most frequently detected (60%). A number of novel structures, such as A42867 and A40926, were also discovered.

This approach concentrates initial antibiotic activity from broths, thus greatly increasing sensitivity of detection. At the same time, because of the specificity of the ligand used, the detection of antibiotic activity is indicative of the presence of a glycopeptide.

E. Antibody-Based Assay

In 1983, responding to a need for new antibacterials against Gram-positive infections, especially MRS, we at Eli Lilly and Company initiated a program to search for new glycopeptides. We believed that there was a need for an agent like vancomycin with reduced toxicity and improved pharmacokinetics, as well as better efficacy. Since we had developed an enzyme-linked immunosorbent assay (ELISA) approach to the detection of specific chemical types in fermentation metabolites (90), this procedure was adapted for glycopeptide antibiotics.

ELISA, like other immunoassays, exhibits a high level of sensitivity and specificity. This analytical technique has been established as an extremely valuable tool in both the clinical and research environment. Its application in the detection and discovery of fermentation metabolites has been well documented (90,91). Using polyclonal antibodies for gentamicin (aminoglycoside) and 23-amino-O-mycaminosyltylonolide (macrolide), antibiotics belonging to a specific chemical class could be detected in fermentation broths when present at low levels. This chemical approach departs from the traditional methods of detection based on biological activities.

In the glycopeptide immunoassay, vancomycin antibody is immobilized onto the surface of a microtiter well (Fig. 2). A vancomycin-enzyme (alkaline phosphatase), prepared using glutaraldehyde as the cross-linking reagent, was used as a label for direct ELISA. The binding of the enzyme label to the antibody can be quantified by measuring the enzyme activity. Utilizing a competitive assay, the presence of vancomycin in a sample can be detected. Vancomycin molecules in the sample compete with the enzyme label for the limited antibody binding sites on the wall of the microtiter wells. The quantity of enzyme label bound to the antibody decreases with increasing concentration of free competing vancomycin, resulting in reduced absorbance values. The concentration of vancomycin required to displace 50% of the enzyme label, defined as IC_{50}, is 8 ng/ml. If 20% displacement is used as the cutoff for the lower detection limit, vancomycin can be detected at a 0.5 ng/ml level making this assay one of the most sensitive tools available (Fig. 3). This sensitivity is similar to the detection of drugs in body fluids using radioimmunoassay and other immunoassays in clinical laboratories. This ELISA procedure is outlined in Figure 4. The technique provides a sensitivity 1000-fold better than that of the agar diffusion assay.

The specificity of this assay is evidenced by the lack of cross-reactivity with this antibody to antibiotics chemically unrelated to glycopeptides (Table 3). A polyclonal instead of monoclonal antibody was specifically chosen with the hope that this antibody would recognize various members of this chemical class,

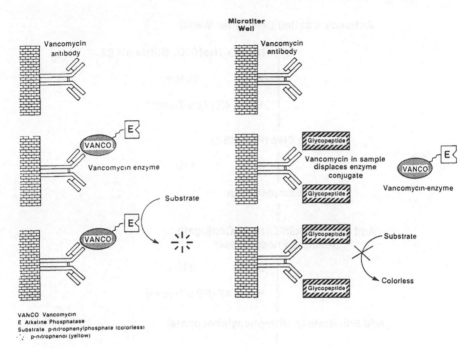

Fig. 2 Competitive vancomycin ELISA.

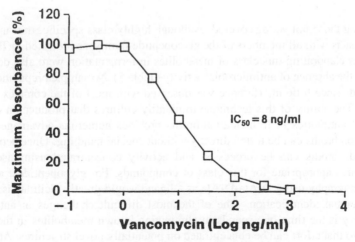

Fig. 3 Competitive vancomycin ELISA dose-response curve.

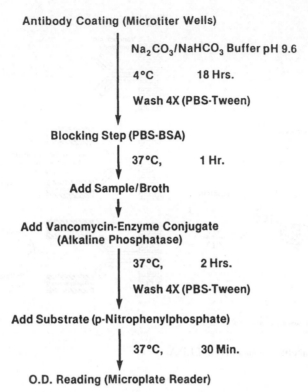

Antibody Coating (Microtiter Wells)

Na$_2$CO$_3$/NaHCO$_3$ Buffer pH 9.6

4°C   18 Hrs.

Wash 4X (PBS-Tween)

Blocking Step (PBS-BSA)

37°C,   1 Hr.

Add Sample/Broth

Add Vancomycin-Enzyme Conjugate
(Alkaline Phosphatase)

37°C,   2 Hrs.

Wash 4X (PBS-Tween)

Add Substrate (p-Nitrophenylphosphate)

37°C,   30 Min.

O.D. Reading (Microplate Reader)

Fig. 4   Competitive vancomycin ELISA procedure.

including those not yet discovered. Although highly class specific, this antibody cross-reacts with all members of the glycopeptide family, as indicated in Table 4. Cultures elaborating this class of metabolites in fermentation were also detected even in the absence of antimicrobial activity (Table 5). No sample preparation was required, since little interference was observed with most of the complex media tested. The ability of this technique to identify cultures that produce an unique class of antibiotics in fermentation broths provides numerous advantages. Fermentation broths can be tested directly without special handling. Once activity is detected, broths can be processed and activity concentrated using isolation procedures appropriate for this class of compounds. For glycopeptides, affinity chromatography using D-Ala-D-Ala as a ligand would greatly facilitate isolation and eventual identification. One of the most difficult challenges in antibiotic discovery is the time one spends in eliminating known metabolites in the early stages so that effort can be concentrated on potentially novel structures. Application of appropriate HLPC systems and FAB mass spectrometry, such as the

Table 3 Cross-reactivity of Vancomycin Antibody with Antibiotics of Various Chemical Classes

Antibiotics (100 μg/ml)	Zone of inhibition (mm) versus *Bacillus subtilis*	ELISA (% reactivity)[a]
Benzylpenicillin	29	0
Tetracycline	24	2
Erythromycin	36	4
Monensin	—	7
Actidione	—	11
Cycloserine	Trace	18
Novobiocin	13	12
Chloramphenicol	14	9
Gentamicin	23	0
Kanamycin	27	0
Neomycin	19	0
Tobramycin	27	0
Streptomycin	26	0
Streptothricin	18	0
Vancomycin		
100 μg/ml	19	100
10 μg/ml	13	98
1 μg/ml	Trace	94
Control[b]	—	0

[a]ELISA response of vancomycin (100 μg/ml) normalized at 100%.
[b]Phosphate-buffered saline + 0.5% bovine serum albumin (BSA).

scheme used by the Smith, Kline and French group (86), would greatly facilitate the identification of a new structure since the number of known metabolites is limited. Since this assay is set up in a 96-well microtiter plate format, the resulting high-volume throughput capability certainly meets one of the major criteria essential for the success of a screening program. This unique approach, which detects metabolites of a specific chemical type rather than biological activity, constitutes another valuable addition to the arsenal of screening and discovery techniques.

Using this approach alone without optimization of any other parameters favoring this class of antibiotics, glycopeptides were detected at the rate of 23 per 5100 cultures screened (0.5%). Of these 23 glycopeptides, 12 were characterized as vancomycin producers (52%); 6 others were also known glycopeptide producers [actinoidin, ristocetin, LL-AM372 (2), A47934, and A35512B], and 5 new structures were detected. It is interesting to note that for every 4.6 glycopeptides detected, 1 turned out to be a new structure (22%), and for every 1020 cultures

Table 4 Cross-reactivity of Vancomycin Antibody with Glycopeptide Antibiotics

Glycopeptide	(µg/ml)	Zone of inhibition (mm) versus *Bacillus subtilis*	ELISA (% reactivity)[a]
Vancomycin	100	20	100
	10	16	99
	1	T	96
Actinoidin	100	17	94
	10	12	75
	1	T	63
LL-AM374	100	17	89
	10	12	68
	1	T	49
Avoparcin	100	16	89
	10	11	65
	1	T	40
Teicoplanin	100	16	95
	10	12	85
	1	0	77
Actaplanin G	100	14	95
	10	12	93
	1	T	48
Ristocetin A	100	15	40
	10	T	26
	1	0	22
A35512B	100	17	33
	10	10	32
	1	0	35
A47934	100	15	80
	10	T	67
	1	0	60
A41030	100	15	73
	10	10	68
	1	0	67
Ardacin	100	—	67
	1	—	35
Control[b]			0

[a]ELISA response of vancomycin (100 µg/ml) normalized at 100%.
[b]Phosphate-buffered saline + 0.5% BSA.

Table 5 ELISA Detection of Glycopeptide-Producing Cultures in Fermentation

Organism[a]	NRRL	Glycopeptide	Zone of inhibition (mm)[b]	ELISA (% reactivity)[c]
Amycolatopsis/Nocardia				
A. orientalis	2,452	Vancomycin	30	75
A. orientalis	2,451	Vancomycin	30	68
A. orientalis	15,232	N-demethyl-vancomycin	26	59
Amycolatopsis sp.	3,218	Avoparcin	20	75
Nocardia sp.	8,156	A35512	25	49
Nocardia sp.	3,582	LL-AM374	28	25
Nocardia sp.	—	Ristocetin	17	55
A. lurida	2,430	Ristocetin	Trace	22
Actinoplanes				
A. teichomyceticus	31,121[c]	Teicoplanin	14	92
A. missouriensis	23,342[c]	Actaplanin	—	34
Streptomyces				
S. toyocaensis	15,009	A47934	19	64
S. virginiae	12,525	A41030	14	53
Control (broth)[d]			—	0

[a]Cultures were grown in 250 ml flasks each containing 50 ml glucose-peptone broth for 5 days at 30°C and 250 rpm.
[b]Agar diffusion assay against *Bacillus subtilis* American Type Culture Collection (ATCC) 6633.
[c]ELISA response of vancomycin (100 μg/ml) normalized at 100%
[d]ATCC.

screened randomly, 1 new glycopeptide can be found (0.01%). Chemical structures of 4 of these new glycopeptides have been elucidated: A82846 and A83850 are vancomycin types (group I), A80407 is a ristocetin type (group III), and A84575 belongs to the lipoglycopeptide class (group IV).

F. Peptide-Binding Assay

Similar to the antibody-based immunoassay, a solid-phase competitive binding assay was reported by both Lepetit Research Laboratories and our laboratory (92,93). Instead of an antibody, a peptide ending in D-Ala-D-Ala coupled to a carrier protein is used. The protein facilitates the attachment of the peptide hydrophobically onto the surface of the polystyrene wells (Fig. 5). Vancomycin-enzyme or teicoplanin-enzyme labels were used in an approach similar to that of an ELISA. The presence of glycopeptide-related antibiotics could be detected in a competitive assay. Since D-Ala-D-Ala is the target binding site, this assay, as

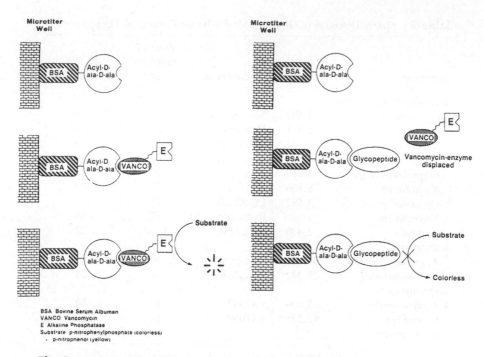

BSA Bovine Serum Albumen
VANCO Vancomycin
E Alkaline Phosphatase
Substrate p-nitrophenylpnosphate (colorless)
  - p-nitrophenol (yellow)

Fig. 5   Detection of glycopeptides using peptide-binding assay.

expected, is highly specific. Different glycopeptides exhibited varying binding kinetics, and in both assays actaplanin showed the strongest competition. Although this assay showed the specificity needed as a discovery tool, its sensitivity is comparable only to that of microbiological agar diffusion assay.

G.   Gene Induction Assay

Gene induction assays of different types have been reported to detect metabolites of interest. Elespuru and collaborators (94) described the use of γ phage lysogen fused to a reporter gene (β-galactosidase) to screen for antitumor agents. Lysogens of phage are induced by a wide variety of compounds that either interact with DNA or interfere with its synthesis. The expression of β-galactosidase can be easily detected on agar plates using chromogenic substrates. Sykes and Wells (95) recently reported the use of a strain of *Bacillus licheniformis* that carries an inducible β-lactamase gene, the induction of which can be monitored by the use of a chromogenic substrate, such as nitrocephin. β-Lactam antibiotics may be detected with this assay at nanogram levels, making this assay one of the most sensitive detection techniques.

Bacterial strains carrying fusions between antibiotic-induced promoters and β-galactosidase have also been employed to search for antibiotics of a specific chemical class. Kirsch and coworkers (96) isolated a vancomycin-induced gene fusion from a Tn903 random gene fusion pool in *Bacillus subtilis*. Again, using chromogenic substrates for β-galactosidase in an agar-based detection system, this strain was induced by vancomycin, ristocetin, A35512B, and, weakly, by teicoplanin. As little as 200 ng vancomycin could be detected with the fusion, but 700 ng vancomycin was required to produce a zone of inhibition with *Staphylococcus aureus* FDA 209P. With several thousand microbial fermentations assayed, a variety of glycopeptides were identified and confirmed using a peptide reversal assay similar to the method described earlier.

Table 6   Recently Discovered Glycopeptide Antibiotics

Antibiotic	Producing organism	Research group
Ardacin	*Kibdelosporangium aridum*	1985, Smith Kline
Chloropolysporin	*Micropolyspora* sp.	1985, Sankyo
Izupeptin	*Nocardia* sp.	1986, Kitasato Research Institute
Kibdelin	*Kibdelosporangium aridum*	1986, Smith Kline
Synmonicin	*Synnemonyces mamnoorii*	1986, Smith Kline
Parvodicin	*Actinomadura parvosata*	1986, Smith Kline
A40926	*Actinomadura* sp.	1986, Dow Lepetit
Eremomycin	*Actinomycetes*	1987, Institute New Antib., Moscow
Chloroorienticin	*Amycolatopsis orientalis*	1987, Shionogi
Orienticin	*Amycolatopsis orientalis*	1987, Shionogi
UK68597	*Actinoplanes* sp.	1988, Pfizer, Ltd.
A82846	*Amycolatopsis orientalis*	1988, Eli Lilly
A42867	*Nocardia* sp.	1988, Dow Lepetit
A84575	*Streptosporangium carneum*	1989, Eli Lilly
A80407	*Kibdelosporangium philippinensis*	1989, Eli Lilly
MM45289	*Amycolatopsis orientalis*	1989, Beecham
MM47766	*Amycolatopsis orientalis*	1989, Beecham
MM47761	*Amycolatopsis orientalis*	1990, Beecham
MM55266	*Amycolatopsis* sp.	1990, Beecham
MM55270	*Amycolatopsis* sp.	1990, Beecham
Decaplanin	*Kibdelosporangium decaensis*	1990, Hoechst
UK72051	*Amycolatopsis orientalis*	1990, Pfizer, Ltd.
Galacardin	*Actinomycetes*	1991, Sankyo
Helvecardin	*Pseudomonocardia compacta*	1991, Sankyo
UK-69542	*Saccharothrix aerocolonigenes*	1991, Pfizer, Ltd.
A83850	*Amycolatopsis albus*	1991, Eli Lilly

V. CONCLUSION

Vancomycin, the first and most important member of the glycopeptide family, has evolved as a useful antibiotic for human medicine since its discovery in 1956. Avoparcin, reported in 1968, is a commercially significant agricultural product. Of the many other glycopeptides, only teicoplanin is currently being pursued. The application of mechanism-based and targeted approaches to screening have resulted in an explosion of new glycopeptides in the last decade (Table 6). Many of these new agents are produced by rare genera of actinomycetes. Factors of the

Table 7   In Vivo Efficacy of A82846 Factors[a]

	A82846 A	X=H; Y=Cl
	A82846 B	X=Y=Cl
	A82846 C	X=Y=H

Antibiotic	Staphylococcus aureus	Streptococcus pyogenes	Streptococcus pneumoniae
A82846A	0.19	0.19	0.17
A82846B	0.19	0.20	0.18
A82846C	2.18	2.71	5.87
Vancomycin	1.30	0.72	1.52

[a]ED_{50} (mg/kg × 2) mice, subcutaneous.

glycopeptide complex designated A82846 at Eli Lilly and Company (Table 7) are possibly the most potent entities of this class detected to date, with A82846B showing the best activity (97) (Table 7). As mentioned in Section II, factors of this complex have been isolated independently by several laboratories. Further utilization of directed approaches to screening can be expected to expedite the discovery of even more useful therapeutic agents.

VI. ADDENDUM

There have been two recent publications of patents reporting new glycopeptides (98). The balhimycin complex belonging to the vancomycin class is produced by an actinomycete, Y-86,21022 (DSM 5908).

MM56597 and MM56598 are two additional members of the MM55266 complex produced by *Amycolatopsis* sp. NCIB 40098 (99). This class of antibiotics has been reviewed by Cavalleri and Parenti (100).

ACKNOWLEDGMENTS

The authors wish to thank D. K. Baisden, D. M. Berry, R. Hamill, D. F. Mahoney, F. P. Mertz, K. H. Michel, and D. W. Norton for their contributions in the discovery of new glycopeptides.

REFERENCES

1. McCormick M.H., Stark W.M., Pittenger G.E., Pittenger R.C., McGuire J.M. (1956). Vancomycin, a new antibiotic. I. Chemical and biological properties. Antibiot. Annu. 1955–1956:606–611.

2. Pittenger R.C., Brigham R.B. (1956). *Streptomyces orientalis*, n.sp., the source of vancomycin. Antibiot. Chemother. 6:642–647.

3. Lechevalier M.P. (1976). The taxonomy of the genus *Nocardia*: Some light at the end of the tunnel? In Goodfellow M., Brownell G.H., Serrano J.A. (Eds.), The Biology of the Nocardiae. Academic Press, New York, pp. 1–38.

4. Lechevalier M.P., Prauser H., Labeda D.P., Ruan J-S (1986). Two new genera of nocardioform actinomycetes: *Amycolata* gen. nov. and *Amycolatopsis* gen. nov. Int. J. Syst. Bacteriol. 36:29–37.

5. Philip J.E., Schenck J.P., Hargie M.P. (1957). Ristocetins A and B, two new antibiotics. Isolation and properties. Antibiot. Annu. 1955–1956:699–705.

6. Lepper M.H., Dowling H.F., Jackson G.G., Moulton B., Spies HW (1953). Effect of antibiotic usage in the hospital on the incidence of antibiotic-resistant strains among personnel carrying staphylococci. J. Lab. Clin. Med. 42:832.

7. Campoli-Richards D.M., Brogden R.N., Faulds D (1990). Teicoplanin: A review of its antibacterial activity, pharmacokinetic properties and therapeutic potential. Drugs 40:449–486.

8.   Howard M.A., Firkin B.G. (1971). Ristocetin—a new tool in the investigation of platelet aggregation. Thromb. Diath. Haemorrh. 26:362–369.

9.   MacGregor R.C. (1988). Growth promoters and their importance in ruminant livestock production. In Haresign W., Cole D.J.A. (Eds.), Recent Advances in Ruminant Nutrition, Vol. 2. Butterworths, London, pp. 308–322.

10. Moughan P.J., Stevens, E.V.J., Reisima I.D., Rendel J. (1989). The effect of avoparcin on the ileal and faecal digestibility of nitrogen and amino acids in the milk-fed calf. Anim. Prod. 49:63–71.

11. Scheifinger C.C., Anderson D.B., McGuffey R.K., Pendlum L.C., Richardson L.F., Wellenreiter R.H. (1983). The use of actaplanin in ruminant and non-ruminant nutrition (abstract 439). 23rd Intersci. Conf. Antimicrob. Agents Chemother. Las Vegas, Nevada.

12. Lancini G.C. (1989). Fermentation and biosynthesis of glycopeptide antibiotics. In Bushell M.E., Graefe U. (Eds.) Bioactive Metabolites from Microorganisms. Elsevier, Amsterdam. Prog. Ind. Microbiol. 27:283–287.

13. Sitrin R.D., Wasserman G.F. (1989). Affinity and HPLC purification of glycopeptide antibiotics. In Wagman G.H., Cooper R. (Eds.) Natural Products Isolation. Elsevier, Amsterdam, p. 119.

14. Riva E., Gastaldo L., Beretta M.G., Ferrari P., Zerilli L.F., Cassani, Selva E., Goldstein B.P., Berti M., Parenti F., Denaro M. (1989). A42867, a novel glycopeptide antibiotic. J. Antibiot. 42:497–505.

15. Hunt A.H., Marconi G.G., Elzey T.K., Hoehn M.M. (1984). A51568A; N-demethylvancomycin. J. Antibiot. 37:917–919.

16. Nagarajan R., Berry D.M., Hunt A.H., Occolowitz J.L., Schabel A.A. (1988). Conversion of antibiotic A82846B to orienticin A and structural relationships of related antibiotics. J. Org. Chem. 54:983–986.

17. Hunt A.H. unpublished results (patent application).

18. Tsuji N., Kamigauchi T., Kobayashi M., Terui Y. (1988). New glycopeptide antibiotics. II. Isolation and structures of chloroorienticins. J. Antibiot. 41:1506–1510.

19. Franco C.M.M., Chatterjee S., Vijakumar E.K.S., Chatterje D.K., Ganguli B.N., Rupp R.H., Fehlhaber H.W., Kogler H., Seibert G., Teetz V. (1990). Glycopeptide antibiotic decaplanin. European Patent Appl. 356894 (assigned to Hoechst AG), March 7, 1990.

20. Lomakina N.N., Berdnikova T.F., Tokareva N.L., Abramova E.A., Dokshina N.Y. (1989). Structure of eremomycin, a novel polycyclic glycopeptide antibiotic. Antibiot. Khimioter. 34:254–258.

21. Athalye M., Elson A., Gilpin M.L., Jeffries L.R. (1989). Production of antibiotics MM 45289 and MM 47756 by fermentation with Amycolatopsis orientalis. European Patent Appl. 309161 (assigned to Beecham Group plc), March 29, 1989.

22. Box S.J., Elson A.L., Gilpin M.L., Winstanley D.J. (1990). MM 47761 and MM 49721, glycopeptide antibiotics produced by a new strain of Amycolatopsis orientalis: Isolation, purification and structure determination. J. Antibiot. 43:931–937.

23. Jeffs P.W., Yellin B., Mueller L., Heald S.L. (1988). Structure of the antibiotic OA-7653. J. Org. Chem. 53:471–477.

24. Ang S.G., Williamson M.P., Williams D.H. (1988). Structure elucidation of a glycopeptide antibiotic, OA-7653. J. Chem. Soc., Perkin Trans. 1:1949–1956.

25. Tsuji N., Kobayshi M., Kamigauchi T., Yoshimura Y., Terui Y. (1988). New glycopeptide antibiotics I. The structures of orienticins. J. Antibiot. 41:819–822.
26. Skelton N.J., Williams D.H., Rance M.J., Ruddock J.C. (1990). Structure elucidation of UK-72051, a novel member of the vancomycin group of antibiotics. J. Chem. Soc., Perkin Trans. 1:77–81.
27. Harris C.M., Harris T.M. (1982). Structure of the glycopeptide antibiotic vancomycin. Evidence for an asparagine residue in the peptide. J. Am. Chem. Soc. 104: 4293–4295.
28. Heald S.L., Mueller L., Jeffs P.W. (1987). Actinoidins A and A2: Structure determination using 2D NMR methods. J. Antibiot. 40:630–645.
29. McGahren W.J., Martin J.H., Morton G.O., Hargreaves R.T., Leese R.A., Lovell F.M., Ellestad G.A., O'Brien E., Holker J.S.E. (1980). Structure of avoparcin compounds. J. Am. Chem. Soc. 102:1671–1684.
30. Takatsu T., Takahashi S., Nakajima M., Haneishi T., Nakamura T., Kuwano H., Kinoshita T. (1987). Chloropolysporins A, B, and C. Novel glycopeptide antibiotics from *Faenia interjecta* sp. nov. III. Structure elucidation of chloropolysporins. J. Antibiot. 40:933–940.
31. Inukai M., Takahashi H., Takeuchi M., Enokida R., Nagaki H., Kagasaki T. (1991). New antibiotics galacardin A and B. Japan Patent Appl. 03-83594 (assigned to Sankyo Co.), April 9, 1991.
32. Takeuchi M., Takahashi S., Inukai M., Nakamura T., Kinoshita T. (1991). Helvecardins A and B, novel glycopeptide antibiotics II. Structural elucidation. J. Antibiot. 44:271–277.
33. Athalye M., Coates N.J., Milner P.H. (1989). Novel glycopeptide antibiotics from *Amycolatopsis orientalis*, European Patent Appl. 339982 (assigned to Beecham Group plc), November 2, 1989.
34. Hunt A.H. (1983). Structure of the pseudoaglycone of A35512B. J. Am. Chem. Soc. 105:4463–4468.
35. Hunt A.H., Dorman D.E., Debono M., Molloy R.M. (1985). Structure of antibiotic A41030A. J. Org. Chem. 50:2031–2035.
36. Hunt A.H., Occolowitz J.C., Debono M., Molloy R.M., Maciak G.M. (1983). A47934 and A41030 factors—new glycopeptides and glycopeptide aglycones: Structure determination. 23rd Interscience Conference on Antimicrobial Agents and Chemotherapy, Las Vegas, Nevada, Abstract 441.
37. Doolin L.E., Gale R.M., Godfrey O.W. Jr., Hamill R.L., Mahoney D.F., Yao R.C.F. (1989). New glycopeptide antibiotic A80407. European Patent Appl. 299707 (assigned to Eli Lilly and Company), January 18, 1991.
38. Hunt A.H., Elzey T.K., Merkel K.E., Debono M. (1984). Structures of the actaplanins. J. Org. Chem. 49:641–645.
39. Harris C.M., Harris T.M. (1982). Structure of ristocetin A: Conformational studies of the peptide. J. Am. Chem. Soc. 104:363–365.
40. Lomakina N.N., Katrukha G.S., Brazhnikova M.G., Silaev A.B., Muravyova L.I., Trifonova Z.P., Tokareva N.L., Diarra B. (1982). Final structure of the glycopeptide antibiotic ristomycin A. Antibiotiki 27:248–252.
41. Skelton N.J., Williams D.H., Monday R.A., Ruddock J.C. (1990). Structure elucidation of the novel glycopeptide antibiotic UK 68597. J. Org. Chem. 55:3718–3723.

42. Skelton N.J., Williams D.H., Rance M.J., Ruddock J.C. (1991). Structure elucida-
     tion of a novel antibiotic of the vancomycin group. The influences of ion-dipole
     interactions on peptide backbone conformation. J. Am. Chem. Soc. 113:3757–
     3765.
43. Waltho J.P., Williams D.H., Selva E., Ferrari P. (1987). Structure elucidation of the
     glycopeptide antibiotic complex A40926. J. Chem. Soc., Perkin Trans. 1:2103–2107.
44. Michel K.H., Yao R.C.F. (1991). New lipoglycopeptide antibiotic A84575. Euro-
     pean Patent Appl. 424051 (assigned to Eli Lilly and Company), April 24, 1991.
45. Jeffs P.W., Muller L., DeBrosse C., Heald S.L., Fisher R. (1986). The structure of
     aridicin A. An integrated approach employing 2D NMR, energy minimization and
     distance constraints. J. Am. Chem. Soc. 108:3063–3075.
46. Folena-Wasserman G., Poehland B.L., Yeung E.W.K., Staiger D., Killmer L.B.,
     Snader K., Dingerdissen J.J., Jeffs P.W. (1986). Kibdelins (AAD-609), novel
     glycopeptide antibiotics. J. Antibiot. 39:1395–1406.
47. Box S.J., Coates N.J., Davis C.J., Gilpin M.L., Houge-Frydrych C.S., Milner P.H.
     (1991). MM 55266 and MM 55268, glycopeptide antibiotics produced by a new
     strain of *Amycolatopsis*. Isolation, purification and structure determination. J.
     Antibiot. 44:807–813.
48. Christensen S.B., Allaudeen H.S., Burke M.R., Carr S.A., Chung S.K., DePhillips
     P., Dingerdissen J.J., Dipaolo M., Giovenella A.J., Heald S.L., Killmer L.B., Mico
     B.A., Mueller L., Pan C.H., Poehland B.L., Rake J.B., Roberts G.D., Shearer
     M.C., Sitrin R.D., Nisbet L.J., Jeffs P.W. (1987). Parvodicin, a novel glycopeptide
     from a new species. *Actinomadura parvosata*: Discovery, taxonomy, activity and
     structure elucidation. J. Antibiot. 40:970–990.
49. Hamill R.L., Haney M.E. Jr., Stark W.M. (1973). Antibiotic A477. U.S. Patent
     3780174 (assigned to Eli Lilly and Company), December 18, 1973.
50. Michel K.H., Shah R.M., Hamill R.L. (1980). A35512, a complex of new
     antibacterial antibiotics produced by *Streptomyces candidus*. I. Isolation and charac-
     terization. J. Antibiot. 33:1397–1406.
51. Goldstein B.P., Selva E., Gastaldo L., Berti M., Pallanza R., Ripamonti F., Ferrari
     P., Denaro M., Ariolo V., Cassani G. (1987). A40926, a new glycopeptide antibiotic
     with anti-*Neisseria* activity. Antimicrob. Agents Chemother. 31:1961–1966.
52. Boeck L.D., Mertz F.P., Clem G.M. (1985). A41030. A complex of novel glycopep-
     tide antibiotics produced by a strain of *Streptomyces virginiae*. Taxonomy and
     fermentation studies. J. Antibiot. 38:1–8.
53. Boeck L.D., Mertz F.P. (1986). A47934, a novel glycopeptide-aglycone antibiotic
     produced by a strain of *Streptomyces toyocaensis*. Taxonomy and fermentation
     studies. J. Antibiot. 39:1533–1540.
54. Doolin L.E., Gale R.M., Godfrey O.W., Hamill R.L., Mahoney D.F., Yao R.C.
     (1991). Production of antibiotics by *Kibdelosporangium philippinensis*. U.S. Patent
     4996148 (assigned to Eli Lilly and Company), February 26, 1991.
55. Hamill R.L., Mabe J.A., Mahoney D.F., Nakatsukasa W.M., Yao R.C. (1988).
     Glycopeptide antibiotics A82846. European Patent Appl. 265071 (assigned to Eli
     Lilly and Company), April 27, 1988.
56. Mertz F.P., Yao R.C. (1989). *Streptosporangium carneum* sp. nov. isolated from soil.

Proceedings of the Society of Industrial Microbiology, Abstract P-80, Seattle, WA, August 13–18.

57. Tamura A., Takeda I. (1975). Antibiotic AB-65, a new antibiotic from *Saccharomonospora viride*. J. Antibiot. 28:395–397.

58. Boeck L.D., Stark W.M. (1984). Actaplanin, a complex of new glycopeptide antibiotics—fermentation. Dev. Ind. Microbiol. 25:505–514.

59. Yurina M.S., Lavrova M.F., Brazhnikova M.G. (1961). Actinoidin and its separation into biologically active variants. Antibiotiki 6:609–618.

60. Shearer M.C., Actor P., Bowie B.A., Grappel S.F., Nash C.H., Newman D.J., Oh Y.K., Pan C.H., Nisbet L.J. (1985). Aridicins, novel glycopeptide antibiotics. I. Taxonomy, production and biological activity. J. Antibiot. 38:555–560.

61. Kunstmann M.P., Mitscher L.A., Porter J.N., Shay A.J., Darken M.A. (1969). LL-AV290, a new antibiotic I. Fermentation, isolation and characterization. Antimicrob. Agents Chemother. 1968:242–245.

62. Omura S., Tanaka H., Tanaka Y., Spiri-Nakagawa P., Oiwa R., Takahashi Y., Matsuyama K., Iwai Y. (1979). Studies on bacterial cell wall inhibitors. VII. Azureomycins A and B, new antibiotics produced by *Pseudonocardia azurea* nov. sp. Taxonomy of the producing organism, isolation, characterization and biological properties. J. Antibiot. 32:985–994.

63. Okazaki T., Enokita R., Miyaoka H., Takatsu T., Torikata A. (1987). Chloropolysporins A, B, and C, novel glycopeptide antibiotics from *Faenia interjecta* sp. nov. I. Taxonomy of producing organism. J. Antibiot. 40:917–923.

64. Boeck L.D., Mertz F.P., Wolter R.K., Higgens C.E. (1984). *N*-demethylvancomycin, a novel antibiotic produced by a strain of *Nocardia orientalis*: Taxonomy and fermentation. J. Antibiot. 37:446–453.

65. Gauze G.F., Brazhinikova M.G., Lomakina N.N., Gol'berg L.E., Laiko A.V., Fedorova G.B., Berdnikova T.F. (1989). Eremomycin, a novel antibiotic of the polycyclic glycopeptide group. Antibiot. Khimioter. 34:348–352.

66. Takeuchi M., Enokita R., Okazaki T., Kagasaki T., Inukai M. (1991). Helvecardins A and B, novel glycopeptide antibiotics I. Taxonomy, fermentation, isolation and physio-chemical properties. J. Antibiot. 44:263–270.

67. Spiri-Nakagawa P., Fukushi Y., Maebashi K., Imamura N., Takahashi Y., Tanaka Y., Tanaka H., Omura S. (1986). Izupeptins A and B, new glycopeptide antibiotics produced by an actinomycete. J. Antibiot. 39:1719–1723.

68. Kunstmann M.P., Porter J.N. (1972). Antibiotic AM374 and the method of production using *Streptomyces eburosporeus*. U.S. Patent 3700768 (assigned to American Cyanamid Company), October 24, 1972.

69. Hayashi T., Harada Y., Ando K. (1975). Isolation and characterization of mannopeptins. J. Antibiot. 28:503–507.

70. Coates N.J., Elson A.L., Athalye M., Curtis L.M., Moores L.V. (1990). New glycopeptide antibiotic complex. European Patent Appl. 375213 (assigned to Beecham Group plc), June 27, 1990.

71. Kamogashira T., Nishida T., Sugawara M. (1983). A new glycopeptide, OA-7653, produced by *Streptomyces hygroscopicus* subsp. *hiwasaensis*. Agric. Biol. Chem. 47:499–506.

72. Brazhinikova M.G., Lomakina N.N., Lavrova M.F., Tolstykh I.V., Yurina M.S.,
     Klyueva L.M. (1963). Isolation and properties of ristomycin. Antibiotiki 8:392–396.
73. Verma A.K., Goel A.K., Rao A., Venkateswarlu A., Sitrin R.D. (1988). Glycopep-
     tide antibiotics. U.S. Patent 4742045 (assigned to Smith Kline Beckman Corpora-
     tion), May 3, 1988.
74. Parenti F., Beretta G., Berti M., Arioli V. (1978). Teichomycins, new antibiotics
     from *Actinoplanes teichomyceticus* nov. sp. J. Antibiot. 31:276–283.
75. Holdom K.S., Maeda H., Ruddock J.C., Tone J. (1988). Manufacture of glycopep-
     tide antibiotic UK-68597 with *Actinoplanes*. European Patent Appl. 265143 (as-
     signed to Pfizer, Ltd.), April 27, 1988.
76. Katayama N., Nozaki Y., Okonogi K., Ono H., Harada S., Okazaki H. (1985).
     Formadicins, new monocyclic β-lactam antibiotics of bacterial origin. I. Taxonomy,
     fermentation and biological activities. J. Antibiot. 38:1117–1127.
77. Nozaki Y., Katayama N., Tsubotani S., Ono H., Okazaki, H. (1985). Cephabacin
     M$_{1-6}$, new 7-methoxycephem antibiotics of bacterial origin. I. Producing organism,
     fermentation, biological activities and mode of action. J. Antibiot. 38:1141–1150.
78. Reynolds P.E. (1961). Studies on the mode of action of vancomycin. Biochim.
     Biophys. Acta 52:403–405.
79. Jordan D.C. (1961). Effect of vancomycin on the synthesis of the cell wall and
     cytoplasmic membrane of *Staphyloccus aureus*. Can. J. Microbiol. 11:390–393.
80. Wallace C., Strominger J.L. (1963). Ristocetin, inhibitors of cell wall synthesis in
     Staphyloccus aureus. J. Biol. Chem. 233:2264–2266.
81. Perkins H.R. (1969). Specificity of combination between mucopeptide precursors
     and vancomycin and ristocetin. Biochem J. 111:195–205.
82. Nieto M., Perkins H.R. (1971). Physicochemical properties of vancomycin and
     iodovancomycin and their complexes with diacetyl-L-lysyl-D-alanyl-D-alanine.
     Biochem. J. 123:773–787.
83. Kalman J.R., Williams D.H. (1980). An NMR study of the structure of the antibiotic
     ristocetin A. The negative nuclear Overhauser effect in structure elucidation. J. Am.
     Chem. Soc. 102:897–905.
84. Williams D.H., Butcher D.W. (1981). Binding site of the antibiotic vancyomycin
     for a cell-wall peptide analogue. J. Am. Chem. Soc. 103:5700–5704.
85. Reynolds P.E. (1989). Structure, biochemistry and mechanism of action of glyco-
     peptide antibiotics. Eur. J. Microbiol. Infect. Dis. 8:943–950.
86. Jeffs P.W., Nisbet L.J. (1988). Glycopeptide antibiotics: A comprehensive approach
     to discovery, isolation, and structure determination. In Actor P., Daneo-Moore L.,
     Higgins M.L., Salton K.R.J., Schockman G.D. (Eds.), Antibiotic Inhibition of
     Bacterial Cell Surface Assembly and Function. American Society of Microbiology,
     Washington, D.C. pp. 509–530.
87. Rake J.B., Gerber R., Mehta R.J., Newman D.J., Oh Y.K., Phelen C., Shearer M.C.,
     Sitrin R.D., Nisbet L.J. (1986). Glycopeptide antibiotics: A mechanism-based
     screen employing a bacterial cell wall receptor mimetic. J. Antibiot. 39:58–67.
88. Nieto M., Perkins H.R., Reynolds P.E. (1972). Reversal by a specific peptide
     (diacetyl-α,γ-L-diaminobutyryl-D-alanyl-D-alanine) of vancyomcyin inhibition in
     intact bacteria and cell-free preparations. Biochem. J. 126:139–149.

89.   Cassini G. (1989). Glycopeptides: Antibiotic discovery and mechanism of action. In Bushell M.E., Grafe U. (Eds.), Bioactive Metabolites from Microorganisms. Elsevier, Amsterdam Prog. Ind. Microbiol. 27:221–235.

90.   Yao R.C., Mahoney D.F. (1984). Enzyme-linked immunosorbent assay for the detection of fermentation metabolites: Aminoglycoside antibiotics. J. Antibiot. 37: 1462–1468.

91.   Yao R.C., Mahoney D.F. (1989). Enzyme immunoassay for macrolide antibiotics: Characterization of an antibody to 23-amino-O-mycaminosyltylonolide. Appl. Environ. Microbiol. 56:1507– 1511.

92.   Corti A., Rurali C., Borghi A., Cassani G. (1985). Solid-phase enzyme-receptor assay (SPERA): A competitive-binding assay for glycopeptide antibiotics of the vancomycin class. Clin. Chem. 31:1606–1610.

93.   Mahoney D.F., Baisden D.K., Yao R.C. (1989). A peptide binding chromogenic assay for detecting glycopeptide antibiotics. J. Ind. Microbiol. 4:43–47.

94.   Elespuru R.K., Yarmolinsky M.W. (1979). A colorimetric assay of lysogenic induction designed for screening potential carcinogenic and carcinostatic agents. Environ. Mutagen. 1:65–78.

95.   Sykes R.B., Wells J.S. (1982). Screening for beta-lactam antibiotics in nature. J. Antibiot. 38:119–121.

96.   Kirsch D.R., Lai M.H., McCullough J.M., Gillum A.M. (1991). The use of beta-galactosidase gene fusions to screen for antibacterial antibiotics. J. Antibiot. 44: 210–217.

97.   Counter F.T. Unpublished results.

98.   Vertesy L., Betz J., Fehlhaber H.W., Limbert M. (1993). New glycopeptide antibiotic balhimycin. European Patent Appl. 521408 (assigned to Hoechst AG), January 7, 1993.

99.   Coates N.J., Davis C.J., Curtis L.M., Sykes R. (1991). Glycopeptide antibiotics and their manufacture with *Amycolatopsis*. PCT Int. Appl. WO 9116346 (assigned to Beecham Group plc), October 31, 1991.

100. Cavalleri B., Parenti F. (1992). Glycopeptides (dalbaheptides). In Kirk-Othmer Encyclopedia of Chemical Technology, 4th Ed., Vol. 2. John Wiley & Sons, New York, pp. 995–1018.

2
Separation Methodology

ROBERT D. SITRIN
Merck Research Laboratories, West Point, Pennsylvania

GAIL FOLENA-WASSERMAN
MedImmune, Inc., Gaithersburg, Maryland

I. INTRODUCTION

The glycopeptides offer an interesting challenge to the isolation chemist in that they are very polar, highly charged molecules that cannot be extracted into organic solvents. The reported isolation methodology for this class of antibiotics spans the extreme of classic methods involving ion exchange, carbon adsorption, and selective precipitation and the use of modern affinity and high-performance liquid chromatography (HPLC) techniques. This chapter focuses on chromatographic methodology, both classic and modern, used to isolate and purify members of this class. In particular, the use of ligand affinity chromatography for the isolation of this class of antibiotics offers a first in the antibiotics isolation area, a generic isolation scheme applicable without exception to every member of this class regardless of its structural perturbations.

A. Overview of Glycopeptide Class

The first two glycopeptides to be discovered, ristocetin and vancomycin, are the two most extensively studied members of this class. Degradation experiments coupled with high-field nuclear magnetic resonance (NMR) and x-ray crystallography data, respectively, led to the assignment of structures for these two antibiotics (1–6). Table 1 lists in alphabetical order the currently known members of this class, producing organisms, date, and literature reference for the publication describing the first or most complete isolation procedure. Structures have been determined for most of the antibiotics listed and can be found elsewhere in this volume. Although they all contain a similar central heptapeptide core of

Table 1   Glycopeptide Antibiotics

Antibiotic	Producing organism	Date	Reference
A477	*Actinoplanes* sp.	1973	7
A35512	*Streptomyces candidus*	1980	8
A41030	*Streptomyces virginiae*	1986	9
A42867	*Nocardia* sp.	1989	10
A47934	*Streptomyces toyocaensis*	1983	11
A80407	*Kibdelosporangium philippinensis*	1990	12
AB65	*Saccharomonospora viride*	1975	13
A82846	*Nocardia orientalis*	1989	14
AM374	*Streptomyces eburosporeus*	1974	15
Actaplanin	*Actinoplanes missouriensis*	1984	16
Actinoidin	*Proactinomyces actinoides*	1963	17
Ardacin	*Kibdelosporangium aridum*	1985	18
Avoparcin	*Streptomyces candidus*	1969	19
Chloroorienticin	*Amycolatopsis orientalis*	1988	20
Chloropolysporins	*Faenia interjecta*	1987	21
Decaplanin	*Kibdelosporangium deccaensis*	1990	22
Eremomycin	*Actinomycete* sp.	1989	23
Helvecardin	*Pseudonocardia compacta*	1991	24
Kibdelins	*Kibdelosporangium aridum*	1986	25
Mannopeptin	*Streptomyces platensis*	1975	26
MM45289	*Amycolatopsis orientalis*	1990	27
MM47761	*Amycolatopsis orientalis*	1990	28
MM47766	*Amycolatopsis orientalis*	1990	29
MM49728	*Amycolatopsis orientalis*	1991	30
MM55270	*Amycolatopsis orientalis*	1990	31
OA7653	*Streptomyces hygroscopicus*	1983	32
Orienticins	*Nocardia orientalis*	1988	33
PA45052	*Nocardia orientalis*	1990	34
Parvodicin	*Actinomadura parvosata*	1986	35
Ristocetin	*Nocardia lurida*	1956	36
Synmonicin	*Synnemomyces mamnoorii*	1986	37
Teicoplanin	*Actinoplanes teichomyceticus*	1978	38
UK-68,597	*Actinoplanes* sp.	1990	39
UK-72,051	*Amycolatopsis orientalis*	1990	40
Vancomycin	*Streptomyces orientalis*	1956	41

aromatic and aliphatic amino acids terminating in actinoidinic acid, as a class they differ in the extent of chlorination of the nucleus, types of amino acids located elsewhere in the chain, extent of glycosylation, and presence or absence of fatty acid amides.

B. Mode of Action

Although the isolation schemes used for an antibiotic class are usually quite independent of its mode of action, the unique binding properties of the glycopeptides to bacterial cell wall moieties provide a suggestion for an alternative general isolation scheme. In general, all glycopeptide antibiotics have Gram-positive bacteriocidal activity against strains of *Staphyloccocus* and *Streptococcus*. Like the β-lactams, the glycopeptides are inhibitors of bacterial cell wall synthesis. Studies by Perkins and coworkers (42–46) led to the conclusion that vancomycin and the related antibiotic ristocetin both killed bacteria by binding to murein pentapeptides terminating in D-Ala-D-Ala, thereby inhibiting the extracellular transpeptidases that cross-link the bacterial cell wall. Without this final step the bacteria became susceptible to lysis. This particular affinity for peptides terminating in D-Ala-D-Ala was used to design a screen for novel glycopeptides (47) and as a basis for the affinity column isolation procedure described later in this chapter.

II. CLASSIC METHODS FOR EXTRACTION AND PURIFICATION

A. Vancomycin

When first discovered in the 1950s, vancomycin and ristocetin posed a difficult problem to the isolation chemist since they were neither solvent extractable, like the macrolides, nor readily isolable on an ion-exchange resin, like the aminoglycosides. As shown in the following isolation scheme from the original vancomycin publication and patent (41,48,49), most isolation schemes used charcoal or ion-exchange matrices in what appears to be a reversed-phase mode:

For clarified broth, follow these steps:

1. Adsorption at pH 7 onto Permutit DR resin or at pH 8.5 onto IRC-50 (H⁺)
2. Elution with acidic (or basic) aqueous acetone
3. Chromatography on Norite with acidic aqueous methanol
4. Precipitation as picrate salt
5. Conversion to hydrochloride salt
6. Purification by ion exchange on CM Sephadex with ammonium bicarbonate or precipitation as zinc salt

Since the elution from the ion-exchange column required organic solvent, the interaction was more hydrophobic than ionic in nature. This use of an ion-exchange resin was a forebear to the use of the HP-20 (Mitsubishi Industries) and XAD (Rohm and Haas) resins, currently the matrices of choice for the extraction of glycopeptides and many other antibiotics. The success of this original isolation was in part due to the reasonably high titers (0.2 mg/ml) of antibiotic in the fermentation broth (48).

B. Ristocetin

Ristocetin was isolated as shown here, again using charcoal as well as chromatography on acidic alumina (36). For clarified broth, follow these steps:

1. Adsorption at pH 7 onto Darco G-60
2. Elution with acidic aqueous acetone
3. Chromatography on a Darco G-60 column with acidic aqueous acetone
4. Chromatography on acidic alumina with aqueous methanol
5. Conversion to free base on mixed-bed ion-exchange column
6. Recrystallization from aqueous methanol

C. Actinoidin

Actinoidin, also first studied in the late 1950s, was isolated and purified by ion-exchange chromatography (17). For clarified broth, follow these steps:

1. Absorption at pH 3 onto Dowex 50X2
2. Elution with 1 N ammonia
3. Chromatography on Dowex 50 with ammonium acetate

D. Other Glycopeptides

Similar schemes were used to isolate the remaining glycopeptides listed in Table 1 using permutations of charcoal, acid alumina, and ion-exchange supports. Other laboratories have reported the use of polyamide with water as eluant for the purification of actaplanin (16), A477 (7), A35512 (8), and chloropolysporin (21); the use of an ion pair extraction for the isolation of avoparcin (50); and the use of butanol extraction and chromatography on LH-20 (Pharmacia) for the isolation and purification of teicoplanin (38).

E. Use of Macroreticular Resins XAD and HP-20

By far the most effective universal adsorbents for the extraction of glycopeptide antibiotics from crude fermentation broths, especially on a large scale, are the XAD (Rohm and Haas) and Dianion HP-20 (Mitsubishi Chemicals) macroreticular polymeric resins. These porous hydrophobic resins carry out a solid-phase extraction by a reversed-phase mode. Although silica-based reversed-phase adsorbents could in theory be used to extract these antibiotics from clarified broths, the polymeric resins are more robust and considerably less expensive for large-scale use. Polymeric resins have been used as the first step in the isolation of A35512 (8), A41030 (9), A47934 (11), A80407 (12), AB2846 (14), actaplanin (51), actinoidin A$_2$ (52), ardacin (18), chloroorienticins (20), chloropolysporins (21), decaplanin (22), helvecardin (24), kibdelin (23), MM47761 (28), orienticins (33), PA-45052 (34), parvodicin (35), synmonicins (37), UK-68,597 (39), and vancomycin (51,53). Although these resins are very effective in extracting glycopeptides from crude fermentation broths, the eluted product is often only 50–70% pure and requires further chromatography to separate individual components from each other and from broth contaminants.

III. AFFINITY CHROMATOGRAPHY

The glycopeptide antibiotics all contain a highly cross-linked aromatic heptapeptide nucleus that provides sufficient hydrophobicity for the solid-phase extraction onto XAD or HP-20 resins. However, the structural diversity of the members of this class precludes the use of other generic isolation and purification techniques based on a single physicochemical property. The glycopeptides are all too polar to extract into organic solvents, and they differ too widely in charge characteristics, from the negatively charged ardacins (18,54) to the positively charged ristocetin (54). Further diversity is evident in the number and size of attached carbohydrates and fatty acids. Thus, it is impossible to predict a single ion-exchange extraction procedure useful for all glycopeptides. However, one major parameter common to all glycopeptide antibiotics is their ability to bind to peptides terminating in D-Ala-D-Ala. It is this property that can be used to develop a universal isolation adsorbent for this class of antibiotics using affinity chromatography.

Affinity chromatography is a commonly used tool for the purification of proteins wherein the material to be isolated binds to a specific receptor or ligand that is immobilized on an insoluble support (55,56). After washing the support with a suitable solvent to remove weakly bound contaminants, the desired protein is desorbed from the column by elution with the ligand or with a solution that disrupts the interaction. The receptor or ligand is chosen to mimic the natural affinity of the protein to be isolated for a specific substrate or cofactor. Almost all reported examples of affinity chromatography are those in which at least one of the interacting species is a protein or large peptide. Examples in which both the ligate and ligand are of low molecular weight are rare, limited at present to chiral columns (57). The structures posed by two potentially interacting low-molecular-weight ligates have only a small number of sites available to form a tight binding complex. Glycopeptide antibiotics are an important exception to this since they inhibit peptidoglycan biosynthesis in the bacterial cell wall by specifically binding to pentapeptide precursors terminating with L-Lys-D-Ala-D-Ala (1,58) and are themselves less than 2000 molecular weight. The binding of these low-molecular-weight glycopeptides to cell wall precursors is analogous to the same phenomenon observed for protein-ligand interactions and suggests it can also be exploited using the technique of affinity chromatography.

For affinity chromatography to be effective, binding constants between the ligand and the protein should optimally be at least 10^4 L/mol (55,56), there should be a minimum of nonspecific binding and there should be an effective desorption procedure. Solution binding studies by Nieto and Perkins previously showed that the glycopeptide antibiotics vancomycin and ristocetin bind to the tripeptide Ac_2-L-Lys-D-Ala-D-Ala with a binding constant of 10^6 L/mol and to several dipeptide derivatives containing Ac-D-Ala-D-Ala with binding constants between 10^4 and 10^5 L/mol (1,2,42–46,58). Model studies of solid-state interactions performed with an insoluble peptide made by linking α-N-Ac-L-Lys-D-Ala-D-Ala to carboxymethylcellulose demonstrated that adsorption of vancomycin mimics its natural binding to cell wall preparations (46). Unfortunately, desorption of bound vancomycin from this support was difficult and recovery of more than 50% of the antibiotic was never achieved. Corti and Cassani (59) described an affinity technique for isolating teicoplanin from fermentation broths on Sepharose 4B-D-Ala-D-Ala. A 90% recovery of antibiotic was reported, but the product purity was only 65%. In the next section, the development and optimization of our affinity support are described, along with experimental protocols for its use and examples of purifications run on the micro- and macroscale. More detailed discussions of the development of the matrix can be found in our publication (60).

A. Selection of Ligands and Support

The choice of ligands used in this study was based on the original solution binding studies carried out on the glycopeptides vancomycin and ristocetin B by Nieto and Perkins (43,45). As noted, the strongest binding was observed for the tripeptide Ac_2-L-Lys-D-Ala-D-Ala, with almost 10-fold lower affinity observed for the dipeptide Ac-D-Ala-D-Ala. In all cases, a D configuration is required in both alanines for binding to occur. Longer peptides approaching the length of the natural pentapeptide (L-Ala-D-γ-Glu-L-Lys-D-Ala-D-Ala) failed to exhibit tighter binding. Studies by Williams and others (61–64) using NMR techniques have led to the proposed model shown in Figure 1 for the interaction of vancomycin with the tripeptide Ac_2-L-Lys-D-Ala-D-Ala. Although several of the ligands studied effectively mimicked the natural pentapeptide with binding constants in excess of 10^4 L/mol, studies were undertaken to see whether the full tripeptide sequence was required for effective purification of glycopeptide antibiotics or whether a smaller fragment would suffice. The ligands selected for immobilization included the tripeptide (with a free ϵ-amino group as an attachment site), the dipeptide D-Ala-D-Ala, and the amino acids D-Ala and glycine

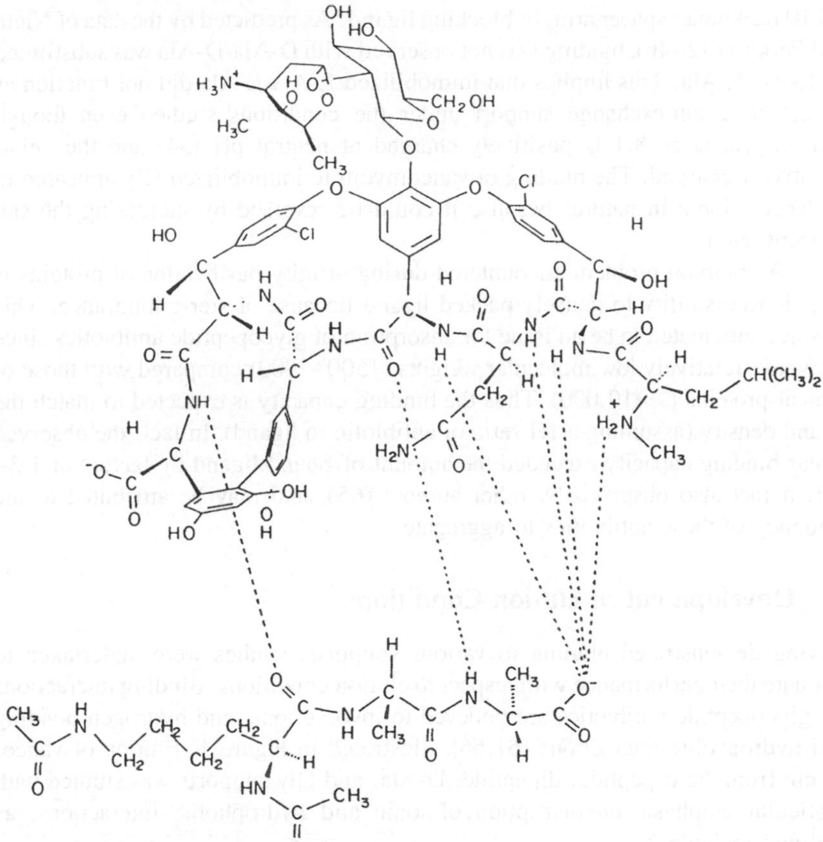

Fig. 1 Proposed interactions between vancomycin and Ac$_2$-L-Lys-D-Ala-D-Ala. [From Williams et al. (60).]

(with free α-amino groups as attachment sites). A commercially available activated support containing a 10-carbon spacer terminating in an N-hydroxysuccinimide ester, Affi-gel 10 (Bio-Rad Laboratories, Richmond, CA), was used for these studies. Immobilization was carried out by reacting a twofold excess of ligand to active ester followed by blocking of unreacted active groups with 1 M ethanolamine adjusted to pH 9. Typically, this procedure yielded a gel containing 5–8 μmol ligand per ml as measured by amino acid analysis.

Vancomycin was observed to bind to all the immobilized ligands. It did not bind to a control gel that was completely blocked by reaction with ethanolamine, suggesting there are no nonspecific interactions between vancomycin and the Affi-gel 10 backbone, spacer arm, or blocking ligand. As predicted by the data of Nieto and Perkins (42–46), binding was not observed with D-Ala-D-Ala was substituted by L-Ala-L-Ala. This implies that immobilized L-Ala-L-Ala did not function as an effective ion-exchange support under the conditions studied even though vancomycin at p*I* 8.1 is positively charged at neutral pH (54) and the gel is negatively charged. The binding of vancomycin to immobilized Gly appeared to be largely ionic in nature, because it could be reversed by increasing the salt concentration.

A common problem encountered during affinity purification of proteins is their inaccessibility to densely packed ligand because of steric hindrance. This was not anticipated to be an issue for adsorption of glycopeptide antibiotics since they have relatively low molecular weights (1500–2000) compared with those of typical proteins ($\gg$10,000). Thus the binding capacity is expected to match the ligand density (assuming a 1:1 ratio of antibiotic to ligand). In fact, the observed molar binding capacity exceeded the amount of bound ligand by factors of 1.3–1.7, a fact also observed by other authors (65), and may be attributed to the tendency of these antibiotics to aggregate.

B. Development of Elution Conditions

Having demonstrated binding to various supports, studies were undertaken to evaluate their performance with respect to elution conditions. Binding interactions for glycopeptide antibiotics are believed to involve ionic and hydrogen bonding and hydrophobic interactions (61,66), illustrated in Figure 1. Elution of vancomycin from the tripeptide, dipeptide, D-Ala, and Gly supports was studied with particular emphasis on disruption of ionic and hydrophobic interactions, as outlined in Table 2.

Solution studies predict (43,45) that the binding constants of the immobilized ligands should increase going from Gly to D-Ala to α-N-Ac-L-Lys-D-Ala-D-Ala. As expected, elution of vancomycin by disruption of ionic forces alone followed this trend. For example, with 0.5 M NaCl at pH 8, near the isoelectric point of vancomycin, 66% of bound antibiotic was eluted from the Gly support, 38% was recovered from the D-Ala support, and 29% from the dipeptide support, and no antibiotic was eluted from the tripeptide support. Using 0.4 M sodium carbonate buffer at pH 9.5 (above the isoelectric point of vancomycin), almost complete recovery could be obtained from the immobilized Gly. The same conditions yielded only 61% recovery from the dipeptide gel. Since a further increase in pH could lead to air oxidation or degradation of the antibiotic, other means of elution were evaluated. The use of acetonitrile-water mixtures to disrupt hydrophobic interactions without pH or ionic strength adjustments gave poor recoveries (20–27%, Table 2) from the amino acid and dipeptide supports. By

Table 2 Development of Elution Conditions[a]

		% Recovery			
Interaction	Eluant	Gly	D-Ala	D-Ala-D-Ala	Ac-L-Lys-D-Ala-D-Ala
Ionic	0.5 M NaCl, pH 8	66	38	29	0
	0.4 M carbonate, pH 9.5	82	73	61	—
Hydrophobic	30% Acetonitrile	27	20	23	—
Ionic + hydrophobic	0.4 M carbonate + 30% acetonitrile, pH 9.5	104	100	100	40
	0.1 M ammonia + 50% acetonitrile	100	100	100	—

[a]A solution containing 16 mg vancomycin in 0.02 M phosphate, pH 7, was batched onto 0.5 ml of each support and eluted with 4×4 ml of each eluant. Vancomycin was determined by UV spectroscopy.

using both high pH and acetonitrile, however, full recovery of vancomycin could be achieved for the amino acid and dipeptide gels.

The support containing the tripeptide ligand was considered undesirable because of the poor recovery of antibiotic on elution ($\sim$40%) and the need for synthesis of the commercially unavailable ligand. However, it was unclear whether the binding to the amino acid or dipeptide gels would be tight enough to allow purification of antibiotics from crude fermentation broths containing many contaminants. Therefore, studies were undertaken to determine if binding to the other gels would be tight enough to allow a series of selective wash steps to remove nonspecifically bound contaminants before elution of pure product. As shown in Table 3, the results on affinity chromatography of a vancomycin fermentation broth sample were similar to those previously observed for elution of the pure standard. Although vancomycin from a fermentation broth binds to the Gly gel, over 90% was washed off the gel in the low and high ionic strength washes (phosphate and ammonium acetate), along with adsorbed colored broth contaminants. In contrast, only 19% of the antibiotic was eluted from the D-Ala gel with the washes, and recovery was 81% in the final ammonia-acetonitrile elution. The dipeptide gel had the greatest specificity in that 95% of the bound vancomycin was recovered in the final eluant and only 5% was lost in the washing steps. Because removal of broth contaminants requires both high ionic strength (0.5 M acetate) and organic solvent (acetonitrile) washes, the D-Ala-D-Ala ligand was the smallest peptide of those studied that maintained sufficient selectivity between nonspecifically adsorbed materials and the antibiotic.

Table 3   Modification of Elution Conditions for Purification from Fermentation Broths[a]

		% Recovery on sequential elution			
Antibiotic	Ligand	0.02 M Phosphate	0.5 M Am Ac, pH 7.8	10% MeCN	0.1 M NH$_3$ and 50% MeCN
Vancomycin	Gly	32	59	8	0
	D-Ala	0	15	4	81
	D-Ala-D-Ala	0	5	0	95
Ardacin	Gly	84	5	11	0
complex	D-Ala	7	2	85	6
	D-Ala-D-Ala	1	0	72	27

[a]Fermentation broths containing 8 mg of each antibiotic were loaded onto 2 ml of each support, and the gels sequentially eluted with 10 ml of each buffer or solvent. Sequential recoveries were measured by HPLC analysis.

C.   Applicability to Other Glycopeptide Antibiotics

To test the generality of this observation, the same series of experiments were repeated using the ardacin complex since these antibiotics show significant variations in both physical and chemical properties compared with vancomycin. The ardacins are novel glycopeptide antibiotics (Fig. 2) that differ from vancomycin in that they contain $C_{10}-C_{12}$ lipid side chains and a carboxyl group instead of an amino group on their carbohydrates, giving them a net negative charge at neutral pH (18,67–69). The difference in isoelectric points (8.1 versus 3.8) between vancomycin and the ardacin complex (18,54) offered an opportunity to test whether the negatively charged ardacins could be purified on a negatively charged matrix. Elution studies of the ardacin complex from the Gly, D-Ala, and dipeptide gels are also presented in Table 3. Binding to the Gly support was very weak and largely ionic in nature: the bulk of antibiotics appeared in the early washes, presumably as a result of charge repulsion. For the D-Ala- and D-Ala-D-Ala-containing gels, an additional interaction was evident in that minimal breakthrough of the complex occurred at pH 7.8 despite a net negative charge on both the matrix and the antibiotic. Since over 70% of the ardacin complex could be recovered by elution with 10% acetonitrile-water for both gels, the extra interaction was presumably hydrophobic in nature. Although the leakage of the antibiotic complex from the dipeptide gel was substantial, effective purification could be achieved by washing with one column volume of 1% acetonitrile-water to remove contaminants and then eluting bound complex with 0.1 M ammonia and 50% acetonitrile in high recovery (see the large-scale isolation described later).

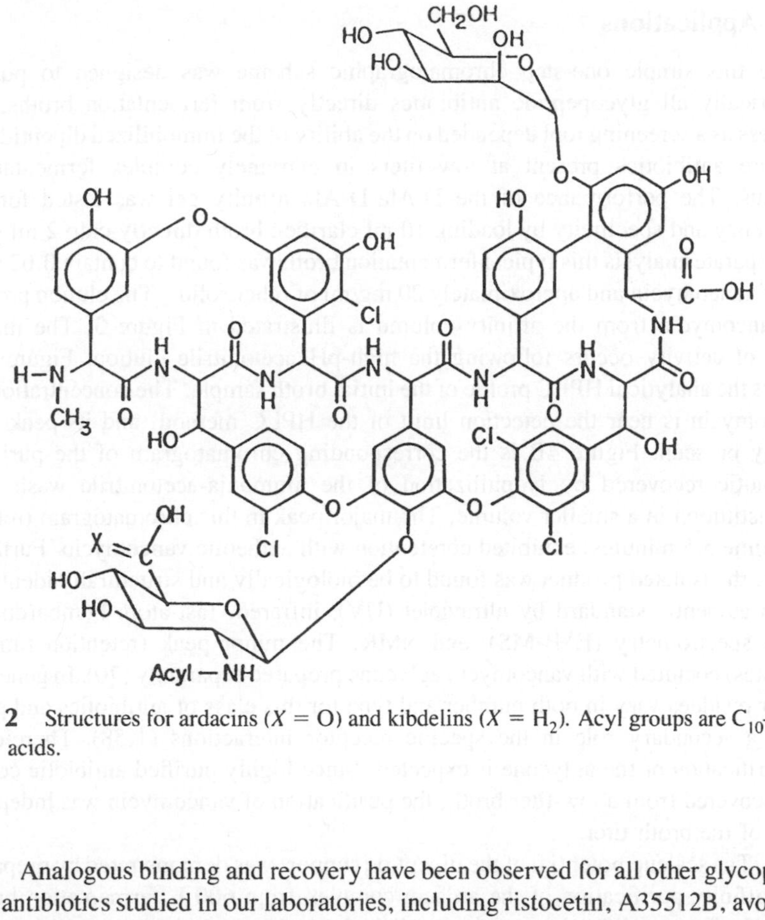

Fig. 2 Structures for ardacins ($X = O$) and kibdelins ($X = H_2$). Acyl groups are C_{10}–C_{12} fatty acids.

Analogous binding and recovery have been observed for all other glycopeptide antibiotics studied in our laboratories, including ristocetin, A35512B, avoparcin, OA-7653, A477, parvodicin, actaplanin, actinoidin, kibdelin, synmonicin, and teicoplanin. The D-Ala-D-Ala affinity support could thus be predicted to serve as a generic affinity support for isolating similar glycopeptide antibiotics operating by the same mechanism. This class-specific binding and elution became the basis for the use of this gel in a screen for glycopeptide antibiotics (47) efficiently to isolate and purify any member of this class, including previously unreported novel entities. Several of the more recently published glycopeptides, including A42867 (10), helvecardin (24), MM45289 (27), MM47761 (28), MM47766 (29), MM55220 (31), UK-68,597 (39), and UK-72,051 (40) have used affinity chromatography as an isolation step. Affinity chromatography has been used by one group as a primary screen for glycopeptide antibiotics to minimize the possibility of interference in a disk competition assay by the presence of other classes of antibiotics (27).

D. Applications

Since this simple one-step chromatographic scheme was designed to purify
generically all glycopeptide antibiotics directly from fermentation broths, its
success as a screening tool depended on the ability of the immobilized dipeptide to
capture antibiotics present at low titers in extremely complex fermentation
milieus. The performance of the D-Ala-D-Ala affinity gel was tested for its
efficiency and specificity by loading 10 ml clarified broth directly onto 2 ml gel.
On separate analysis this typical fermentation broth was found to contain 0.02 mg/
ml of vancomycin and approximately 20 mg/ml of other solids. The elution profile
of vancomycin from the affinity column is illustrated in Figure 3. The major
peak of activity occurs following the high-pH acetonitrile elution. Figure 4A
shows the analytical HPLC profile of the initial broth sample. The concentration of
vancomycin is near the detection limit of the HPLC method, and its peak can
barely be seen. Figure 4B is the corresponding chromatogram of the purified
antibiotic recovered by lyophilization of the ammonia-acetonitrile wash and
reconstitution in a smaller volume. The major peak in this chromatogram (retention time 5.5 minutes) exhibited coretention with authentic vancomycin. Furthermore, the isolated product was found to be biologically and structurally identical
to an authentic standard by ultraviolet (UV), infrared, fast-atom bombardment
mass spectrometry (FAB-MS), and NMR. The minor peak (retention time 8
minutes) coeluted with vancomycin aglycone prepared separately (70). In general,
sugar residues vary in both number and type for this class of antibiotics and play
only a secondary role in the specific receptor interactions (1,58). Therefore,
copurification of the aglycone is expected. Since highly purified antibiotic could
be recovered from a low-titer broth, the purification of vancomycin was independent of the broth titer.

The scaleup potential of the dipeptide support was demonstrated by preparative affinity purification of the ardacin complex from 600 L fermentation broth
(Fig. 5). Two modifications of the small-scale studies were incorporated into the
final large-scale process. An initial precipitation and a chromatography step on
XAD-7 were added to reduce the initial 600 L volume to 4 L and to remove broth
contaminants, which tended to contaminate the affinity gel. Second, after the
XAD-7 concentrate was loaded onto the affinity gel, the column was washed with
0.5 M, pH 7.8, acetate buffer and 1% acetonitrile before elution with 0.1 M
ammonia containing 50% acetonitrile. This acetonitrile wash step using 1%
instead of 10% solvent resulted in less than 5% loss of bound ardacins and
eliminated the bulk of colored contaminants. The 36 g purified product obtained
after direct lyophilization was identical in purity to an 8 mg sample purified in a
small-scale experiment. The product was a white powder that contained less than
1% inorganic material on combustion analysis. The analytical HPLC of this
product (Fig. 6) shows at least 28 peaks, all of which were later identified as

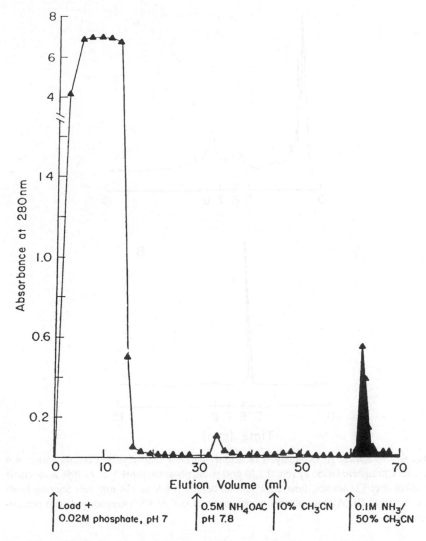

Fig. 3 Affinity purification of vancomycin from clarified fermentation broth. Clarified broth (10 ml) was chromatographed on a 10 × 1 cm ID column containing 2 ml Affi-gel 10–D-Ala-D-Ala. The effluent was monitored by UV at 280 nm. Elution of vancomycin is indicated by the shaded region. [Reproduced with permission (60).]

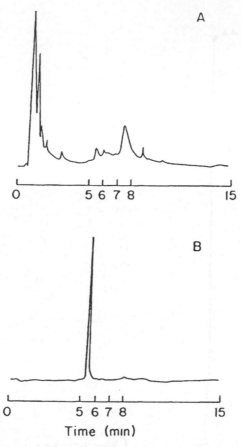

Fig. 4 Analytical HPLC analysis of affinity-purified vancomycin. Column 150 × 4.6 mm ID Ultrasphere ODS; solvent 0.1 M potassium phosphate, pH 3.2, 7–30% acetonitrile gradient over 13 minutes; flow 2 ml/minute; detection UV at 254 nm. (A) Starting broth at 0.2 AUFS; (B) Affinity-purified vancomycin at 0.1 AUFS. [Reproduced with permission (60).]

ardacin-like glycopeptides (71). The components were found to vary only in their glycolipid moieties as determined by UV analyses, isoelectric points, FAB mass spectra, and carbohydrate and lipid compositions. The major components, ardacins A, B, and C, were identical to those obtained by conventional chromatography (18).

Almost all reported glycopeptide antibiotics are, like the ardacins, produced

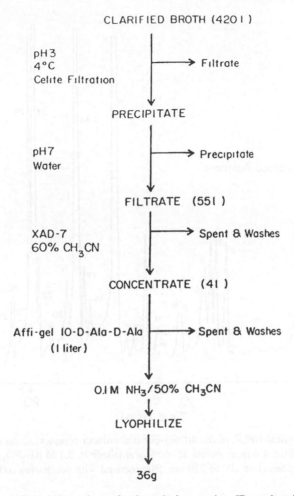

CLARIFIED BROTH (420 l)

pH 3
4°C
Celite Filtration
→ Filtrate

PRECIPITATE

pH 7
Water
→ Precipitate

FILTRATE (55 l)

XAD-7
60% CH₃CN
→ Spent & Washes

CONCENTRATE (4 l)

Affi-gel IO-D-Ala-D-Ala
(I liter)
→ Spent & Washes

0.1 M NH₃/50% CH₃CN

LYOPHILIZE

36g

Fig. 5 Large-scale isolation scheme for the ardacin complex. [Reproduced with permission (60).]

as mixtures that differ in carbohydrate, lipid, and chlorine content. Since all members of this class function by the same mechanism, the affinity gel is expected to yield the entire mixture of antibiotic products. By careful design of the elution conditions, however, selective desorption from the column can be achieved. For example, during the fermentation of the kibdelin complex significant amounts of the corresponding ardacins are produced (25,72). The two antibiotics (Fig. 2) differ only in the oxidation state of the glycolipid side chains. The kibdelins with

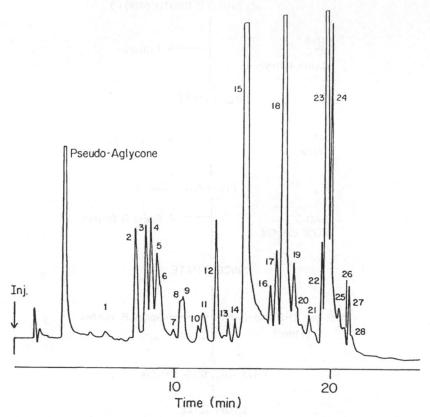

Fig. 6   Analytical HPLC of the affinity-purified ardacin complex. Column Ultrasphere ODS, 5 μm 150 × 4.6 mm; solvent 25–60% acetonitrile in 0.1 M KH_2PO_4, pH 3.2; flow 1.5 ml/minute; detection UV at 220 nm. [Reproduced with permission (60).]

their 2-acylamino glucose moieties, are the biosynthetic precursors of the corresponding ardacins with their 2-acylamino glucuronic acids. Because of this subtle structural difference, the two complexes elute very closely on analytical HPLC. Thus, it could be predicted that these two complexes would be difficult to separate by preparative HPLC. Fortunately, complete separation of the two complexes from each other could be effected by careful design of the elution conditions. Since the ardacins are highly anionic at neutral pH, charge repulsion forces weaken their binding to the affinity matrix. Thus, all the ardacins can be selectively eluted off the gel using 5% acetonitrile before complete elution of the kibdelins using the normal 50% acetonitrile ammonia eluant. The high-pH eluate contained only the kibdelins and was suitable for separation by preparative HPLC.

E. General Utility of the Immobilized D-Ala-D-Ala Supports

Affinity chromatography with Affi-gel 10 and D-Ala-D-Ala can be used either in a batch procedure with a 1:5 or 1:10 volume ratio of gel to sample or by direct flow through a column. The kinetics of binding at room temperature are generally fast, with quantitative binding of antibiotic at flow rates up to 120 cm/h until the capacity of the column is reached. Similarly, a 10–15 minute contact time in batch mode is sufficient. Binding occurred under a variety of conditions typically encountered in fermentation broths and is not sensitive to batch-to-batch variations. All glycopeptides we have examined bind to the gel at neutral pH. Of the several elution conditions examined for vancomycin, the 0.1 M ammonia and 50% acetonitrile mixture offered both high recovery of bound antibiotic and the ability to isolate pure product by lyophilization without needing a subsequent chromatography step to remove nonvolatile salt components. Replacement of acetonitrile by methanol or ethylene glycol or elution at pH values between 2.5 and 9.5 using 4 M NaCl without acetonitrile resulted in lower product recovery.

An interesting modification of the affinity technique was recently described (73). In place of a rigid agarose support D-Ala-D-Ala was attached to a polyethylene glycol (PEG) polymer, which was used to prepare an affinity two-phase PEG-dextran partitioning system. The technique offers some potential use on scaleup because the liquid phases are in theory easier to handle.

IV. PREPARATIVE REVERSED-PHASE HPLC

The affinity chromatography techniques described earlier can be considered a general isolation step for all glycopeptide antibiotics since the binding mechanism is directly related to their mode of action. Since every reported glycopeptide antibiotic is produced as a complex mixture of analogs, the affinity products by definition are mixtures. To separate these mixtures into their components for spectral and biological evaluation, additional chromatographic separations are required. Just as analytical HPLC is now the standard accepted tool for the analysis of pharmaceuticals (74,75), preparative HPLC is also becoming a vital tool for the isolation and purification of natural products (for examples see references listed in Refs. 76–79). The reversed-phase mode is particularly effective in separating glycopeptides since most glycopeptides exist as complexes differing in carbohydrate, lipid, or chlorine content, aspects predicted to affect hydrophobicity. When, as is the case for the ardacins, the factors contain homologous fatty acids (18,69,71), the resulting lipophilicity differences are sufficient to separate the antibiotics easily on a reversed-phase column (see later). Preparative reversed-phase HPLC has been used in the isolation of most of the glycopeptides listed in Table 1.

A fundamental question arises when choosing to run preparative HPLC:

what column size, column loading, and particle size are most appropriate for a given separation? A strategic approach to answering these questions based on α values and plate counts has been presented in review articles on the subject (76–79). Briefly, there appear to be two types of preparative HPLC modes: a "semi-preparative" run at relatively low loading on columns packed with 5–10 μm supports and a "large-scale" run on large inner diameter (ID) columns packed with larger particle (40–60 μm) supports run at high loading.

A.   Small-Scale Separations on 5–10 μm Supports

From a strategic point of view, there are few decisions to be made in choosing the appropriate column to isolate the milligram quantities of a new antibiotic to obtain in vitro biological data and structural information by FAB-MS and NMR techniques. A commercially packed semipreparative column with a 5 or 10 μm support attached to an ordinary analytical HPLC system suffices to separate milligram amounts of an antibiotic in a few hours time. This technique has been widely used in the natural products area (see Refs. 67–92 in Ref. 76) and in particular for the purification of the glycopeptides OA-7653 (32), teicoplanin (80), chloropolysporins (21), and the ardacin complex. In practice, an analytical solvent system is developed that separates the mixture into its components and then the separation is scaled up onto a larger diameter column packed with the same analytical (small particle) support. Figure 6 shows an analytical separation of several members of the affinity-purified ardacin complex on a 4.6 mm ID column packed with 5 μm reversed-phase support (71). To scale up this separation to a preparative level while maintaining efficiency and resolution, loading for the corresponding preparative run had to be limited to 200 mg per injection on a 25 mm ID column, as shown in Figure 7. The quantities obtained from this column were sufficient for obtaining biological data as well as FAB-MS and NMR spectra. In many cases, the separate antibiotic factors were isomers and differed only in fatty acid content. For example, the factors labeled 21, 22, and 23 differed in that 21 and 22 were found to contain branched C_{12} fatty acids and factor 23 contained a straight-chain C_{12} fatty acid. The reversed-phase support was capable of distinguishing between isomers of molecular weight 1800 differing in the position of a single carbon atom on a fatty acid side chain (71).

In this example, the sample was loaded as a large volume of dilute solution in an organic-free solvent. The antibiotic adsorbed to the head of the column, and polar contaminants washed through the column with the elution solvent. Antibiotic was eluted with a high ionic strength (0.1–0.2 M) aqueous buffer using a continuous shallow gradient of acetonitrile. The highly buffered aqueous solvent resulted in sharper peaks and provided strict control of pH during the chromatography. Fractions containing the desired product as detected by UV absorption were desalted by passing over a column of XAD-7 or HP-20 followed by elution with 50% acetonitrile and lyophilization. The affinity purification procedure described

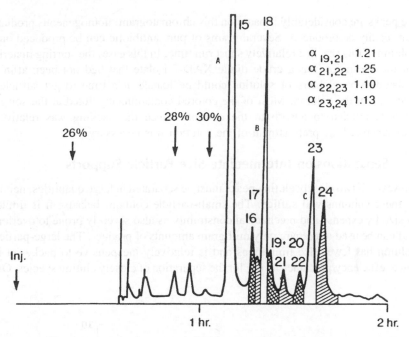

Fig. 7 Semipreparative separation of the ardacin complex. Sample 200 mg: column Whatman Magnum 20, Partisil ODS-3, 10 μm, 2.2 × 50 cm; solvent 26–30% acetonitrile in 0.1 M KH_2PO_4, pH 6.0; flow 15 ml/minute; detection UV at 254 nm (71).

earlier is an excellent pretreatment step in that all nonglycopeptide contaminant products are removed from the sample before HPLC separation.

B.   Separations on Large-Particle Supports

Unfortunately, the high cost and limited availability of small-particle semipreparative columns precludes their use in the preparation of larger (multigram) amounts of antibiotic needed for extensive biological, chemical, spectroscopic, or semisynthetic studies. One approach to the scaleup of preparative HPLC is the use of larger particle (40–60 μm) packing in 2 inch or larger diameter columns. These reversed-phase supports are widely available from several manufacturers, are relatively inexpensive, and are easily user packed by a dry-packing procedure. Furthermore, because of their larger particle size, lower instrument backpressures are required to obtain reasonable flow rates, typically under 1000 psi at 500 ml/minute for a 2 inch diameter column. Unfortunately, because of the large particle size, these columns are relatively inefficient and can be used only where peaks are widely separated. Figure 8 shows a separation of the ardacin antibiotics on a column packed with a 40–60 μm reversed-phase support (18). Note that although the peaks are considerably broader on this chromatogram, homogeneous products can readily be produced. Several grams of pure antibiotic can be produced on a column of this size in a relatively short run time. In this case, the starting material for the separation was a crude dilute XAD-7 isolate that had not been affinity isolated. Several liters of solution could be loaded at a time to get sufficient material on the column. Most of the colored contaminants eluted at the solvent front and failed to adsorb to the column. Since the packing was relatively inexpensive, little pretreatment of the solution was necessary.

C. Separations on Intermediate-Size Particle Supports

However, if two closely eluting peaks must be separated in large quantities, neither of these columns will suffice. The small-particle column, because it is limited in size by expense and mechanical constraints, is also severely prone to overload and can be used to prepare only milligram amounts of product. The large-particle column has fewer size limitations and is relatively inexpensive to pack, yet its lower efficiency precludes its use for the separation of closely eluting species. One

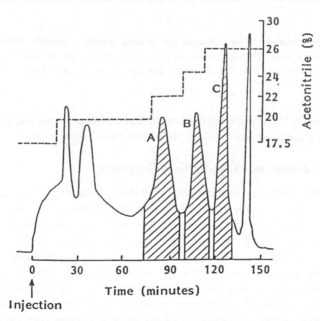

Fig. 8 Preparative HPLC separation of ardacins A, B, and C on a large particle support. Sample 25 g crude ardacin complex containing 2.2 g ardacin A, 1.8 g ardacin B, and 1.4 g ardacin C; column Whatman Partisil Prep 40 ODS-3, 37–60 μm, 4.8 × 50 cm; solvent 17–26% acetonitrile in 0.1 M KH_2PO_4, pH 6.0; flow 250 ml/minute; detection UV at 210 nm. [Reproduced with permission (18).]

approach to solving this problem is the use of an intermediate (15–20 μm) particle size. Columns 2 inches in diameter or larger can readily be dry packed with these supports and yield efficiencies in excess of 4000 plates per meter. An example from our laboratories is the separation of the parvodicin antibiotics (35). In this case, only limited amounts of pure antibiotics could be obtained on a 10 μm column, and no separation could be observed on a column packed with a 40–60 μm support. Figure 9 shows an analytical separation of an affinity chromatography-purified parvodicin complex in which the separation of factors C_1 and C_2 was desired. Figure 10 shows the corresponding preparative separation on a 2 inch diameter column packed with a 15–20 μm support. Structurally these factors are isomers that differ only in that one contains a branched C_{12} fatty acid and the other a straight-chain fatty acid, in a manner analogous to the structural differences in ardacins 21–23. Several hundred milligrams of each isomer could be prepared in high purity using three injections to minimize overload.

The development of technology for packing large columns with 15 μm or smaller supports (81) enables this technique to be scaled up even further for

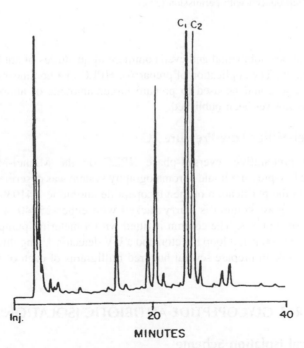

Fig. 9 Analytical separations of the parvodicin complex. Column Vydac C-18, 5 μm, 250 × 4.6 mm; solvent 20–30% acetonitrile in 0.1 M KH_2PO_4, pH 6.0; flow 1.0 ml/minute; detection UV at 225 nm. [Reproduced with permission (35).]

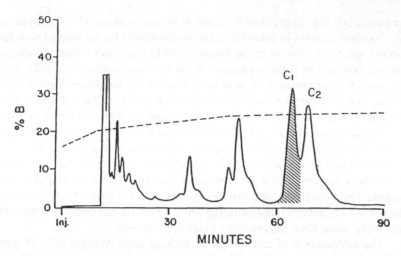

Fig. 10 Preparative separation of the parvodicin complex. Sample affinity-purified parvodicin complex (650 mg); column Vydac C-18, 15–20 μm, 300 Å, 500 × 51 mm; solvent 15–20% acetonitrile in 0.1 M KH₂PO₄, pH 6.0; flow 100 ml/minute; detection UV at 280 nm. [Reproduced with permission (35).]

preparation of developmental and even commercial quantities of antibiotics with similar retention. The application of prepartive HPLC in a nonlinear or displacement mode also could be used to prepare larger amounts of antibiotics. This approach has not yet been published.

D.   Michel-Miller Low-Pressure LC

The use of preparative reversed-phase HPLC on the Michel-Miller high-performance low-pressure liquid chromatography system was described by Eggert and Michel in the purification of the glycopeptide antibiotic A41030 (9). In this case a tapered glass column is slurry packed with either 25–40 or 10–20 μm reversed-phase supports. The column is fitted with a metering pump capable of 25 ml/minute at 100 psi, a loop injector, and a UV detector. Using this system the authors were able to prepare several hundred milligrams of each of the A41030 factors.

V.   GENERAL GLYCOPEPTIDE ANTIBIOTIC ISOLATION SCHEME

A.   General Isolation Scheme

From the previous discussions it is apparent that the three techniques just described, HP-20, affinity chromatography, and preparative reversed-phase

HPLC, offer a general purification scheme for all glycopeptide antibiotics. For clarified broth, follow these steps:

1. Adsorption onto HP-20
2. Elution with aqueous methanol
3. Adsorb onto immobilized D-Ala-D-Ala
4. Elute with ammonia-acetonitrile
5. Neutralize and concentrate
6. Preparative reversed-phase HPLC
7. Desalt on HP-20

This procedure can be used on a preparative scale for the production of multi-gram quantities of pure products for structural, semisynthetic, and biological evaluation and, conceivably, for the manufacture of highly purified antibiotic preparations. This scheme is effective for all glycopeptide antibiotics, provided they have the same mode of action, and has been used in the recent literature for the isolation of most of the newly described members of this class.

B. Microversion for Isolation from Screening Broths

This procedure can also be applied on a microscale for analysis of new fermentation broths for structural, semi-synthetic, and biological evaluation. The success in developing the Affi-gel 10–D-Ala-D-Ala resin led to its application for screening crude fermentation broths for novel glycopeptides based on antagonism of *B. subtilis* activity by the tripeptide di-N-Acetyl-L-Lys-D-Ala-D-Ala (47). Since only members of the glycopeptide class bind to peptides terminating in D-Ala-D-Ala, this screen is expected to be very specific, with no false positives.

A broth sample (10 ml) from the primary screen can be passed through a small column of affinity gel to yield a salt-free residue of high purity for detailed microcharacterization. Because of the unique method used to isolate the antibiotic from the original primary screen sample, the product is of sufficient purity to obtain reliable data regarding HPLC retention, p*I*, FAB-MS, and antimicrobial activity. These tests can all be run within 2 weeks from the discovery of the original culture and are more than sufficient to decide whether the culture warrants further work.

C. Manufacturing Considerations

Currently three glycopeptides are manufactured for commercial use: vancomycin, avoparcin, and teicoplanin. Although the general isolation scheme described earlier works for relatively small scale preparations, it is interesting to speculate about which (if any) of the steps are useful (or are being used) for full-scale manufacture.

Vancomycin was originally prepared as a very crude preparation by the scheme described earlier (41,48,49). In recent years the quality of the drug has improved. The manufacturer has published at least two patents on the use of HP-20 resin for the isolation of vancomycin and has commented on the yield and purity improvements achieved with the use of these resins (51,53). Since HP-20 is widely available in bulk, this is presumably used as the capture column. However, further purification is needed to reach a pharmaceutically acceptable purity level. Since the same manufacturer has not published on the use of affinity gels but has presented work on the preparative HPLC of insulin on a very large scale (81), one can presume that preparative HPLC on a reversed-phase column is used for the final purification step. Clearly, the value of this lifesaving drug allows the manufacturer to use rather costly and sophisticated technology for its manufacture.

Avoparcin is sold as a growth promoter in Europe. Because of the scale needed for veterinary use, there are fewer opportunities for expensive and complicated technology and severe regulatory limitations on implementing newer alternative isolation procedures. Presumably the method of manufacture is by the ion-paired extraction procedure described in the original literature (50) or the use of a dried whole-broth preparation.

Teicoplanin is marketed in Europe as a human pharmaceutical. Since the manufacturer has published extensively on the use of affinity chromatography, one can speculate that this step is used in the manufacture of the drug, although it is unclear whether this is a cost-effective process on a large scale (10,59). One patent application has appeared describing the immobilization of D-alanine onto relatively inexpensive silica supports. (82).

VI. ANALYTICAL SEPARATION TECHNIQUES

Two separation techniques, analytical reversed-phase HPLC and isoelectric focusing, warrant further discussion because they provide excellent data to determine whether an entity isolated in a glycopeptide screen is novel.

A. HPLC

Reversed-phase HPLC is an excellent discrimination tool for antibiotics in general and in particular for glycopeptide antibiotics for which the structural diversity in chlorine, carbohydrate, and lipids contributes to differences in lipophilicity. In many cases, the use of HPLC to discriminate known antibiotics carries with it some risk in that there is no guarantee that a particular peak in a chromatogram is associated with the bioactive entity or that the antibiotic is eluting within the chromatographic "window" being monitored. If a D-Ala-D-Ala support is used to isolate an antibiotic complex, however, it can be assumed that every (UV-adsorbing) peak observed in a subsequent HPLC chromatogram comes from an glycopeptide antibiotic, not an extraneous contaminant. Table 4 lists the HPLC

Table 4 HPLC Retention Times for Various Glycopeptide Antibiotics

Antibiotic	Retention time (minutes)[a]	
	System A	System B
A-477	11.5, *11.7*[b], 12.1	8.4, *8.9*, 10.3
A35512B	5.2, 6.2	*6.1*, 7.2
Actaplanin	6.1, *6.3*, 6.6, 10.7	7.5, 7.9,17.2
Actinoidins A, B	*5.5*, 5.9, 6.1	6.5, 7.1, *8.1*
Actinoidin A$_2$	4.4	6.9
Ardacins A, B, C	15.3, 15.9, 17.2	14.5, 15.2, 16.1
Avoparcin	4.8, *5.8*, 6.0	7.6, 7.7
Kibdelins A, B, C	15.0, 15.8, 16.7	14.6, 15.6, 16.4
LL-AM374	2.0, *5.1*, 6.6	6.8, 10.3
OA-7653	10.3	—
Parvodicins	15.8, 16.5, 17.2, 17.7	—
Ristocetin	3.6	5.4
Synmonicin B	*5.4*, 7.5	—
Teicoplanin	13.1, *13.6*, 13.8	*13.8*, 14
Vancomycin	6.0	7.4

[a]Column: ultrasphere ODS. 5 μm, 4.6 × 150 mm; flow 1.5 ml/minute; detection UV 220 nm; system A. 7–34% acetonitrile (7% for 1 minute, then ramp to 34% over 13 minutes) in pH 3.2, 0.1 M phosphate; system B, 5–35% acetonitrile (5% for 1 minute, then ramp to 35% over 13 minutes) in pH 6.0, 0.025 M phosphate.

[b]Retention times in italics are the major peak in a multicomponent mixture.

retention times for various glycopeptide antibiotics in two different HPLC solvent systems (18). The long gradients were chosen to assure the elution of every known glycopeptide antibiotic.

Several pieces of critical information can be obtained from this HPLC analysis on a potentially novel glycopeptide isolate. Examination of the HPLC profile will suggest whether the product is primarily one factor or a mixture. In most cases, the new isolate is a complex and gives several peaks, as did many of the reference samples analyzed earlier. Since finding a match in retention times, when there are several peaks, can be accidental, coelution in both pH solvent systems and the appearance of a similar profile can suggest that the antibiotic complex is known. If samples are available, coinjection experiments in both systems should be performed to confirm novelty. Unfortunately, the lack of a coincidentally eluting peak does not rule out the possibility that the glycopeptide being tested contains the same nucleus as a known glycopeptide but differs in a subtle aspect, such as having a different carbohydrate composition. Unfortunately, the use of a diode array detector in this case offers little help in distinguishing members of this class since they all have the same UV chromophore. Differentiation can be carried out after further analysis by such techniques as mass spectrometry and aglycone analysis.

Structural features of the new antibiotic can also be discerned from the HPLC data. The glycolipid-containing antibiotics (ardacins, kibdelins, parvocidins, and teicoplanin) all elute relatively late in the HPLC systems, with retention times in excess of 13 minutes, whereas all the other glycopeptides elute before 13 minutes. It has been shown that the glycolipid-containing glycopeptides have longer serum half-lives than the more polar species, such as vancomycin (83). For example, the ardacins and teicoplanin had serum half-lives of 150–277 minutes in mice, whereas vancomycin and ristocetin had half-lives of 20 and 62 minutes, respectively (83). Since a longer serum half-life offers a significant therapeutic advantage, the preselection of glycolipid-containing glycopeptides at this stage is highly advantageous.

B. Isoelectric Focusing

Another structural feature potentially useful for discrimination of the known glycopeptide antibiotics are their charge characteristics, which can be determined by electrophoretic techniques. Glycopeptide antibiotics are amphoteric molecules containing charged amino, carboxyl, and phenolic groups. The pK_a of a typical glycopeptide can be determined using titration procedures with computer analysis of the titration curve (18). Vancomycin, for example, has several phenolic groups with pK_a 10.6, 10.3, and 9.4; two amino groups (N-terminal and amino sugar) with pK_a 8.6 and 6.8, and one C-terminal carboxylate with pK_a 2.5 (18). Unfortunately, determination of the pK_a values by titrate requires complicated equipment and significant amounts (100 mg) of material. A far simpler method involves the use of isoelectric focusing to determine the isoelectric point (pI) using B. subtilis bioautographic detection (54). The pI of a molecule is the pH at which there is no net charge: that is, the negative and positive charges cancel out. For vancomycin the pI is expected to be between 6.8 and 8.6, where the carboxyl would have a charge of -1, the amine with pK_a 6.8 would be deprotonated and have no charge, and the amine with pK_a 8.6 would still have a charge of 1. Thus, the net charge at this pH would be zero. Experimentally, the pI of vancomycin was found to be 8.1 (54).

To determine the pI of an antibiotic, small amounts (5–10 μl of an approximately 1 mg/ml solution) are deposited on a gel containing ampholines and the gel focused in an isoelectric focusing apparatus. The antibiotics are detected by blotting the gel on inoculated agar through a piece of filter paper pretreated with high ionic strength phosphate buffer to solubilize the antibiotic and neutralize the ampholines. After overnight incubation, the antibiotics are detected as spots of no growth of the organism. Antibiotics, such as vancomycin, are used as standards. Table 5 lists the pI of the glycopeptide antibiotics. Note that values vary significantly from the highly acid ardacins (pI 3.8), with two carboxyl and one amino group, to the highly basic ristocetin (pI 8.4), with two amino groups and no carboxyl function. It is apparent that significant structural details can be obtained from this procedure in that glycopeptides with high pI (ristocetin, A35512B) contain amino sugars and methyl ester functions, whereas those with low pI (ardacins and parvodicin) contain an extra carboxylic acid function. Those with pI near neutrality (teicoplanin and OA-7653) contain one amino and one carboxyl group. This technique is not limited to glycopeptide antibiotics or even amphoteric antibiotics, since even the aminoglycoside gentamicin can be focused with pI 9.1. In this case the negative charge comes from the ionization of a carbohydrate hydroxyl group. The use of capillary electrophoresis has not yet been reported for the glycopeptides but should be relatively straightforward because the only UV-adsorbing peaks in an affinity gel isolate would again be attributable to members of the class, not to contaminants.

Table 5 Isoelectric Points of the Glycopeptide Antibiotics

Antibiotic	pI	Number of ionic groups	
		COOH	NH$_2$
A-477	6.5		
A35512B	8.1	0	2
A41030	5.1[a]	1	1
A47934	3.0[a]	2 (SO$_4^-$)	1
AM374	8.1		
Actaplanin	8.5	0	1
Ardacin	3.8	2	1
Actinoidin	8.5, 8.0	1	3
Actinoidin A$_2$	6.9	1	2
Avoparcin	8.5, 8.0	1	2
Kibdelins	5.1	1	1
OA-7653	4.9	1	1
Parvodicin	3.8	2	1
Ristocetin	8.4	0	2
Synmonicin	8.0	1	2
Teicoplanin	5.1	1	1
Vancomycin	8.1	1	2

[a]Estimated.

VII.   CONCLUSIONS

The screening of microorganisms for the production of novel natural products is a highly competitive enterprise that has been carried out for decades. Traditionally, the many months required to purify a newly discovered antibiotic were relatively insignificant compared with the months or years needed to evaluate the structure of a new entity. With the advent of modern spectroscopic techniques, the limiting factor in discovering and evaluating a new entity is often the time required for its isolation and the effectiveness of its purification. The application of affinity chromatography for the facile isolation of glycopeptide antibiotics is truly notable, as is the use of preparative reversed-phase HPLC to provide homogeneous entities for structure determination. Similar approaches to the purification of other antibiotics using ligand affinity chromatography or antibodies to antibiotic classes may be an effective tool for the discovery of other novel fermentation products.

ACKNOWLEDGMENTS

The authors acknowledge the many chemists and biologists who participated in the antibiotic screening program at Smith Kline and French Laboratories. It was their joint efforts that created such a productive and enthusiastic environment and enabled the successful execution of the glycopeptide antibiotic screen. In particular, many thanks go to Louis Nisbet, Peter Jeffs, Ken Snader, John Dingerdissen, Marcia Shearer, George Chan, Peter DePhillips, George Udowenko, Jim Rake, Claire Phelen, Charles Pan, Sue Spaeth, David Newman, Fred Chapin, Donald Pitkin, Raj Mehta, Sarah Grappel, Yong Oh, Gerald Roberts, Steven Carr, Al Giovenella, and Charles DeBrosse. In addition many thanks to our coworkers V. Arjuna Rao, A.K. Goel, A.K. Verma, A.K. Sadhukhan, D. Kandasamy, and A. Venkateswarlu at the ESKAYEF Limited Research Labs in Bangalore, India.

REFERENCES

1.   Williams D., Rajananda V., Williamson M., Bojesen G. (1980). The vancomycin and ristocetin group of antibiotics. Top. Antibiot. Chem. 5:119–158.
2.   Barna J., Williams D. (1984). The structure and mode of action of glycopeptide antibiotics of vancomycin group. Annu. Rev. Microbiol. 38:339–357.
3.   Katrukha G., Silaev A. (1986). The chemistry of glycopeptide antibiotics of the vancomycin group. Chem. Peptides Proteins 3:289–306.
4.   Sztaricskai F., Bognar R. (1984). The chemistry of the vancomycin group of antibiotics. In Szantay, C. (Ed.), Recent Developments in the Chemistry of Natural Carbon Compounds. Akademiai Kiado, Budapest, 10:91–201.
5.   Williamson M., Williams D. (1983). Structure revision of the antibiotic vancomycin.

The use of nuclear Overhauser effect difference spectroscopy. J. Am. Chem. Soc. 103:6580–6585.

6. Harris C., Harris T. (1982). Structure of the glycopeptide antibiotic vancomycin. Evidence for an asparagine residue in the peptide. J. Am. Chem. Soc. 104:4293–4295.

7. Hamill R., Haney M., Stark W. (1973). Antibiotic A477 and process for preparation thereof, USP 3780174.

8. Michel K., Shah R., Hamill R. (1980). A complex of new antibacterial antibiotics produced by *Streptomyces candidus*. I. Isolation and characterization. J. Antibiot. 33: 1397–1406.

9. Eggert J., Michel K. (1986). Isolation and characterization of A41030, a complex of novel glycopeptide antibiotics: Application of the Michel-Miller high performance low pressure liquid chromatography system. J. Antibiot. 39:792–799.

10. Riva E., Gastaldo L., Beretta M., Ferrari P., Zerilli L., Cassani G., Selva E., Goldstein B., Berti M., Parenti F., Denaro N. (1989). A42867, a novel glycopeptide antibiotic. J. Antibiot. 42:497–503.

11. Hamill R., Boeck L., Kostner R., Gale R. (1983). A novel glycopeptide-aglycone antibiotic: Fermentation, isolation and characterization. 23rd Interscience Conference on Antimicrobial Agents Chemotherapy. Abstract 443.

12. Doolin L., Hamill R., Mahoney D., Yao R. (1990). Production of glycopeptide antibiotics by *Kibdelosporangium philippinensis*. CA 112:6072.

13. Tamura A., Takeda I. (1975). Antibiotic AB-65 a new antibiotic from *Sacchararomonospora viride*. J. Antibiot. 28:395–397.

14. Hamill R., Mabe J., Mahoney D., Nakatsukasa W., Yao R. (1989). Glycopeptide antibiotics AB2846, their manufacture with *Nocardia orientalis*, and pharmaceuticals containing them. CA 111:76518.

15. Kunstmann M., Porter J. (1974). Animal feed composition containing antibiotic AM374 as a growth promoter. USP 3803306.

16. Debono M., Merkel K., Molloy R., Barnhart M., Presti E., Hunt A., Hamill R. (1984). Actaplanin, new glycopeptide antibiotics produced by *Actinoplanes missouriensis*. The isolation and preliminary chemical characterization of actaplanin. J. Antibiot. 37:85–95.

17. Brazhnikova M., Lomakina N., Yurina M., Lavrova M. (1963). Antibiotic actinoidin: Its isolation and chemical properties. II Int. Symp. Chemotherapy. Naples, 1961, pp. 253–260.

18. Sitrin R., Chan G., Dingerdissen J., Holl W., Hoover J., Valenta J., Webb L., Snader K. (1985). Aridicins, novel glycopeptide antibiotics. II. Isolation and characterization. J. Antibiot. 38:561–571.

19. Kunstmann M., Mitscher L., Porter J., Shay A., Darken M. (1969). LL-AV290, a new antibiotic I. Fermentation, isolation, and characterization. Antimicrob. Agents Chemother. 1968:242–245.

20. Tsuji N., Kamigauchi T., Kobayashi M., Terui Y. (1988). New glycopeptide antibiotics. I. The isolation and structures of chloroorienticins. J. Antibiot. 41:1506–1510.

21. Takatsu T., Nakajima M., Oyajima S., Itoh Y., Sakaida Y., Takahashi S., Haneishi T. (1987). Chloropolysporins A, B and C, novel glycopeptide antibiotics from *Faeinia*

interjecta sp. nov. II. Fermentation, isolation and physico-chemical characterization. J. Antibiot. 40:924–932.

22. Franco C., Chatterjee S., Vijakumar E., Chatterjee D., Gangulim B., Rupp R., Fehlhaber H., Kogler H., Seibert G., Teetz V. (1990). Glycopeptide antibiotic decaplanin. CA 113:76615.

23. Gause G., Brazhnikova M., Lomakina N., Berdnikova T., Fedorova G., Tokareva N., Borisova V., Batta G. (1989). Eremomycin—new glycopeptide antibiotic: Chemical properties and structure. J. Antibiot. 42:1790–1799.

24. Takeuchi M., Enokita R., Okazaki T., Kagasaki T., Inukai M. (1991). Helvecardins A and B, novel glycopeptide antibiotics. I. Taxonomy, fermentation, isolation and physico-chemical properties. J. Antibiot. 44:263–270.

25. Folena-Wasserman G., Poehland B., Yeung E., Staiger D., Killmer L., Snader K., Dingerdissen J., Jeffs P. (1986). Kibdelins (AAD-609), novel glycopeptide antibiotics. II. Isolation, purification and structure. J. Antibiot. 39:1395–1406.

26. Hayashi T., Harada Y., Ando K. (1975). Isolation and characterization of mannopeptins, new antibiotics. J. Antibiot. 28:503–507.

27. Good V., Gwynn M., Knowles D. (1990). MM 45289, a potent glycopeptide antibiotic which interacts weakly with diacetyl-l-lysyl-d-alanyl-d-alanine. J. Antibiot. 42:550–555.

28. Box S., Elson A., Gilpin M., Winstanley D. (1990). MM47761 and MM 49721, glycopeptide antibiotics produced by a new strain of *amycolatopsis orientalis*: Isolation, purification and structure determination. J. Antibiot. 43:931–937.

29. Athalye M., Coates N., Milner P. (1990). Novel glycopeptide antibiotics from *Amycolatopsis orientalis*. CA 113:170474.

30. Coates N., Davis C., Curtis L., Sykes R. (1991). Isolation of glycopeptide antibiotic complex MM49728 from *Amycolatopsis* NCIB 40089. CA 114:99959.

31. Coates N., Elson A., Athalye M., Curtis L., Moores L. (1990). Manufacture of glycopeptide antibiotics MM55270, MM55271 and MM55272 with *Amycolatopsis*. CA 113:189780.

32. Kamogashira T., Nishida T., Sugawara M. (1983). A new glycopeptide antibiotic, OA-7653, produced by *Streptomyces hygroscopicus* subsp. *hiwasaensis*. Agr. Biol. Chem. 47:499–506.

33. Tsuji N., Kobayashi M., Kamigauchi T., Yoshimura Y., Terui Y. (1988). New glycopeptide antibiotics. I. The structures of orienticins. J. Antibiot. 41:819–822.

34. Kondo E., Kawamura Y., Kamigauchji T., Hayashi Y., Konishi T. (1990). Manufacture of glycopeptide antibiotics PA-45052 with *Nocardia* and its use as animal growth stimulators. CA 113:113843.

35. Christensen S., Allaudeen H., Burke M., Carr S., Chung S., DePhillips P., Dingerdissen J., DiPaolo M., Giovenella A., Heald S., Killmer L., Mico B., Mueller L., Pan C., Poehland B., Rake J., Roberts G., Shearer M., Sitrin R., Nisbet L., Jeffs P. (1987). Parvodicin, a novel glycopeptide from a new species, *Actinomadura parvosata*: Discovery, taxonomy, activity and structure elucidation. J. Antibiot. 40:970–990.

36. Philip J., Schenck J., Hargie M. (1956). Ristocetins A and B, two new antibiotics. Antibiot. Annu. 1956–1957:699–705.

37. Verma A., Prakash R., Carr S., Roberts G., Sitrin R. (1986). Synmonicins: A novel

antibiotic complex. II. Isolation and preliminary characterization. 26th Interscience Conference on Antimicrobial Agents Chemotherapy, Abstract 940.

38. Bardone M., Paternoster M., Coronelli C. (1978). Teichomycins, new antibiotics from *Actinoplanes teichomyceticus* nov. sp. II. Extraction and chemical characterization. J. Antibiot. 31:170–177.

39. Skelton N., Williams D., Monday R., Ruddock J. (1990). Structure elucidation of the novel glycopeptide antibiotic UK-68,597. J. Org. Chem. 55:3718–3723.

40. Skelton N., Williams D., Rance M., Ruddock J. (1990). Structure elucidation of UK-72,051, a novel member of the vancomycin group of antibiotics. J. Chem. Soc., Perkin Trans. I:77–81.

41. McCormick M., Stark W., Pittenger G., Pittenger R., McGuire J. (1956). Vancomycin, a new antibiotic. I. Chemical and biological properties. Antibiot. Annu. 1955–1956:606–611.

42. Nieto M., Perkins H. (1971). Physiocochemical properties of vancomycin and iodovancomycin and their complexes with diacetyl-l-lysyl-d-alanyl-d-alanine. Biochem. J. 123:773–787.

43. Nieto M., Perkins H. (1971). Modifications of the acyl-d-alanyl-d-alanine terminus affecting complex formation with vancomycin. Biochem. J. 123:789–803.

44. Nieto M., Perkins H., Reynolds P. (1972). Reversal by a specific peptide (diacetyl-α-γ-L-diaminobutyryl-d-alanyl-d-alanine) of vancomycin inhibition in intact bacteria and cell-free preparations. Biochem. J. 126:139–149.

45. Nieto M., Perkins H. (1971). The specificity of combination between ristocetins and peptides related to bacterial cell wall mucopeptide precursors. Biochem. J. 124: 845–852.

46. Perkins H., Nieto M. (1974). Part IV. Inhibitors of the synthesis of peptidoglycans and other wall components: The chemical basis for the action of the vancomycin group of antibiotics. Ann. N.Y. Acad. Sci. 235:348–363.

47. Rake J., Gerber R., Mehta R., Newman D., Oh Y., Phelen C., Shearer M., Sitrin R., Nisbet L. (1986). Glycopeptide antibiotics: A mechanism-based screen employing a bacterial cell wall receptor mimetic. J. Antibiot. 39:58–67.

48. Higgens H., Harrison W., Wild G., Bungay H., McCormick M. (1958). Vancomycin, a new antibiotic. VI. Purification and properties of vancomycin. Antibiot. Annu. 1957–1958:906.

49. McCormick M., McGuire L., McGuire J. (1962). Vancomycin and method for its preparation. USP 3,067,099.

50. Shu P., Daun M. (1974). Method for the production and isolation of the antibiotic AV290 sulfate. USP 3819836.

51. McCormick M., Wild G. (1984). Purification of glycopeptide antibiotics using nonfunctional resins. USP 4,440,753.

52. Dingerdissen J., Sitrin R., DePhillips P., Giovenella A., Grappel S., Mehta R., Oh Y., Pan C., Roberts G., Shearer M., Nisbet L. (1987). Actinoidin A₂, a novel glycopeptide: Production, preparative HPLC separation and characterization. J. Antibiot. 40: 165–172.

53. Glass S., Johnson C., Spencer J. (1989). Recovery of glycopeptide antibiotics from fermentation media using polystyrene divinylbenzene resins. CA 111:37970.

54. Henner J., Sitrin R. (1984). Isoelectric focusing and electrophoretic titration of antibiotics using bioautographic detection. J. Antibiot. 37:1475–1478.

55. Scouten W. (1983). Introduction of affinity purification of biopolymers. Solid Phase Biochemistry. John Wiley and Sons, New York, pp. 17–78.

56. Lowe C., Dean P. (1974). Affinity Chromatography. John Wiley and Sons, New York.

57. Pirkle W., Hyun M., Bank B. (1984). A rational approach to the design of highly-effective chiral stationary phases. J. Chromatogr. 316:585–604.

58. Gale E., Cundliffe E., Reynolds P., Richmond M., Waring M. (1981). The Molecular Basis of Antibiotic Action. John Wiley and Sons, New York, pp. 144–161.

59. Corti A., Cassani G. (1985). Synthesis and characterization of d- alanyl-d-alanine-agarose: A new bioselective adsorbent for affinity chromatography of glycopeptide antibiotics. Appl. Biochem. Biotechnol. 11:101–109.

60. Folena-Wasserman G., Sitrin R., Chapin F., Snader K. (1987). Affinity chromatography of glycopeptide antibiotics. J Chromatogr. 392:225–238.

61. Williams D., Williamson M., Butcher D., Hammond S. (1983). Detailed binding sites of the antibiotics vancomycin and ristocetin A: Determination of intermolecular distances in antibiotic/substrate complexes by use of the time-dependent NOE. J. Am. Chem. Soc. 105:1332–1339.

62. Harris C., Kopecka H., Harris T. (1983). Vancomycin: Structure and transformation to CDP-1. J. Am. Chem. Soc. 105:6915–6922.

63. Kalman J., Williams D. (1980). An NMR study of the structure of the antibiotic ristocetin A. The negative nuclear Overhauser effect in structure elucidation. J. Am. Chem. Soc. 102:897–905.

64. Williams D., Rajananda V., Bojesen G., Williamson M. (1979). Structure of the antibiotic ristocetin A. J. Chem. Soc. Chem. Commun. 1979:906–908.

65. Corti A., Soffientini A., Cassani G. (1985). Binding of the glycopeptide antibiotic teicoplanin to d-alanyl-d-alanine-agarose: The effect of micellar aggregates. J. Appl. Biochem. 7:133–137.

66. Williamson M., Williams D. (1984). Hydrophobic interactions affect hydrogen bond strengths in complexes between peptides and vancomycin or ristocetin. Eur. J. Biochem. 138:345–348.

67. Sitrin R., Chan G., Chapin F., Giovenella A., Grappel S., Jeffs P., Phillips L., Snader K., Nisbet L. (1986). Aricidins, novel glycopeptide antibiotics. III. Preparation, characterization, and biological activities of aglycone derivatives. J. Antibiot. 39:68.

68. Jeffs P., Mueller L., DeBrosse C., Heald S., Fisher R. (1986). Structure of aridicin A. An integrated approach employing 2D NMR, energy minimization, and distance constraints. J. Am. Chem. Soc. 108:3063–3075.

69. Jeffs P., Chan G., Sitrin R., Holder N., Roberts G., DeBrosse C. (1985). The structure of the glycolipid components of the aridicin antibiotic complex. J. Org. Chem. 50:1726.

70. Marshall F. (1965). Structure studies on vancomycin. J. Med. Chem. 8:18–22.

71. DePhillips P., Giovenella A., Grappel S., Hedde R., Jeffs P., Killmer L., Lindsey T., Nisbet L., Roberts G., Snader K., Sitrin R. (1985). Isolation, structural studies and biological evaluation of the major and minor components of the Aridicin (AAD-216)

complex. 25th Interscience Conference on Antimicrobial Agents and Chemotherapy, Abstract 790.

72. Folena-Wasserman G.. Poehland B.. Killmer L.. Yeung E.. Jeffs P.. Dingerdissen J.. Shearer M.. Grappel S.. Pan C.. Nisbet L. (1985). Affinity isolation and characterization of new glycopeptide antibiotics from SK&F AAD-609. 25th Interscience Conference on Antimicrobial Agents and Chemotherapy. Abstract 793.

73. Lee C-K.. Sandler S. (1990). Vancomycin partitioning in aqueous two-phase systems: Effects of pH. salts. and an affinity ligand. Biotechnol. Bioeng. 35:408–416.

74. Wainer I. (1985). Liquid Chromatography in Pharmaceutical Development: An Introduction. Aster Publishing. Springfield, Oregon.

75. Mierzwa R.. Marquez J.. Patel M.. Cooper R. (1989). HPLC detection methods for microbial products in fermentation broth. In Wagman G.. Cooper R. (Eds.) Natural Products Isolation. Elsevier. New York, pp. 55–109.

76. Sitrin R.. DePhillips P.. Dingerdissen J.. Erhard K.. Filan J. (1986). Preparative liquid chromatography. LC/GC 4:530–550.

77. Sitrin R.. DePhillips P.. Dingerdissen J.. Erhard K.. Filan J. (1985). Liquid Chromatography in Pharmaceutical Development: An Introduction. Wainer I. (Ed.), pp. 265–304. Aster Publishing. Springfield, Oregon.

78. Sitrin R.. DePhillips P.. Dingerdissen J. (1987). Preparative reversed-phase HPLC of polar fermentation products. Dev. Ind. Microbiol. 27:65–75.

79. Sitrin R.. Chan G.. DePhillips P.. Dingerdissen J.. Valenta J.. Snader K. (1985). Purification of fermentation products: Applications to large-scale process. In LeRoith D.. et al. (Eds.). ACS Symposium Series. No. 271 Am. Chem. Soc.. Washington DC. pp. 71–89.

80. Borghi A.. Coronelli C.. Faniuolo L.. Allieve G.. Pallanza R.. Gallo G. (1984). Teichomycins, new antibiotics from *Actinoplanes teichomyceticus* nov. sp. IV. Separation and characterization of the components of teichomycin (teicoplanin). J. Antibiot. 37:615–620.

81. Kroeff E.. Owens R.. Campbell E.. Johnson R.. Marks H. (1989). Production scale purification of biosynthetic human insulin by reversed-phase high performance liquid chromatography. J. Chromatogr. 461:45.

82. Cantwell A.. Hill J. (1987). Preparation of guanidine-containing dipeptides as ACE inhibitors. CA 110:8698.

83. Pitkin D.. Mico B.. Sitrin R.. Nisbet L. (1986). Charge and lipophilicity govern the pharmacokinetics of glycopeptide antibiotics. Antimicrob. Agents Chemother. 29: 440–444.

3
Approaches to the Synthesis of the Vancomycin Aglycones

DAVID A. EVANS and KEITH M. DeVRIES*
Harvard University, Cambridge, Massachusetts

I. INTRODUCTION

A. History of Structure Proof and Aspartate-Isoaspartate Rearrangement

In 1956 vancomycin was isolated from *Streptomyces orientalis* (1) and has been in clinical use for nearly three decades. Vancomycin was the first glycopeptide antibiotic to be isolated and is one member of a large family of antibiotics that are becoming increasingly important pharmaceutical agents in the treatment of severe staphlococcal infections (2).

In 1978, a proposal for the structure of vancomycin was extended by Williams et al. The structural assignment was compiled from the combined use of nuclear magnetic resonance (NMR) spectroscopic data, degradation studies, and x-ray diffraction on the degradation product CDP-1 (CDP-1M; Fig. 1) (3). In 1981, Williamson and Williams amended their original structure with reassignment of the stereochemical disposition of the chlorine substituent on the biphenyl ether moiety associated with amino acid residue 2, which they deduced from NMR studies was *different* in vancomycin and the CDP-1 degradation product (4).

In 1982, the Williams structure was modified by Harris et al. to the now generally accepted structure (Fig. 2) (5,6). Harris proposed that vancomycin had undergone an aspartate-isoaspartate rearrangement (7) to provide CDP-1M (64%), the intermediate isolated by Williams from which the x-ray structure was determined, and an isomeric compound CDP-1m (36%; Fig. 2). This arrangement

---

Present affiliation: Pfizer, Inc., Groton, Connecticut

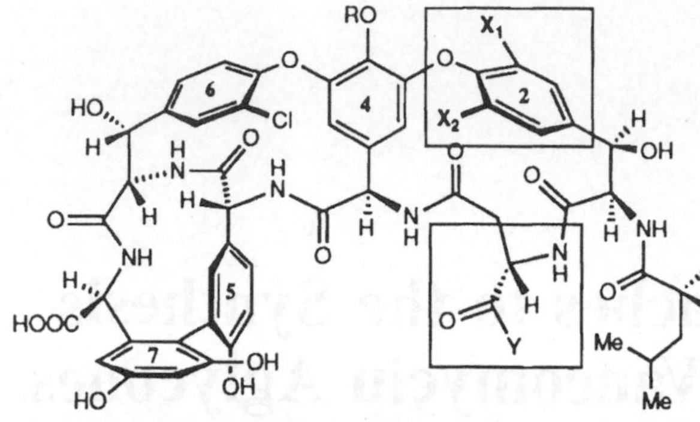

CDP-1 (CDP-1M) $X_1 = Cl, X_2 = H, Y = OH$
Initial Williams Structure $X_1 = Cl, X_2 = H, Y = NH_2$
Second Williams Structure $X_1 = H, X_2 = Cl, Y = NH_2$

Fig. 1 Initial vancomycin structure assignments by Williams and colleagues (2–4).

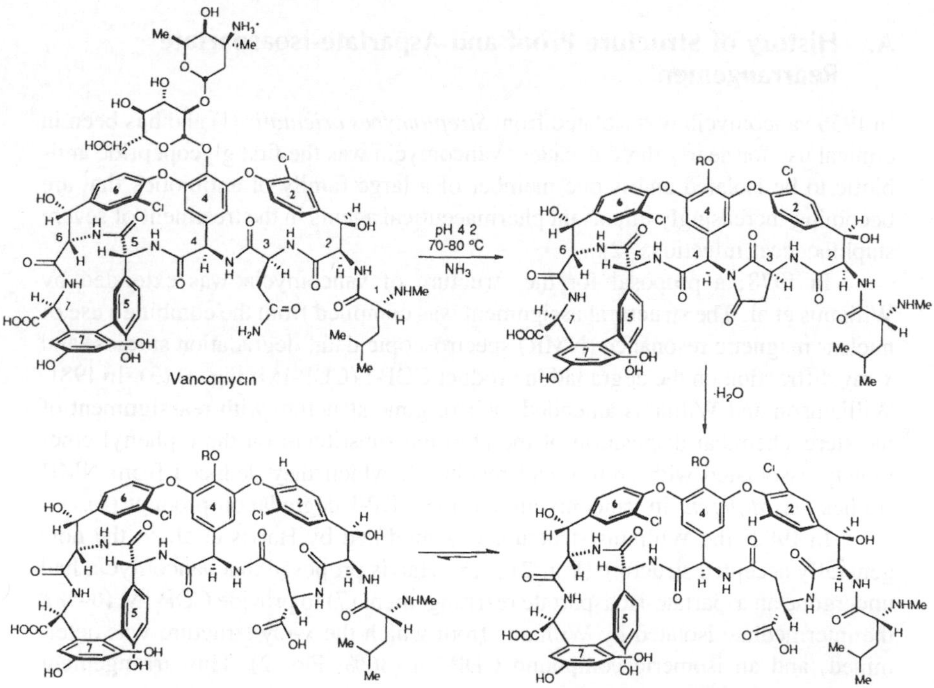

CDP-1 M (64%) CDP 1 m (36%)

Fig. 2 Aspartyl rearrangement of vancomycin to CDP-1.

proceeds with loss of ammonia through the illustrated cyclic imide, which is then cleaved to furnish an isoaspartic acid at residue 3. Upon rearrangement, the cyclic peptide moiety containing rings 2 and 4, henceforth referred to as macrocycle M(2–4), is expanded by one methylene unit, allowing the aryl ring 2 to rotate through the macrocycle to provide the mixture of chlorine-derived atropisomers CDP-1M and CDP-1m. [To clarify the discussion pertaining to the individual macrocycles contained within the vancomycin aglycone, we use the descriptor $M(X-Y)$, where X and Y are the numbers of the *aromatic amino acid residues* contained within the ring, to identify that cycle.] In comparison, vancomycin itself exists as a single atropisomer corresponding to CDP-1m. The error in the original Williams structure was associated with the fact that the aspartate-isoaspartate rearrangement had escaped detection, and although NMR studies eventually revealed the chlorine atropisomerization of residue 2 during the degradation, the aspartate isomerization was overlooked.

B. Representative Structures in the Vancomycin Family

Another subgroup of the vancomycin family of antibiotics is represented by ristocetin (Fig. 3; for a general discussion and leading references see Ref. 8). In ristocetin, residues 1 and 3 are interconnected by an additional diphenyl ether linkage. This more rigid structure possesses different binding selectivities (see later) compared with vancomycin, and the additional level of complexity offers yet another challenge to any projected synthesis. Approximately 100 glycopeptide antibiotics have been isolated to date (9), and vancomycin and ristocetin are representative of the two basic structural variants. These vancomycin congeners, although bearing close similarities, exhibit the following structural variations: (1) the nature and degree of cross-linking between residues 1 and 3; (2) the number and point of attachment of the sugars; (3) the degree of chlorination on rings 2 and 6; and (4) the presence or absence of a benzylic hydroxyl group in residue 2.

C. Mode of Action

The antibiotic activity of vancomycin is proposed to be the result of formation of a noncovalent 1:1 complex between the natural product and the D-Ala-D-Ala terminus of the growing peptidoglycan of the bacterial cell wall (Fig. 4). Using NMR techniques, Waltho and Williams concluded that the amide protons of residues 2, 3, and 4 form hydrogen bonds to the carboxyl terminus of the bound D-Ala-D-Ala residue (9). As initially postulated by Perkins with Nieto (10,109), this binding to the D-Ala-D-Ala terminus of the growing peptidoglycan of the bacterial cell wall is proposed to inhibit the cross-linking that gives the bacterial cell wall its rigidity. Much of the synthetic activity in the vancomycin area (see later) has been directed toward constructing the M(2–4) macrocycle, referred to by Williams as the "vancomycin binding pocket." Because of the relatively small

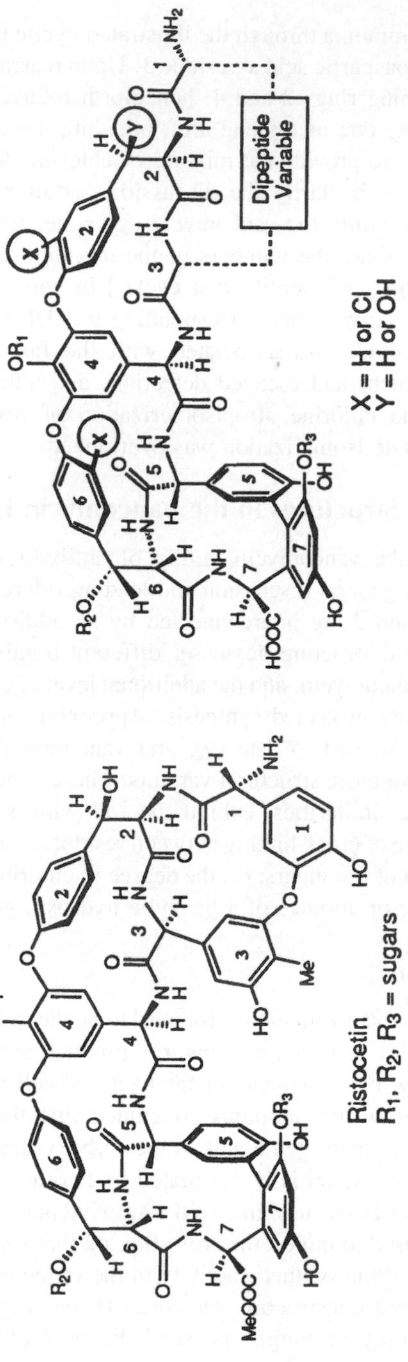
Ristocetin
R1, R2, R3 = sugars
X = H or Cl
Y = H or OH
Dipeptide Variable

Fig. 3 Representative vancomycin structural variants.

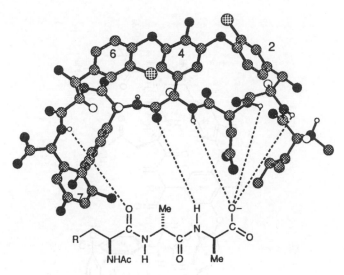

Fig. 4 Vancomycin D-Ala-D-Ala complex.

size of vancomycin in comparison with enzymes and larger proteins, this has been viewed as an ideal system for studying the subtleties of ligand-receptor interactions. Interestingly, residues 1 and 3, which are part of the vancomycin binding pocket, show the most structural variation among the vancomycin family members. In comparison, the vancomycin antibiotics show almost no structural variability in the biaryl portion of the molecule.

D. Summary of Relevant Studies of Biosynthesis

Origin of the Amino Acid Constituents

Although the synthetic organic chemist cannot reproduce nature's efficiency in building complex natural products, consideration of a molecule's biosynthesis can often be inspirational to the development of a synthesis plan. Unfortunately, limited work has been published to date with regard to the biosynthesis of vancomycin. Williams and coworkers have added [13]C- and [2]H-labeled acetate to the fermentation of *S. orientalis* and determined that residue 7 is derived from the cyclization of a polyketide. Feeding experiments with [13]C- and [2]H-labeled tyrosine added to the fermentation broth resulted in the incorporation of label at the residue 4 and 5 arylglycines as well as both the residue 2 and 6 β-hydroxy-α-amino acids (11). To summarize Williams' conclusions, of the five nonproteinogenic residues present in vancomycin, one is derived from a polyacetate cyclization and four are derived from tyrosine.

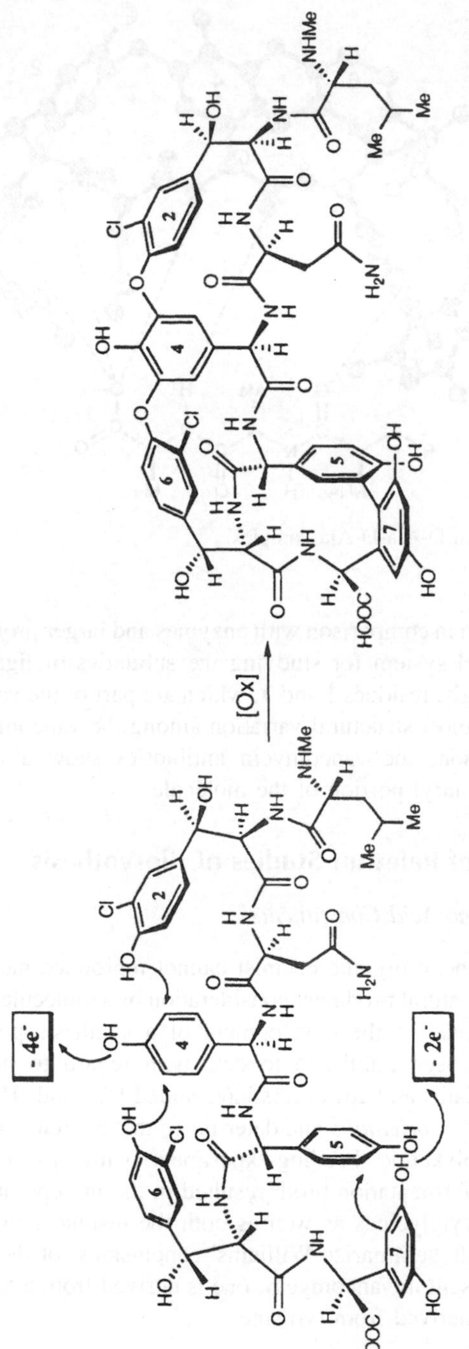

Fig. 5 Phenolic coupling steps in the biosynthesis of vancomycin.

Phenolic Coupling Steps

Although no studies on the biosynthesis of the diphenyl ethers and the biaryl bond present in vancomycin have been reported, Taylor and Battersby extensively surveyed the oxidative coupling of phenols and discussed the biosynthesis of diphenyl ethers and biaryl bonds (12). It is likely that the assemblage of the peptide backbone precedes oxidative cross-coupling through the diphenyl ether and biaryl linkages (Fig. 5). The development of a laboratory mimic to duplicate nature's approach to these bond constructions in vancomycin might arguably provide the pivotal methodology for a convergent laboratory synthesis of this family of natural products.

II. DEGRADATION STUDIES: FORMATION OF VANCOMYCIN AGLYCONE AND FURTHER DEGRADATION TO AMINO ACID CONSTITUENTS

It is important to note that, although x-ray diffraction studies have been carried out by Williams for the rearranged CDP-1M, no x-ray structures for any of the vancomycin natural products are available, and structural assignments of every member of this family of natural products are based on spectroscopic methods combined with chemical degradation [Workers at Eli Lilly & Co. have also published the x-ray structure for the rearranged CDP-1 product of a vancomycin analog (13).]

The disaccharide appended to the phenol of residue 4 can be hydrolyzed under mild conditions to provide vancomycin aglycone and the amino sugar vancosamine (see Fig. 1). Under more stringent hydrolysis conditions, three other vancomycin subunits were also isolated, although with extensive racemization (14). Recently, Jeffs and coworkers were able to obtain actinoidinic acid and a derivative of ristomycinic acid from the degradation of aridicin (Fig. 6) with no racemization of the arylglycine residues. Actinoidinic acid was obtained, however, as a mixture of biaryl atropisomers (see later). Although vancomycinic acid did not survive the harsh hydrolysis conditions, a degradation product containing the upper diphenyl ether network was also isolated (15). In addition to providing insight on the stability of vancomycin, structural subunits derived from degradation often define key bond disconnections for a projected total synthesis and represent synthetic targets in their own right (see later).

III. SYNTHESIS CHALLENGE

A. Atropisomerism

Compounds that do not contain a stereogenic center can still be chiral if there is hindered rotation about certain key bonds. Unsymmetrically substituted biaryls

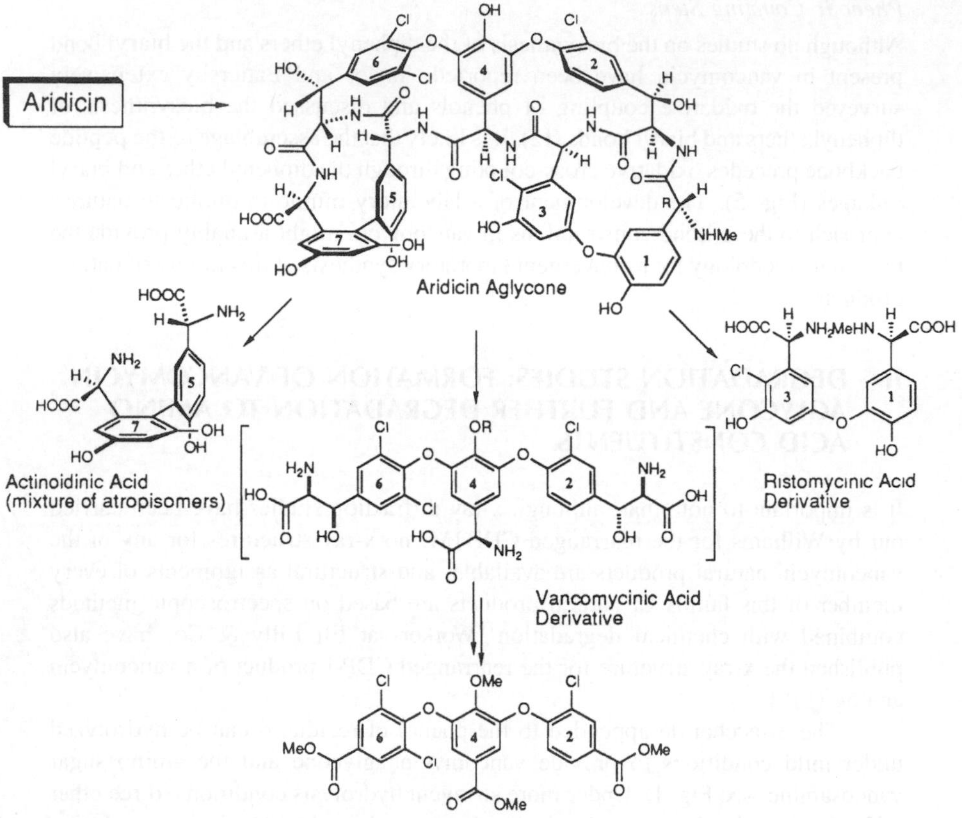

Fig. 6 Aridicin degradation (From Ref. 14, with permission.)

containing ortho substituents that are sufficiently large to inhibit rotation about the aryl-aryl bond represent one example of atropisomerism (16). The examples in equations 1 and 2 (17) illustrate the concept of chirality resulting from hindered rotation about a biaryl bond. The vancomycin structure arguably provides one of nature's most elegant illustrations of atropisomerism.

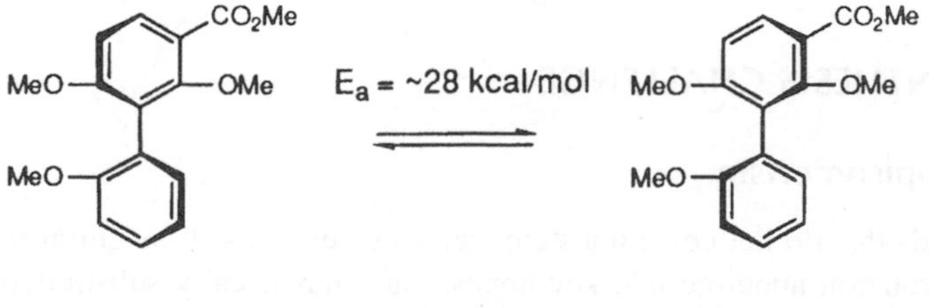

(1)

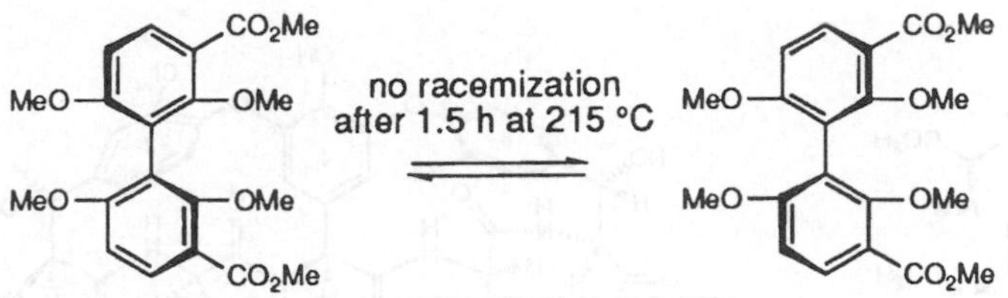

$$(2)$$

Atropisomerism of the Chlorinated Rings

Perhaps the most subtle synthetic challenge present in vancomycin is the atropisomerism contained in the chlorinated aromatic rings associated with residues 2 and 6 (Fig. 7). There is restricted rotation of each of these rings through the 16-membered macrocycles M(2–4) and M(4–6). Accordingly, the presence of a single chlorine substituent on each aromatic ring creates the potential for four vancomycin atropodiastereomers. Since vancomycin exists as a single atropisomer, the selective formation of the correct atropisomer presents a formidable challenge. Harris and coworkers have dehalogenated vancomycin to provide didechlorovancomycin (18), which was recently isolated as the natural product orienticin C (19). Since orienticin C lacks the complicating feature of atropisomerism of the 2 and 6 rings, this structure will probably represent a preliminary objective for total synthesis as a prelude to synthesis of the more elaborate vancomycin target.

Biaryl Atropisomerism

The biaryl bond linking residues 5 and 7 in vancomycin possess three ortho substituents (Fig. 8); consequently, the barrier to rotation is high enough to restrict facile conformational interconversion at ambient temperatures. Accordingly, this conformation restriction introduces yet another element of chirality (for a discus-

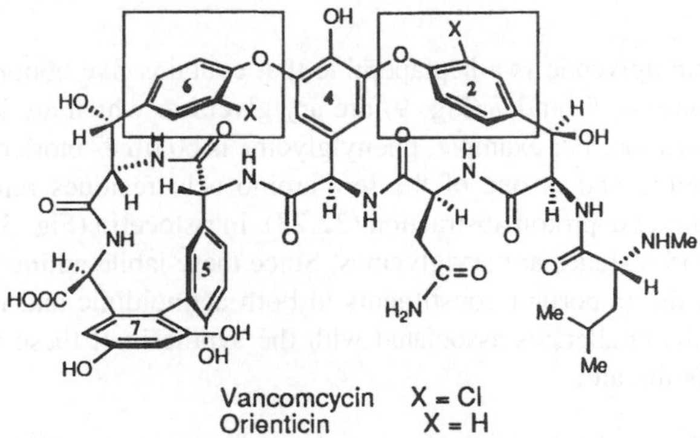

Fig. 7 Atropisomerism of the chlorinated rings and orienticin C aglycone.

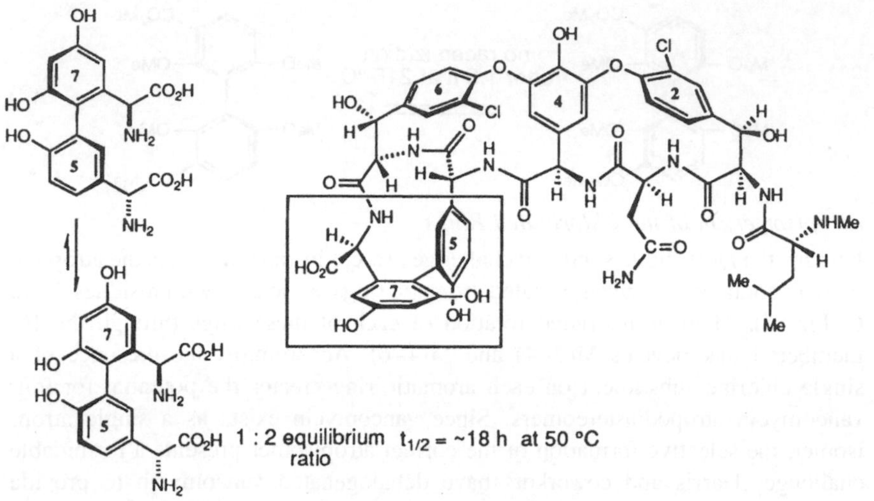

Fig. 8 Biaryl atropisomer and actinoidinic acid atropisomers.

sion of biaryl atropisomerism, see Ref. 20). The actinoidinic acid isolated during the degradation work by Jeffs was obtained as a mixture of atropisomers, the minor atropisomer of the equilibrium mixture corresponding to that found in vancomycin. The biaryl atropisomers of actinoidinic acid were assigned using circular dichroism (21). In comparison, the natural product itself exists as a single atropisomer. It is presumed that the constraints of the M(5–7) macrocycle containing actinoidinic acid dictate the configuration about the biaryl moiety.

B. Nonproteinogenic Amino Acids

Arylglycines

The vancomycin aglycone is a heptapeptide that contains five nonproteinogenic residues. Residues 4, 5, and 7 (Fig. 9) are arylglycines, which are known to be prone to racemization. For example, phenylglycine is 60 times more racemization prone than alanine and is one of the few amino acid residues reported to be racemized by direct α-proton abstraction (22,23). In ristocetin (Fig. 3), five of the seven amino acid residues are arylglycines. Since these labile amino acid derivatives represent the important constituents in both actinoidinic and ristomycinic acid (Fig. 6), the challenges associated with the synthesis of these vancomycin subunits are significant.

β-Hydroxy-α-Amino Acids

Two additional nonproteinogenic amino acid types are also present in vancomycin. Residue 2 is a syn-β-hydroxy-α-amino acid, and residue 6 is an anti-β-hydroxy-α-

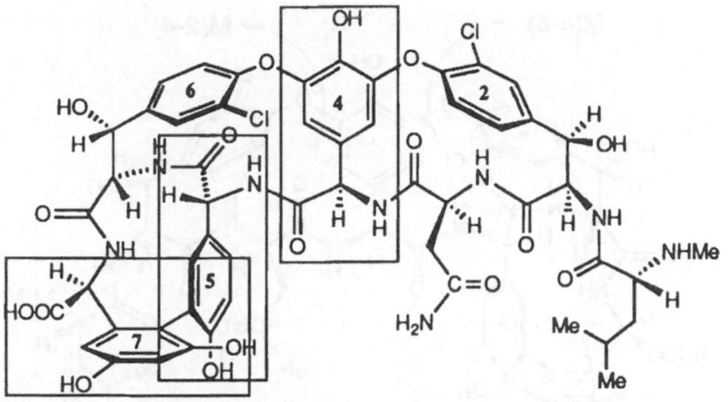

Fig. 9 Arylglycines present in vancomycin aglycone.

amino acid (Fig. 10). The interest in vancomycin as a target for synthesis has been the impetus for developing efficient and reliable methods for obtaining these unnatural amino acids, whose generalized structures are shown in Figure 10.

C. Macrocyclic Ring Syntheses

Vancomycin presents the unique problem of having the amino acid side chains oxidatively cross-linked to provide a tricyclic structure (Fig. 11). The two 16-membered rings M(2–4) and M(4–6) are oxidatively coupled through diphenyl ether linkages. The third macrocycle M(5–7) is a highly strained 12-membered ring joined through a biaryl carbon-carbon bond. Two strategies are evident for

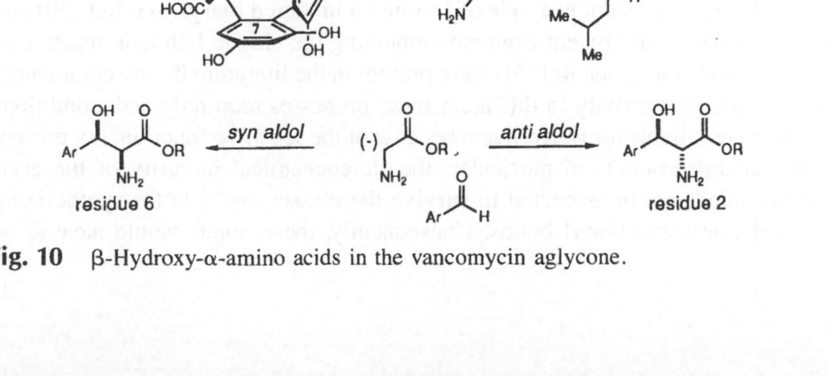

Fig. 10 β-Hydroxy-α-amino acids in the vancomycin aglycone.

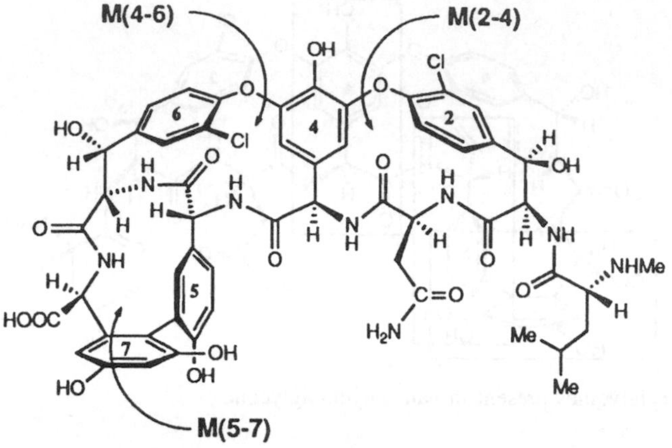

Fig. 11   Macrocyclic rings of vancomycin.

forming the macrocycles present in vancomycin. In the more realistic approach, one might consider constructing those bonds formed via phenolic coupling (Fig. 5) and then utilize well-precedented macrolactamization reactions to assemble the peptide aglycone. Alternatively, one might develop the relevant oxidative macro-cyclization reactions necessary for the synthesis. The main attribute of the latter strategy is that of convergency. The main liability is associated with the fact that little precedent exists for the needed bond constructions. The appeal associated with the phenolic coupling approach is further dampened by the checkered history that such reactions have recorded in the area of alkaloid synthesis.

Conventional Condensation Methodology

In the absence of oxidative routes for the formation of diphenyl ethers and biaryls (see later), any approach to the synthesis of vancomycin would involve the consecutive closure of two 16-membered rings containing the diphenyl ethers and the 12-membered ring containing the biaryl through macrolactamization using well-precedented peptide cyclization reactions (for recent examples, see Refs. 24–26). Although methods for synthesizing biaryl bonds (for recent reviews, see Refs. 27–29; for a more recent example of forming a hindered biaryl, see Ref. 30) and diphenyl ethers (for a recent example employing the classic Ullmann reaction to form diphenyl ethers, see Ref. 31) were present in the literature before commence-ment of synthesis activity in this area, these processes required harsh conditions and were not suitable for vancomycin because of the sensitive functionality present in the natural product. In particular, the stereochemical integrity of the aryl-glycines might not be expected to survive the classic methods for synthesizing diphenyl ethers and biaryl bonds. Consequently, these bonds would have to be formed before incorporation of the amino acid side chains. Recently Schmidt and Boger and their colleagues (32,33) reported milder variants of the Ullmann reaction for forming diphenyl ethers so that the chirality of the amino acids is not destroyed during formation of the desired bond construction (for intramolecular Ullmann couplings see Refs. 34 and 35). It is unlikely, however, that even these conditions are mild enough for those bond constructions that might be constrained by the lability of arylglycine amino acid constituents.

A retrosynthetic scheme for the introduction of the labile arylglycine amine-substituted stereocenters after formation of the diphenyl ether and biaryl bonds is illustrated in Figure 12. In one approach, vancomycin could be dissected to provide vancomycinic acid, which contains the two diphenyl ethers, and acti-noidinic acid, which contains the biaryl bond. These two fragments might then be joined through their amide linkages to provide vancomycin. The closure of the 12-membered ring containing the biaryl might be anticipated to be a nontrivial peptide cyclocondensation.

Oxidative Macrocyclization

An alternative strategy to macrolactamization could be based on vancomycin's proposed biosynthesis. In this approach, after first assembling the peptide back-bone the relevant oxidative coupling methodology might be applied to the sequential construction of the diphenyl ether and biaryl bonds (Fig. 13). Although only one permutation of the ring assemblage sequence is illustrated, the others may be readily visualized.

Quite recently, methods have been developed (see later) so that such a

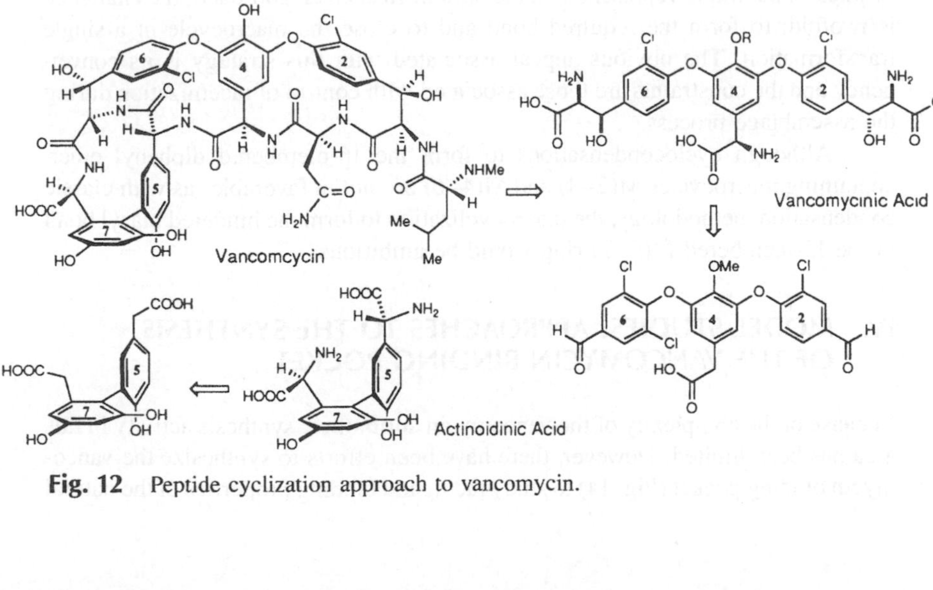

Fig. 12 Peptide cyclization approach to vancomycin.

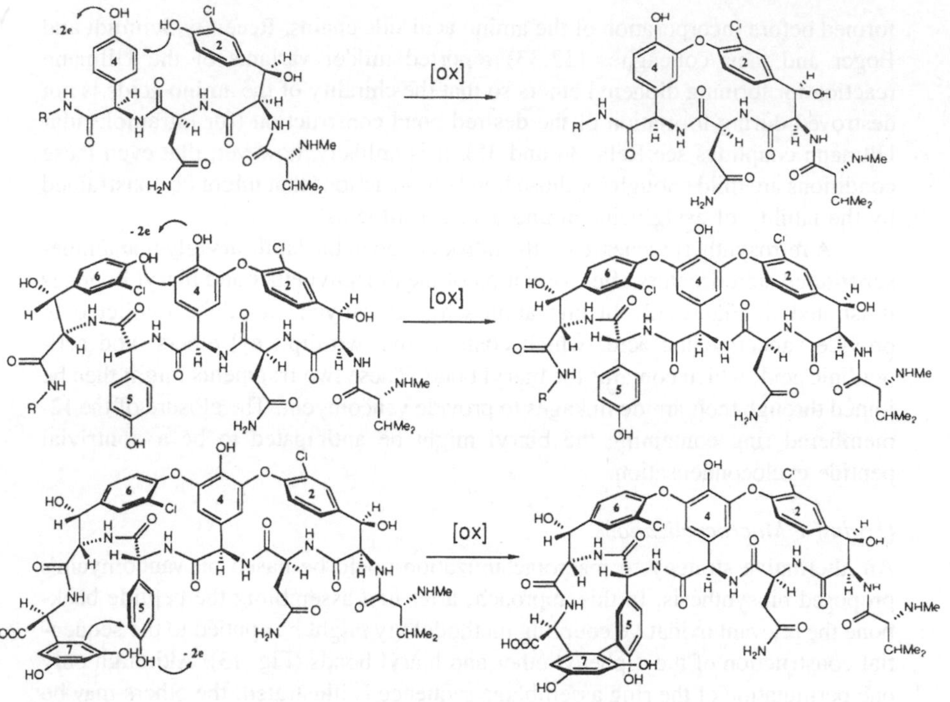

Fig. 13 Oxidative macrocyclization approach to vancomycin aglycone.

synthesis plan might represent a viable option. In such an approach, the challenge is twofold: to form the required bond and to close the macrocycle in a single transformation. The obvious appeal associated with this strategy is its convergency and the constraints are those associated with control of racemization during the assemblage process.

Although cyclocondensations to form the 16-membered diphenyl ether-containing macrocycles M(2–4) and M(4–6) are not unfavorable, as with classic condensation methodology, the macrocyclization to form the hindered biaryl bond in the 12-membered M(5–7) ring would be ambitious.

IV. MODEL STUDIES: APPROACHES TO THE SYNTHESIS OF THE VANCOMYCIN BINDING POCKET

Because of the complexity of the vancomycin antibiotics, synthesis activity in this area has been limited. However, there have been efforts to synthesize the vancomycin binding pocket (Fig. 14) to study the ligand binding properties of the natural

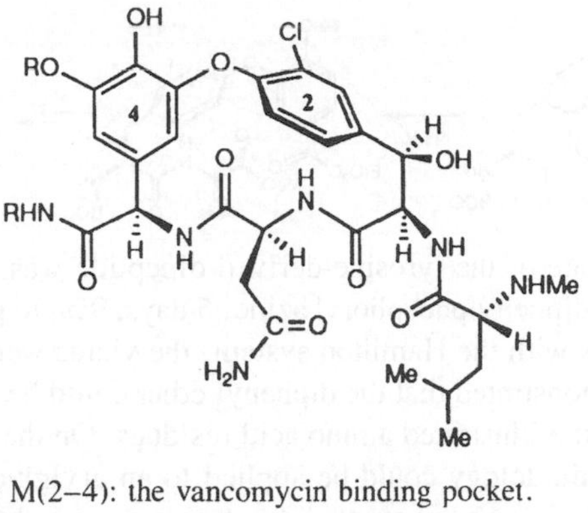

Fig. 14 Macrocycle M(2–4): the vancomycin binding pocket.

product. According to the Williams model, the M(2–4) macrocycle contains the key contact points for binding of the D-Ala-D-Ala terminus of the bacterial cell wall peptidoglycan (9).

A. Hamilton's Studies

Hamilton reported the first synthesis of an analog binding pocket, albeit in low yield (eq. 3) (36,37). The key step relies on nucleophilic aromatic substitution to

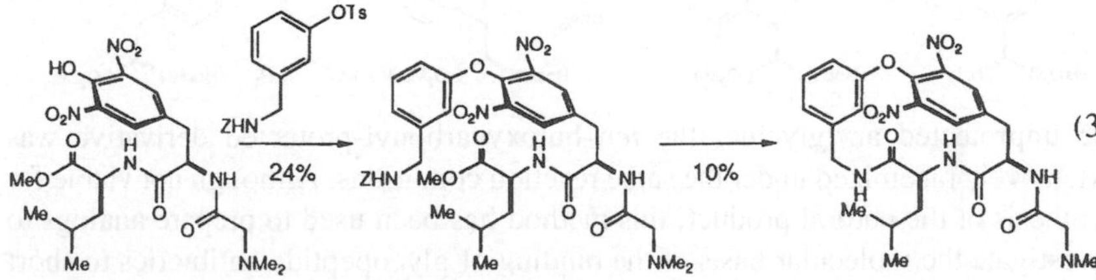

$$(3)$$

assemble the requisite biaryl ether. In this reaction, the union of the illustrated aromatic tosylate and phenolate anion to give the acyclic peptide proceeded in 24% yield.

The subsequent cyclocondensation of the derived amino acid (N,N-bis[2-oxo-3-oxazolidinyl]phosphorodiamidic chloride, 5 days) then proceeded to give the synthetic binding pocket in low yield. Although the yields were less than optimal, the chirality present in the molecule before the key bond construction was not compromised.

B. Brown's Studies

Brown recently used a different phenol displacement reaction to form an analog binding pocket (eq. 4) (38). Displacement of the illustrated aryliodonium salt with

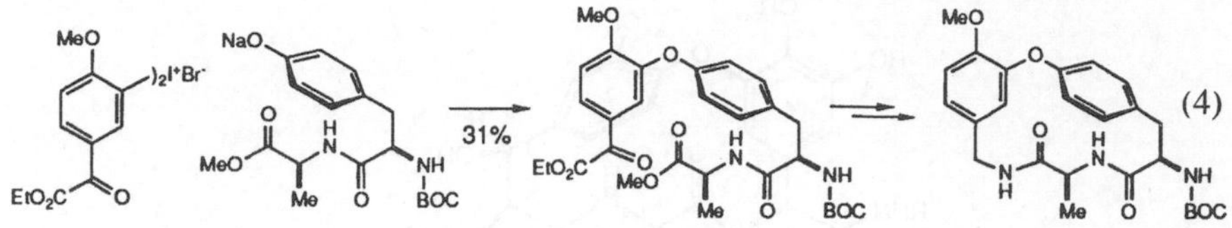

(4)

the sodium phenolate of the tyrosine-derived dipeptide was followed by peptide macrocyclization (diphenylphosphoryl azide, 5 days, 9%) to provide the synthetic binding pocket. As with the Hamilton system, the yields were modest; nevertheless, the study demonstrated that the diphenyl ether could be formed successfully in the presence of the illustrated amino acid residues. On the other hand, it is not clear that such methodology could be applied to an arylglycine-containing substrate. Although the Brown substrate did not give measurable binding to D-Ala-D-Ala, it exhibited some in vivo antibiotic activity.

C. Still's Studies

Hobbs and Still employed a mild photochemical substitution process to form a thiophenyl ether analog of vancomycinic acid in greater than 95% yield and with less than 5% racemization of the arylglycine moiety (eq. 5) (39,40). In contrast to

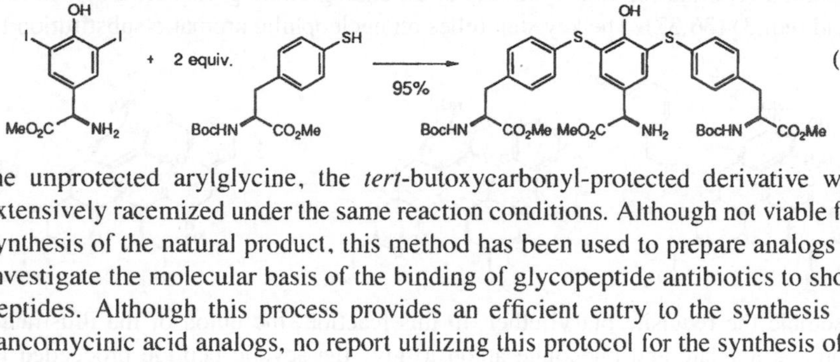

(5)

the unprotected arylglycine, the *tert*-butoxycarbonyl-protected derivative was extensively racemized under the same reaction conditions. Although not viable for synthesis of the natural product, this method has been used to prepare analogs to investigate the molecular basis of the binding of glycopeptide antibiotics to short peptides. Although this process provides an efficient entry to the synthesis of vancomycinic acid analogs, no report utilizing this protocol for the synthesis of a vancomycin binding pocket analog has been reported.

V. SYNTHESIS OF THE AMINO ACID SUBSTITUENTS

The nonproteinogenic amino acids present in vancomycin consist of three arylglycines as well as both syn- and anti-β-hydroxy-α-amino acids (Figs. 9 and 10). Before the commencement of synthetic efforts in the vancomycin area, reliable methods for the synthesis of these amino acid types were unavailable. Only recently have general methods for preparing these unnatural amino acids been reported (for a review of asymmetric amino acid synthesis see Ref. 41).

A. Synthesis of Arylglycines

General Overview

For the obvious reasons associated with reactivity, the application of chiral glycine enolates to the synthesis of arylglycines (eq. 6) has not been pursued. Rather, classic resolution has been commonly used to obtain these amino acids in enantiomerically pure form (42,43). Two other strategies for the formation of arylglycines were recently reported. These strategies exploit the utilization of either electrophilic glycinate (eq. 7) or nitrogen equivalents (eq. 8). In addition

to the review contained in the following discussion, a summary of the various approaches to the synthesis of arylglycines recently appeared (44).

The electrophilic glycinate equivalent developed by Williams and coworkers (45,46) (eq. 9) proceeds with high diastereoselectivity for the C-C bond construction (for another example of a cationic glycine equivalent, see Ref. 47).

As an alternative to nucleophilic and electrophilic glycinate equivalents, this laboratory has reported two electrophilic nitrogen sources that react with the enolate generated from their carboximide chiral auxiliaries selectively to provide the α-azido (48,49) and α-hydrazido (50) derivatives (eqs. 10 and 11). Both

transformations provide general approaches to the asymmetric synthesis of amino acids and proceed with high diastereoselectivity, even for arylglycines. The minor

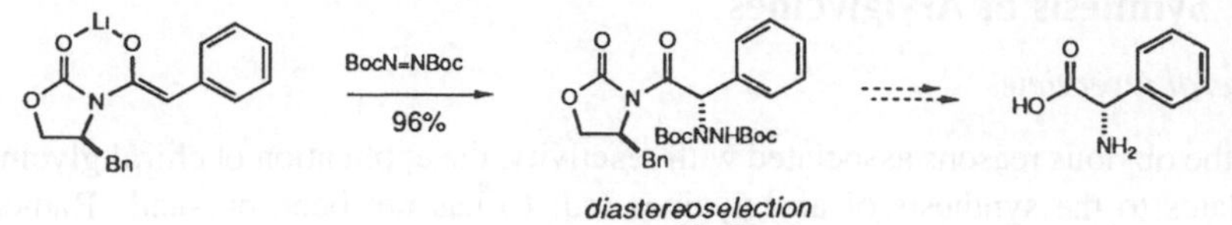

$$diastereoselection$$
$$97:3$$

(11)

diastereomer formed in these reactions can be removed chromatographically, and the α-azido or α-hydrazino groups can be further functionalized to provide protected amino acids of high enantiomeric purity (for references on racemization-free removal of the chiral auxiliary, see Refs. 51 and 52).

Applications to Vancomycin Substrates

The carboximide chiral auxiliaries developed in these laboratories have been used to synthesize the aryglycine constituents found in both vancomycin and risto-mycinic acid. Syntheses of residue 7 of vancomycin and the methylated dihy-droxyphenylglycine of ristomycinic acid are illustrated in equations 12 and 13.

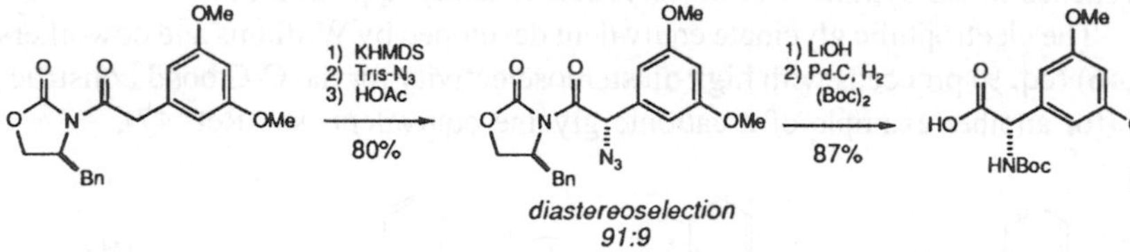

$$diastereoselection$$
$$91:9$$

(12)

$$diastereoselection$$
$$88:12$$

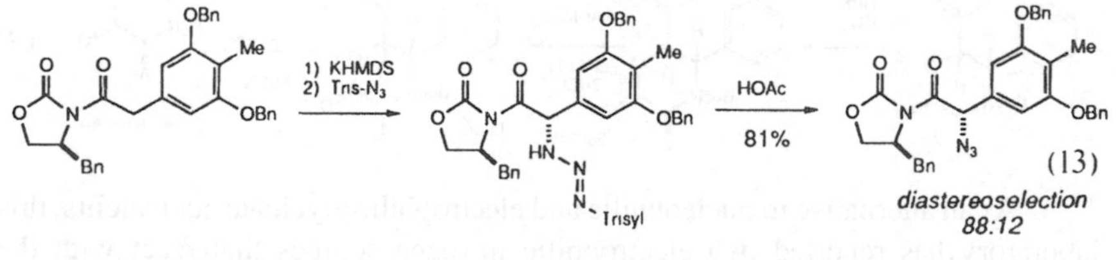

(13)

Enolization (potassium hexamethyldisilylazide) followed by treatment with trisyl azide and an acetic acid quench resulted in formation of the desired azido carbox-imide both in high yield and with high selectivity (53). The minor diastereomer formed in these reactions was readily removed by flash chromatography. One noteworthy point with regard to these transformations is that the triazine inter-mediate (eq. 13) is slow to break down to the azide. In contrast to a report by Williams and coworkers that the triazine intermediate failed to break down to the azide (54), the direct azide transfer protocol has worked well for the synthesis of the majority of the unnatural arylglycine constituents in this family of molecules. Cleavage of the chiral imide auxiliary with lithium hydroxide followed by reduc-tion and acylation provides a general approach to these N-protected amino acids.

The only other vancomycin-related arylglycine synthesis reported was an application of the Schöllkopf (55–59) bis-lactim ether methodology used by Pearson and coworkers in the synthesis of a ristomycinic acid derivative (eq. 14) (60,61).

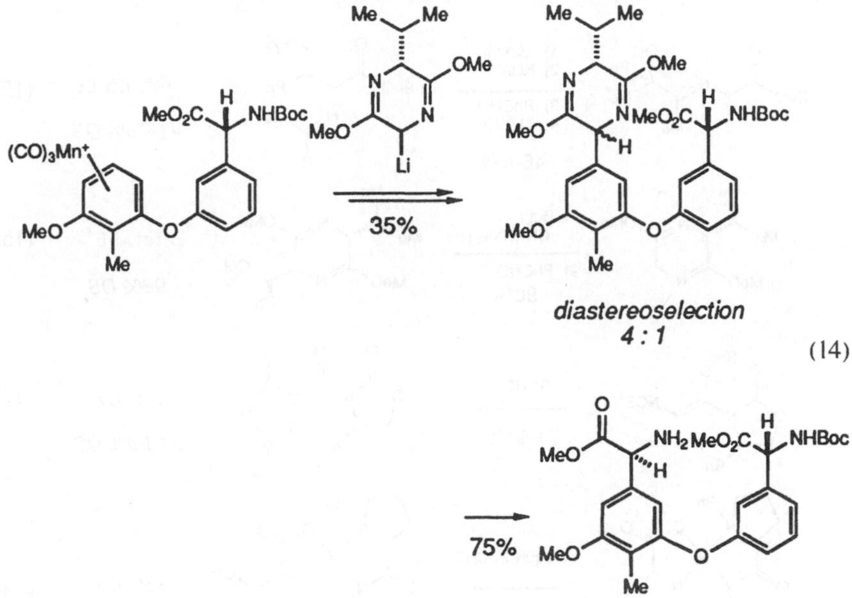

diastereoselection
4 : 1

(14)

B. Synthesis of Syn-β-Hydroxy-α-Amino Acids

General Overview

A large number of peptide natural products, including vancomycin, bouvardin (62), echinochandin (63), and cyclosporin (64), contain β-hydroxy-α-amino acids (Fig. 15) as unnatural amino acid constituents. Interest in these natural products as targets for synthesis has resulted in the development of new methods that efficiently deliver these residues in enantiomerically pure form. The obvious

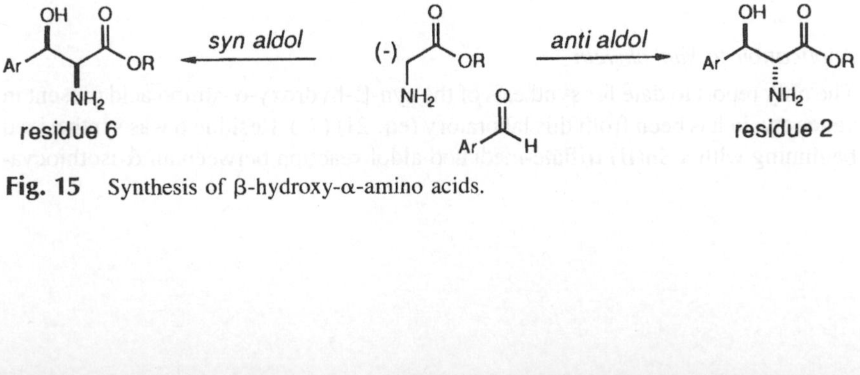

Fig. 15 Synthesis of β-hydroxy-α-amino acids.

approach to this class of compounds is the controlled aldol union of a chiral glycine enolate equivalent and an appropriate aldehyde.

Equations 15–20 illustrate some of the currently available methods, all of which proceed to give the syn product (55–59, 65–70). The method of Hayashi (eq. 20) is noteworthy in that this aldol process is catalytic.

Application to Vancomycin

The only report to date for synthesis of the syn-β-hydroxy-α-amino acid present in vancomycin has been from this laboratory (eq. 21) (71). Residue 6 was synthesized beginning with a Sn(II) triflate-mediated aldol reaction between an α-isothiocya-

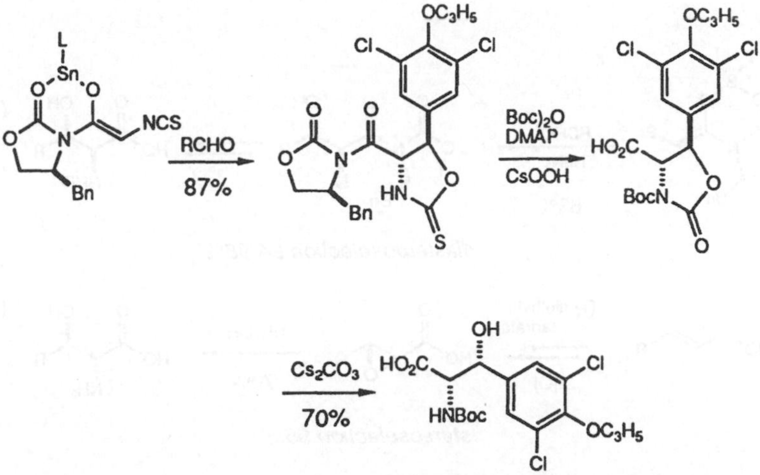

(21)

nate carboximide and the aromatic aldehyde to afford an initial adduct that spontaneously cyclizes to provide the thiooxazolidinine intermediate. Acylation and hydrolysis (72) of this intermediate provided the protected syn-β-hydroxy-α-amino acid in good yield and with high selectivity.

C. Synthesis of Anti-β-Hydroxy-α-Amino Acids

General Overview

Although the chiral glycine enolates are efficient for forming syn-β-hydroxy-α-amino acids, synthesis of residue 2 in vancomycin would require the equivalent of an antialdol addition. Miller et al. used the Williams auxiliary, which provides an antialdol isomer selectively (eq. 22) (73). An alternative approach developed in these laboratories utilizes a synaldol followed by S_N2 displacement of an α-leaving group to access the desired antiisomer (eq. 23) (74). Another method for obtaining these β-hydroxy-α-amino acids utilizes the selectivity of the Sharpless epoxidation (eq. 24) (75).

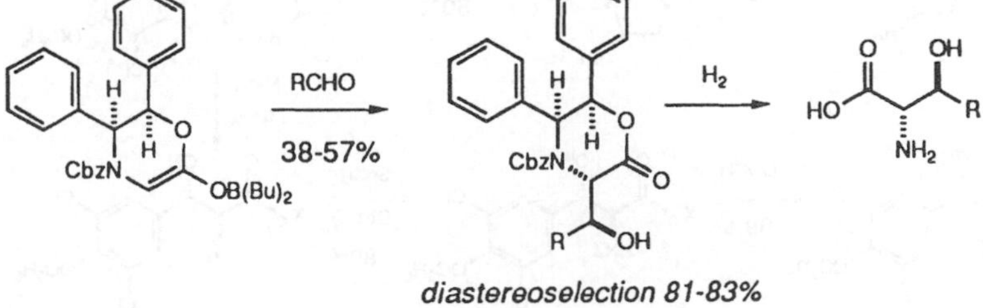

diastereoselection 81-83%

(22)

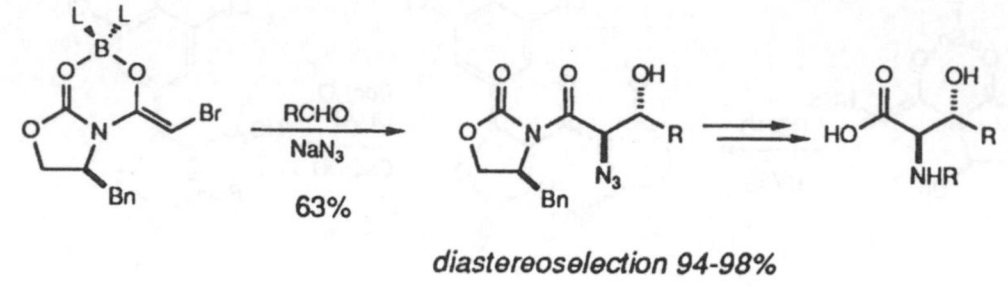

$$(23)$$

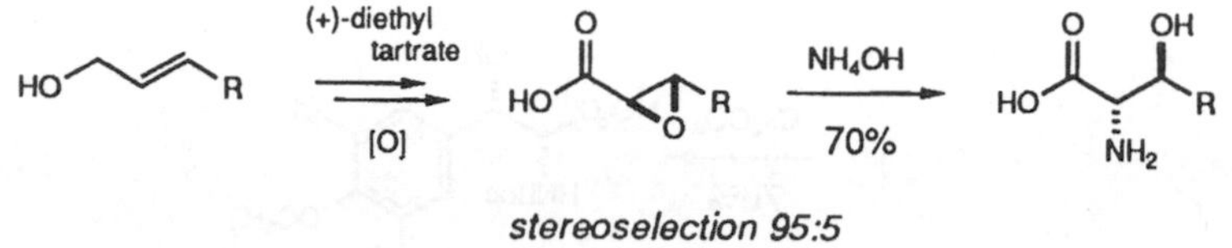

diastereoselection 94-98%

$$(24)$$

stereoselection 95:5

Application to Vancomycin

This laboratory has used chiral carboximides to obtain the anti-β-hydroxy-α-amino acid present in vancomycin (Fig. 16). (55–59). Residue 2 was synthesized by an aldol addition between the boron enolate of the α-bromocarboximide and the aromatic aldehyde to provide the illustrated aldol adduct in reasonable yield and diastereoselectivity. Displacement of the bromide with sodium azide cleanly provided the α-azidocarboximide. Reduction of the azido functionality, acylation, and cleavage of the chiral auxiliary provided the desired protected amino acid.

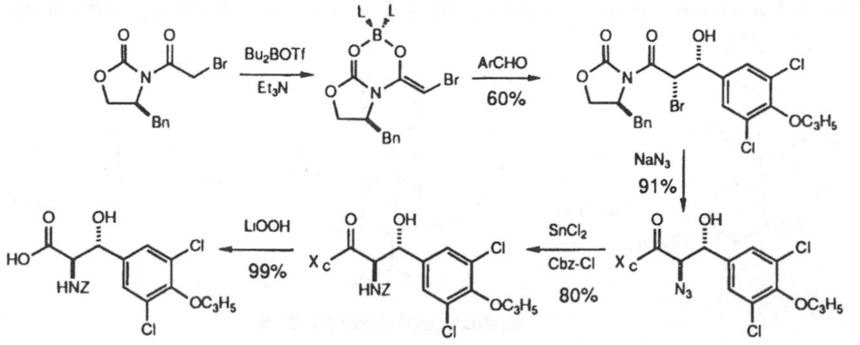

Fig. 16   Synthesis of anti-β-hydroxy-α-amino acids.

VI. SYNTHESIS OF VANCOMYCIN AND RISTOCETIN SUBUNITS

A. Approaches to Ristomycinic Acid

Pearson Route

Pearson and coworkers used iron and manganese-arene species to form diaryl ethers and applied this methodology to the synthesis of a deoxyristomycinic acid derivative (Figs. 3 and 17) (60,61). The yields for forming the metal-arene complexes and their coupling to form diaryl ethers are high and the coupling is reported to proceed with no racemization of the arylglycine moiety. Because of the harsh conditions necessary to form these transition metal-arene complexes, however, the stereogenic center associated with the second amino acid must be introduced *after* formation of the diphenyl ether. As illustrated in Figure 17, the Schöllkopf auxiliary was employed to introduce the second arylglycine moiety. Oxidative cleavage of the product metal-arene complex and hydrolysis of the auxiliary is efficient and proceeds with no racemization.

Evans Route

This laboratory is in the process of developing a route to ristomycinic acid based on the addition of the phenolate anion of a protected arylglycine derivative to a

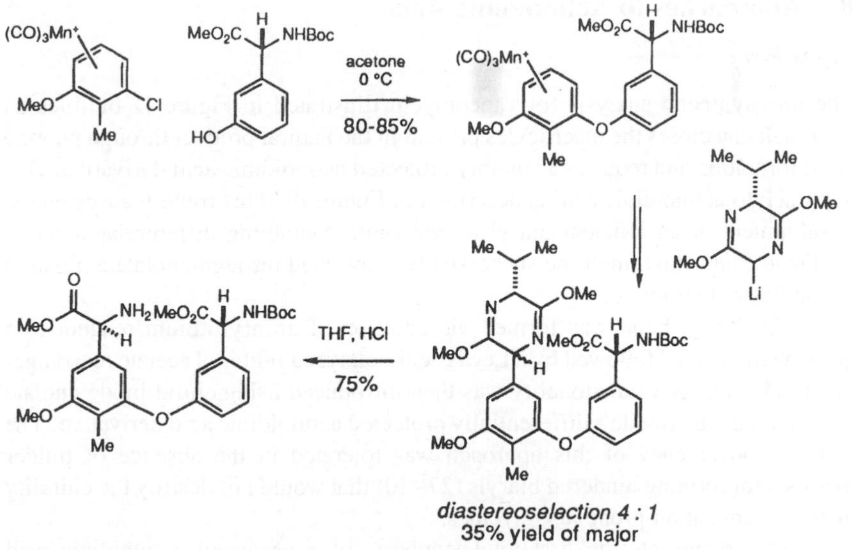

diastereoselection 4 : 1
35% yield of major

Fig. 17 Pearson approach to ristomycinic acid.

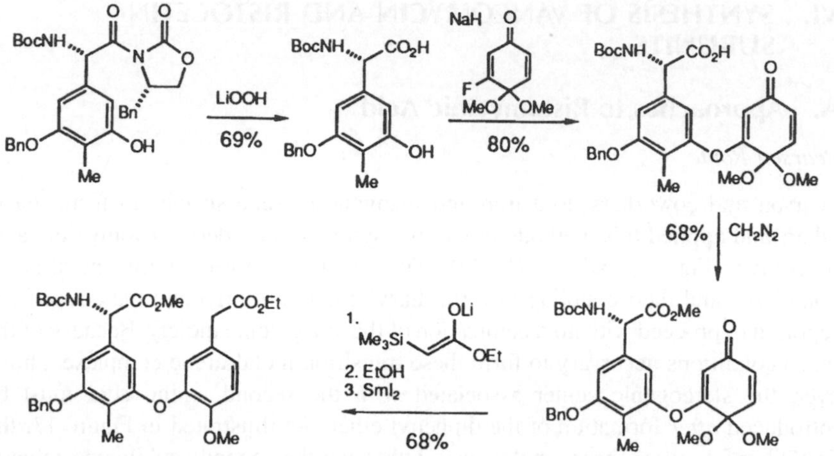

Fig. 18 Evans approach to ristomycinic acid.

quinone monoketal (Fig. 18) (76). The phenol partner in this coupling was obtained using the direct azide transfer protocol developed in these laboratories. Incorporation of the second nitrogen-bearing stereocenter would complete a synthesis of ristomycinic acid.

B. Approaches to Actinoidinic Acid

Evans Route

The retrosynthetic analysis for vancomycin illustrated in Figure 12 outlines an approach that closes the macrocyles present in the natural product through peptide bond formation and requires a suitably protected actinoidinic acid derivative. One approach to actinoidinic acid is described in Figure 19. This route focuses on the development of an efficient biaryl intermediate containing differentiated acetic acid side chains that might be successively elaborated through enolate amination to the desired structure.

The biaryl bond was formed via addition of an aryllithium reagent to a quinone monoketal followed by a Lewis acid-catalyzed p-quinol acetate rearrangement. The nitrogen functionality was then introduced using chiral imide enolate methodology to provide a differentially protected actinoidinic acid derivative. The lack of convergency of this approach was tolerated in the absence of milder methods for forming hindered biaryls (27–30) that would not destroy the chirality of the racemization-prone arylglycines.

This represents the first total synthesis of a protected actinoidinic acid derivative (77,78). As expected for a biaryl moiety with three ortho substituents,

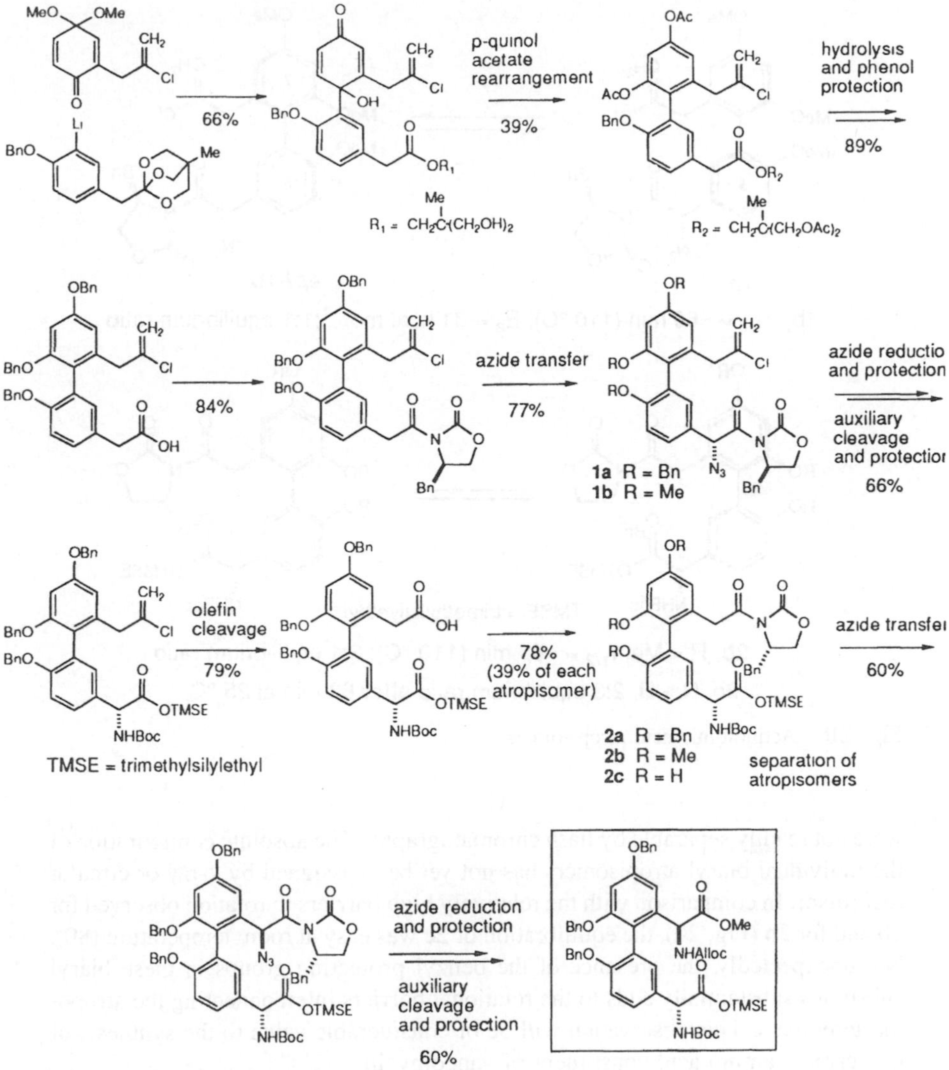

Fig. 19 Synthesis of actinoidinic acid.

there is hindered rotation about this bond with a measured barrier to rotation for **1b** of 31 kcal/mol (Fig. 20) (79).

The 1:1 mixture of atropisomers formed in this biaryl synthesis could be cleanly separated after attachment of the chiral auxiliary to provide imide (**2a**). Before this point, although the atropisomers were distinguishable by NMR, they

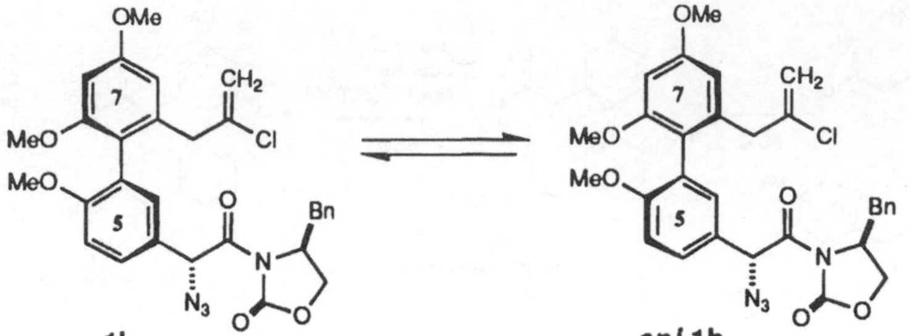

1b, $t_{1/2}$ = ~60 min (110 °C), E_a = 31 kcal mol^{-1}; 1:1 equilibrium ratio

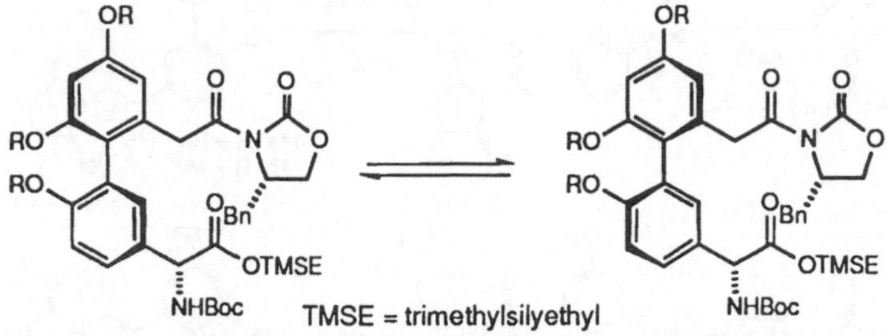

2b, R = Me, $t_{1/2}$ = ~60 min (110 °C); 3:1 equilibrium ratio

2c, R = H, 2:3 equilibrium ratio after 60 min at 25 °C

Fig. 20   Actinoidinic acid atropisomers.

were not readily separable by flash chromatography. The absolute configuration of the individual biaryl atropisomers has not yet been assigned by x-ray or circular dichroism. In comparison with the relatively high barriers to rotation observed for **1b** and for **2b** (Fig. 20), the equilibration of **2c** was easy at room temperature (80). Not unexpectedly, the presence of the benzyl protecting groups in these biaryl substrates substantially adds to the rotational barriers interconnecting the atropodiastereomers. This observation will be of considerable value to the synthesis of this critical amino acid constituent of vancomycin.

C.   Approaches to Vancomycinic Acid

Pearson Route

Pearson et al. also used metal-arene complexes to form a derivative of vancomycinic acid (81). Both diphenyl ethers were formed in high yield, and the protected tyrosine residues, which were used in place of the β-hydroxy-α-amino acids present in vancomycin itself, were not racemized (Fig. 21). As with the

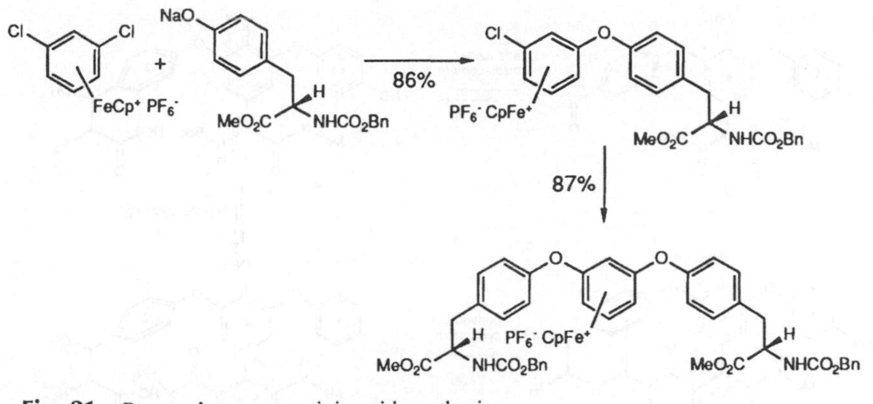

Fig. 21 Pearson's vancomycinic acid synthesis.

Pearson ristomycinic acid synthesis (Fig. 17), however, the harsh conditions necessary to form the metal-arene complexes would require that the arylglycine side chain be introduced after diphenyl ether formation.

Evans Route

The approach to vancomycin outlined in Figure 12 also requires a differentiated vancomycinic acid derivative. The harsh conditions of the Ullmann reaction, the classic method for forming diphenyl ethers, preclude any chirality during execution of these bond constructions. The plan was to form an achiral, differentiated diphenyl ether network and to elaborate this fragment "tridirectionally" using chiral enolate methodology. The synthesis of an achiral vancomycinic acid derivative employing the classic Ullmann reaction is illustrated in Figure 22 (82).

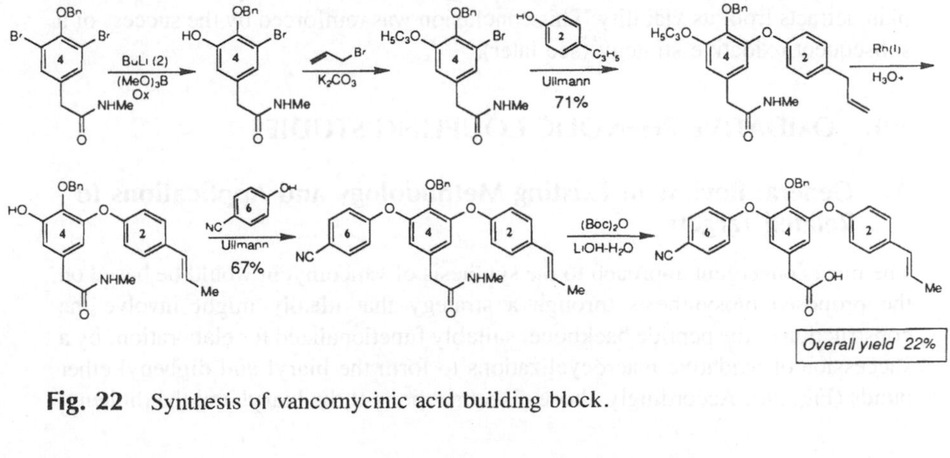

Fig. 22 Synthesis of vancomycinic acid building block.

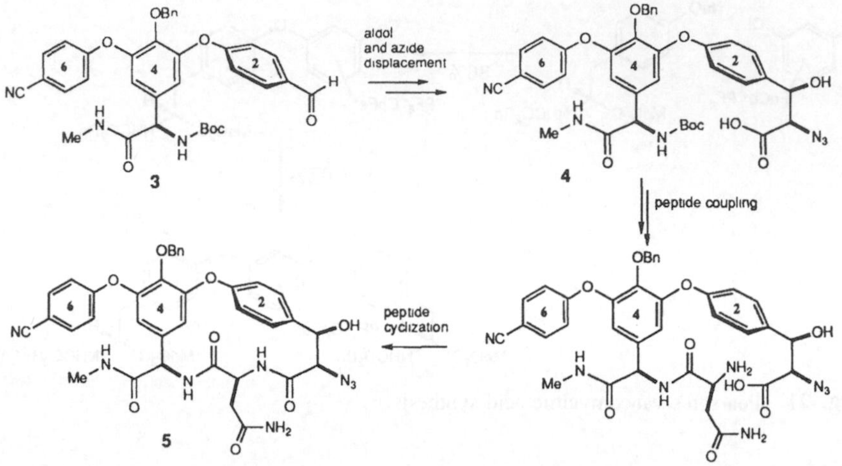

Fig. 23 Formation of the M(2–4) monocycle by peptide cyclization.

The direct azide transfer methodology (see eq. 10) was employed to introduce the nitrogen functionality of residue 4 (Fig. 23) to provide **3** (61%, 12:1 diastereoselection). This transformation was followed by a bromoacetate aldol addition and subsequent azide displacement (see Fig. 16) to form the anti-β-hydroxy-α-azido acid of residue 2 to furnish **4** (58%, >10:1 diastereoselection). The asparagine residue was then introduced, and the M(2–4) macrocycle was formed using conventional peptide cyclization methodology (diisopropylcarbodiimide, pentafluorophenol) to provide **5** [21% overall from (**4**)].

Although this route provides access to a differentiated vancomycinic acid derivative that could be coupled to an appropriately functionalized actinoidinic acid derivative prepared as described in Figure 19, the linearity of the synthesis plan detracts from its viability. This conclusion was reinforced by the success of a subsequent oxidative strategy (see later).

VII. OXIDATIVE PHENOLIC COUPLING STUDIES

A. General Review of Existing Methodology and Applications to Related Targets

The most convergent approach to the synthesis of vancomycin would be based on the proposed biosynthesis through a strategy that ideally might involve the construction of the peptide backbone, suitably functionalized for elaboration, by a succession of oxidative macrocyclizations to form the biaryl and diphenyl ether bonds (Fig. 24). Accordingly, those disconnections at the biaryl and the diphenyl

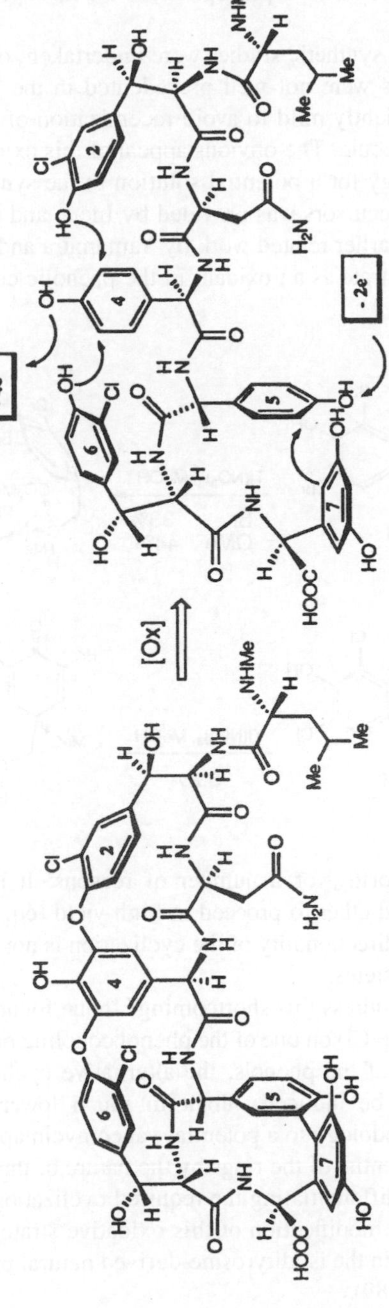

Fig. 24   Oxidative macrocyclization approach to vancomycin.

ether bonds to provide the linear heptapeptide are the most powerfully simplifying transforms.

At the time initial synthetic studies were undertaken, oxidative cyclizations to form diphenyl ethers were not well precedented in the literature, especially under conditions sufficiently mild to avoid racemization of the arylglycine residues present in the molecule. The obvious appeal of this oxidative approach is its convergency. An analogy for a potential solution to the synthesis of such biaryl ethers from phenolic precursors was provided by Inoue and coworkers (83). This study, an extension of earlier related work by Yamamura and coworkers (84–87), employed thallium trinitrate as an oxidant in the phenolic coupling step (eqs. 25 and 26).

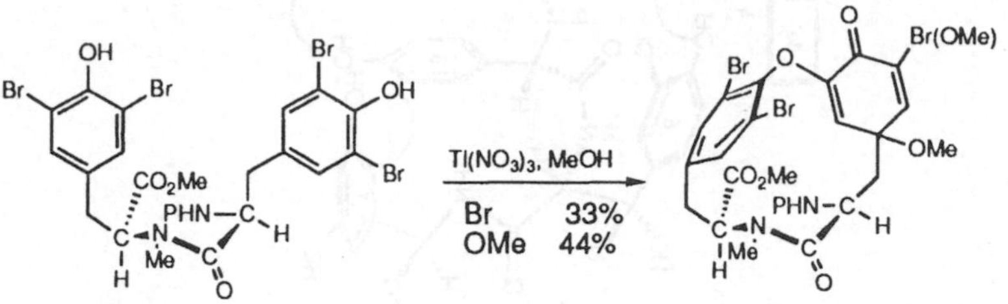

$$Tl(NO_3)_3, MeOH$$

$$\begin{array}{ll} Br & 33\% \\ OMe & 44\% \end{array}$$

(25)

$$Tl(NO_3)_3, MeOH$$

$$5.2\%$$

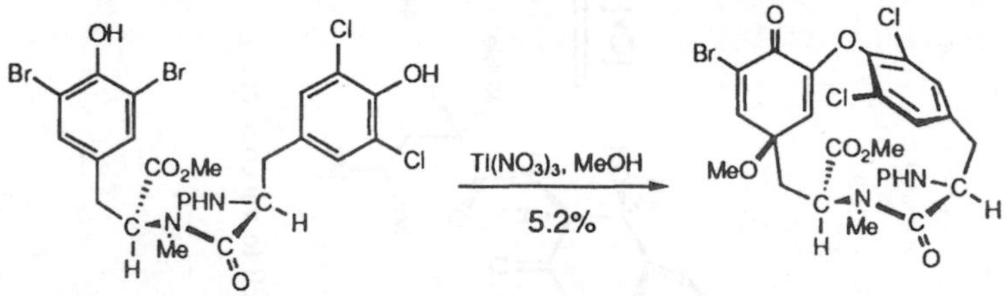

(26)

This report was noteworthy for a number of reasons. It is the first oxidative cyclization to a diphenyl ether to proceed in high yield (eq. 25). The liability of this reaction is that the directionality of the cyclization is not readily controlled in such unsymmetrical systems.

In attempting to address this shortcoming, Inoue found that by altering the halogen substituent (Br→Cl) on one of the phenol coupling partners, thus altering the oxidation potential of the phenols, the alternative cyclization mode for the desired product could be induced, albeit in much lower yield (eq. 26). In extrapolating this methodology to a potential vancomycin application, the ability to control the redox potential of the rings by the nature of the halogen substituent would be essential for differentiating the required cyclization modes. Yamamura recently used the Inoue modification of this oxidative strategy to synthesize the diphenyl ethers present in the isodityrosine-derived natural products OF-4949-III and K-13 (Fig. 25) (88,89).

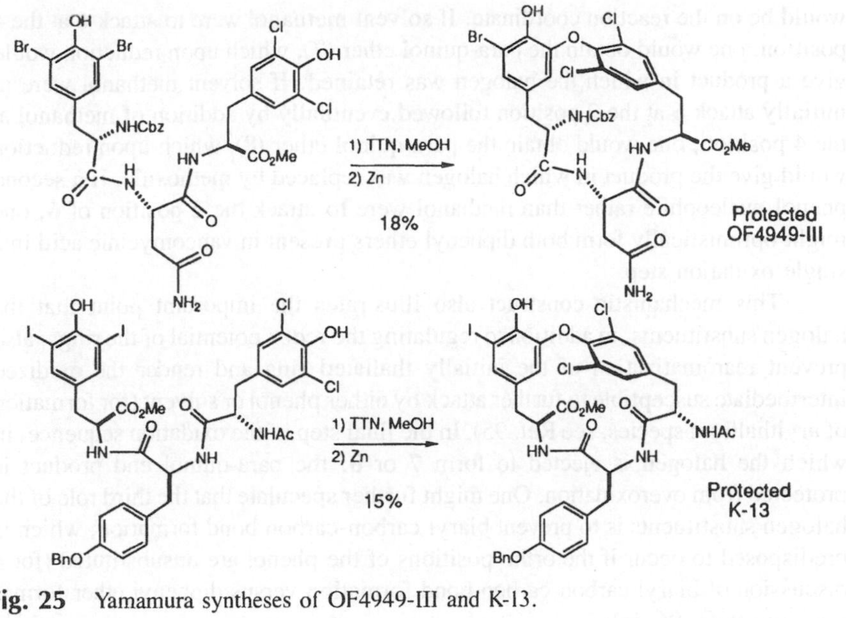

Fig. 25 Yamamura syntheses of OF4949-III and K-13.

The following mechanistic construct provides an explanation for the two products observed by Inoue in equation 25. A discrete arylthallium intermediate may not be formed initially, although Taylor proposed its existence (Fig. 26) (90). The biosynthesis of diphenyl ethers is proposed to proceed through phenoxy radicals (for a discussion, see Refs. 91–94). Regardless of the mechanism, compound **6**

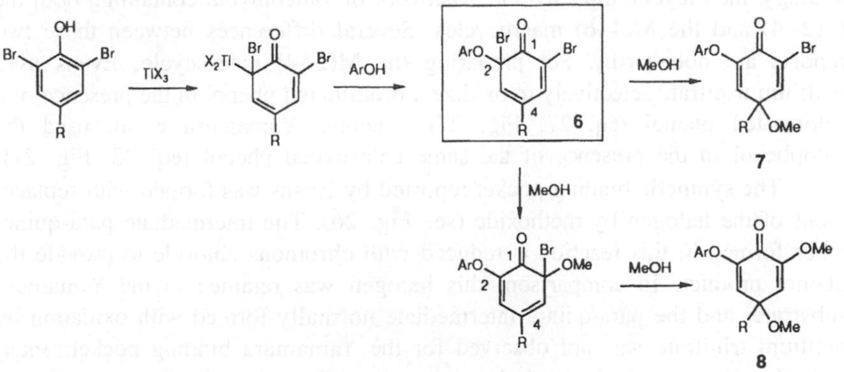

Fig. 26 Mechanistic construct for oxidative cyclizations.

would be on the reaction coordinate. If solvent methanol were to attack **6** at the 4 position, one would obtain the para-quinol ether (**7**), which upon reduction would give a product in which the halogen was retained. If solvent methanol were to initially attack **6** at the 2 position followed eventually by addition of methanol at the 4 position, one would obtain the para-quinol ether (**8**), which upon reduction would give the product in which halogen was replaced by methoxide. If a second phenol nucleophile rather than methanol were to attack the 2 position of **6**, one might optimistically form both diphenyl ethers present in vancomycinic acid in a single oxidation step.

This mechanistic construct also illustrates the important point that the halogen substituents, in addition to regulating the redox potential of the rings, also prevent rearomatization of the initially thallated ring and render the oxidized intermediate susceptible to further attack by either phenol or solvent (for formation of arylthallium species, see Ref. 95). In the final step of the oxidation sequence, in which the halogen is ejected to form **7** or **8**, the para-quinol end product is protected from overoxidation. One might further speculate that the third role of the halogen substituents is to prevent biaryl carbon-carbon bond formation, which is predisposed to occur if the ortho positions of the phenol are unsubstituted (for a discussion of biaryl carbon-carbon bond formation versus diphenyl ether formation, see Refs. 96–99).

B. The Evans and Yamamura Studies

This mild diphenyl ether synthesis allows one to effect macrocyclization in conjunction with the diphenyl ether bond construction with no attendant racemization of the arylglycine residues. This approach was employed by Evans et al. (71) and subsequently by Yamamura and colleagues (100,101) to form the vancomycin binding pocket, the M(4–6) macrocycle, and, with a sequential cyclization strategy, the bicyclic diphenyl ether network of vancomycin containing both the M(2–4) and the M(4–6) macrocycles. Several differences between these two reports are noteworthy. For preparing the M(2–4) macrocycle, Evans used thallium trinitrate selectively to oxidize a brominated phenol in the presence of a chlorinated phenol (eq. 27, Fig. 27), whereas Yamamura et al. used the iodophenol in the presence of the same chlorinated phenol (eq. 28, Fig. 27).

The synthetic binding pocket reported by Evans was formed with replacement of the halogen by methoxide (see Fig. 26). The intermediate para-quinol ether formed in this reaction is reduced with chromous chloride to provide the phenol product. In comparison, this halogen was retained in the Yamamura substrate, and the para-quinol intermediate normally formed with oxidation by thallium trinitrate was not observed for the Yamamura binding pocket. Interestingly, Yamamura obtained the reduced product directly. Incorporation of methanol in the Evans cyclization made a dicyclization strategy outlined in

Evans Route

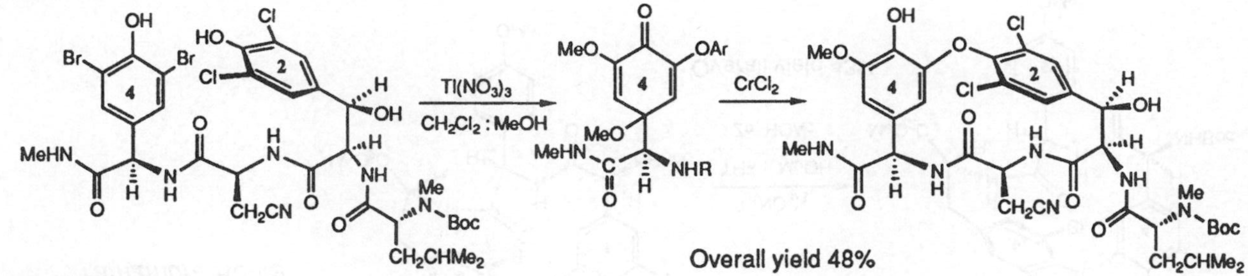

(27)

Yamamura Route

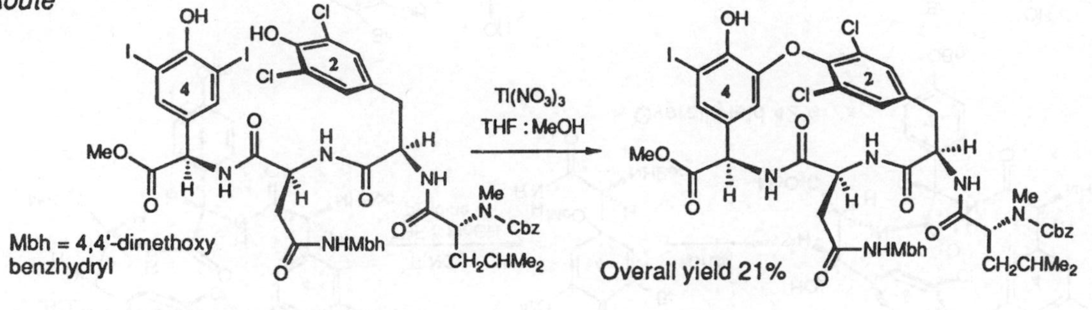

(28)

Fig. 27 Syntheses of the M(2–4) macrocyclic fragment.

Evans Route

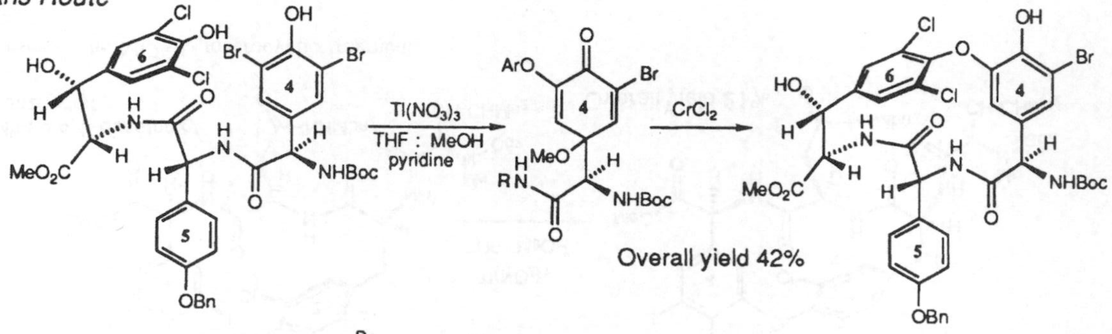

(30)

Yamamura Route

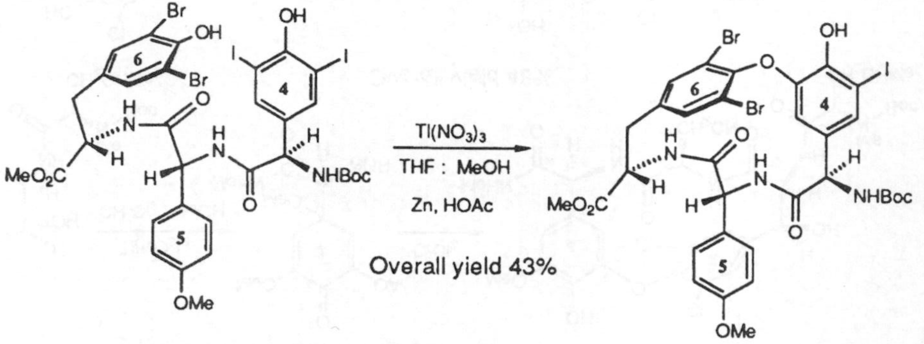

(31)

Fig. 28 Syntheses of the M(4–6) macrocyclic fragment.

equation 29 look potentially viable. For this transformation to be realized, a second phenol nucleophile rather than methanol must add into the ortho position of (6) in Figure 26.

$$(29)$$

The M(4–6) macrocylization for both the Evans (eq. 30, Fig. 28) and Yamamura substrates (eq. 31, Fig. 28) proceeded with retention of the halogen substituent, and in both cases the intermediate para-quinol ether was observed. The selectivity of this reagent is quite remarkable since the phenolic ether on residue 5 is not oxidized under these conditions.

Although incorporation of methanol in the Evans binding pocket (eq. 27, Fig. 27) made a dicyclization strategy look potentially attractive, a more conservative strategy of forming the second diphenyl ether in a sequential manner was pursued. Since only the M(4–6) macrocyclization proceeded with retention of halogen in the Evans system and this halogen was required for the second diphenyl ether construction, this dictated the order of assembly for the sequential cyclization. After coupling the remaining tripeptide fragment and deprotection of the residue 2 phenol, oxidative cyclization was effected by thallium trinitrate oxidation to afford the intermediate para-quinol ether, which was reduced to provide the bicyclic M(2–4), M(4–6)-vancomycin diphenyl ether network (Fig. 29, eq. 32). Yamamura has also reported the synthesis of a similar fragment that lacks the benzylic hydroxyl groups in residues 2 and 6 present in the natural product (Fig. 29, eq. 33).

VIII. FUTURE PROSPECTS

The remaining formidable obstacle to completion of the synthesis of didechloro-vancomycin must address the construction of the 12-membered M(5–7) ring containing the actinoidinic acid residue. Indeed, even with efficient synthesis of actinoidinic acid in hand, the subsequent construction of the highly strained 12-membered ring containing the two racemization-prone arylglycines may prove to be the most challenging aspect of any projected syntheses of this family of natural products. As is evident from the illustrated partial x-ray structure of CDP-1M containing the biaryl moiety, the amide bond joining residues 5 and 6 has adopted a cis geometry, presumably to alleviate the inherent strain associated with the M(5–7) macrocycle (Fig. 30).

The absence of efficient redox methodology for the formation of hindered biaryl bonds (27–30), especially under conditions mild enough to preclude arylglycine racemization of residues 5 and 7, may dictate that the nitrogen

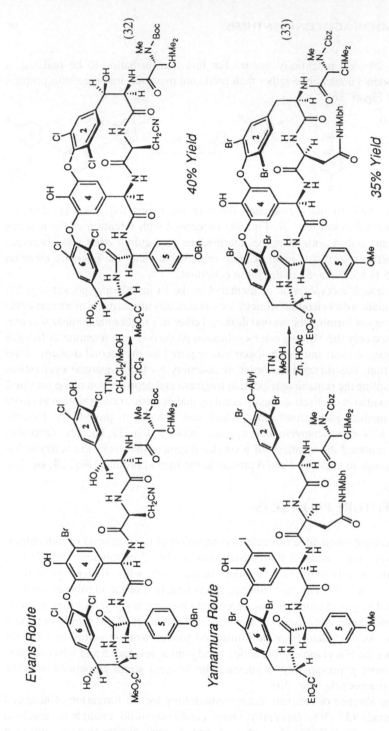

Fig. 29 Syntheses of the M(2–4):M(4–6) bicyclic diphenyl ether fragment.

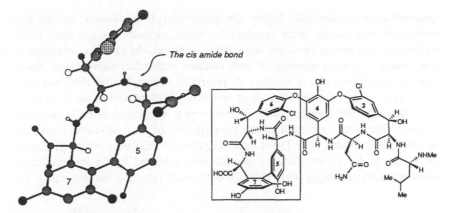

Fig. 30   Biaryl-containing vancomycin macrocycle M(5–7).

functionality be stereoselectively incorporated after formation of the biaryl bond construction, as described in Figure 19 for the Evans synthesis of actinoidinic acid. Indeed, within the last year the first published synthesis of actinoidinic acid was reported by Rao and coworkers by just such a route (102).

Certainly the most efficient route to the construction of the biaryl bond in vancomycin would be through oxidative phenolic coupling, ideally from the appropriate tripeptide residue found in the natural product. A recent report by Magnus et al. (103) provides an encouraging analogy for the use of thallium(III) in a relevant bond construction (eq. 34; for the primary study on such reactions, see Ref. 104, and for a related oxidative cyclization, see Ref. 100 and references therein). In comparison with other biaryl bond-forming reactions, this oxidative approach appears quite mild, highly substituted biaryls are formed, and the aryl halide and aryl metals often required for other methods do not need to be regioselectively incorporated into the starting materials. In addition, dehalogenation, which is often a problematic side reaction with other conventional biaryl bond-forming protocols, is avoided. A projected intramolecular oxidative coupling approach to an actinoidinic acid derivative is illustrated in equation 35.

Recently, the complete M(5–7) macrocycle was constructed in the most obvious biomimetic approach to the construction of the biaryl-containing 12-membered tripeptide (106). The macrocyclization of the linear tripeptide precursor to the cyclic 12-membered tripeptide was achieved by an intramolecular oxidative biaryl coupling with VOF₃ in 64% yield (eq. 36). As illustrated, the

DCB = 3,4-dichlorobenzyl   (36)

R = DCB
R = H

kinetic product of this transformation was found to be the undesired biaryl atropodiastereomer. The key to effecting this critical reaction was the presence of a second differentiated phenolic residue on the aromatic ring in position 5, which serves either to lower its oxidation potential or to increase aromatic ring nucleophilicity. After removal of the requisite ortho phenolic residue and subsequent demethylation of this intermediate to the unnatural atropisomer, atropisomeriza-

equilibrium ratio
11 : 89   (37)

tion to the M(5–7) macrocyclic tripeptide vancomycin subunit was achieved (eq. 37). The rotational barrier for this conformational change was determined to be 22 kcal/mol. In contrast, the trimethylated analogs were found to be resistant to isomerization at temperatures in excess of 100°C.

IX. CONCLUSION

Targets of the complexity of vancomycin invariably provide the impetus for the development of important new reaction methodology. The present instance is no exception. Important advances have already been made in the development of new approaches to the synthesis of α amino acids, including arylgylcines and β-hydroxy-α-amino acids, suitably derivatized to be practical building blocks for a synthesis venture. Significant progress has also been made in the development of reliable peptide macrocyclization reagents and strategies. Finally, new insights into the design of phenol oxidative coupling have been identified. As yet, the challenge posed by the vancomycin structure has yet to be met. Nevertheless, from the chemistry described in this review, considerable progress toward this objective has been made.

X. EPILOGUE

Since the completion of this review other significant articles have appeared (107,108).

REFERENCES

1. McCormick M.H., Stark W.M., Pittenger G.E., Pittenger R.C., McGuire G.M. (1955–1956). Antibiot. Annu. 606.
2. Barna J.C.J., Williams D.H. (1984). Annu. Rev. Microbiol. 38:339–357.
3. Sheldrick G.M., Jones P.G., Kennard O., Williams D.H., Smith G.A. (1978). Nature 271:223–225.
4. Williamson M.P., Williams D.H. (1981). J. Am. Chem. Soc. 103:6580–6585.
5. Harris C.M., Harris T.M. (1982). J. Am. Chem. Soc. 104:4293–4295.
6. Harris C.M., Kopecka H., Harris T.M. (1983). J. Am. Chem. Soc. 105:6915–6922.
7. Barany G., Merrifield R.B. (1979). In Gross E., Meienhofer J. (Eds.), The Peptides, Vol. 2. Academic Press, New York, pp. 192–208.
8. Waltho J.P., Williams D.H. (1989). J. Am. Chem. Soc. 111:2475–2480.
9. Williams D.H., Waltho J.P. (1988). Biochem. Pharmacol. 37:133–141.
10. Perkins H.R. (1969). Biochem. J. 111:195–205.
11. Hammond S.J., Williamson M.P., Williams D.H., Boeck L.D., Marconi G.G. (1982). J. Chem. Soc., Chem. Commun. 344–346.
12. Taylor W.I., Battersby A.R. (Eds.), (1967). Oxidative Coupling of Phenols. Marcel Dekker, New York.

13. Nagarajan R., Merkel K.E., Michel K.H., Higgins H.M. Jr, Hoehn M.M., Hunt A.H., Jones N.D., Occolowitz J.L., Schabel A.A., Swartzendruber J.K. (1989). J. Am. Chem. Soc. 111:7896–7997.

14. Harris C.M., Kibby J.J., Fehlner J.R., Raabe A.B., Barber T.A., Harris T.M. (1979). J. Am. Chem. Soc. 101:437–445.

15. Jeffs P.W., Chan G., Mueller L., DeBrosse C., Webb L., Sitrin R. (1986). J. Org. Chem. 51:4272–4278.

16. Oki M. (1983). Top. Stereochem. 14:1–81.

17. Kawano N., Okigawa M., Hasaka N., Kouno I., Kawahara Y., Fujita Y. (1981). J. Org. Chem. 46:389–392.

18. Harris C.M., Kannan R., Kopecka H., Harris T.M. (1985). J. Am. Chem. Soc. 107: 6652–6658.

19. Tsuji N., Kobayashi M., Kamigauchi T., Yoshimura Y., Terui Y. (1988). J. Antibiot. 41:819–822.

20. Bringmann G., Walter R., Weirich R. (1990). Angew. Chem. Int. Ed. (Engl.) 29: 977–991.

21. Harada N., Nakanishi K. (1983). Circular Dichroic Spectroscopy—Exciton Coupling in Organic Stereo Chemistry. University Science Books, Mill Valley, CA.

22. Smith G.G., Sivakua T. (1983). J. Org. Chem. 48:627–634.

23. Bodanszky M., Bodanszky A. (1967). J. Chem. Soc., Chem. Commun. 591–593.

24. Evans D.A., Ellman J.E. (1989). J. Am. Chem. Soc. 111:1063–1072.

25. Boger D.L., Yohannes D. (1991). J. Am. Chem. Soc. 113:1427–1429.

26. Brady S.F., Freidinger R.M., Paleveda W.J., Colton C.D., Homnick C.F., Whitter W.L., Curley P., Nutt R.F., Veber D.F. (1987). J. Org. Chem. 52:764–769.

27. Bringmann G., Walter R., Weirich R. (1990). Angew. Chem. Int. Ed. (Engl.) 29: 977–991.

28. Sainsbury M. (1980). Tetrahedron 36:3327–3359.

29. Semmelhack M.F., Helquist P., Jones L.D., Keller L., Mendelson L., Ryono L.S., Smith J.G., Stauffer R.D. (1981). J. Am. Chem. Soc. 103:6460–6471.

30. Muller D., Fleury J-P. (1991). Tetrahedron Lett. 32:2229–2232.

31. Evans D.A., Ellman J.E. (1989). J. Am. Chem. Soc. 111:1063–1072.

32. Schmidt U., Weller D., Holder A., Lieberknecht A. (1988). Tetrahedron Lett. 26: 3227–3230.

33. Boger D.L., Yohannes D. (1990). J. Org. Chem. 55:6000–6017.

34. Boger D.L., Yohannes D. (1991). J. Am. Chem. Soc. 113:1427–1429.

35. Boger D.L., Yohannes D. (1991). J. Org. Chem. 56:1763–1767.

36. Mann M.J., Pant N., Hamilton A.D. (1986). J. Chem. Soc., Chem. Commun. 158–160.

37. Pant N., Hamilton A.D. (1988). J. Am. Chem. Soc. 110:2002–2003.

38. Crimmin M.J., Brown A.G. (1990). Tetrahedron Lett. 31:2017–2020, 2021–2024.

39. Hobbs D.W., Still W.C. (1987). Tetrahedron Lett. 28:2805–2808.

40. Hobbs D.W., Still W.C. (1989). Tetrahedron Lett. 30:5405–5408.

41. Britton T.C. (1988). Ph.D. Dissertation, Harvard University.

42. Clark J.C., Phillips G.H., Steer M.R., Stephenson L., Cooksey A.R. (1976). J. Chem. Soc., Perkin Trans. 1:471–474.

43. Clark J.C., Phillips G.H., Steer M.R. (1976). J. Chem. Soc., Perkin Trans. I: 475–481.
44. Williams R.M., Hendrix J.A. (1992). Chem. Rev. 92:889–917.
45. Williams R.M., Im M.N. (1991). J. Am. Chem. Soc. 113:9276–9286.
46. Williams R.M., Hendrix J.A. (1990). J. Org. Chem. 55:3723–3728.
47. Yamamoto Y., Ito W., Maruyama K. (1985). J. Chem. Soc., Chem. Commun. 1131–1132.
48. Evans D.A., Britton T.C. (1987). J. Am. Chem. Soc. 109:6881–6883.
49. Evans D.A., Britton T.C., Ellman J.A., Dorow R.L. (1990). J. Am. Chem. Soc. 112: 4011–4030.
50. Evans D.A., Britton T.C., Dorow R.L., Dellaria J.F. (1988). Tetrahedron 44:5525–5540.
51. Evans D.A., Britton T.C., Ellman J.A. (1987). Tetrahedron Lett. 28:6141–6144.
52. Evans D.A., Ellman J.A., Dorow R.L. (1987). Tetrahedron Lett. 28:1123–1126.
53. Evans D.A., Evrard D.A., Rychnovsky S.D., Früh T., Whittingham W.G., DeVries K.M. (1992). Tetrahedron Lett. 33:1189–1192.
54. Stone M.J., Maplestone R.A., Rahman S.K., Williams D.H. (1991). Tetrahedron Lett. 32:2663–2666.
55. Schöllkopf U., Groth U., Nozulak J. (1983). Liebigs Ann. Chem. 1133–1151.
56. Schöllkopf U. (1983). Tetrahedron 39:2085–2091.
57. Schöllkopf U. (1983). Pure Appl. Chem. 55:1799–1806.
58. Schöllkopf U., Nozulak J., Grauert M. (1985). Synthesis 55–56.
59. Grauert M., Schöllkopf U. (1985). Liebigs Ann. Chem. 1817–1824.
60. Pearson A.J., Shin H. (1992). Tetrahedron 48:7527–7538.
61. Pearson A.J., Lee S-H., Gouzoules F. (1990). J. Chem. Soc., Perkin Trans. I:2251–2254.
62. Jolad S.D., Hoffman J.J., Torrance S.J., Wiedhopf R.M., Cole J.R., Arora S.K., Bates R.B., Gargiulo R.L., Kriek G.R. (1977). J. Am. Chem. Soc. 99:8040–8044.
63. Traber R., Keller-Juslen C., Loosli H-R., Kuhn M., von Wartberg A. (1979). Helv. Chim. Acta 62:1252–1267.
64. White D.J.G. (Ed.) (1982). Cyclosporin A. Biomedical, Amsterdam.
65. Nakatsuka T., Miwa T., Mukaiyama T. (1981). Chem. Lett. 279–282.
66. Nakatsuka T., Miwa T., Mukaiyama T. (1982). Chem. Lett. 145–148.
67. Evans D.A., Weber A.E. (1986). J. Am. Chem. Soc. 108:6757–6761.
68. Belekon Y.N., Bulychev A.G., Vitt S.V., Struchkov Y.T., Batsanov AS. Timofeeva T.V., Tsyryapkin V.A., Ryzhov M.G., Lysova L.A., Bakhmutov V.I., Belikov V.M. (1985). J. Am. Chem. Soc. 107:4252–4259.
69. Seebach D., Juaristi E., Miller D.D., Schlickli C., Weber T. (1987). Helv. Chim. Acta 70:237–261.
70. Ito Y., Sawamura M., Hayashi T. (1986). J. Am. Chem. Soc. 108:6405–6406.
71. Evans D.A., Ellman J.E., DeVries K.M. (1989). J. Am. Chem. Soc. 111:8912–8914.
72. Evans D.A., DeVries K.M. Harvard University, unpublished results.
73. Reno D.S., Lotz B.T., Miller M.J. (1990). Tetrahedron Lett. 31:827–830.
74. Evans D.A., Sjogren E.B., Weber A.E., Conn R.E. (1987). Tetrahedron Lett. 28: 39–42.

75. Kurokawa N., Ohfune Y. (1986). J. Am. Chem. Soc. 108:6041–6043.
76. Whittingham W.G. (1990). Postdoctoral research report, Harvard University.
77. Morrissey M. (1986). Ph.D. Dissertation, Harvard University.
78. Fruh T. (1988). Postdoctoral research report, Harvard University.
79. Evans D.A., Morrissey M.M. Harvard University, unpublished results.
80. Evans D.A., Fruh T. Harvard University, unpublished results.
81. Pearson A.J., Park J.G., Zhu P.Y. (1992). J. Org. Chem. 57:3583–3589.
82. Rychnovsky S.D. (1987). Postdoctoral research report, Harvard University.
83. Inaba T., Umezawa I., Yuasa M., Inoue T., Mihashi S., Itokawa H., Ogura K. (1987). J. Org. Chem. 52:2957–2958.
84. Noda H., Niwa M., Yamamura S. (1981). Tetrahedron Lett. 22:3247–3248.
85. Nishiyama S., Yamamura S. (1982). Tetrahedron Lett. 23:1281–1284.
86. Nishiyama S., Suzuki T., Yamamura S. (1982). Tetrahedron Lett. 23:3699–3702.
87. Nishiyama S., Nakamura K., Suzuki T., Yamamura S. (1986). Tetrahedron Lett. 27: 4481–4484.
88. Nishiyama S., Suzuki Y., Yamamura S. ((1988). Tetrahedron Lett. 29:559–562.
89. Nishiyama S., Suzuki Y., Yamamura S. (1989). Tetrahedron Lett. 30:379–382.
90. McKillop A., Perry D.H., Edwards M., Antus S., Farkas L., Nogradi M., Taylor E.C. (1976). J. Org. Chem. 41:282–287.
91. Lewis N., Wallbank P. (1987). Synthesis 1103–1106.
92. Dreher E.L. (1979). Houben-Weyl, 4th ed., Vol. VII/3b. Georg Thieme Verlag, Stuttgart, p. 580.
93. Altwicker E.R. (1967). Chem. Rev. 67:475–531.
94. Taylor W.I., Battersby A.R. (Eds.) (1967). Oxidative Coupling of Phenols. Marcel Dekker, New York.
95. McKillop A., Hunt J.D., Zelesko M.J., Fowler J.S., Taylor E.C., McGillivray G., Kienzle F. (1980). J. Am. Chem. Soc. 93:4841–4844.
96. Scott A.I. (1967). In Taylor W.I., Battersby A.R. (Eds.). Oxidative Coupling of Phenols. Marcel Dekker, New York, p. 107.
97. Nishiyama S., Yamamura S. (1982). Tetrahedron Lett. 23:1281–1284.
98. Nishiyama S., Suzuki T., Yamamura S. (1982). Tetrahedron Lett. 23:3699–3702.
99. Rotermund G.W. (1975). Houben-Weyl, 4th ed., Vol. IV/lb. Georg Thieme Verlag, Stuttgart, p. 761.
100. Nishiyama S., Suzuki Y., Yamamura S. (1989). Tetrahedron Lett. 29:6043–6046.
101. Nishiyama S., Suzuki Y., Yamamura S. (1990). Tetrahedron Lett. 31:4053–4056.
102. Rao A.V.R., Chakraborty T.K., Joshi S.P. (1992). Tetrahedron Lett. 33:4045–4048.
103. Magnus P., Schultz J., Gallagher T. (1985). J. Am. Chem. Soc. 107:4984–4988.
104. McKillop A., Turrell A.G., Taylor E.C. (1977). J. Org. Chem. 42:764–765.
105. Robin J-P., Landais Y. (1988). Org. Chem. 53:224–226.
106. Evans D.A., Dinsmore D.J., Evrard D.A., DeVries K.M. (1993). J. Am. Chem. Soc. 115:6426–6427; Evans, D.A., Dinsmore, D.J. (1993). Tetrahedron Lett. 34:6029–6032.
107. Rao A.V.R., Gurjar M.K., Kaiwar V., Khare V.B. (1993). Tetrahedron Lett. 34: 1661–1664.
108. Rao A.V.R., Chakraborty T.K., Reddy K.L., Rao A.S. (1992). Tetrahedron Lett. 33: 4799–4802.
109. Nieto M., Perkins H.R. (1971). Biochem. J. 124:845–852.

4
Chemistry of Carbohydrate Components

FERENC SZTARICSKAI and ISTVÁN PELYVÁS-FERENCZIK
Hungarian Academy of Sciences, Lajos Kossuth University, Debrecen, Hungary

I. INTRODUCTION

Vancomycin and related antibiotics significantly differ from other glycopeptide antibiotics in both structural features and biological effects. Differentiation can also be made between such antibiotics on the basis of the carbohydrate components, linked singly or in a combined form to the tri- and tetracyclic heptapeptide aglycones. The nature and quantity of these sugars, as well as their substitution pattern and mode of linkage, are probably important influencing factors (1) on the pharmacokinetic and pharmacodynamic properties and biological action of these antibiotics. Recent knowledge emphasizes the essential role of carbohydrates in the recognition and communication processes between living cells (2,3) by invariable structural arrangement.

Structural investigations of the vancomycin group of antibiotics have shown (4,5) that of the currently known neutral sugars, D-glucose, D-mannose, D-galactose, D-arabinose, L-rhamnose, and L-fucose occur in the molecules of these natural compounds. The data in Table I indicate clearly that D-glucose, D-mannose, and L-rhamnose are the most frequent structural elements, whereas D-galactose, D-arabinose, and L-fucose appear only in a few cases.

For carbohydrate chemists the structural elucidation of the glycopeptide antibiotics turned out to be quite challenging because of the rare 3-amino-2,3,6-trideoxyhexopyranose building units (4–7), most of which became known as structural units of glycopeptide antibiotics.

The structures of these aminodeoxy sugars, each belonging to the L-sugar series and present in the antibiotic molecules in form of the $^1C_4$ conformer, are shown in Figure 1. The theoretically possible four stereoisomers and the corre-

105

Table 1 Occurrence of Neutral Carbohydrates in the Vancomycin Antibiotics

Monosaccharide	Antibiotic
D-arabinose	Ristocetin (Ristomycin) A
L-fucose	A35512B
(6-deoxy-L-galactose)	
D-galactose	A41030C, F, G
D-glucose	Actaplanins A, B$_{1-3}$, G, D$_1$, actinoidins A, B, avoparcins α, β, A35512B, A51568, A42867, A82846A, B, C, chloropolysporins B, C, synmonicins (CWI-785) B, C, decaplanin, eremomycin, helvecardins A, B, M43 A, B, C, D, MM45289, MM47756, MM47761, MM47767, MM49721, MM55256, OA-7653, MM55266, MM55268, orienticin A, ristocetins, (ristomycins) A, B, UK-72,051, vancomycin, UK-68,597
D-mannose	Actaplanins A, B$_{1-3}$, C$_{1-3}$, D$_1$, actinoidins A, B, aricidins (ardacins) A, B, C, avoparcins α, β, A35512B, A40926, chloropolysporins B, C, synmonicins (CWI-785) A, M, helvecardins A, B, MM55266, MM55268, kibdelins A, B, C$_{1-2}$, C$_{1-4}$, X, ristocetin (ristomycin) A, teicoplanins A-2-1-5, A-3-1
L-rhamnose	Actaplanins B1, C1, avoparcins α, β, A35512B, A42867, synmonicins (CWI-785) B, C, decaplanin, helvecardins A, B, MM47761, MM49721, ristocetins (ristomycins) A, B

sponding C$_3$ methyl-branched analogs are listed. The occurrence of the 3-amino-2,3,6-trideoxy-L-hexoses among the representatives of the vancomycin group of antibiotics is summarized in Table 2. The trivial names of these aminodeoxy sugars were derived from the name of the antibiotic from which the amino sugars were isolated. Thus, acosamine and actinosamine were obtained (8,9) from the actinoidins, ristosamine was isolated (10) from the ristomycins (ristocetins), and vancosamine from (11–13) vancomycin. 3-*Epi*- and 4-*epi*-L-vancosamine (with structures closely related to that of vancosamine) were found in antibiotics A35512B (14) and A82846B (15), respectively.

The molecules of teicoplanins (16,17) and related antibiotics [aridicins (18–20), kibdelins (21), parvodicins (22), and others] are the only exceptions, carrying D-glucosamine and D-glucosamine uronic acid instead of 3-aminodeoxyhexoses.

The members of the vancomycin group of antibiotics possess significant variety in the place and mode of attachment of carbohydrate units. Accordingly, differentiation can be made on the basis of the mono-, bis-, tris-, and tetra-glycoside subgroups sugar attachments to the heptapeptide aglycone portion, and the bis- and trisglycosides are the most common (Table 3).

Neither C- and N-glycosidic bondings, nor a glycosylamine moiety are found in these antibiotics. All the neutral carbohydrates (Table 1) and the basic

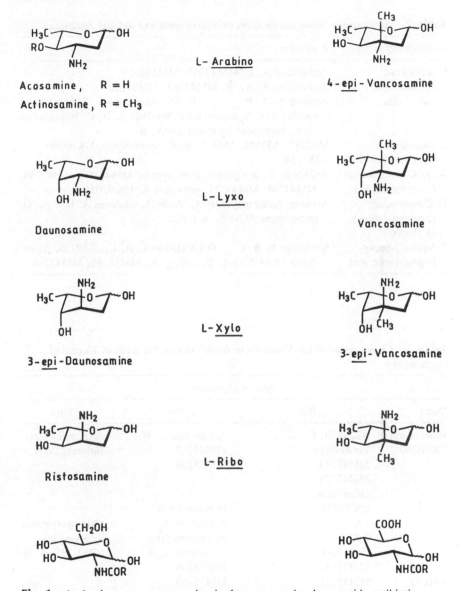

Fig. 1 Aminodeoxy sugars occurring in the vancomycin glycopeptide antibiotics.

Table 2 Aminodeoxy Sugar Components of Vancomycin and Related Antibiotics

Aminodeoxy sugar	Antibiotic
L-acosamine	Actinoidins A, B; MM47767, MM55256
L-actinosamine	Actinoidins A, A_2, B; MM47767; MM55256
L-ristosamine	Actaplanins A, B_{1-3}, C_{1-3}, G, D_2, avoparcins α, β, chloropoly-sporins B, C, synmonicins (CWl-785) A, B, C, helvecardins A, B, ristocetins (ristomycins) A, B
L-vancosamine	M42867, A51568, M43 A, B, C, vancomycin, UK-68597
3-*Epi*-L-vancosamine	A35512B
4-*Epi*-L-vancosamine (L-eremosamine)	A82846B, C, decaplanin, eremomycin, MM45289, MM47756, MM47761, MM49721, orienticin A, UK-72051
D-Glucosamine (2-amino-2-deoxy-D-glucose)	Aridicins (ardacins) A, B, C, A40926, kibdelins A, B, C_{1-2}, D, teicoplanins A-2-1-5, A-3-1-2
2-Amino-2-deoxy-D-glucuronic acid	Kibdelins A, B, C_{1-2}, D (AAD-609A, B, C), A40926, parvo-dicins (AAJ-271) A, $B_{1,2}$, C_{1-4}, X, MM55266, MM55268

Table 3 Classification of the Vancomycin Antibiotics on the Basis of Degree of Glycosidation

Type of glycoside			
Mono	Bis	Tris	Tetra
Vancomycin	A82846B, C	Actinoidins A, B	Avoparcins α, β
UK-68597	Eremomycin	MM47767	Helvecardin A
	MM45289	MM55256	
	MM47756		
	Orienticins A–D		
	UK-72051	Helvecardin B	
		Actinoidin A_2	Chloropolysporin
	A42867	Ristocetins (risto-	Synmonicins
	Decaplanin	mycins) A, B	(CWl-785) A, B
	MM47761	MM55266	
A41030	MM49721	MM55268	
OA7653		Chloropolysporin C	
	A40926 A, B	Teicoplanins A-2-1-5	
	Aridicins (ardacins) A–C	Teicoplanins A-3-1-3	
	Kibdelins A–D		
	(AAD-609 A–C)		
	Parvodicins		

aminodeoxy sugars (Table 2 and Fig. 1) are linked to the aromatic amino acids, via O-glycosidic linkages. The carbohydrate units are most frequently attached either to one of the phenolic hydroxyl groups of vancomycinic acid, actinoidinic acid, and ristomycinic acid or the β-hydroxyl group of the phenylserine unit of vancomycinic acid (or its mono or didechloro analog) in these antibiotics.

D-Glucose is generally present in a β-D-glucopyranosyl unit linked to the phenolic hydroxyl group of the central p-hydroxy-D-phenylglycine portion of vancomycinic acid derivatives. However, there is a β-N-acyl-D-glucosaminyl group (17,21) in the teicoplanins and kibdelins, whereas aridicins, antibiotics A40926A and B, and parvodicins carry a β-N-acyl-D-glucosamine uronic acid (20,22,23) function. In the monoglycoside vancomycin and antibiotic UK-68597 the C_2 hydroxyl group of the D-glucopyranosyl unit is substituted (24,25) via α-glycosidic linkage, with L-vancosamine (Fig. 2).

Most of the bisglycosides (Fig. 2) contain a 2-O-(3-amino-2,3,6-trideoxy-α-L-hexopyranosyl)-β-D-glucopyranosyl disaccharide side chain, and the aminodeoxy sugar constituent is L-vancosamine, L-ristosamine, 3-*epi*-L-vancosamine, or 4-*epi*-L-vancosamine. In many cases, such as antibiotics A35512B (14), MM45298 and MM47756 (27), UK-72051 (28), orienticins (29), and eremomycin (30), an additional molecule of these amino sugars is always α-glycosidically attached to the hydroxyl group of phenylserine situated at the C terminus of the heptapeptide aglycone.

In the disaccharide portion of decaplanin (31), A42867 (32), MM47761, and MM49721 (33), the aminodeoxy sugar is replaced by L-rhamnose, but the place and type of the interglycosidic linkage are the same.

The trisglycosides are represented by the actinoidins (34,35) MM47767 and MM55256 (Fig. 2) (36). As in bisglycosides, the aglycone moiety of these antibiotics is glycosylated with a heterodisaccharide side chain and with one molecule of an aminodeoxy sugar. In addition, these antibiotics carry an α-D-mannopyranosyl unit linked to one of the phenolic hydroxyl group of the tetra-substituted aromatic ring of actinoidinic acid. D-Mannose is attached to the same place in actinoidin A_2 (37,38), ristocetins (ristomycins) (39,40) and teicoplanins (17). The only difference is that the previously mentioned heterodisaccharide side chain is replaced by a 2-O-(α-L-rhamnopyranosyl)-β-D-glucopyranosyl group in actinoidin A_2, whereas in ristocetin B and ristocetin A a rutinosyl and a ristotetraosyl moiety (42), respectively, are present. Instead of related carbohydrate units the teicoplanins carry only an N-acyl-β-D-glucosaminyl portion.

A reversed arrangement of the carbohydrate units was proposed (43) only for the tris-glycosides MM55266 and MM55268, but no convincing proofs were offered.

The first tetraglycoside representative of the glycopeptide antibiotics is avoparcin α and β (44,45). Similarly to the bisglycosides, the aminodeoxy sugar components in these molecules are present both in the heterodisaccharide side chain and linked to one of the phenylserine units. At the same time, an α-D-

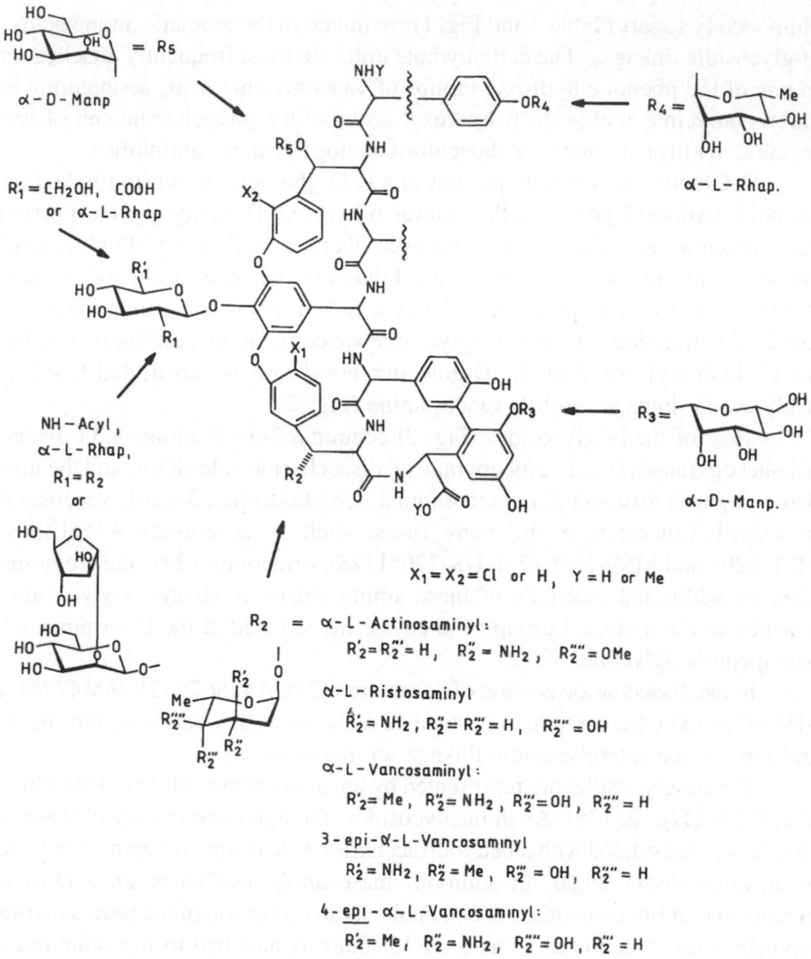

Fig. 2　Site of linkages of the carbohydrates to the aglycone moiety.

mannopyranosyl group is attached to the other phenylserine moiety of monode-chlorovancomycinic acid.

　　As the fourth most frequently occurring sugar found in these antibiotics, L-rhamnose is linked to the phenolic hydroxyl group of the N-terminal amino acid. In an analogous manner, the D-mannose and L-rhamnose are connected in chloropolysporin B (41) and helvecardin A (26).

　　In these antibiotics, the monosaccharide units are linked to the aglycone portion or to each other via $\alpha(1 \rightarrow 2)$ linkages. However, attachment of D-glucose is an exception to this generally observed rule.

II. COMPOSITION AND STRUCTURAL ELUCIDATION OF THE CARBOHYDRATE UNITS BY CHEMICAL METHODS

A. The Neutral Sugars

In glycopeptide antibiotics the neutral sugars are linked to the aglycone either monomerically or in the form of a heterooligosaccharide side chain. Most of these sugars belong to the D-aldohexopyranose series (Table 1), except L-rhamnose, L-fucose, and D-arabinose.

The free reducing sugars, liberated upon mild acid hydrolysis of the antibiotics, give positive anthrone and Folin-Ciocalten phenolic probes and Fehling reaction. For qualitative detection and quantitative determination of the freed neutral sugars, a variety of methods have been applied. Earlier, paper chromatography was frequently employed for such purposes, allowing the detection of D-glucose and D-mannose in the actinoidins (46) and D-glucose, D-mannose, D-arabinose, and L-rhamnose in ristocetins (47,48) and ristomycins (49). In vancomycin, D-glucose was also identified (50) as penta-O-acetyl-β-D-glucopyranose (1) derived by acid hydrolysis followed by acetylation (Fig. 3). Alternatively, the glycosidic linkages of per-O-acetylristomycin A were split with asymmetric dichlorodimethyl ether and the acetochloro sugars were treated with silver acetate and then deacetylated. This carbohydrate fraction was compared (51) with authentic neutral sugars by thin-layer chromatography.

Cleavage of the glycosidic linkages can be effected with trifluoroacetic acid instead of mineral acids, without decomposition or rearrangement of the aglycone portion. Thus, the D-glucose unit of orienticin A can readily be split off by treatment (52) with 80% trifluoroacetic acid. An additional methodology involves the reduction of the neutral sugars, liberated by acid hydrolysis of the antibiotics, with sodium borohydride, and after acetylation the sugar components are identified in the form of the corresponding per-O-acetyl alditols. This method was followed for the conversion (Fig. 3) of D-mannose, present in the avoparcins (44) and kibdelins (21), into per-O-acetyl-D-mannitol (2). In this case the alditol acetate (3), produced from D-glucosamine, was also obtained. For the detection and identification of these alditol derivatives gas chromatography and the GC/CI-MS technique were also employed. Silylation (TMSCl-BSTFA) of the split-off neutral sugar components results in the volatile persilyl derivatives, which can be detected by thin-layer chromatography or gas chromatography–mass spectrometry. A related procedure was applied for the GC-MS detection (53) of per-O-trimethylsilyl-D-mannose (4) derived from antibiotic A16686 (Fig. 3) and for the gas chromatographic determination (54) of glucose, mannose, rhamnose, and fucose in antibiotic A35512B.

In a modified procedure the methyl glycosides produced by methanolysis are converted into the volatile silyl derivatives suitable for gas chromatographic

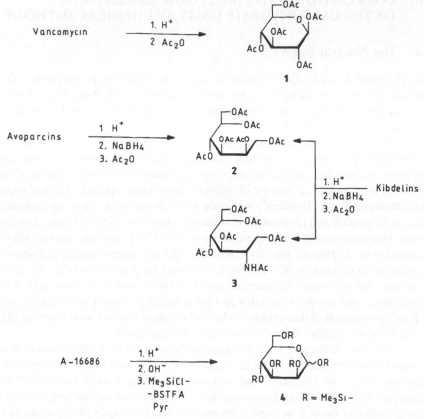

Fig. 3 Methods employed for the identification of the neutral carbohydrate components of the glycopeptide antibiotics.

analysis. The glucose, mannose, and rhamnose components of chloropolysporin B were determined (41) according to this methodology. For antibiotic A42867 the sugars (glucose and rhamnose), liberated upon acid hydrolysis, were transformed into the corresponding dansyl derivatives, and examined by HPLC (55).

Enzymatic hydrolysis is also useful in some cases. For example, treatment (41) of chlorosporin B and C with rhamnosidase (naringinase) or α-mannosidase resulted in the selective hydrolysis of the L-rhamnose and D-mannose units, respectively. A similar methodology was also successful in the structural studies (56) on avoparcin α and β.

These examples give a brief survey of the most important chemical methods employed for the detection and identification of the neutral sugar components present in the glycopeptide antibiotics. The related ^{1}H- and ^{13}C-NMR, as well as the mass spectroscopic methods are used extensively in the structural studies of glycopeptide antibiotics.

B. The Aminodeoxy Sugars

Although D-glucosamine (chitosamine) and D-galactosamine (chondrosamine) have long been known as natural aminodeoxy sugars, the chemistry of related carbohydrate derivatives has become a challenging research field only in connection with structural studies on antibiotic substances. The discovery of daunosamine (3-amino-2,3,6-trideoxy-L-*lyxo*-hexopyranose), as the first representative of this type of rare aminodeoxy sugars (6,7) in the anticancer anthracycline antibiotics daunomycin (daunorubicin), doxorubicin, and carminomycin, opened up new perspectives in the field of carbohydrate research. Also, recognition of the role of related sugars in biochemical processes and the mode of action of glycosidic antibiotics has contributed significantly to correct understanding of structure-activity relationships.

In the glycopeptide antibiotics, the 3-amino-2,3,6-trideoxy-L-hexopyranoses (Fig. 1) are the most characteristic building elements (Table 2). Because of the 2-deoxy function, these aminodeoxy sugars can be split off by extremely mild mineral acid hydrolysis or by methanolysis.

Acid hydrolysis gives the amorphous free aminodeoxy sugars, whereas methanolysis yields the crystalline methyl glycosides. Since the molecular mass of these carbohydrate derivatives amounts at most, to a hundredth part of the molecular weight of the antibiotics, their isolation calls for the most advanced techniques. In addition, besides chemical degradation methodology, the structural identification of these compounds requires both classic and more sophisticated spectroscopic studies (NMR, CD, and mass spectrometric investigations). This section gives a brief survey of related methodologies through the following selected examples.

Treatment of actinoidin A and B with 0.2 N hydrochloric acid in methanol afforded methyl α-L-acosaminide (5) and the Ψ-aglycone. Hydrolysis under more vigorous conditions gave methyl α-L-actinosaminide (6) (Fig. 4). The significant difference in the ability of these two aminodeoxy sugars to hydrolyze indicates their attachment to different structural units of the aglycone in the antibiotic. Based on elemental analytical data, Spiridonova et al. (9) suggested the presence of one primary amino group and a C-CH$_3$ function in both amino sugars, and it was shown that 6 contains an extra methoxy group compared with 5. Application of the reaction sequence 5 → 7 → 8 then clearly proved that L-actinosamine is actually the 4-O-methyl derivative of L-acosamine. The most valuable information for the composition and structure of these sugars was obtained from investigations of the amino acids obtained from N-acetyl-L-acosamine (9). Thus, oxidation

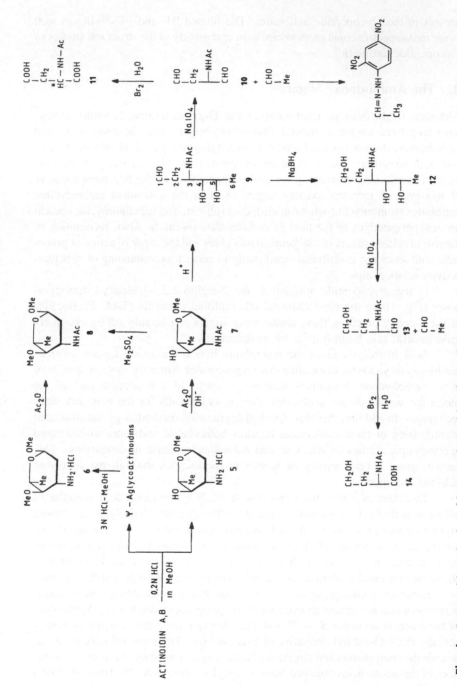

Fig. 4   Structural investigation of L-acosamine and L-actinosamine

of 9 with sodium periodate furnished 10 and acetaldehyde. Oxidation of 10 with bromine-water yielded N-acetyl-L-aspartic acid (11). This amino acid carries the C_1–C_4 carbon skeleton of N-acetyl-L-acosamine (9), whereas acetaldehyde represents carbons C_5 and C_6 of the amino sugar. The structure of these degradation products showed that in 9 the acetamido group is linked to C_3 and C_4 and C_5 are substituted with hydroxyl groups.

All these findings were further substantiated by the isolation of homoserine (14) upon the reduction of 9 with sodium borohydride and subsequent successive oxidation with sodium periodate and bromine-water (Fig. 4, 9 → 12 → 13 → 14).

The ^{1}H-NMR spectra of 5 and 6 showed essential similarities, with the only difference that the spectrum of 5 contained one methoxy signal but two such signals were assigned for 6. In both amino sugar molecules H_1 is equatorial and the protons H_3, H_4, and H_5 are axially disposed. Based on these data 5 and 6 are of 3-amino-2,3,6-trideoxy-*arabino* configuration with either α-L-($^1C_4$) or β-D-($^4C_1$) conformation (56). The observed negative specific optical rotation data then unequivocally proved that the L-*arabino* configuration should be assigned to both acosamine and actinosamine.

Although ristocetin and ristomycin were proved to be identical (38) only in 1980 L-ristosamine was first isolated by the Hungarian chemists (10,57) from ristomycin which was discovered later. This amino sugar was then also found in other glycopeptide antibiotics in the last 15 years (see Table 2).

Mild hydrochloric acid hydrolysis of ristomycin gave the Ψ-aglycoristocetin (see Sec II.D), and further acid hydrolysis of this compound furnished amorphous L-ristosamine (15, Fig. 5).

The sugar 15 gave positive Fehling, Tollens, Keller-Kiliani, and iodoform reactions and xanthidrol color probe, characteristic of aldoses and deoxy sugars. The crystalline methyl glycoside (16) of 15 could only be obtained upon methanolysis of the intact antibiotic or the Ψ-aglycone. When subjected to periodate oxidation, 15 consumed 2 mol oxidizing agent and acetaldehyde was detected. Acetylation and benzoylation of 16 afforded the N,O-diacetyl- (17) and the N-benzoyl derivatives (18), respectively. Sequential oxidation of 19, obtained by the hydrolysis of 18, with sodium periodate and bromine-water resulted in N-benzoyl-D-aspartic acid (20).

Comparison of the molecular rotation value ($[M]_D = -62.7°$) of L-ristosamine (15) with those of the α, β equilibrium mixtures of the possible eight stereoisomeric 2,6-dideoxyhexoses suggested that similar data (in both direction and magnitude) was only found (58) for the L-ribo compound (L-digitoxose, $[M]_D = -68°$). This structural assignment was substantiated (57) by the ^{1}H-NMR data obtained for 17, since the $J_{3,4}$ and $J_{4,5}$ coupling constants indicated the *cis* arrangement of H_3 and H_4, as well as the transdiaxial disposition of H_4 and H_5. Such a steric arrangement can only be possible in the α-L-*ribo* ($^1C_4$) or β-D-*ribo* ($^4C_1$) configuration of the 3-amino-2,3,6-trideoxyhexose. However, on the basis of

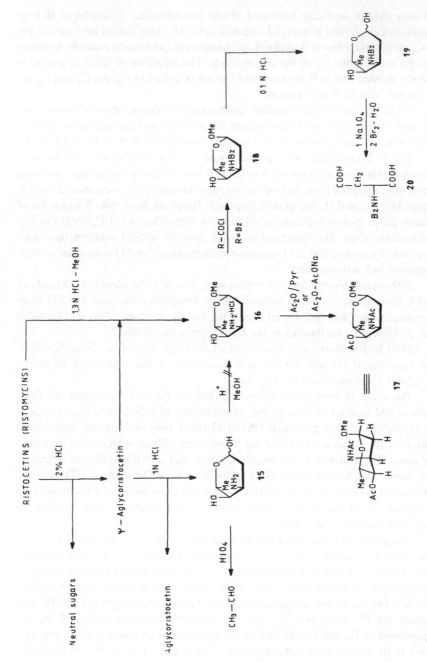

Fig. 5   Determination of the structure of L-ristosamine.

molecular rotation data the D configuration for ristosamine can be unequivocally excluded, and thus the structure of **17** was assigned as methyl 3-acetamido-4-O-acetyl-2,3,6-trideoxy-α-L-*ribo*-hexopyranoside, with a preferred $^1C_4$ conformation in chloroform solution.

A few years later, during the isolation of L-ristosamine from avoparcins α and β, the American chemists (59) observed a novel phenomenon. When processing the hydrolysate of the antibiotic, formation of the tricyclic compound **21** (Fig. 6), composed of three molecules of L-ristosamine and one molecule of ammonia, was recognized. However, the development of such a condensation product does not contradict the previously assigned structure of the aminodeoxy sugar.

Structural investigation (11,13) of the C_3 methyl-branched aminodeoxy sugar, L-vancosamine (**22**), was independently accomplished by two English research groups. Since the instrumental examinations (CD, NMR, and mass spectrometry) of the derivatives prepared from the parent sugar provided sufficient information for identification of the structure, these studies did not involve chemical degradation methodology (Fig. 7).

Treatment of vancosamine (**22**), isolated upon mild acid hydrolysis of vancomycin, with hydrochloric acid in methanol gave the corresponding methyl glycoside **23**. The negative absorption observed at 600 nm in the CD spectrum of **23** indicated a cis arrangement of the hydroxyl and amino functions of different chiralities, and thus the amino sugar was suggested (60) to belong to the L-sugar series. The vicinal position of the amino and hydroxyl groups was also supported by the observed N $\rightarrow$ O acyl migration reaction, induced by means of acetylation of **22** with acetic anhydride in methanol, followed by successive methanolysis of the product and reacylation with benzene sulfonyl chloride in pyridine to obtain methyl N-benzenesulfonyl-O-acetyl-α-L-vancosaminide (**24**). Benzoylation of vancomycin and subsequent methanolysis resulted in methyl N,O-dibenzoyl-α-

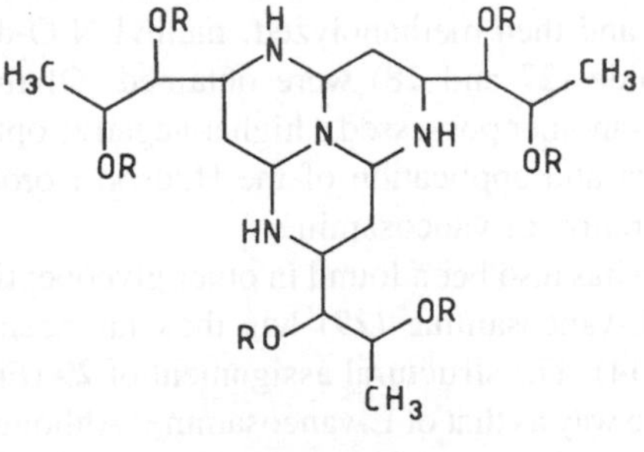

21 R = H or Ac

Fig. 6 Structure of the tricyclic compound produced from L-ristosamine.

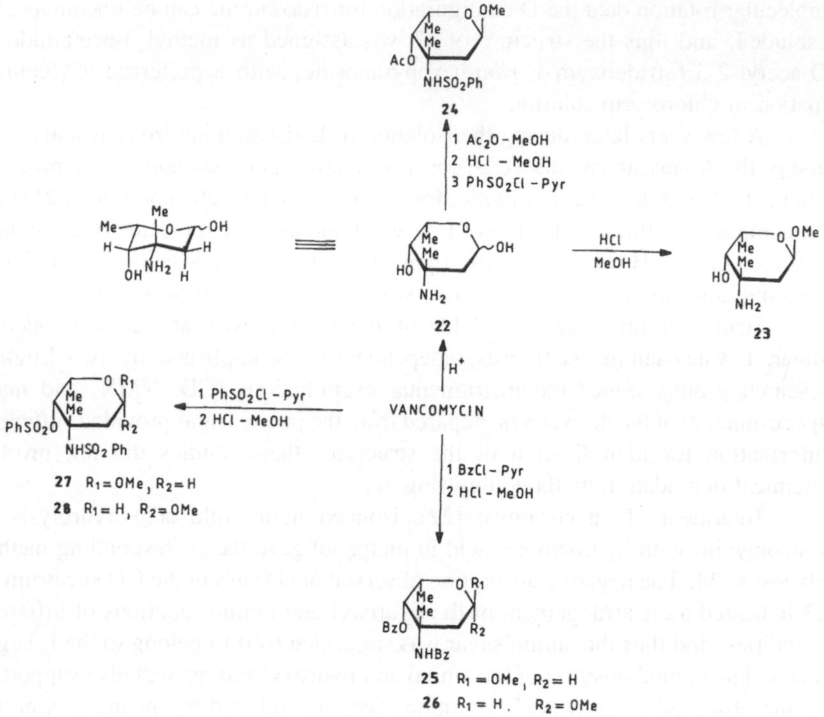

Fig. 7 Structure of L-vancosamine and its derivatives.

and -β-L-vancosaminide (25 and 26, respectively). ^{1}H-NMR studies of 25 proved clearly the 1,3-*cis*-diaxial orientation of H_5 and the C_3-methyl group.

When vancomycin was first reacted with benzene sulfonyl chloride (instead of benzoyl chloride) and then methanolyzed, methyl N,O-dibenzenesulfonyl-α- and β-L-vancosaminide (27 and 28) were obtained. Of the isolated N,O-acyl anomeric pairs, the α-anomer possessed a higher negative optical rotation value in each case. These data and application of the Hudson isorotation rule also supported the L configuration of vancosamine.

L-Vancosamine has also been found in other glycopeptide antibiotics (Table 2). However, 3-*epi*-L-vancosamine (29) has thus far been isolated only from antibiotic A35512B (14). The structural assignment of 29 (Fig. 8) was performed in essentially the same way as that of L-vancosamine. Although the mass spectrum of 30 was identical to that of methyl α-L-vancosaminide (23), the thin-layer R_f value of its N,O-dibenzoyl derivative (31) was different from that of the methyl glycoside 25. Finally, ^{1}H-NMR and CD studies of the dibenzoate 31 proved that

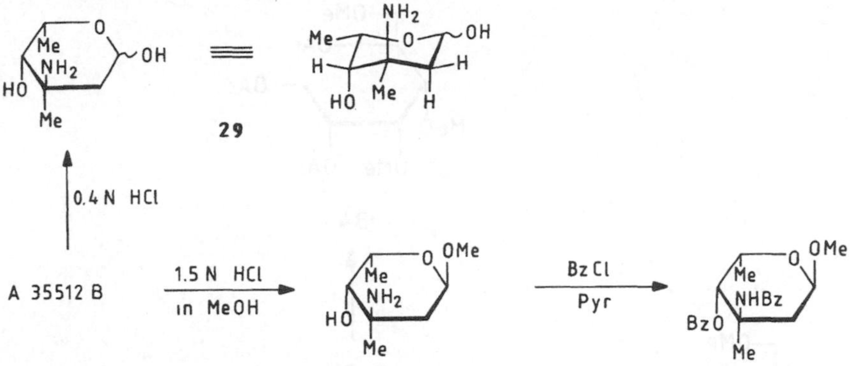

Fig. 8 Structural investigation of 3-*epi*-L-vancosamine.

the structure of the parent sugar is 3-amino-3-C-methyl-2,3,6-trideoxy-L-*xylo*-hexopyranose (**29**).

4-*Epi*-L-vancosamine (**32**, eremosamine) with the 3-amino-3-C-methyl-2,3,6-trideoxy-L-*arabino*-hexopyranose structure was recently found in several glycopeptide antibiotics. The structure of this sugar was determined (28,30, 52,61) by comparison of the NMR and optical rotation properties with those of the previously known C_3 methyl amino sugar derivatives of the glycopeptide antibiotics.

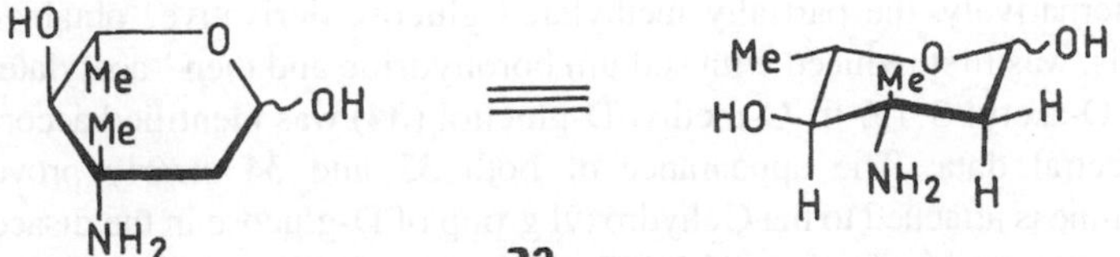

C. Sequence of the Carbohydrate Units

For determination of the sequence of attachment of the monosaccharide units in the oligosaccharide components of vancomycin and related antibiotics, the well-known methods of carbohydrate chemistry have been applied. The most frequently employed procedure is "analysis of the linkages with methylation." For structural proof of the 2-O-(L-vancosaminyl-D-glucosyl)disaccharide unit of vancomycin, the following strategy was used by Roberts and coworkers (63).

The antibiotic was first methylated according to the Hakomori method and then methanolyzed. After deuteriomethylation of the partially methylated prod-

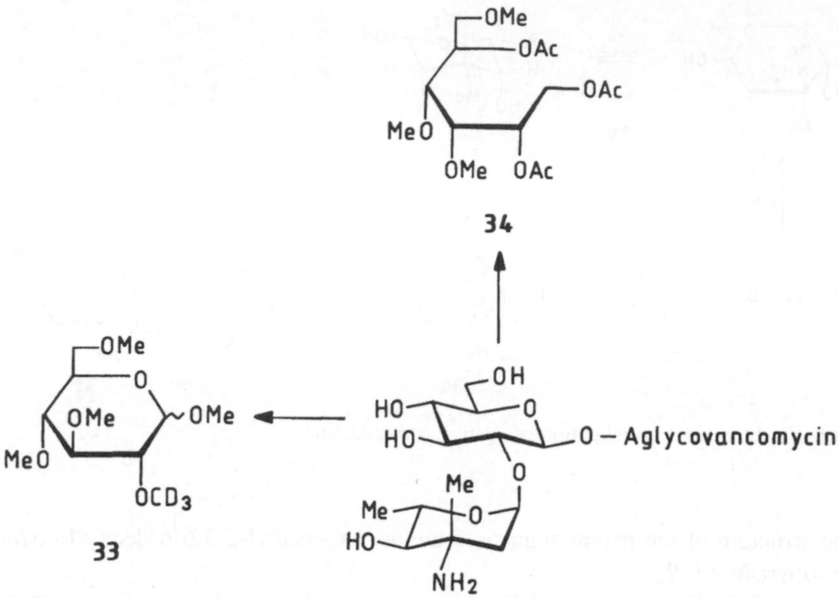

Fig. 9   Sequence of the carbohydrates in vancomycin.

ucts, the structure of 2-O-$^2H_3$-permethyl-D-glucose (**33**) was determined by mass spectrometric measurements (Fig. 9).

Alternatively, the partially methylated glucose derivative, obtained upon hydrolysis, was first reduced with sodium borohydride and then "acetylated" and 1,2,5-tri-O-acetyl-3,4,6-tri-O-methyl-D-glucitol (**34**) was identified according to mass spectral data. The appearance of both **33** and **34** clearly proved that vancosamine is attached to the C_2 hydroxyl group of D-glucose in the disaccharide unit of vancomycin. Further ^{1}H-NMR studies by Williams and Kalman (64) showed that a 2-O-(α-L-vancosaminyl)-β-D-glucopyranosyl side chain is attached to aglycovancomycin.

Determination of the sequence of the sugars is more complicated when, in contrast to vancomycin, the antibiotic contains more than two monosaccharide components as bis-, tris-, or tetrasaccharides. In such cases the results of methylation analysis are thoroughly evaluated in comparison with those of hydrolysis and periodate oxidation studies. The structures of the disaccharide moieties of actinoidins (9) and eremomycin (30) were determined by related complex examinations. Hydrolysis of the antibiotics for different reaction times with acids in various concentrations showed the following order of splitting off of the sugar components:

Actinoidin A,B :

L-acosamine   D-glucose   D-mannose   L-actinos-amine

Thus, in actinoidin, L-acosamine is the most sensitive and L-actinosamine is the most resistant to acid hydrolysis.

Eremomycin:

4-Epi-L-vancosamine (32)   D-glucose   4-Epi-L-vancosamine (32)

The rates of hydrolysis of the glycosidic linkages of the two 4-*epi*-L-vancosamine (32) molecules, present in eremomycin, proved to be quite different. Relevant studies have shown that first the amino deoxy sugar incorporated in the hetero-disaccharide unit is split off, following by liberation of the β-glycosidic D-glucose attached to the central ring of vancomycinic acid (or an analog thereof). The neutral sugars linked to the phenolic hydroxyl groups of actinoidinic acid or another structural portion of the aglycone then appear in the hydrolysate. Finally, hydrolysis of the amino sugar linked to the hydroxyl group of the phenylserine unit occurs. These observations accord well with earlier experience (67) with the different stabilities and rates of the acid-catalyzed hydrolysis of the O-glycosidic bonds, significantly influenced by the ring size of the monosaccharide and the nature of the aglycone (aliphatic or aromatic), as well as by the configuration of the glycosidic linkage.

These general features were also substantiated by observations in connection with the structural elucidation (9) of the actinoidins. Upon oxidation of the antibiotic with periodic acid, only L-actinosamine (35) could be detected, since this sugar component does not contain free hydroxyl groups (Fig. 10). Meth-

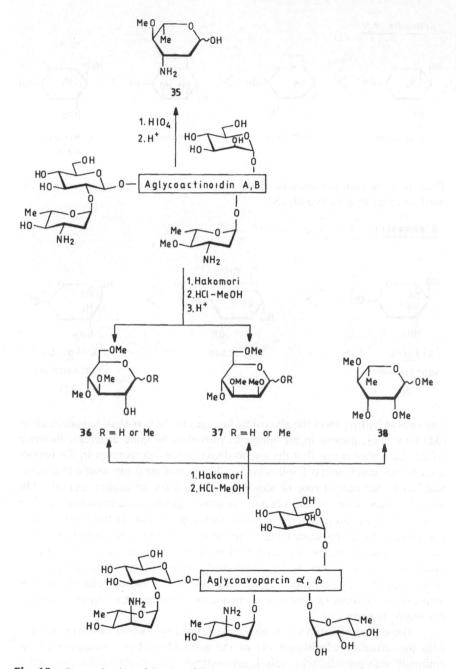

Fig. 10   Determination of the site of linkages of the sugar components in the actinoidins and avoparcins.

anolysis of the permethylated actinoidin and subsequent hydrolysis led to the isolation of 3,4,6-tri-O-methyl-D-glucose (**36**) and 2,3,4,6-tetra-O-methyl-D-mannose (**37**). This experimental fact and the previously shown order of hydrolytic splitting off of the carbohydrate units suggested that the D-mannose molecule was separately attached to the aglycone and that L-acosamine and the D-glucose molecule constitute a 2-O-(α-L-acosaminyl)-β-D-glucopyranosyl disaccharide side chain. The configuration of the glycosidic linkages was later determined by Batta et al. (35) with the aid of ^{1}H-NMR spectroscopic methods.

By hydrolysis studies of eremomycin and permethylation experiments and ^{1}H- and ^{13}C-NMR investigations, Russian and Hungarian researchers (30,66) identified the heterodisaccharide of eremomycin as a 2-O-(4-*epi*-α-L-vancosaminyl)-β-D-glucopyranosyl unit. This disaccharide was later found in the orienticins and in the antibiotic complex A82846. The American chemists followed similar methods for the structural investigation (44,45) of the 2-O-(α-L-ristosaminyl)-β-D-glucopyranosyl side chain of avoparcin α and β (Fig. 10). In the methanolysate of the permethylated avoparcin (prepared by the Hakomori method), methyl 3,4,6-tri-O-methyl-D-glucopyranoside (**36**), methyl 2,3,4,6-tetra-O-methyl-D-mannopyranoside (**37**), and methyl 2,3,4-tri-O-methyl-L-rhamnopyranoside (**38**) were detected, showing that the only nonterminal neutral sugar present in the antibiotic is D-glucose, which is glycosylated with L-ristosamine at the C$_2$ hydroxyl group. This was also supported by spectral studies and by biogenetic analogy to the antibiotics vancomycin and actinoidin, as well as by determination of the place of attachment of L-rhamnose and D-mannose. This heterodisaccharide structural unit (avobiose) of the avoparcin α and β was later found also in helvecardin A and B. Identification (26) of this disaccharide was performed with the aid of mass spectrometric methods.

In the structural elucidation of the oligosaccharide components of the glycopeptide antibiotics the most complicated problem was associated with studies of the heterotetrasaccharide side chain of ristomycin (ristocetin) A and B. The first supposition for the sequence of carbohydrates and for the attachment of the tetrasaccharide to the aglycone was drawn from the established composition of the carbohydrate-containing fragments produced upon mild acid hydrolysis and acetolysis of the antibiotic.

Upon acid hydrolysis the following glycosides and oligosaccharides were obtained (49,68) from ristomycin A: D-Man-aglycone, D-Man-aglycone-D-Glu, aglycone-D-Glu-L-Rha, D-Glu-D-Man, D-Glu-L-Rha, L-Rha-D-Glu-D-Man, and D-Glu-D-Man-D-Ara. From ristomycin B the following related substances were produced: aglycone-D-Man, D-Glu-aglycone-D-Man, aglycone-D-Glu-L-Rha, and D-Glu-L-Rha.

These results clearly showed that in both ristomycin A and ristomycin B the neutral sugars are attached to the aglycone in two different places (Fig. 11). Structural studies on the hydrolysis products were performed by Sztaricskai et al. (69), and of the substances produced upon partial hydrolysis of ristomycin A two

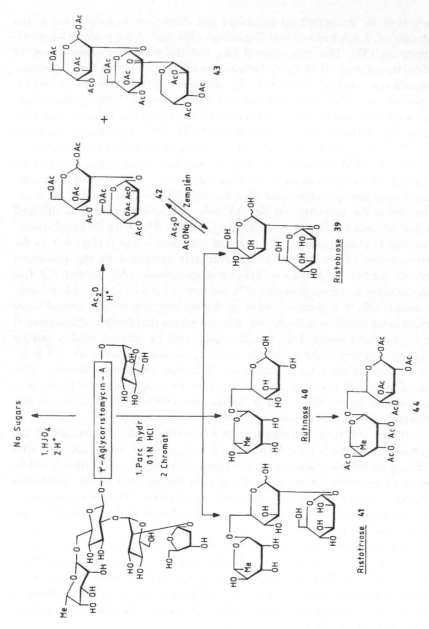

Fig. 11 Identification of the hydrolysis products formed upon mild partial acid hydrolysis of ristomycin (ristocetin).

disaccharides (**39** and **40**) and a trisaccharide (**41**) could be separated. After acetolysis of the antibiotic an octaacetyldisaccharide (**42**) and a decaacetyltrisaccharide (**43**) were isolated. Zemplén saponification of **42** led to a disaccharide that proved to be identical to **39**, ristobiose. Acetylation of **40** resulted in **44**, identified as hepta-O-acetylrutinose (peracetyl-6-O-α-L-rhamnopyranosyl-D-glucopyranose). The structure of this disaccharide provided valuable information for the location and mode of attachment of the two monosaccharides L-rhamnose and D-glucose in the heterotetrasaccharide side chain.

Following the periodate oxidation of the antibiotic no monosaccharides could be detected, indicating that even the D-glucopyranosyl unit, representing the branching unit in the tetrasaccharide, contains a vicinal diol function. In addition, in **44** D-mannose can only be linked to D-glucose at either C_2 or C_4. Subsequent enzymatic hydrolysis and methylation analysis showed that the foregoing compound has a 2-O-α-D-mannopyranosyl-D-glucose structure. The trisaccharide **41**, ristotriose, incorporates the structure of both **39** and **40**, and methylation studies permitted its identification as O-α-L-rhamnopyranosyl-(1→6)-O-α-D-mannopyranosyl-(1→2)-D-glucose. Consequently, D-mannose is linked through a 1 → 2 bonding to D-glucose in each **39** and **41** and in the heterotetrasaccharide side chain of the antibiotic.

The other trisaccharide (ristriose), however, could be obtained only in extremely low quantities. This finding is in good accord with the substantial instability of the D-arabinofuranosides. At the same time, compound **43**, isolated upon acetolysis, was found to contain D-arabinopyranoside. The only explanation for this experimental fact is that the α-D-arabinofuranosyl unit is transformed, via ring enlargement and inversion of the configuration of the anomeric center, into a β-D-arabinopyranosyl moiety. This was finally confirmed by Hungarian and American chemists (40) by means of the mild reductive-alkaline degradation of ristomycin A (ristocetin A).

The antibiotic, protected at the phenolic hydroxyl groups, was subjected to hydrolysis with barium hydroxide in the presence of sodium borohydride to afford a fragment (**45**) that carried the intact heterotetrasaccharide side chain (Fig. 12). By comparison of the ^{13}C-NMR spectra of the pure crystalline ristomycin A, compound **43**, and that of methyl α-D-arabinofuranoside, Sztaricskai et al. (39,40) unambiguously demonstrated that the terminal arabinose unit in the tetrasaccharide is present in α-D-arabinofuranosyl form. Based on these studies the structure of ristotetraose in ristomycin A (ristocetin A) is O-α-D-Araf-(1→2)-O-α-D-Manp-(1→2)-O-|α-L-Rhap|-(1→6)-O-β-D-Glcp.

The results of comparative studies on ristomycin B, involving determination of the molar ratio of the neutral carbohydrates and structural investigation of the hydrolysis products, led to the recognition that ristomycin B differs (69) from ristomycin A only by the lack of the 2-O-α-D-arabinofuranosyl-α-D-mannopyranosyl unit. Later independent studies of Williamson and Williams (70) confirmed

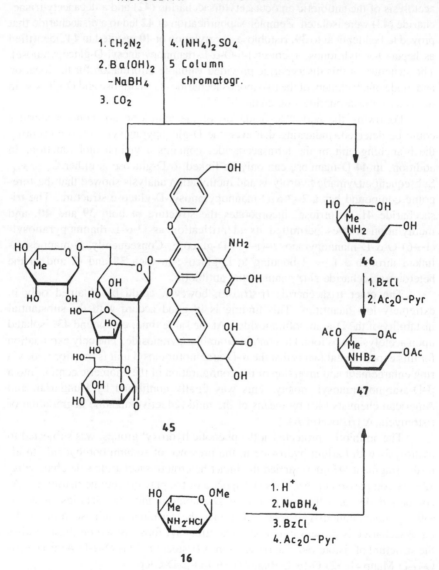

Fig. 12 Investigation of the heterotetrasaccharide side chain by the reductive alkaline degradation of ristomycin (ristocetin).

these results. NMR studies with phenyl α- and β-D-glucopyranoside derivatives proved the β configuration of the glycosidic linkage between D-glucose and the aglycone moiety.

The other product of the reductive alkaline degradation procedure was L-ristosaminitol (46) produced from L-ristosamine, linked to one of the β-hydroxythyrosine unit of the antibiotic via β elimination under alkaline conditions. The structure of 46 and its N-benzoyl-tri-O-acetyl derivative 47 was confirmed by independent synthesis using methyl α-L-ristosaminide (16).

Finally, ¹H-NMR spectroscopic examination (71,73) of the Ψ-aglycone of ristocetin and its derivatives clearly demonstrated (see Sec II.D) that the α-L-ristosamine unit is O-glycosidically attached to the hydroxyl group of didechloro-vancomycinic acid close to the C terminus.

Apart from analysis of the sequence of the sugar units, all these results provided valuable information for the exact determination of the site of the aglycone-sugar linkages in the glycopeptide antibiotics (see Sec. II.D).

D. Determination of the Site of the Aglycone-Sugar Linkages

To determine the site of the aglycone-sugar linkages, the isolation of a fragment, still carrying the sugar components in a simple or more combined form after partial degradation of the antibiotic, offers the most valuable information. The Ψ-aglycones have been shown to be most suitable for structural elucidation. Because of the different rate of hydrolysis of the aliphatic and phenolic glycosides, the Ψ-aglycones are readily obtained. These substances carry exclusively the aminodeoxy sugar unit, linked to the phenyserine portion, since this glycosidic bonding is the most resistant to proton-catalyzed hydrolysis (see Chap. 3).

Detailed ¹H-NMR spectroscopic studies of English (71,73) and American researchers (74) on Ψ-aglycoristocetin and Ψ-aglycoactaplanin showed that L-ristosamine is α-glycosidically linked, in both molecules, to the hydroxyl group of the phenylserine unit close to the C terminus (Fig. 13). The Ψ-aglycones of ristomycins (ristocetins) and actaplanins, representing a common structural composition, possess a considerable part of the biological activity of the parent antibiotic, but they are more toxic (75).

An alternative methodology for determination of the place of the aglycone-sugar linkages is based on the isolation of 45 (see Sec. II.C) or related fragments. Such a procedure is primarily applied to the structural investigation (5) of glycopeptides containing serine and threonine.

In addition, the peptide structure of the aglycones of the glycopeptide antibiotics permits the application of the sequential Edman degradation method, leading, in fortunate cases, to the isolation of thiohydantoin derivatives glycosylated with the carbohydrate components of the antibiotics. Such a methodology was first applied by Hlavka et al. (72) for the degradation of avoparcin, and the

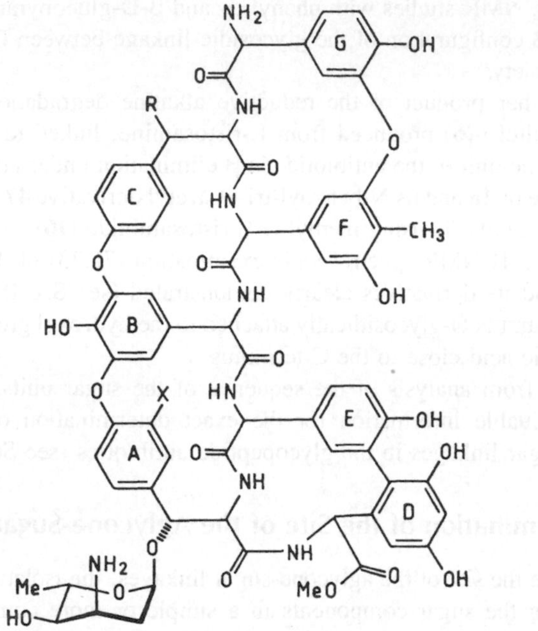

ψ–Aglycoristocetin (R = OH, X = H)
ψ–Aglylcoactaplanin (R = H, X = Cl)

Fig. 13   Structure of ψ-aglycoristocetin and ψ-aglycoactaplanin.

L-rhamnoside of the phenylthiohydantoin derivative of the N-terminal *p*-hydroxy-phenylsarcosine (**48**) could be isolated (Fig. 14), together with the phenylisocyanate derivative (**49**) of L-ristosamine. However, the composition of **49** did not yield information about the site of attachment of L-ristosamine to the aglycone.

Later, this procedure was successfully employed for the structural elucidation of the helvecardins (26) and chloropolysporins (40), closely related to avoparcins.

In most of the cases, however, only indirect methods can be applied to the determination of the site of aglycone-carbohydrate linkages. Many of the amino acids incorporated into the aglycone moiety are substituted with phenolic hydroxyl groups, but only some of these are glycosylated with the sugar components. The free phenolic hydroxyl functions can be readily transformed into phenol ethers with diazoalkanes, but those linked to carbohydrates do not react. In this way the introduced alkyl ether groups show the number and location of the phenolic hydroxyl groups present in the intact antibiotic. This method is usually completed with further derivatization of those phenolic hydroxyls, which are liberated only upon splitting off of the sugar units by hydrolysis. The subsequently introduced novel alkyl ether functions should be different from those generated previously to ensure exact differentiation (with the aid of spectroscopic methods).

Following this "labeling" technique, the antibiotics are subjected to combined degradation processes to allow the isolation of smaller fragments carrying the introduced groups unchanged, thus diagnostically indicating the place of attachment of the carbohydrate units to the aglycone.

The scope of the strategy is illustrated by the following classic examples.

Acid hydrolysis of a sample of avoparcin, which was previously treated with an excess of diazomethane, gave a partially methylated aglycone that was first ethylated with diazoethane and then oxidized with potassium permanganate. Repeated methylation with diazomethane furnished compound **50** (Fig. 14), produced from the monodechlorovancomycinic acid unit of the parent antibiotic.

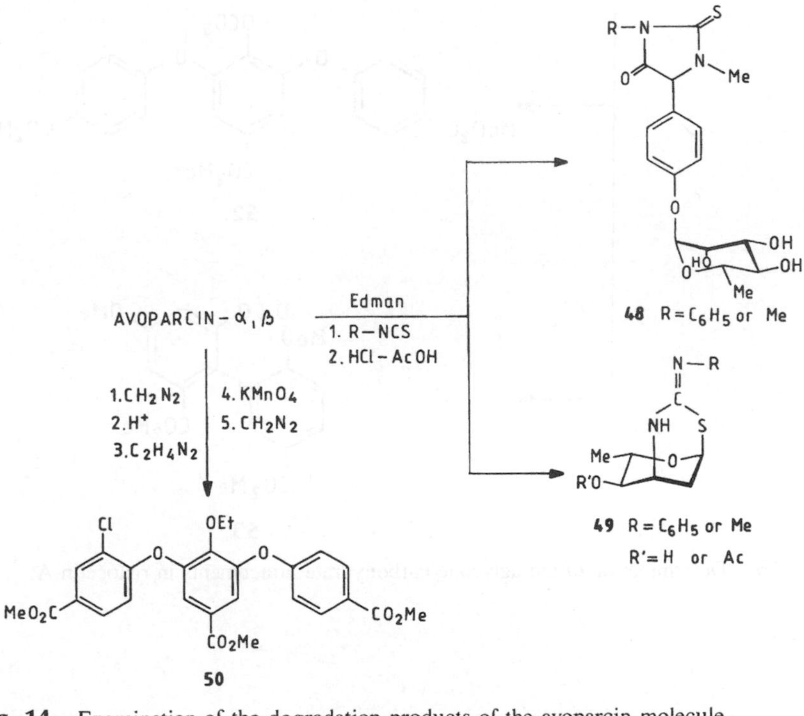

Fig. 14 Examination of the degradation products of the avoparcin molecule.

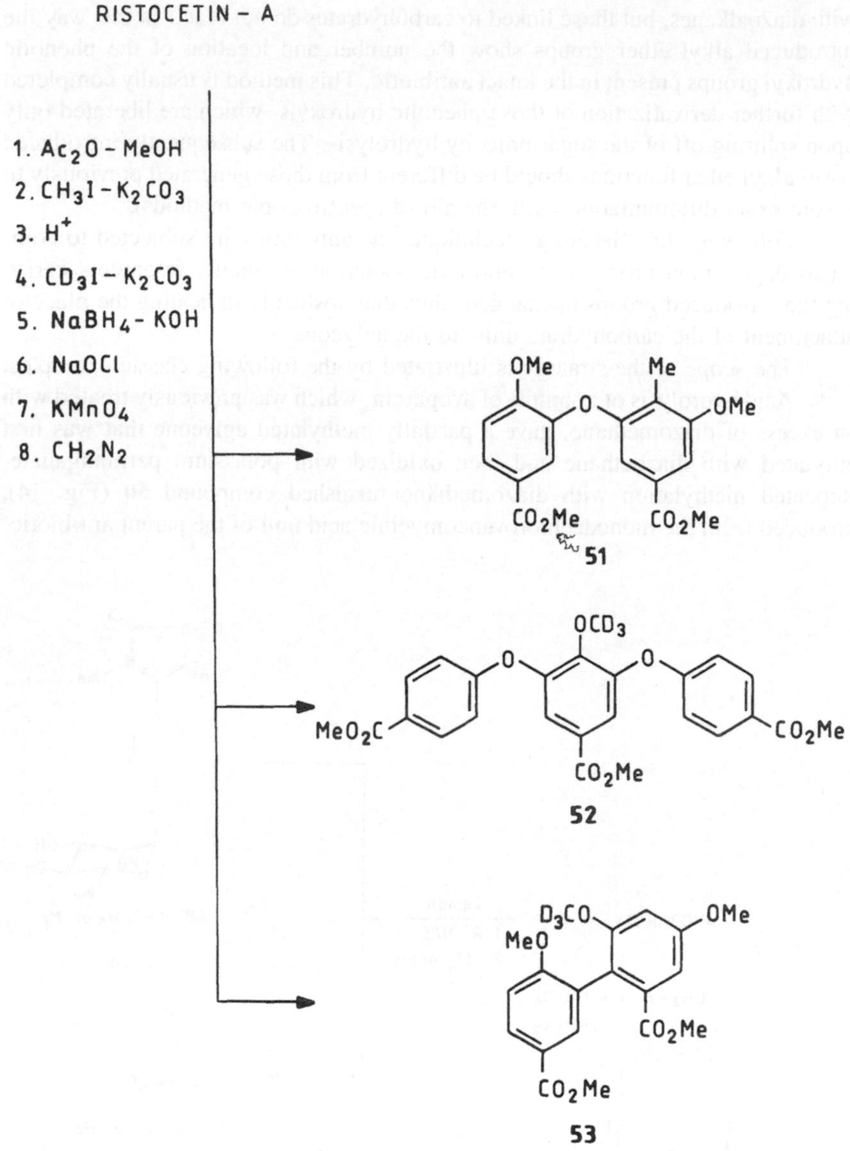
RISTOCETIN – A
1. Ac2O – MeOH
2. CH3I – K2CO3
3. H+
4. CD3I – K2CO3
5. NaBH4 – KOH
6. NaOCl
7. KMnO4
8. CH2N2
OMe
Me
O
OMe
CO2Me
CO2Me
51
OCD3
O
O
MeO2C
CO2Me
CO2Me
52
D3CO
OMe
MeO
CO2Me
CO2Me
53

Fig. 15 Determination of the aglycone carbohydrate attachments in ristocetin-A.

The substitution pattern of **50** demonstrated that the ethyl-substituted phenolic hydroxyl group originally carried the 2-O-(α-L-ristosaminyl)-β-D-glucopyranosyl moiety or another monosaccharide molecule (see Sec. II.C). Introduction of an ethyl function was necessary to make a distinction from the methyl groups upon ^{1}H-NMR spectroscopic investigations.

During the structural studies of ristocetin A, Sztaricskai et al. (40) found that the introduction of the CD_3 group instead of the ethyl substituent was more suitable (Fig. 15). Thus, the N-acetylated O-methylated ristocetin A was first hydrolyzed (to remove the carbohydrate moieties) and the liberated phenolic hydroxyl groups were blocked by treatment with deuteriomethyl iodide. The aglycone derivative obtained this way was subjected to alkaline hydrolysis in the presence of sodium borohydride. The O-methylated amino acids were degraded upon treatment with hypochlorite, followed by oxidation with potassium permanganate, and then methylated again with diazomethane to isolate the three compounds **51–53**. The deuteriomethyl group appeared in **52** and **53**. Since the triester **52** could be produced only from the didechlorovancomycinic acid portion and previous studies (see Sec. II.C) have shown that the heterotetrasaccharide side chain is attached to the didechlorovancomycinic acid, the placement of the C_3' CD_3 group in the diester **53** (produced from actinoidinic acid) requires that the other D-mannose should be originally attached at this phenolic moiety. The α configuration of the glycosidic linkage was finally determined (39) by comparative ^{13}C-NMR studies of the antibiotic and phenyl α-D-mannopyranoside as a model compound.

The deuteriomethylation procedure was successfully applied (40) for the structural elucidation of the chloropolysporins.

III. CHEMICAL SYNTHESIS OF THE AMINODEOXY SUGARS

A. The Glucosamine Derivatives

2-Amino-2-deoxy-D-glucose (**54a**, D-glucosamine), the first amino sugar to be found in nature, is one of the most abundant monosaccharides. Apart from the occurrence of this sugar in the hard shells of many living organisms and in numerous glycosaminoglycans, certain derivatives of D-glucosamine have become known (76) as antibiotics, such as streptozotocin and diumycin. 2-Amino-2-deoxy-D-glucuronic acid (**55a**) has also been found (76) in many bacterial polysaccharides.

As shown by the data in Table 2 (see Sec. I), both **54** and **55** were recently shown to be the aminodeoxy sugar components of glycopeptide antibiotics. The two amino hexoses are present (Fig. 16) in these antibiotics in N-acetylated form (i.e., **54b**) or are substituted at the amino function with a longer saturated or unsaturated fatty acid chain (i.e., **54a–e** and **55a–d**).

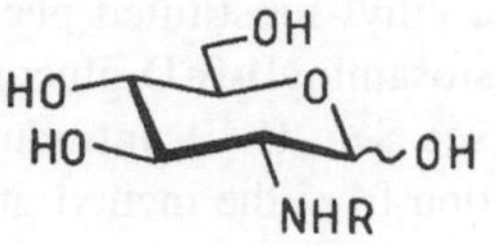

54

a R = H

b R = Ac

c R = Y

d R = Q

e R = Z

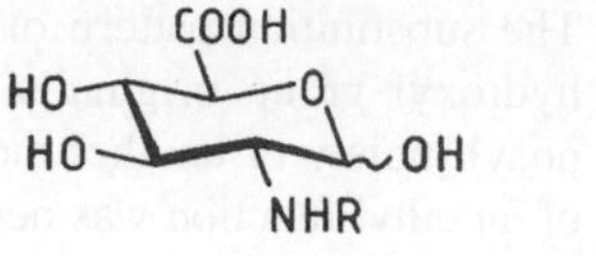

55

a R = H

b R = Y

c R = Q

d R = Z

$$Y = -\underset{\underset{O}{\|}}{C} - (CH_2)_8CH\begin{matrix} CH_3 \\ CH_3 \end{matrix}$$

$$Q = -\underset{\underset{O}{\|}}{C} - (CH_2)_{10} - CH_3$$

$$Z = -\underset{\underset{O}{\|}}{C} - (CH_2)_2CH{=}CH(CH_2)_4 - CH_3$$

Fig. 16 The glucosamine derivatives occurring in the glycopeptide antibiotics.

Although no syntheses have been reported for these functionalized amino sugars, both parent sugars (**54a** and **55a**) are commercially available, and N-acylation can be readily accomplished by means of the well-known Schotten-Baumann procedure to furnish these N-acyl analogs.

B. The 3-Amino-2,3,6-trideoxyhexoses

Based on literature reports concerning the synthesis of L-acosamine, L-actinosamine, L-ristosamine, L-vancosamine, 3-*epi*-L-vancosamine, and 4-*epi*-L-vancosamine (eremosamine), the trideoxyamino sugar components of the glycopeptide antibiotics, acosamine and ristosamine are the most frequently chosen synthetic targets (6,7,77–82). This is clearly because L-acosamine derivatives often serve as intermediates (7) to L-daunosamine and also because of the continuing interest of organic chemists in the structural modification of the antitumor anthracycline glycoside antibiotics.

In the mid-1970s Arcamone et al. (83,84) recognized that replacement of the L-daunosamine (3-amino-2,3,6-trideoxy-L-*lyxo*-hexopyranose) moiety of daunorubicin and doxorubicin with L-acosamine results in reducing the cardiotoxicity of the parent antibiotics, but the new 4'-*epi* analogs retain the favorable anticancer activity. Semi-synthetic anthracyclines carrying an L-ristosamine sugar portion were much less active anticancer agents than daunorubicin or doxorubicin.

Recent results from the authors' laboratory have shown (85) that an azidothymidine (AZT) analog bearing the C_3 azido derivative of L-ristosamine as the sugar moiety possesses considerable activity against the human immunodeficiency virus.

Further, the growing utilization of 3-amino-2,3,6-trideoxyhexoses and the intermediates of their syntheses (7) for the production of other biologically active compounds [nucleosides (85,86), aminocyclitols (87,88), β-lactam (7) and negamycin (7) derivatives, and others], encourage synthetic chemists to undertake research on these types of aminodeoxy sugars.

During the past two decades almost 100 papers have been aimed at the synthesis of these six 3-amino-2,3,6-trideoxyhexose components of the glycopeptide antibiotics, starting from either carbohydrates or nonsugar precursors. The following summarizes the most important synthetic approaches to these challenging amino sugars.

Syntheses from Carbohydrates

L-Acosamine (3-Amino-2,3,6-trideoxy-L-*arabino*-hexopyranose)

Syntheses from O_3 Sulfonyl Derivatives. In connection with the first synthesis of L-daunosamine, the Goodman group (89) prepared a derivative (methyl 3-acetamido-4-O-methanesulfonyl-2,3,6-trideoxy-α-L-*arabino*-hexopyranoside) of L-acosamine as early as 1967, 6 years before the structure (8,9) of this amino sugar became known. Then, in 1974, Gupta (90) described a complete synthetic route to methyl α-L-acosaminide (**5**) from L-rhamnal, based on the Goodman method (89), without the knowledge that the parent sugar had been isolated from the antibiotic actinoidin. The structure of L-acosamine was independently confirmed by utilization of the original procedure (89) for a definitive synthesis (91). These two substantially similar approaches (90,91) to L-acosamine are summarized in Figure 17.

Methoxymercuration of L-rhamnal (**56**) and subsequent reductive demercuration with sodium borohydride led to methyl 2,6-dideoxy-α-L-*arabino*-hexopyranoside (**57**). Tosylation gave mainly the 3-O-tosylate **58**, which was converted into the 3,4-anhydro sugar **59**. Nucleophilic opening of the epoxide function of **59** with sodium azide afforded 88% of the arabino C_3 azide **60**, which

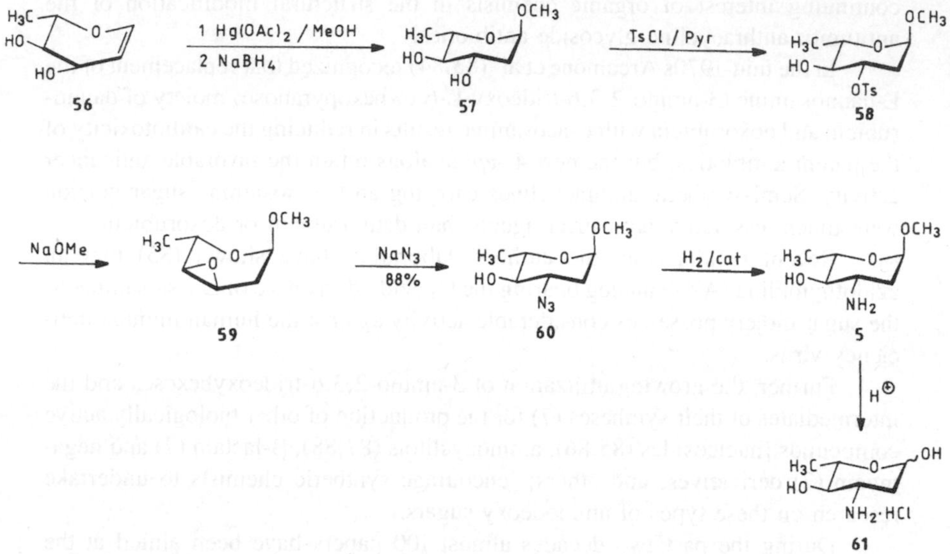

Fig. 17   The first synthesis of L-acosamine derivatives.

was hydrogenated to methyl α-L-acosaminide **5**, also identified (91) as the corresponding N-acetate (**7**) and N-benzoate. The methyl glycoside **5** was then subjected to mild acid hydrolysis, resulting in the hydrochloride salt (**61**) of L-acosamine.

The product distributions in the methoxymercuration and tosylation reactions were later described in a communication (92) from the authors' laboratory in connection with the utilization of the sulfonate **58** for the first definitive synthesis of L-ristosamine (see later in this section). A substantial improvement in the regioselectivity of the O-3 tosylation step (**57** → **58**) via stannylidene acetal derivatives was reported by Monneret et al. (93).

In a similar approach by Grethe et al. (94) to L-acosamine and L-daunosamine derivatives, the 2,6-dideoxyglycoside **57** was synthesized from L-arabinose by the well-known Sowden-Fischer one-carbon enlargement procedure. The same group (95) elaborated an ingenious method for the production of methyl 3,4-anhydro-2,6-dideoxy-α-L-*ribo*-hexopyranoside (**59**) from methylcyclohexadiene. The steric outcome of opening of the anhydro ring of **59** and related 3,4-anhydro-hexopyranosides with various nitrogen nucleophiles to obtain L-acosamine and its regio and stereoisomers was investigated in detail by Monneret et al. (96).

Application of the Allylic-Azide Rearrangement of Glycals.   Another carbohydrate-based approach to L-acosamine (and L-ristosamine) derivatives

involves the utilization of the allylic-azide rearrangement reaction of 3,4-di-
O-acetyl-L-rhamnal (62). In the first trial (Fig. 18), Heyns and coworkers (97)
treated 62 with sodium azide in the presence of boron trifluoride etherate, and the
resulting C_3 epimeric mixture (63) of 3-azidoglycals (60%) was sequentially
iodomethoxylated and hydrogenated. After acetylation and separation 38.6% of
methyl N,O-diacetyl-α-L-acosaminide (64) and 24.6% of the corresponding
L-*ribo* isomer 17 (methyl N,O-diacetyl-α-L-ristosaminide) were obtained.

Subsequent detailed studies by Monneret et al. (98,99) and Thiem et al.
(100,102) showed that the "azide mixture" 63 represents an equilibrium system
(Fig. 18) consisting of the α- and β-1-azido-hex-2-enopyranosides (65 and 66) and
the *ribo* (67) and *arabino* (68) 3-azidoglycals. The composition of this equilibrium
is dependent on the reaction conditions, and the separated azidoglycals (67 and 68)
quickly rearrange into the equilibrium mixtures (65 ⇌ 67 and 66 ⇌ 68,
respectively, indicating the intramolecular character of the allylic-azide rearrangement (98,103).

Reduction of this four-component mixture with lithium aluminum hydride
followed by O-deacetylation gave the *arabino* (69) and *ribo* (70) aminoglycals,
which were transformed (98) into methyl N-trifluoroacetyl-α-L-acosaminide- (71)
and α-L-ristosaminide (72) derivatives, respectively.

Alternatively, the equilibrium mixture (65–68) was treated with *N*-iodosuc-
cinimide and benzyl alcohol (101,102) [or cyclohexanol (100)] to give the corresponding 3-azido-2-iodoglycosides. Sequential (102) or simultaneous (100) reduction of the iodo- and azido functions (into 2-deoxy and 3-amino, respectively) then
afforded the glycosides of the aminodeoxy sugars L-acosamine and L-ristosamine.

Recently Monneret et al. (104,105) recognized that conjugate addition of
hydrazoic acid to the hex-2-enopyranose 73 ("pseudorhamnal") proceeds with a
predominant *equatorial* incorporation of the azido group at C_3 (Fig. 19) to obtain,
after acetylation, the *arabino* azide 74 as the major component of the product
mixture. Glycosylation of this latter with methanol in the presence of K-10
montmorillonite allowed the isolation of 48% methyl 4-O-acetyl-3-azido-2,3,6-
trideoxy-α-L-*arabino*-hexopyranoside (75) along with insignificant amounts (1–
10%) of the corresponding β-L-*arabino*- and α- and β-L-*ribo*-glycosides.
Zemplén transesterification and subsequent hydrogenation then gave (104,105)
methyl α-L-acosaminide (5) in an overall yield of 35% from the rhamnal ester 62.

Syntheses from Hexopyranosid-uloses.   Hexopyranosid C_3 uloses are frequent starting materials for the preparation (7) of 3-amino-2,3,6-trideoxyhexoses
and their C_3 methyl-branched derivatives. A convenient procedure, elaborated in
the authors' laboratory for the production of all four stereoisomeric 3-amino-
2,3,6-trideoxy-L-hexoses, is based on the chemoselective oxidation of L-rhamnal
(56) with silver carbonate (106) or, preferably, with barium manganate to obtain
the enone 76 (Fig. 20). Michael addition of methanol onto 76 gave methyl

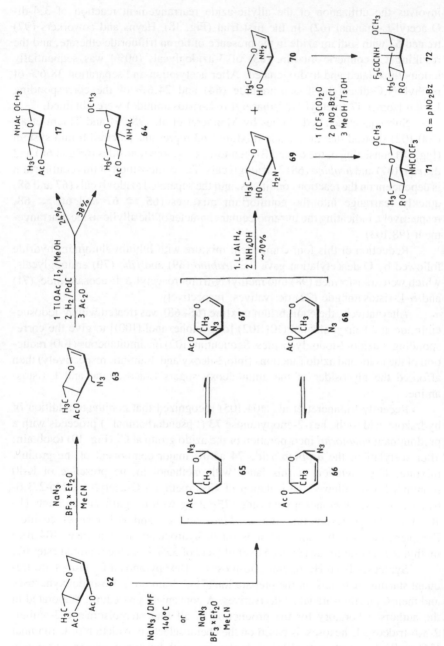

Fig. 18 The allylic-azide rearrangement approach to L-acosamine and L-ristosamine.

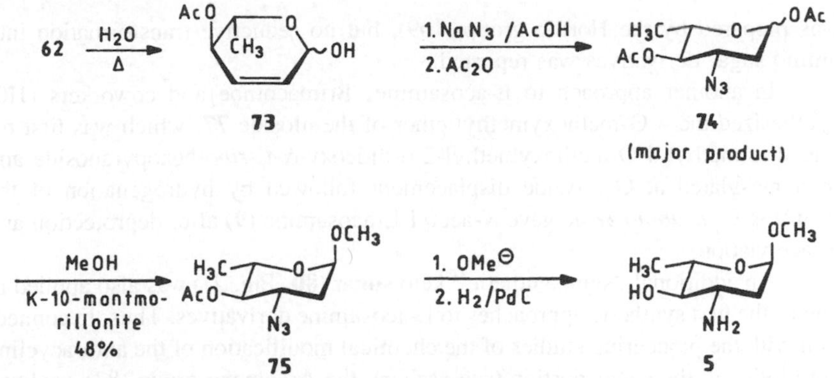

Fig. 19 Synthesis of L-acosamine from the "pseudoglycal" **73**.

2,6-dideoxy-α-L-*erythro*-hexopyranosid-3-ulose (**77**) contaminated with approximately 6% of the corresponding β-anomer. The ketone **77** was converted into the acetylated oxime **78**, which was then reduced (107,108) with the borane-tetrahydrofuran complex. After N-trifluoroacetylation of the crude reaction mixture, methyl N-trifluoroacetyl-α-L-acosaminide (**79**) was the only isolable product (with a 41% overall yield from **56**), and only traces of the L-*ribo* (i.e., ristosaminide) epimer could be detected. Note that catalytic hydrogenation (106) of the non-acetylated analog of the oxime **78** provided the latter amino sugar as the major product (Fig. 28). The S-benzylglycoside derivative of **79** was synthesized (107, 108) in an essentially similar way from **76**. The 3-O-methyl ether of the oxime **78**

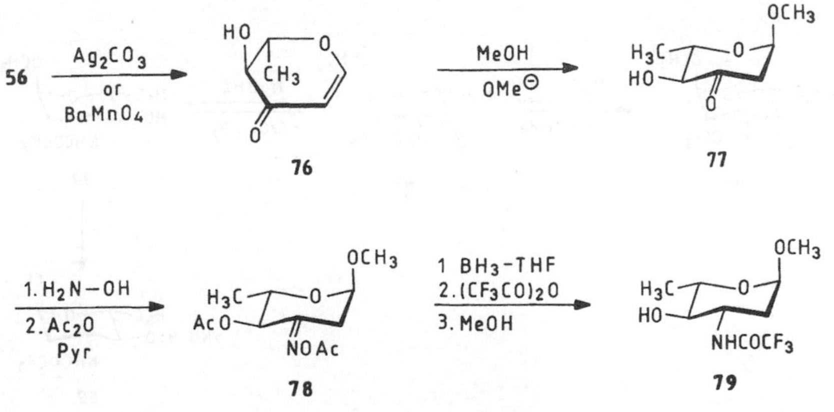

Fig. 20 The hexopyranosid-3-ulose route to L-acosamine derivatives.

was prepared by the Horton group (109), but no reductive transformation into amino sugar derivatives was reported.

In another approach to L-acosamine, Brimacombe and coworkers (110) synthesized the 4-O-methoxymethyl ether of the uloside 77, which was first reduced to methyl 4-O-methoxymethyl-2,6-dideoxy-α-L-*ribo*-hexopyranoside and then mesylated at O_3. Azide displacement followed by hydrogenation of the resulting C_3 *arabino* azide gave N-acetyl-L-acosamine (9) after deprotection and N-acetylation.

An additional "semisynthetic" keto sugar (80, Fig. 21) was also applied in one of the first synthetic approaches to L-acosamine derivatives. Thus, in connection with the pioneering studies of the chemical modification of the anthracycline antibiotics in the sugar portion (see earlier), the Arcamone group (83) oxidized methyl N-trifluoroacetyl-α-L-daunosaminide (81) with ruthenium tetroxide. The resulting semisynthetic C_4 L-*threo*-ketone (80) was stereoselectively reduced with sodium borohydride into methyl N-trifluoroacetyl-α-L-acosaminide (79). By further transformations 79 was converted into the protected 1-chlorosugar 82, and this was applied (83) to the first synthesis of the therapeutically most useful 4'-*epi*-daunorubicin and 4'-*epi*-doxorubicin.

In the course of a rather economical carbohydrate-based synthesis (108) of L-daunosamine derivatives, the C_4 ketone (80) was (reversely) prepared from 79 (Figs. 20 and 21) by oxidation with the chromium trioxide-dipyridyl complex. Subsequent reduction of 80 with L-Selectride afforded exclusively methyl N-trifluoroacetyl-α-L-daunosaminide (81), demonstrating that sodium borohydride and lithium tri-*sec*-butylborohydride (L-Selectride) preferentially generate (7) *equatorial* and *axial* hydroxyl groups, respectively, from carbohydrate C_4 uloses. Based on this experience a straightforward methodology was developed by

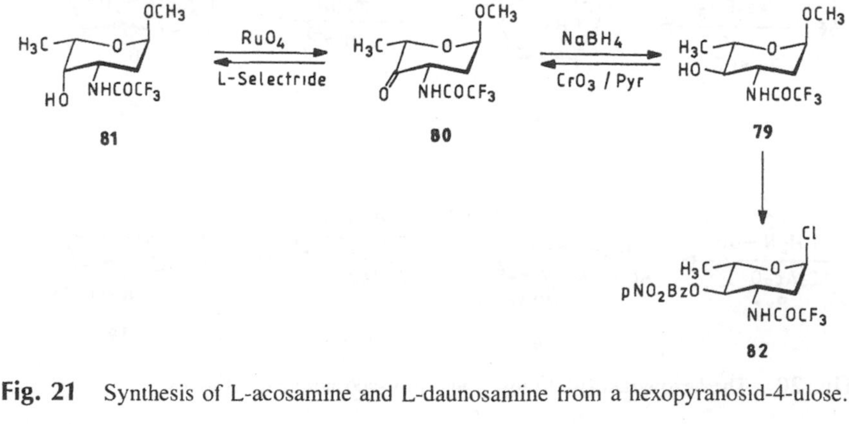

Fig. 21   Synthesis of L-acosamine and L-daunosamine from a hexopyranosid-4-ulose.

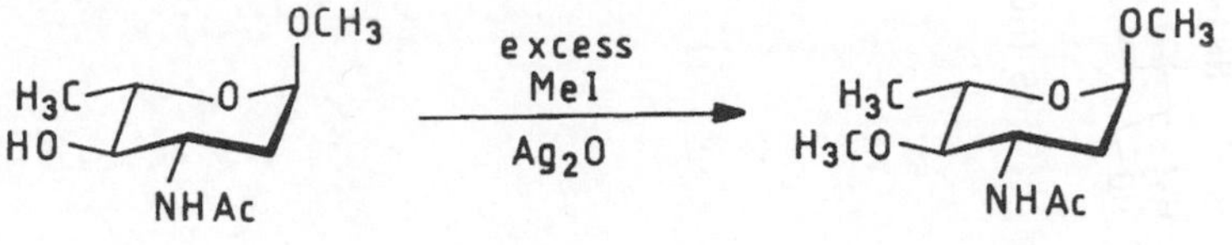

Fig. 22 Definitive synthesis of the amino sugar L-actinosamine.

the Arcamone group (111) for the direct interconversion of daunorubicin to 4'-*epi*-daunorubicin.

L-Actinosamine (3-Amino-4-O-methyl-2,3,6-trideoxy-L-*arabino*-hexopyranose). The first synthesis of actinosamine was reported by the Goodman group (91) in connection with the preparation of L-acosamine derivatives (see the last section). Thus, methyl N-acetyl-α-L-acosaminide (7) was methylated with a large excess of methyl iodide in the presence of silver oxide (Fig. 22) to give methyl N-acetyl-α-L-actinosaminide (8). The physical and spectral data for 8 were in good agreement with those of a sample derived by the methanolysis (8) of actinoidin.

During their early work on the structural modification of anthracycline antibiotics, Arcamone et al. (84) described an independent synthesis of N-trifluoroacetyl-L-actinosamine (83), as shown in Figure 23. Methylation of 79 (Fig. 21) was performed by boron trifluoride-catalyzed treatment with diazomethane, and the resulting actinosaminide 84 was hydrolyzed into the reducing sugar 83. This was then converted to the corresponding 1-chlorosugar, suitable for glycosylation of anthracycline aglycones.

L-Ristosamine (3-Amino-2,3,6-trideoxy-L-*ribo*-hexopyranose)

Syntheses from 3-O-Sulfonyl Derivatives. Following structural elucidation studies (10), the Bognár group proved the structure of L-ristosamine (15) by a definitive synthesis (112) of N-benzoyl-L-ristosamine (19) and additional derivatives (92) of this aminodeoxy sugar (Fig. 24). The Goodman team (113) not much later published exactly the same procedure for the structural proof of L-ristosamine.

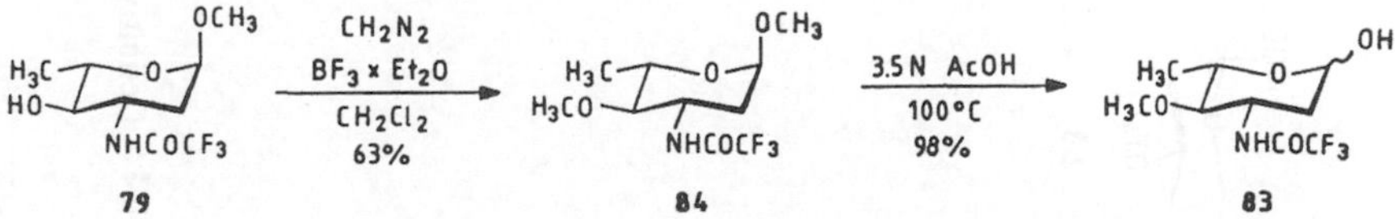

Fig. 23 A preparative approach to L-actinosamine derivatives.

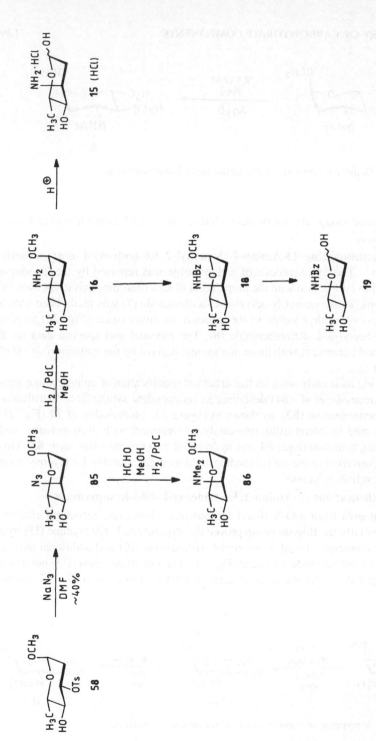

Fig. 24 Definitive synthesis of L-ristosamine.

Methyl 3-O-(p-toluenesulfonyl)-2,6-dideoxy-α-L-*arabino*-hexopyranoside (**58**), the key intermediate of the first synthesis (Fig. 17) of L-acosamine, was treated with sodium azide to give approximately 40% of the C_3 *ribo*azide **85**. The relatively low yield may be attributed to the β-transaxial effect (7) due to the axially disposed glycosidic methoxy group in **58**. Catalytic hydrogenation of **85** gave methyl α-L-ristosaminide (**16**), identified as the hydrochloride salt, which upon mild acid hydrolysis furnished (91,112,113) L-ristosamine hydrochloride (**15**). Schotten-Baumann benzoylation of **16** led to N-benzoyl glycoside (**18**), and this was converted (92,112) to N-benzoyl-L-ristosamine (**19**), a stable crystalline derivative of the amino sugar suitable for comparison with samples originating from the antibiotic ristomycin (ristocetin).

Note that the azido glycoside **85** was utilized by Bartner et al. (114) for the synthesis of methyl 3-dimethylamino-2,3,6-trideoxy-α-L-*ribo*-hexopyranoside (**86**), the methyl glycoside of the aminodeoxy sugar component (L-megosamine) of the macrolide antibiotic megalomycin.

Because of the high cost of L-sugar starting materials, 6-deoxy-L-hexoses (including aminodeoxy sugars) are often synthesized from their C_5 epimeric diastereoisomers, obtainable from commercially available, cheap D-sugars by standard chemical transformations. Many examples based on such a methodology have been described, most particularly for the preparation (7) of L-daunosamine and L-vancosamine.

Accordingly, Figure 25 illustrates the strategy (115) for obtaining L-ristos-amine derivatives from D-glucose. This procedure involves a successive triple configurational inversion, first at C_3, then at C_5, and finally at C_4. (Note that either C_4 or C_5 configurational inversion is a procedure that is followed quite often (7) in syntheses of naturally occurring rare deoxy and aminodeoxy sugars.)

Thus, the 3-O-mesylester **87** [readily available from D-glucose in six steps (116)] was first converted into the D-*arabino* C_3 azide **88**. Hanessian-Hullar opening of the benzylidene acetal of **88**, followed by dehydrobromination with silver fluoride, resulted in the 3-azido-C_5-exomethylene sugar **89** (for recent material on the application of this methodology in aminodedeoxy sugar syntheses, see Ref. 7). Simultaneous reduction of the azido group and the double bond in **89** gave exclusively the 3-acetamido-β-L-*xylo*-glycoside (**90**) after acetylation, which was subjected to C_4 configurational interconversion (**90** → **91**) via the intermediary 4-O-mesylate to yield methyl N-acetyl-β-L-ristosaminide (**91**).

Adaption of the Allylic-Azide Rearrangement of Glycal Esters. In an earlier section (Fig. 18) we discussed related methods (97–105) that usually give reaction condition-dependent mixtures of L-acosamine and L-ristosamine derivatives. However, in an attempted synthesis of L-daunosamine from the equilibrium mixture (**65–68**, Fig. 18), Thiem and Springer (101,102) recognized that the pyridinium dichromate oxidation of the 3-azido-2-iodoglycoside **92** (obtained upon treatment with N-iodosuccinimide/benzyl alcohol) resulted in a simulta-

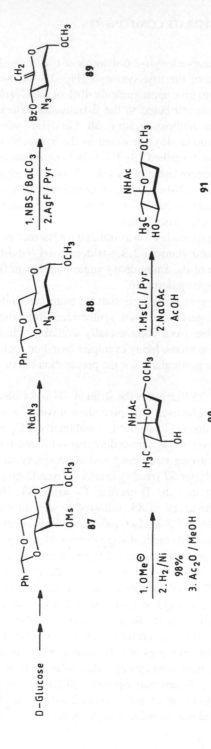

Fig. 25 Synthesis of L-ristosamine derivatives from D-glucose.

neous *trans* elimination of hydrogen iodide, and 65% of the 3-azido-hex-2-enopyranosid-4-ulose **93** was isolated (Fig. 26). Reduction of **93** with sodium borohydride at −20°C gave exclusively the 3-azido-α-L-*erythro*-hex-2-enopyranoside **94**, which could be converted into benzyl N,O-diacetyl-α-L-ristosaminide (**95**) by borohydride reduction at elevated temperature and subsequent acetylation.

Syntheses from Hexopyranosid-3-Uloses.   Arcamone et al. (117) reported that treatment of the oxime **97** (Fig. 27), derived from methyl 4,6-O-benzylidene-2-deoxy-α-L-*erythro*-hexopyranosid-3-ulose (**96**), with lithium aluminum hydride and subsequent removal of the acetal function and N-trifluoroacetylation yielded the 6-hydroxy derivative (**98**) of methyl N-trifluoroacetyl-α-L-ristosaminide. Using the Hanessian method for conversion of **98** to the 6-bromo-6-deoxy derivative **99**, followed by catalytic hydrogenation, methyl N-trifluoroacetyl-α-L-ristosaminide (**72**) was obtained. Note that although this procedure appears to be rather straightforward, the protected C_3 uloside **96** was prepared from the extremely expensive 2-deoxy-L-glucose.

This drawback could be eliminated by employing the authors' procedure (Fig. 20). Thus, the uloside **77** (derived from the cheaper L-rhamnose) was converted (106) into the oxime **100** (Fig. 28), whose hydrogenation over Raney nickel afforded a readily separable, approximately 8:1.5 mixture of methyl N,O-diacetyl-α-L-ristosaminide (**17**) and methyl N,O-diacetyl-α-L-acosaminide (**64**) after acetylation. The reduction of the acetylated analog (**78**) of oxime **100** with the borane-tetrahydrofuran complex gave exclusively the acosamine glycoside (Fig. 20).

Alternatively, Brimacombe et al. (118, 119) applied (Fig. 28) the 4-O-(2-methoxyethoxymethyl) ether derivative **101** of the uloside **77**, derived from methyl 2,3-O-benzylidene-4-O-(2-methoxyethoxymethyl)-α-L-rhamnopyranoside, by

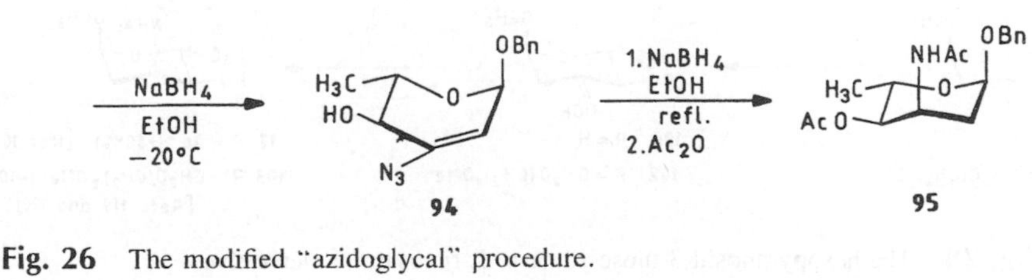

Fig. 26 The modified "azidoglycal" procedure.

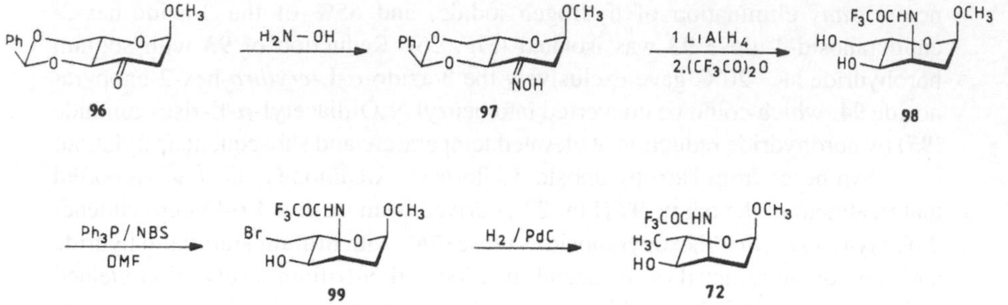

Fig. 27 Preparation of L-ristosamine derivatives from 2-deoxy-L-glucose.

employing the well-known Klemer-Rodemeyer procedure. After conversion of **101** into the oxime (**102**), this was hydrogenated over platinum oxide to ensure quantitative formation of the L-ristosamine glycoside (**103**).

cis-Oxyamination of Hex-2-Enopyranoside Derivatives (Pseudoglycals). A very simple and convenient regio- and stereoselective construction of the molecules of 3-amino-2,3,6-trideoxyhexoses carrying the C_3 amino and C_4 hydroxyl functions in a *cis* relationship (i.e., ristosamine and daunosamine) was elaborated by the groups of Cardillo (120), Fraser-Reid (121,122), and Sammes (123). For the preparation of L-ristosamine derivatives the trichloroacetimidate ester **104** of methyl 2,3,6-trideoxy-α-L-*erythro*-hex-2-enopyranoside was employed (Fig. 29) for the *cis*-oxyamination with N-bromosuccinimide (120), iodomium dicollidine perchlorate (121,122), or N-iodosuccinimide (120), resulting in the respective 2-bromo- (**105**) or 2-iodooxazoline derivative (**106**) in excellent yield (95–100%). Radical dehalogenation of **105** and **106** with 1.5–5 equivalents of tributyltin hydride, followed by hydrolysis, then allowed the isolation of methyl ristosaminide hydrochloride (**16**) (120) or methyl N-acetyl-α-L-ristosaminide (**107**) (121,122). Racemic **16** was synthesized in a similar way by Sammes and Thetford (123,124).

The L-*threo* isomer of the trichloroacetimidate [**104**, prepared by C_4 inversion of the starting α-L-*erythro*-hex-2-enopyranoside with the aid of the Mitsu-

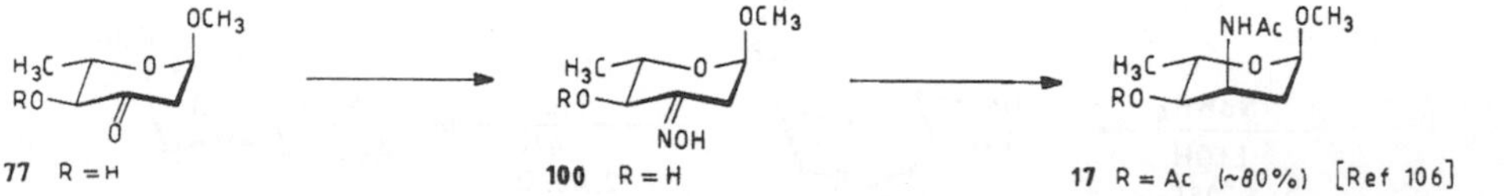

77 R = H
101 R = $CH_2O(CH_2)_2OMe$

100 R = H
102 R = $CH_2O(CH_2)_2OMe$

17 R = Ac (~80%) [Ref 106]
103 R = $CH_2O(CH_2)_2OMe$ (~100%) [Refs 119 and 120]

Fig. 28 The hexopyranosid-3-ulose route to L-ristosamine derivatives.

Fig. 29 *cis*-Oxyamination procedure for obtaining L-ristosamine derivatives.

nobu method (122,123,125) or via nucleophilic displacement (102, 126) of the corresponding 4-O-methanesulfonate| was applied for the hitherto most efficient carbohydrate-based syntheses of L-daunosamine derivatives. This simple route was recently extended (127) to the production of the C_3 methyl-branched 3-amino and 3-nitro analogs (rubranitrose) of the foregoing trideoxyhexoses in the D-*ribo* series.

L-Vancosamine (3-Amino-3-C-methyl-2,3,6-trideoxy-L-*lyxo*-hexopyranose). Since the identification (11–13) of the structure of L-vancosamine (**22**, the C_3 methyl derivative of L-daunosamine) in 1972, considerable effort has been devoted to produce this amino sugar by chemical synthesis. However, during the first trials the preparation of only the C_4 epimer (**32**) of vancosamine (occurring in the glycopeptide antibiotic eremomycin) was obtained (see later) (128). Nevertheless, these pioneering studies dealing with the generation (129) and transformation (130,131) of the C_3 cyanomesylates derived from carbohydrate C_3 uloses into C_3 spiroaziridines have essentially contributed to the development of synthetic strategies (7) for obtaining various C_3 methyl-branched amino and nitro sugars, the latter (L-evernitrose and D-rubranitrose) being the carbohydrate constituents of the antibiotics everninomycin B, C, and D and rubradirin, respectively.

Finally, the first stereospecific synthesis (Fig. 30) of derivatives (**114** and **115**) of L-vancosamine (**22**) was elaborated by the Lukacs group (132) in 1979. Later, a series of papers by the Brimacombe team (133,134), Lukacs et al. (135), and Klemer et al. (136) gave a more extended contribution to this field (7).

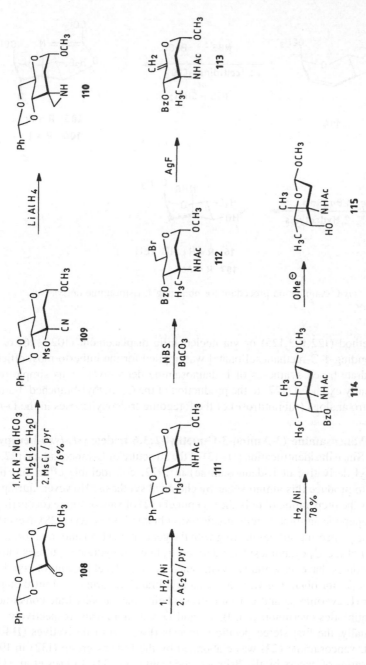

Fig. 30 The first stereospecific synthesis of L-vancosamine based on the cyanomesylation route.

Syntheses from Cyanomesylates Derived from Hexopyranosid-3-Uloses. For the first stereospecific synthesis (132) of L-vancosamine derivatives (Fig. 30), the keto sugar 108 [a popular starting material in aminodeoxyhexose syntheses (7)] was converted into the D-*arabino* C_3 cyanomesylate 109 upon addition of alkaline potassium cyanide under thermodynamic conditions, followed by mesylation of the cyanohydrin intermediate. Lithium aluminum hydride reduction of 109 generated the spiroaziridine 110 via configurational inversion at C_3, and opening of the aziridine ring by catalytic hydrogenation followed by acetylation afforded the C_3 methyl-branched amino sugar 111 with D-*ribo* configuration. Hanessian-Hullar acetal ring opening readily furnished the 6-bromo-6-deoxy derivative 112, which was converted into the C_5 exomethylene compound 113, suitable for a configurational stepover into the L-sugar series. Thus, saturation of the double bond in 113 by catalytic hydrogenation yielded 78% of the desired methyl N-acetyl-4-O-benzoyl-β-L-vancosaminide (114) and ~8% of the corresponding C_5 epimeric α-D-*ribo*-glycoside. The major product was then transformed into methyl N-acetyl-β-L-vancosaminide (115) upon O-debenzoylation. In an essentially similar way the N-trifluoroacetyl analog of 114 was also prepared (135) and employed for the glycosylation of anthracycline antibiotic aglycones by the well-known "aminoglycal procedure."

We already mentioned that such a C_5 configurational interchange as 89 → 90 (Fig. 25) or 113 → 114 (Fig. 30) is nowadays a quite common methodology (6,7) for obtaining rare deoxy (aminodeoxy) L-sugars from the respective C_5 epimeric D-enantiomers, functionalized according to the stereochemistry of the desired product. Note, however, that a related D → L C_5 configurational stepover can only be achieved in the α-D-hex-5-enopyranoside series, since experimental evidence unambiguously demonstrates (7,137) that the β-anomeric glycosides of hex-5-enopyranoses give mainly (or exclusively) the 6-deoxy derivative of the starting D-hexopyranosides upon catalytic saturation of the ethylenic double bond.

Following the "cyanomesylate route" outlined earlier, the Brimacombe team (134) employed the cyanomesylate 117 (Fig. 31) derived from methyl 4-O-methyl-2,6-dideoxy-α-L-*threo*-hexopyranosid-3-ulose 116 under thermodynamic (equilibrating) conditions. Compound 117 was transformed into the spiroaziridine 118 and then into methyl N-acetyl-4-O-methyl-α-L-vancosaminide (119) by a sequence (Fig. 31) essentially the same as that reported by Lukacs et al. (Fig. 30) (132,135).

Note that addition of hydrogen cyanide in pyridine to the ulose 116 (under kinetic control) and subsequent mesylation provided the C_3 epimeric cyanomesylate of 117, which was employed (134) for the preparation of 3-*epi*-vancosamine derivatives (see later).

Syntheses from Hexopyranosid-4-Uloses.   One of these procedures, described by Brimacombe et al. (138), is essentially based on the preparation of methyl 3-acetamido-C_3-methyl-2,3,6-trideoxy-α-L-*arabino*-hexopyranoside

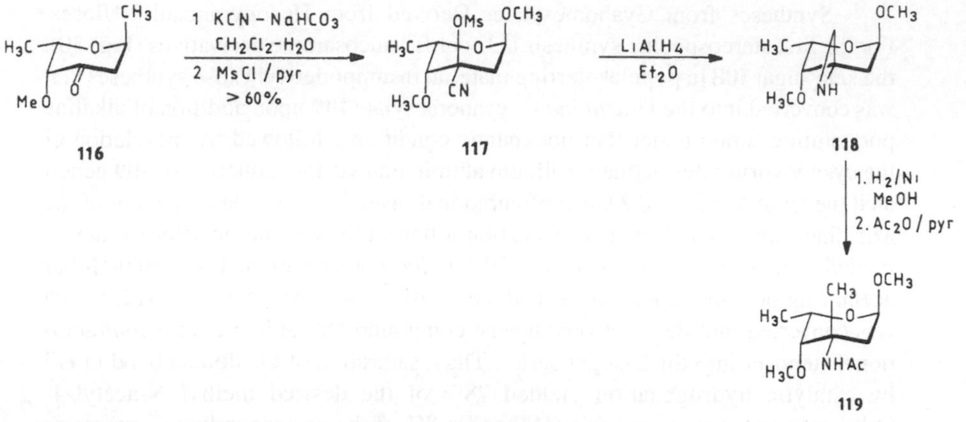

Fig. 31   The Brimacombe approach to L-vancosamine.

(**120**, identified (30) as methyl N-acetyl-α-L-eremosaminide; methyl N-acetyl-4-*epi*-α-L-vancosaminide). The synthesis of **120** is discussed in a later section in connection with obtaining (139) 4-*epi*-vancosamine derivatives from carbohydrate starting materials (Fig. 36).

Configurational inversion at C_4 (Fig. 32) of **120** was performed (138) according to the oxidation (**120** → **121**) and reduction (**121** → **122**) sequences, previously discussed, (Fig. 21) of obtaining the daunosamine glycoside **81** from the C_4 epimeric acosamine derivative **79**. Methyl N-acetyl-α-L-vancosaminide (**122**) was then transformed into the corresponding N,O-diacetate, methyl α-L-vancosaminide (**23**), and methyl N,O-dibenzoyl-α-L-vancosaminide (**25**) by means of conventional chemical transformations.

In 1987, Klemer and Wilbers (136) reported an interesting procedure involving the stereoselective α-C-methylation of N-protected methyl 3-amino-2,3,6-trideoxy-α-hexopyranosid-4-uloses, followed by reduction of the C_4 keto function with L-Selectride or with sodium borohydride to generate axially or equatorially disposed hydroxyl groups. Although both L-vancosamine and 3-*epi*-

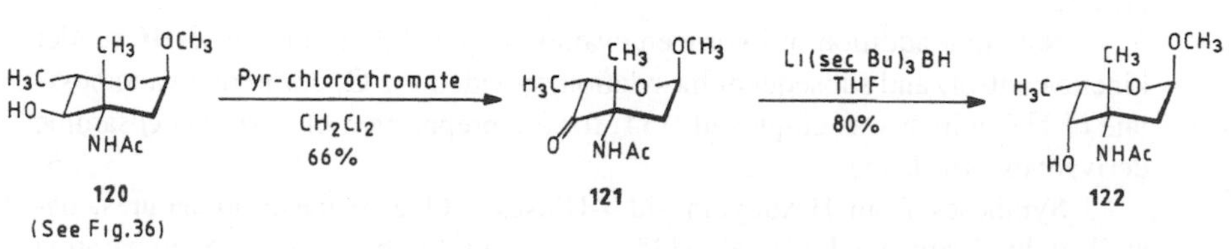

Fig. 32   Synthesis of L-vancosamine by configurational inversion at C_4.

L-vancosamine derivatives (as well as L-rubranitrose and precursors to L-decilo-nitrose and D-kijanose) have been obtained with the aid of this method, each step of the synthetic routes involves tedious separation of the product mixtures, and the yields for the daunosamine derivatives are particularly low.

3-*epi*-L-Vancosamine (3-Amino-3-C-methyl-2,3,6-trideoxy-L-*xylo*-hexo-pyranose). The first derivatives of 3-*epi*-L-vancosamine (**29**) were synthesized (134,140) in connection with the preparation of L-rubranitrose (**123**), the enan-tiomer of 2,3,6-trideoxy-3-C-methyl-4-O-methyl-3-nitro-D-*xylo*-hexopyranose (rubranitrose) isolated from the antibiotic rubradirin. The amino sugar itself then became the target of the syntheses elaborated by the groups of Yoshimura (141) and Brimacombe (142), and all these procedures are based on the generation of the NH_2-C-CH_3 branch by means of the cyanomesylation of carbohydrate C_3 uloses.

In 1984 Yoshimura et al. (140) subjected the β-L-*threo*-hexopyranosid-3-ulose **124** (the β-anomer of **116** prepared from D-glucose in a multistep sequence) to cyanomesylation (Fig. 33) under kinetic control to obtain the L-*lyxo* cyano-

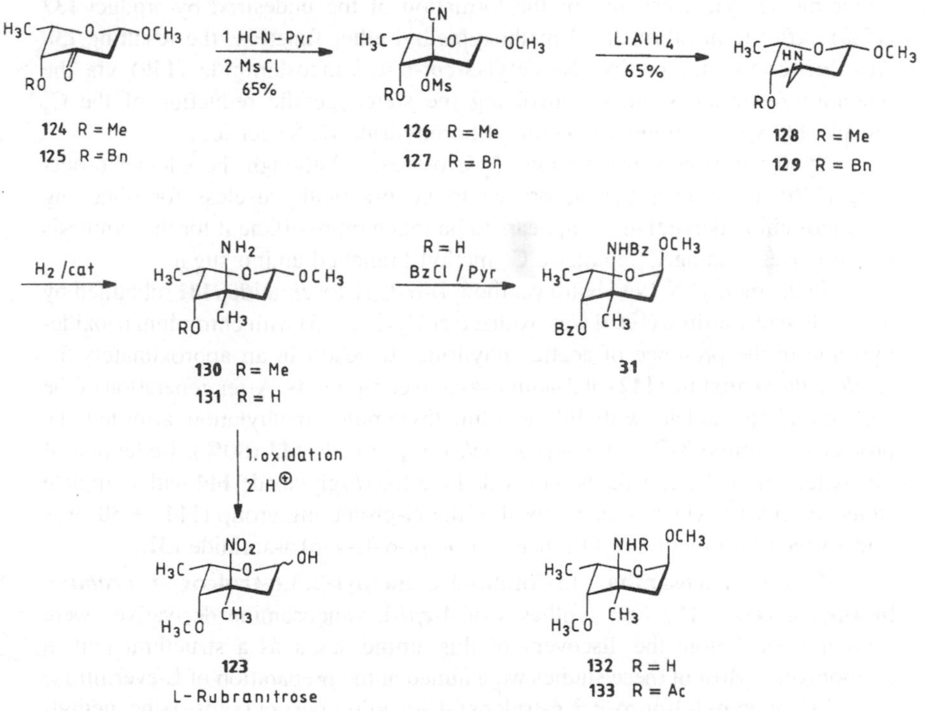

Fig. 33 Synthesis of 3-*epi*-vancosamine and L-rubranitrose derivatives.

mesylate **126** (65%). Under thermodynamic conditions a 1:4 mixture of **126** and the C_3 epimeric L-*xylo* isomer was produced. The rest of the synthetic route (generation of spiroaziridine **128**, its hydrogenolytic ring opening, and further derivatization of the resulting amino sugar **130**) is essentially identical with that (134) outlined in an earlier section (Fig. 31) when discussing the synthesis of L-vancosamine derivatives. This whole procedure was then also performed (141) with a 4-O-benzyl protecting group (**125** → **127** → **129** → **131**) and the β-methyl glycosides of 3-*epi*-vancosamine (**131**) and methyl 3,4-dibenzoyl-3-*epi*-α-L-vancosaminide (**31**) were prepared. (In the latter case an unexpected anomerization occurred during benzoylation of **131** under standard conditions.)

In 1985 the Brimacombe team (134) reported a completely identical route to the methyl 3-*epi*-α-L-vancosaminide derivatives **132** and **133** starting with the ulose **116**.

Another procedure (142) utilizing the previous cyanomesylate → spiroaziridine methodology gave α-L-*arabino*-cyanomesylate (**135**) upon treatment of the O_4-protected α-L-*erythro*-hexopyranosid-3-ulose (**134**) with potassium cyanide under equilibrating conditions (Fig. 34). One-pot conversion of **135** into the L-*ribo* branched-chain amino sugar **136** (via the riboaziridine) could be realized with a rather modest yield because of the formation of the undesired by-product **137** (15%). After removal of the O-methoxymethyl ether function, the resulting **138** was converted into methyl N-acetyl-3-*epi*-α-L-vancosaminide (**140**) via the oxidation-reduction sequence involving the stereospecific reduction of the C_4 uloside **139** with lithium-(tri-*sec*-butyl)borohydride (L-Selectride).

Synthesis from Carbohydrate C_4 Ulosides.   Although the Klemer procedure (136), mentioned earlier, proved to be practically useless for obtaining L-vancosamine derivatives, it appears to be much more efficient for the synthesis of the corresponding C_3 epimeric C_3 methyl-branched amino sugar.

Thus, methyl N-(*tert*-butoxycarbonyl)-α-L-ristosaminide (**141**, obtained by the method of Cardillo (120)] was oxidized at C_4 (Fig. 35) with chromium trioxide-pyridine in the presence of acetic anhydride, to result in an approximately 5:1 *erythro-threo* mixture (**142**) of 3-amino-4-keto compounds. After generation of the corresponding enolate with lithium dimethylamide, methylation afforded the protected 3-amino-3-C-methyl-α-L-*erythro* C_4 uloside **143** (80%). Reduction of this latter with L-Selectride then provided the L-*xylo*glycoside **144** with complete stereoselectivity, which after removal of the N-protecting group (**144** → **30**) was transformed into methyl N,O-dibenzoyl-3-*epi*-α-L-vancosaminide (**31**).

4-*epi*-L-vancosamine   (3-Amino-3-C-methyl-2,3,6-trideoxy-L-*arabino*-hexopyranose).   The first syntheses of 4-*epi*-L-vancosamine derivatives were accomplished before the discovery of this amino sugar as a structural unit in eremomycin. Most of these studies were aimed at the preparation of L-evernitrose (3-C,4-O-dimethyl-3-nitro-2,3,6-trideoxy-L-*arabino*-hexopyranose), the methyl-branched nitro sugar component of everninomycins B, C, and D. The methyl 3,4-

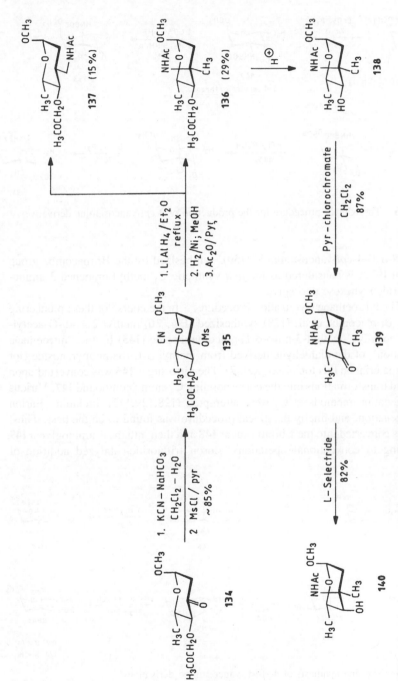

Fig. 34 Synthesis of methyl N-acetyl-3-*epi*-L-vancosaminide.

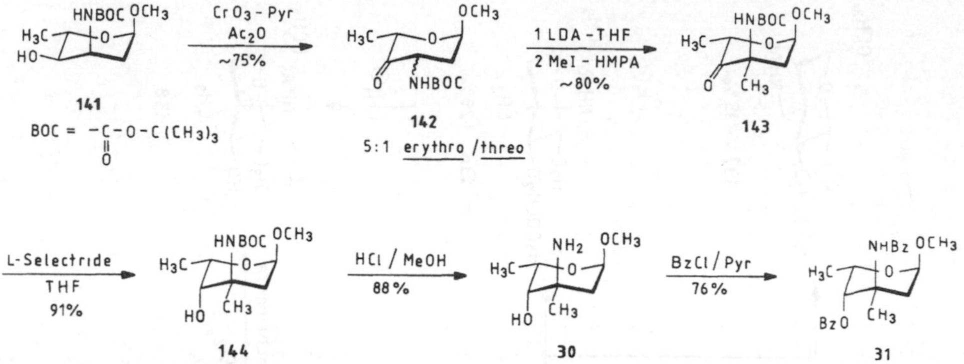

Fig. 35   The Klemer procedure for the production of 3-*epi*-vancosamine derivatives.

diacetyl-α,β-4-*epi*-vancosaminide (**150**), synthesized by the Brimacombe group
(128) in 1972, is considered as the *first* synthetic C_3 methyl-branched 3-amino-
2,3,6-trideoxyhexose derivative.

The Nitroethane Cyclization Procedure.   In the course of these pioneering
studies Brimacombe et al. (128) synthesized (Fig. 36) methyl 2,4,-di-O-acetyl-
3,6-dideoxy-3-C-methyl-3-nitro-α-L-glucopyranoside (**145**) by the "nitroethane
cyclization" of the dialdehyde derived from methyl α-L-rhamnopyranoside (for
recent material on this topic, see Ref. 7). The nitro sugar **145** was converted upon
standard transformations into the corresponding C_3 amino compound **147**. Various
deoxygenation methods at C_2 were attempted (128,129,143), including Barton
deoxygenation, and finally the glycal procedure was found to be the best. Thus,
147 was converted into the 1-bromo sugar **148** and then into the 3-aminoglycal **149**
according to conventional operations. Boron trifluoride-catalyzed addition of

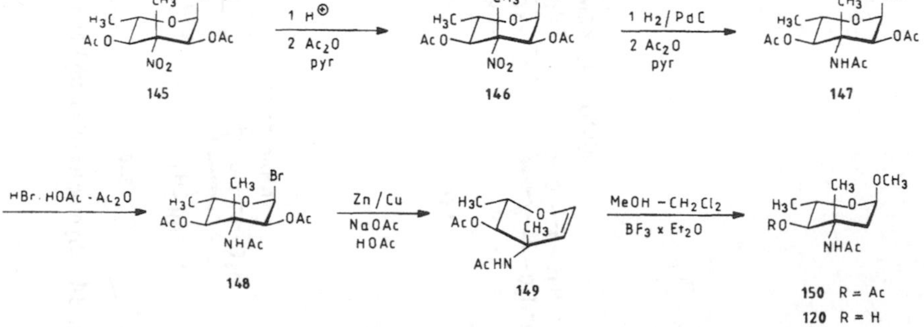

Fig. 36   The first synthesis of 4-*epi*-L-vancosamine derivatives.

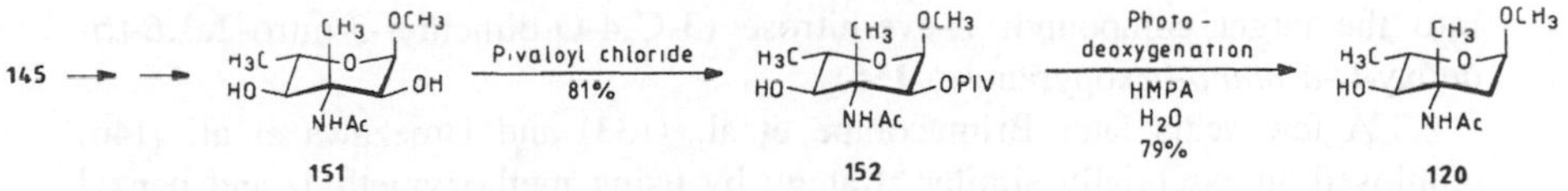

Fig. 37   An efficient C_2 deoxygenation procedure for the preparation of methyl N-acetyleremosaminide.

methanol gave a 3:1 mixture of methyl N,O-diacety-α-L-eremosaminide (**150**) and the corresponding β-anomer.

Most recently Jütten and Scharf (144) succeeded in improving this Brimacombe procedure by avoiding generation of the glycal **149** (Fig. 36) aimed at the development of the 2-deoxy function. Instead, the 3,6-dideoxyamino sugar **151** was prepared (Fig. 37) by the reduction of the nitro derivative **145**, followed by regioselective pivaloylation (at low temperature) to obtain the 2-O-pivaloyl ester **152** in good yield (81%). Photolytic removal of this ester group at 254 nm then ensured smooth deoxygenation at C_2, to afford 79% of methyl N-acetyleremosaminide (**120**).

Synthesis by the Cyanomesylation Method.   By the cyanomesylation procedure, Yoshimura et al. (145) treated (Fig. 38) methyl 2,6-di-deoxy-4-O-methyl-α-L-*erythro*-hexopyranosid-3-ulose (**153**) with hydrogen cyanide (under kinetic control), and subsequent mesylation afforded exclusively the L-*ribo*-cyanomesylate **154**. (Note that the β-anomer of **153** provided a 1:2 mixture of **154** and the corresponding L-*arabino* product under analogous conditions.) Conversion of **154** into methyl 4-O-methyl-α-L-eremosaminide (**155**) was then accomplished according to the methodology previously discussed. The amine **155** was converted

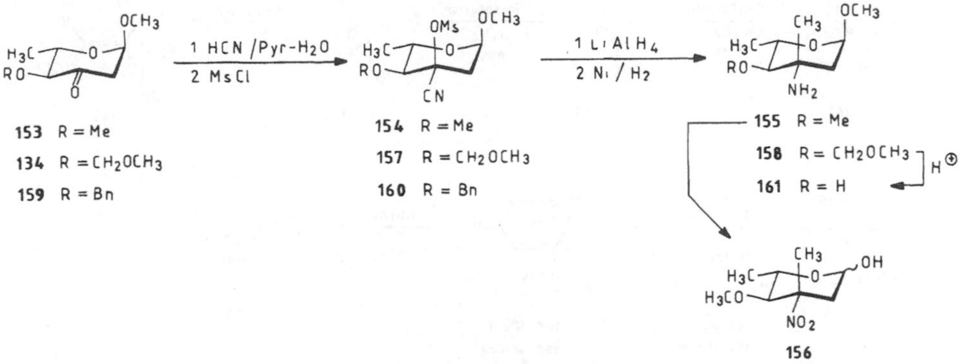

Fig. 38   The cyanomesylation route to eremosamine derivatives.

into the target compound; L-evernitrose (3-C,4-O-dimethyl-3-nitro-2,3,6-tri-deoxy-L-*arabino*-hexopyranose, **156**).

A few years later Brimacombe et al. (133) and Umezawa et al. (146) employed an essentially similar strategy by using methoxymethyl- and benzyl 4-O-protecting groups (Fig. 38), thus ensuring the preparation (146) of methyl α-L-eremosaminide (**161**) or its N-acetyl analog (133); before isolation of the parent aminodeoxy sugar (eremosamine, **32**) from eremomycin.

Syntheses from Noncarbohydrate Precursors

L-Acosamine (3-Amino-2,3,6-trideoxy-L-*arabino*-hexopyranose).   The very first noncarbohydrate-based synthesis of L- (and D,L) acosamine derivatives was described by Dyong and Bendlin (147) in 1978. Later this procedure was utilized by Welch et al. (148,149) for the production of aminodeoxy sugars with related structures.

In the original Dyong approach (Fig. 39), sorbic acid (**162**) was subjected to epoxidation with peracetic acid and the resulting racemic 4,5-anhydrocarboxylic acid **163** was resolved with L-(−)-α-phenylethylamine, followed by conversion of the L-enantiomer into the methyl ester **164**. Regioselective opening of the anhydro ring in **164** and subsequent protection at OH_5 furnished the methyl *trans*-2,3,6-trideoxy-L-*erythro*-hex-2-enoate **165**. The β addition of ammonia followed by removal of the protecting groups in one step (accompanied by simultaneous cyclization) and subsequent N-acetylation gave a mixture of the 3-acetamido-2,3,6-trideoxyhexono-γ- (**166**) and δ-lactones (**168**). Finally, DIBAL reduction of this lactone mixture allowed the isolation of N-acetyl-L-acosamine (**9**) in 39% yield.

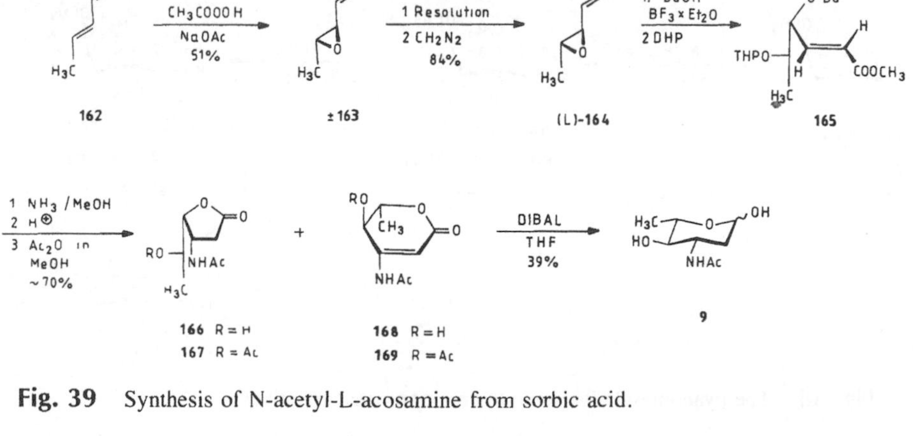

Fig. 39   Synthesis of N-acetyl-L-acosamine from sorbic acid.

By a similar methodology Dyong and Bendlin (147) prepared racemic **9** from methyl sorbate.

Using ethyl sorbate and an analogous epoxidation procedure, Hirama et al. (150) obtained the racemic *erythro*-diol **170** (Fig. 40), which was carbamoylated with chlorosulfonyl isocyanate to yield the biscarbamoyl ester **171**. The key intramolecular Michael addition, induced with potassium *tert*-butoxide, then provided almost exclusively the oxazolidinone **173** with a D,L-*arabino* configuration. Upon alkaline hydrolysis and subsequent acetylation **173** was converted into the lactone mixture **167** + **169** (Fig. 39) and then into N-acetyl-D,L-acosamine (±**9**). When the optically pure methyl ester analog of **171** (with the L-*erythro* configuration) was prepared (151,152) by coupling of O-(*tert*-butyldimethylsilyl)-lactaldehyde and methyl propiolate in the presence of lithium dimethylamine, enantiomerically pure N-acetyl-L-acosamide (**9**) was obtained by employing the same reaction sequence.

The L-*arabino* δ-lactone **169** could also be obtained from (*S*)-(+)-carvone by an ingenious multistep sequence developed by Kametani et al. (153). The Hirama group also utilized this simple procedure for the synthesis of ristosamine derivatives (Figure 51, p. 164).

A closely related structural analog (**176**) of the oxazolidinone **173** (Fig. 40) was prepared in enantiomerically pure form by Trost et al. (154) by the palladium-mediated vicinal hydroxyamination (Fig. 41) of the vinyl epoxide **175** (derived from **174**). Compound **176** was then converted into methyl N-acetyl-α-L-acosaminide (**7**) by employing a conventional hydroboration-oxidation sequence.

In the early 1980s the Fuganti group synthesized numerous aminodeoxy sugars from either natural C_4 carbon compounds (i.e., D-threonine) or from the products of the baker's yeast-mediated coupling reaction of acetaldehyde with

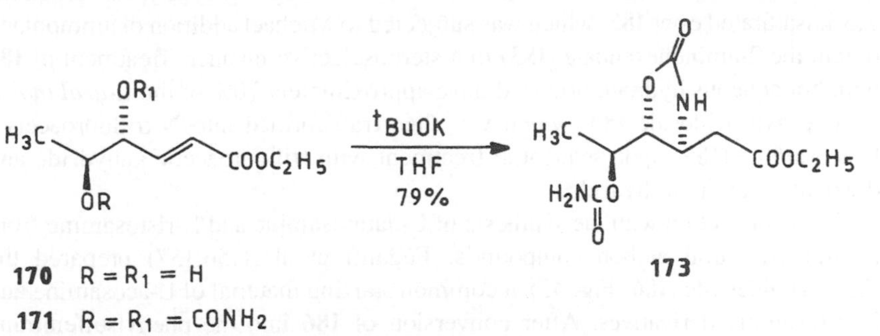

170 $R = R_1 = H$

171 $R = R_1 = CONH_2$

172 $R = CONH_2$, $R_1 = SiEt_3$

Fig. 40 The intramolecular Michael addition methodology.

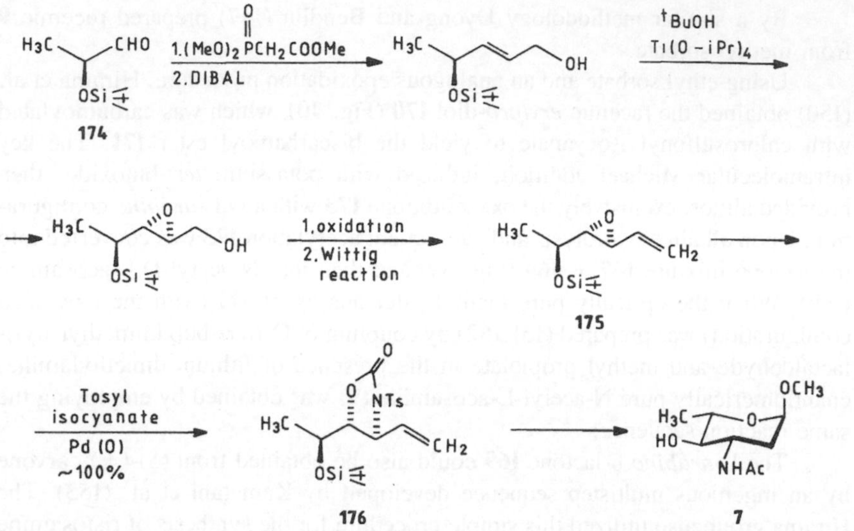

Fig. 41 Synthesis of L-acosamine derivatives by palladium-mediated vicinal hydroxy-amination.

trans-cinnamaldehyde (**177**, Fig. 42). Microbial reduction of the resulting α-hydroxy-ketone **178** (produced upon acyloin condensation) affords the (2S, 3R)-diol **179** in approximately 30% yield, and this versatile chiral substance already carries the C_4 and C_5 hydroxyl groups of the daunosamine type of trideoxyamino-hexoses.

Ozonolytic splitting (155) of the double bond of the isopropylidenated diol **180** followed by Wittig chain elongation of the resulting aldehyde **181** afforded the 2,3-unsaturated ester **182**, which was subjected to Michael addition of ammonia to obtain the 2-aminohexanoate (**183**) in a stereoselective manner. Treatment of **183** with hot aqueous hydrochloric acid gave approximately 70% of the L-*arabino*-γ-lactone hydrochloride **184**, which was then transformed into N-trifluoroacetyl-L-acosamine (**185**) upon sequential treatment with trifluoroacetic anhydride and diisobutylaluminum hydride.

In connection with the synthesis of L-daunosamine and L-ristosamine from natural C_4 chiral carbon compounds, Fuganti et al. (156,157) prepared the (2S, 3S)-aldehyde (**186**, Fig. 43), a common starting material of L-acosamine and L-ristosamine derivatives. After conversion of **186** into the phenylsulfenimine derivative **187**, treatment with dialkyl zinc proceeded with a high *erythro* dia-stereoselectivity to provide the 4,5-*erythro* adduct **188**, a precursor to L-ristos-amine (see Fig. 50, p. 163). On the contrary, addition of allylmagnesium bromide

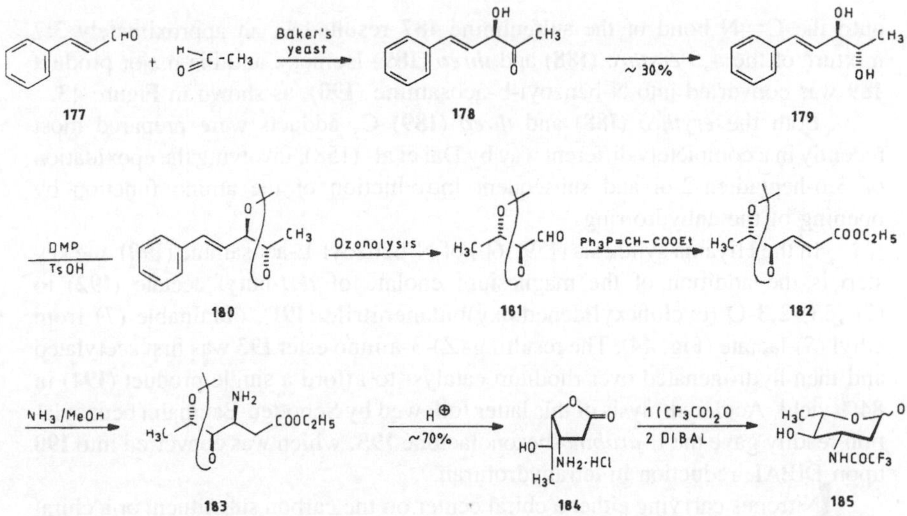

Fig. 42 Synthesis of N-trifluoroacetyl-L-acosamine based on the Baker's yeast-mediated acyloin condensation (Fuganti et al.).

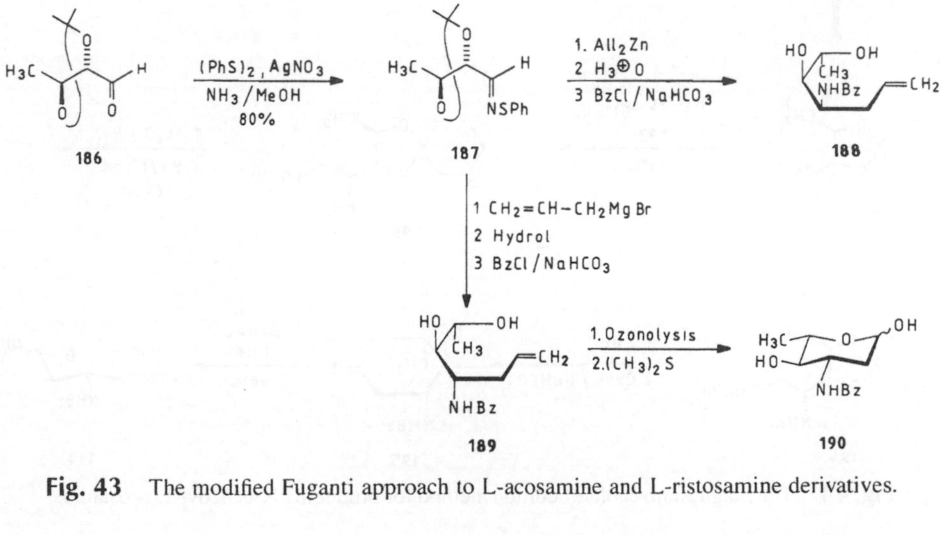

Fig. 43 The modified Fuganti approach to L-acosamine and L-ristosamine derivatives.

onto the C=N bond of the sulfenimine **187** resulted in an approximately 3:7 mixture of the 4,5-*erythro* (**188**) and *threo* (**189**) isomers, and the major product **189** was converted into N-benzoyl-L-acosamine (**190**), as shown in Figure 43.

Both the *erythro* (**188**) and *threo* (**189**) C_7 adducts were prepared most recently in a completely different way by Dai et al. (158), involving the epoxidation of 3,6-heptadien-2-ol and subsequent introduction of the amino function by opening of the anhydro ring.

In the Hiyama synthesis (159,160) of N-benzoyl-L-acosamine (**190**), the key step is the addition of the magnesium enolate of *tert*-butyl acetate (**192**) to (2*R*,3*S*)-2,3-O-(cyclohexylidenedioxy)butanenitrile (**191**), obtainable (7) from ethyl (*S*)-lactate (Fig. 44). The resulting (Z)-3-amino ester **193** was first acetylated and then hydrogenated over rhodium catalyst to afford a single product (**194**) in 84% yield. Acid hydrolysis of this latter followed by Schotten-Baumann benzoylation readily gave the L-*arabino*-hexonolactone **195**, which was converted into **190** upon DIBAL reduction in tetrahydrofuran.

Nitrones carrying either a chiral center on the carbon substituent or a chiral group attached to the nitrogen atom have been reported to display high diastereoselectivity in cycloaddition reactions with achiral dipolarophiles (161). The N—O bonding of these products can be readily split, furnishing derivatives with stereochemistry controlled by factors directing the cycloaddition reaction. This experience was utilized for the production of racemic daunosamine (161), as well as for the enantioselective synthesis (162,163) of L-acosamine and L-daunosamine.

In an intramolecular version of the nitrone cycloaddition reaction, the desired absolute stereochemistry of the three contiguous chiral centers (C_3–C_5) of L-acosamine could be induced (162,163) by the introduction of an (*S*)-α-methyl-

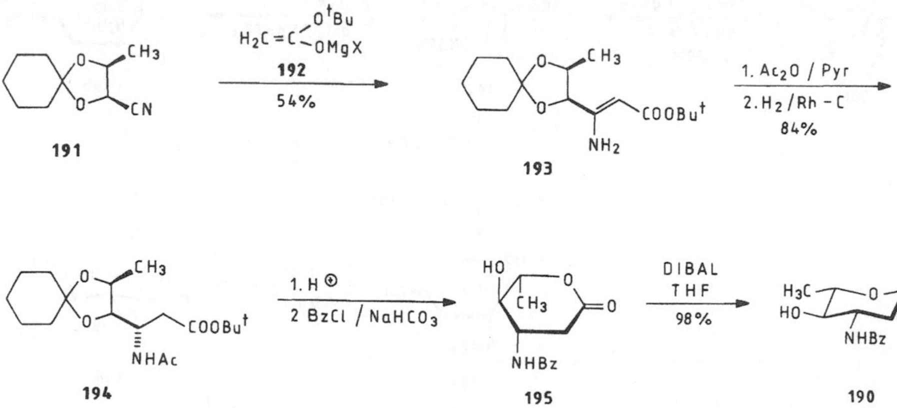

Fig. 44   The magnesium enolate addition method for producing N-benzoyl-L-acosamine.

benzyl chiral substituent to the nitrogen atom of **196**, resulting in the nitrone **197** (Fig. 45). Intramolecular cyclization of **197** gave the isoxazolidine **198** as the major product, with the proper stereochemistry. After reduction of the lactone function, removal of the chiral "directing moiety," and hydrogenolytic cleavage of the N—O bond, the intermediary furanose **199** was converted into methyl α-L-acosaminide **5**.

The Hanessian and Kloss (164) chiral strategy for the synthesis of each of the four stereoisomeric 3-amino-2,3,6-trideoxyhexoses in the L-sugar series, involving the base-catalyzed nitroaldol condensation of O-benzyl-L-lactaldehyde (**200**) with methyl 3-nitropropionate (**201**) in the presence of various catalysts, provided, after equilibration, a separable mixture of each of the four possible β-nitrohexano-ates (**202–205**), from which the desired aminodeoxy sugars, including N-benzoyl-L-acosamine (**190**) and N-benzoyl-L-ristosamine (**19**), were prepared upon sequential reduction, N-benzoylation, and removal of the protecting groups (Fig. 46).

By a similar strategy Suami and coworkers (165) also prepared these aminodeoxy sugars in rather low yields (because of an unnecessarily long (7) and tedious reaction sequence).

Other syntheses of acosamine derivatives from noncarbohydrate precursors include the Hart et al. (166, 167) method for the preparation of this sugar via β-lactam intermediates, and a synthesis by Japanese authors (168) of methyl N-acyl-L-acosaminide analogs by enolate-imine condensation of the lithium dianion of *tert*-butyl-(*S*)-3-hydroxybutanoate with N-acylaldimines. These methods are not discussed here.

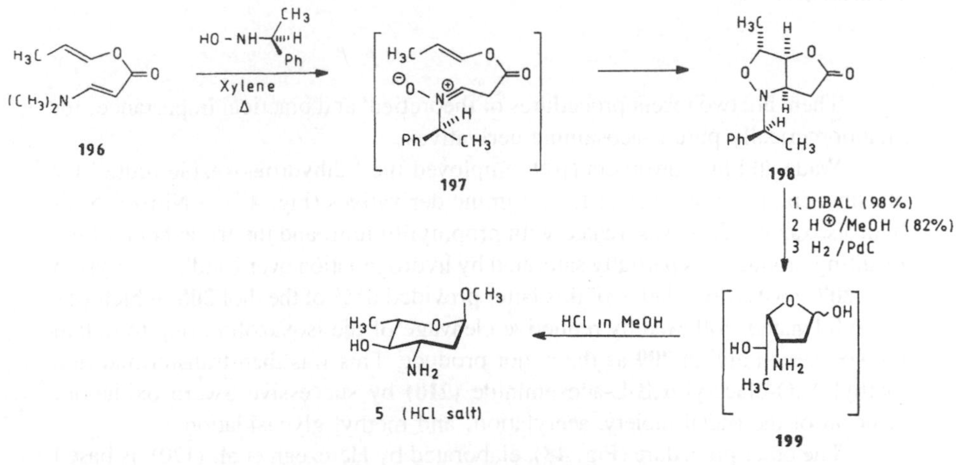

Fig. 45 The nitrone cycloaddition approach to L-acosamine.

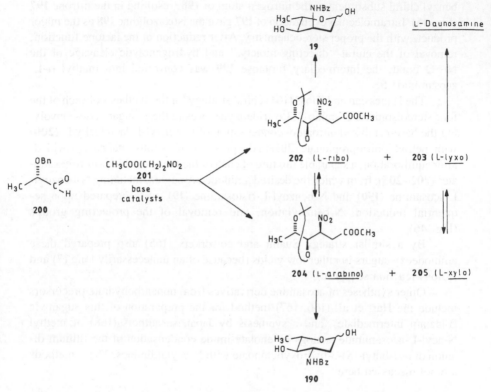

Fig. 46   Synthesis of the 3-amino-2,3,6-trideoxy-L-hexose stereoisomers via nitroaldol condensation.

There are two recent procedures of theoretical and practical importance, for enantiomerically pure L-acosamine derivatives.

Wade and his coworkers (169) employed the "dihydroisoxazole route" for constructing the molecules of L-acosamine derivatives (Fig. 47). 3-Nitro-4,5-dihydroisoxazole (**206**) was treated with propinyllithium, and the triple bond of the resulting product was partially saturated by hydrogenation over Lindlar catalyst to give **207**. cis-Hydroxylation of this latter provided 65% of the diol **208**, which was benzylidenated, followed by reductive cleavage of the isoxazoline ring to obtain the six-carbon alditol **209** as the major product. This was then transformed into methyl N,O-diacetyl-α,β-L-acosaminide (**210**) by successive Swern oxidation, removal of the acetal moiety, acetylation, and methyl glycosidation.

The other procedure (Fig. 48), elaborated by Herczegh et al. (170), is based on the preparation of (2R,3S)-tartraldehyde mercaptal (**212**) by the oxidation

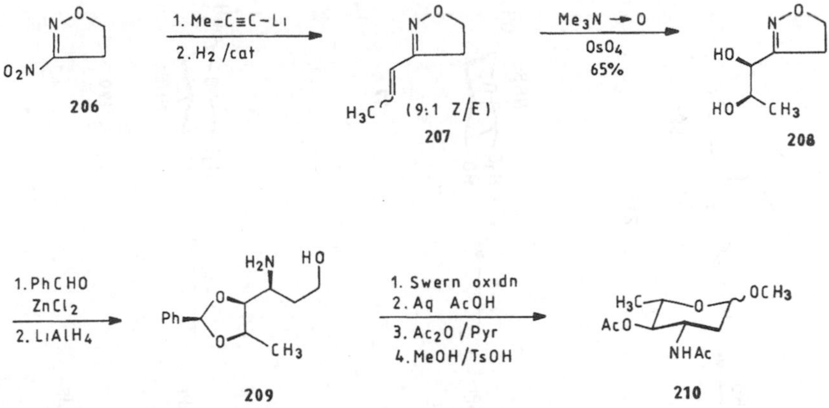

Fig. 47 The dihydroisoxazole route to L-acosamine derivatives.

of 2,3-O-isopropylidene-D-ribose mercaptal (**211**). After a two-carbon chain ascent the resulting α,β-unsaturated aldehyde **213** was subjected to conjugate addition of hydrazoic acid to give a 2:5 mixture (**214**) of the 3,4-*cis* and 3,4-*trans* addition products. Following acetylation and chromatographic separation the pure 3,4-*trans* adduct **216** was converted into N-benzoyl-L-acosamine (**190**) upon reduction of the azido function, reductive desulfurization, and acid hydrolysis. The minor 3,4-*cis* azido compound (**215**) was transformed into N-benzoyl-L-ristosamine (**19**).

L-Ristosamine (3-Amino-2,3,6-trideoxy-L-*ribo*-hexopyranose). The first noncarbohydrate-based syntheses of L-ristosamine derivatives were elaborated by the Fuganti group in 1979–1983.

In one of these approaches (2*R*,3*R*)-tartaric acid was converted (171) into the (2*R*, 3*S*)-aldehyde (**217**, Fig. 49). Wittig chain elongation of this aldehyde gave a rather poor yield of a 13:7 *E* to *Z* mixture (**218**) of hex-2-enoates, and this was converted into the L-*xylo* δ-lactone **219** upon sequential addition of ammonia, acid hydrolysis, and N-benzoylation (Figs. 39 and 42). Inversion of the configuration of this latter at C_4 upon treatment of the 4-O-mesylate **220** with potassium acetate in aqueous medium furnished (172) the N-benzoyl-L-*ribo*-γ-lactone **221**, which was then converted into N-benzoyl-L-ristosamine (**19**) by DIBAL reduction.

By applying a different strategy Fuganti et al. (156,157) prepared the 4*R*, 5*R*,6*S* seven-carbon compound **188** with a 4,5-*erythro* (relative) configuration (Fig. 43). Ozonolysis of **188** and subsequent treatment with dimethylsulfide then gave N-benzoyl-L-ristosamine (**19**, Fig. 50). [Note that **188** was most recently prepared (150) by an alternative route, as mentioned earlier.]

The Hirama group (150–152) extended the intramolecular Michael addition

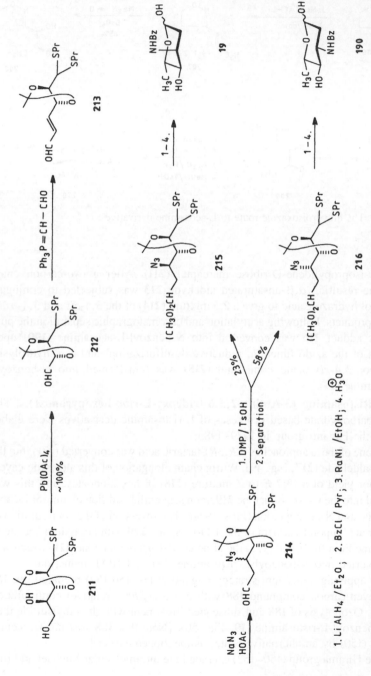

Fig. 48   Synthesis of N-benzoyl-L-acosamine and L-ristosamine from tartraldehyde mercaptals.

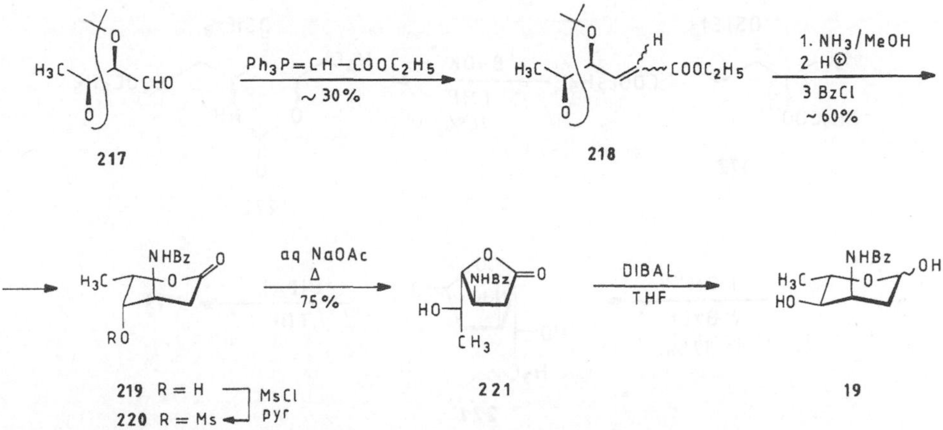

Fig. 49 The first noncarbohydrate-based synthesis of L-ristosamine.

shown in Fig. 40 for the synthesis of L-ristosamine derivatives (Fig. 51). Thus, (4R,5S)-methyl (E)-5-[(carbamoyl)oxy]-4-[(triethylsilyl)oxy]-2-hexenoate (**172**, Fig. 40) was converted (151) into the cyclic carbamate **222**, which upon alkaline hydrolysis and benzoylation gave 77% of the N-benzoyl-L-*ribo*-γ-lactone **221**. Conversion of this latter into N-benzoyl-L-ristosamine (**19**) was performed by DIBAL reduction, as already shown in Figure 49.

Two additional methods were also reviewed earlier in connection with the preparation of L-acosamine derivatives.

N-Benzoyl-L-ristosamine (**19**) could be readily obtained (164) from the L-*ribo* β-nitrohexanoate **202** (Fig. 46), and precursors to L-ristosamine (**15**) were prepared in a parallel work (173) by using an essentially similar method.

The sugar **19** was also synthesized (170) according to the reaction sequence shown in Figure 48 from the masked, functionalized dialdehyde **215** by applying the synthetic operations shown for the transformation of **216** into **190**.

Three additional papers (174–176) described methodologies for the produc-

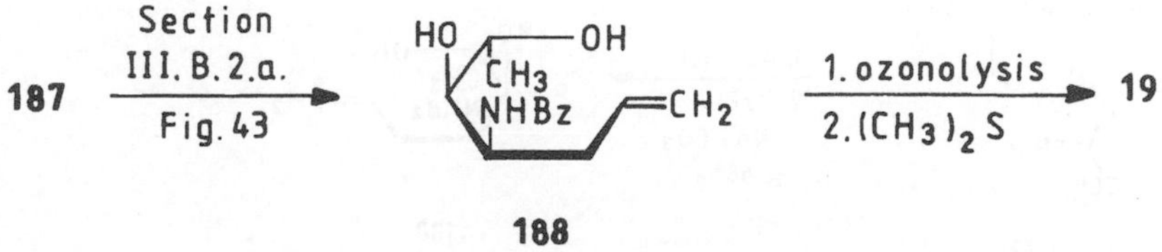

Fig. 50 A modified approach to N-benzoyl-L-ristosamine by Fuganti et al.

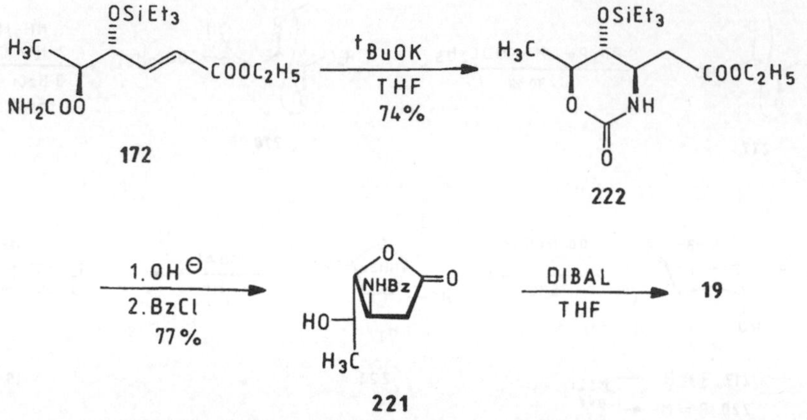

Fig. 51   Synthesis of N-benzoyl-L-ristosamine via intramolecular Michael addition.

tion of racemic ristosamine derivatives. Two procedures of theoretical importance (174,175) are summarized in recent review materials (6,7).

The third, more practical method (Fig. 52) (176) shows that epoxidation of the 4-aminoheptadiene **223**, obtainable in three steps from the readily available racemic 3,6-heptadien-2-ol with 3-chloroperbenzoic acid, is quite regio- and stereoselective, furnishing 42% of N-(*xylo*-5,6-epoxyhept-1-en-4-yl)trichloroacetamide (**224**) along with approximately 10% of each of the corresponding *lyxo*

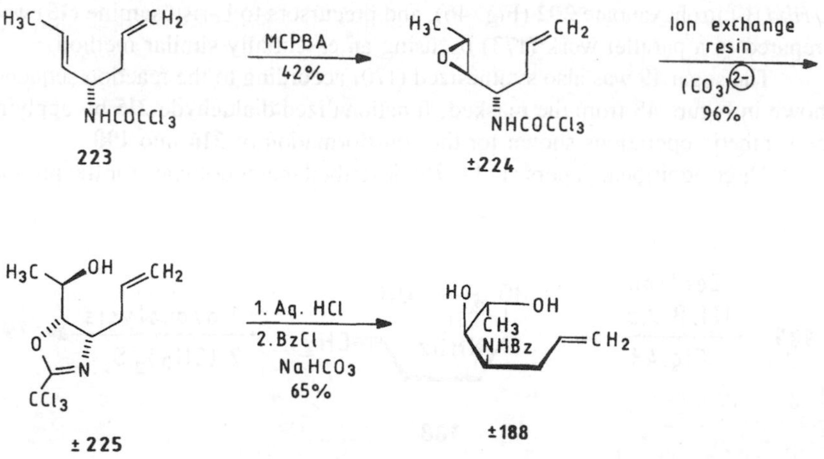

Fig. 52   Synthesis of precursors to racemic ristosamine derivatives.

isomer and a mixture of diepoxides. The α-opening of the anhydro ring of **224** gave the oxazoline **225**, which upon sequential hydrolysis, N-benzoylation, and one-carbon ozonolytic descent (Fig. 50) of the resulting **188** was converted into N-benzoyl-D,L-ristosamine (±**19**).

L-Vancosamine, 3-*epi*-L-Vancosamine, and 4-*epi*-L-Vancosamine (3-Amino-C$_3$-methyl-2,3,6-trideoxy-L-*lyxo*-, -L-*xylo*-, and -L-*arabino*-hexopyranose). An overview of the synthetic procedures and methodologies developed for these three amino sugars from non-carbohydrate precursors are summarized here.

In the course of their pioneering studies aimed at the preparation of C$_3$-methyl-branched aminodeoxy sugars, Dyong and coworkers (177,178) synthesized racemic vancosamine and 3-*epi*-vancosamine derivatives (Fig. 53). Thus, amination of DL-*trans*-3-methyl-4-hexenal-ethylene acetal (**226**) with bis(tosylimino)selenide gave a mixture of the *trans*-3-methyl-3-tosylamino-4-hexenal acetals **227** and **228**. After *cis*-hydroxylation with N-methylmorpholine-N-oxide in the presence of osmium tetroxide, the resulting 2:3 mixture of **229** and **230** was separated by HPLC and the individual components were transformed into methyl

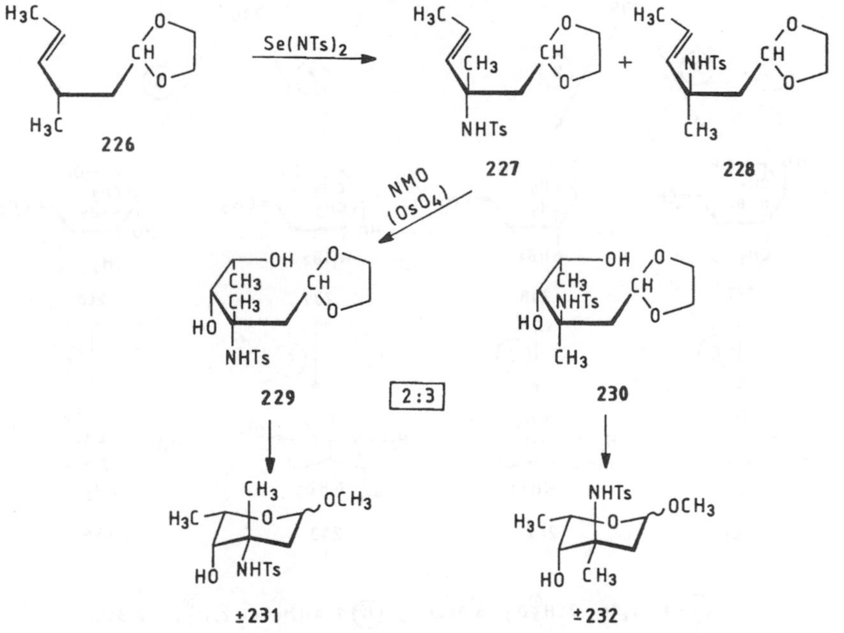

Fig. 53 The classic Dyong synthesis of racemic vancosamine and 3-*epi*-vancosamine derivatives.

N-tosyl-α,β-D,L-vancosaminide (±**231**) and its C_3 epimer (±**232**) upon hydrolysis and methyl glycosidation.

In subsequent studies Fronza et al. (179,180) succeeded in extending the method (Fig. 43) introduced for the preparation of acosamine and ristosamine to produce each of the four 3-amino-C_3 methyl-branched trideoxyhexoses in the L-sugar series (Fig. 54).

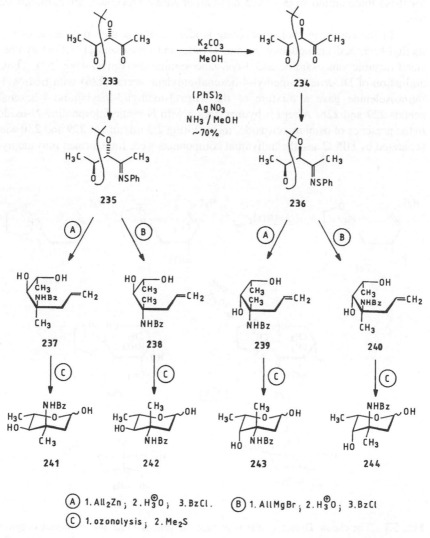

(A) 1.All$_2$Zn; 2.H$_3^{\oplus}$O; 3.BzCl.   (B) 1.AllMgBr; 2.H$_3^{\oplus}$O; 3.BzCl

(C) 1.ozonolysis; 2.Me$_2$S

Fig. 54   Preparation of the four 3-amino-C_3-methyl-2,3,6-trideoxy-L-hexoses.

Thus, the (3S,4S)-methyl ketone (**233**) (obtained by the baker's yeast-mediated acyloin condensation of α-methylcinnamaldehyde with acetaldehyde and subsequent ozonolysis) was isomerized into the C_3 epimer **234**, and both ketones were converted into the corresponding phenylsulfenimines **235** and **236**. As expected on the basis of previous experience (Fig. 43) (156,157), the reaction of **235** and **236** with diallyl zinc (A, Fig. 54) proceeded with high 4,5-*erythro* selectively to afford the C_3 methyl-branched C_7 adducts **237** and **239**, respectively. From the latter N-benzoyl-L-vancosamine (**243**) was obtained upon standard transformations (C).

In contrast, addition of allylmagnesium bromide (B) onto the C═N bond of the sulfenimines **235** and **236** occurred with preferential 4,5-*threo* selectivity, yielding **238** and **240**, which were converted into N-benzoyl-4-*epi*-L-vancosamine (**242**) and N-benzoyl-3-*epi*-L-vancosamine (**244**), respectively, upon ozonolysis and subsequent treatment with dimethylsulfide (C).

Another approach to L-vancosamine derivatives by Hamada et al. (181) is based on the production of an optically active 5-substituted-4-methoxycarbonyl-oxazole derivative (**246**) by the direct C-acylation of methyl isocyanoacetate with optically active carboxylic acids. By employing the protected L-lactic acid **245** (Fig. 55), the oxazole **246** was obtained in approximately 70% yield. Conversion of this latter to the unsaturated lactone **247** was achieved with hydrochloric acid in methanol, and then after N-protection, stereoselective hydrogenation over rhodium on alumina catalyst furnished the protected 2-amino-2,5-dideoxy-L-lyxono-1,4-lactone **248**, a joint intermediate of L-daunosamine and L-vancosamine.

For the production of this latter sugar, **248** was stereoselectively methylated and the resulting C_2-methyl lactone **249** was subjected to sequential lactone reduction and Wittig chain extension to obtain 1,3,4-triacetyl-α,β-L-vancosamine (**250**) after peracetylation.

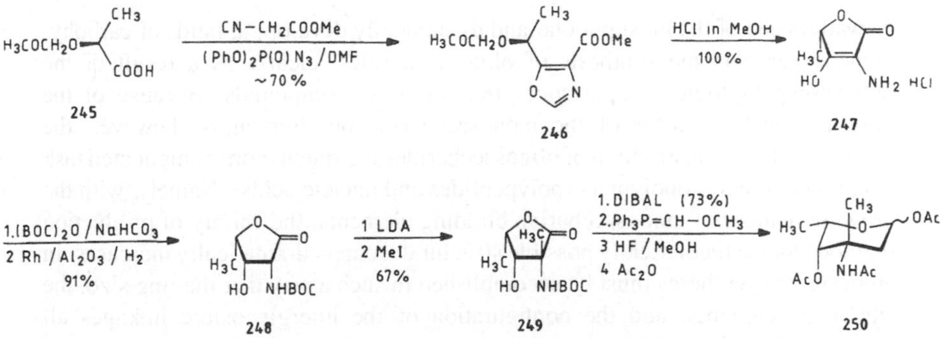

Fig. 55 The oxazole-based route to L-vancosamine derivatives.

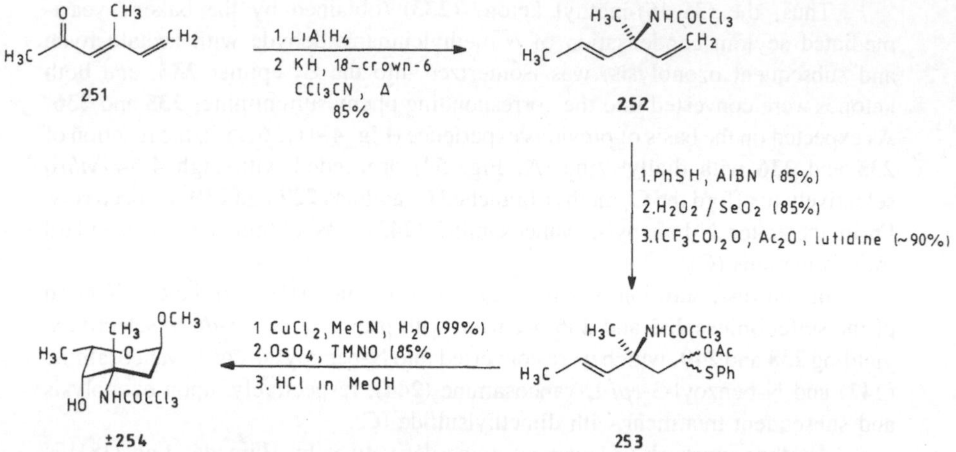

Fig. 56   Synthesis of racemic vancosamine derivatives from acyclic precursors.

In 1986 Hauser et al. (175) extended their earlier work on daunosamine synthesis for the preparation of racemic vancosamine derivatives. The dienone **251** was reduced with lithium aluminum hydride, and subsequent Overman reaction with trichloroacetonitrile furnished 85% of the trichloroacetamide **252** (Fig. 56). Following development of the latent aldehyde function by Pummerer rearrangement of **253**, hydroxylation of the *trans* double bond via the sequence shown in Figure 56 gave methyl N-trichloroacetyl-D,L-vancosaminide (**254**) in an overall yield of approximately 35%.

IV.   SYNTHESIS OF THE OLIGOSACCHARIDE SIDE CHAINS

Nowadays one of the most popular and dynamically developing fields of carbohydrate research is the synthesis of oligosaccharides, clearly as a result of the outstanding biological importance (182) of these compounds. Because of the polyfunctional character of the monosaccharide building units, however, the proper synthetic construction of oligosaccharides is a much more complicated task than that of other biopolymers (polypeptides and nucleic acids). Namely, with the growing number of monosaccharide building elements, the variety of production methods for the theoretically possible structural isomers dramatically increases. In addition, the syntheses must be accomplished in such a way that the ring size, the site of attachments, and the configuration of the interglycosidic linkages all correspond to the desired oligosaccharide structure.

Thus, a successful strategy leading to the required oligosaccharide should involve the following:

1. Preparation of a partially protected glycosyl acceptor, carrying a free hydroxyl group at the position where the glycosylation could be carried out
2. Production of a glycosyl donor with the required ring size and activated at the anomeric center
3. Proper choice of the reaction conditions and the promoter to enable a stereoselective coupling reaction between the glycosyl donor and the acceptor within the range of possibility
4. Removal of the temporary protecting groups of the synthesized oligosaccharide without damage (e.g., hydrolysis or isomerization) to the interglycosidic linkages

Despite numerous efforts in these fields there are still no generally applicable procedures for the synthesis (183) of oligosaccharides, and the production of almost all of them calls for individual synthetic strategies. This is clearly illustrated by the following examples for the synthesis of the oligosaccharide building elements of the antibiotics actinoidins, avoparcins, helvecardins, and ristocetins.

A. Heterodisaccharides

As mentioned in Section I, a common structural feature of certain representatives (such as actinoidins, avoparcins, helvecardins, orienticins, and vancomycin) of the glycopeptide antibiotics is that a heterodisaccharide side chain is linked to the central aromatic ring of the tricyclic vancomycinic acid (or monodechlorovancomycinic acid) portion of the heptapeptide aglycone. For the composition of these disaccharides the 2-O-(3-amino-2,3,6-trideoxy-α-L-hexopyranosyl)-β-D-glucopyranosyl structure is common (5), with the difference that the configuration of the aminodeoxy sugar may vary.

The first synthesis of such a disaccharide, phenyl β-acobioside (the heterodisaccharide component of the actinoidins), was reported by Sztaricskai and coworkers (184,185). For actinoidins A and B (255a and 255b, respectively, in Fig. 57) the aminodeoxy hexose unit of the disaccharide is L-acosamine (3-amino-2,3,6-trideoxy-L-*arabino*-hexopyranose).

For the preparation of the key intermediates of this procedure (Fig. 58), starting with D-glucose and L-rhamnose, the protecting groups were selected so that the target compound could be easily obtained upon deprotection. Thus, the readily available 3-O-benzyl-D-glucose (257) was first converted into 1,2,4,6-tetra-O-acetyl-3-O-benzyl-β-D-glucopyranose (258) and then into phenyl 2,4,6-tri-O-acetyl-3-O-benzyl-α- (259a) and β-D-glucopyranoside (259b) according to the Coleman procedure (186). Compound 259b, used in the next step, could be obtained in pure form by fractional crystallization. Zemplén saponification (187) of this compound afforded phenyl 3-O-benzyl-β-D-glucopyranoside (260), which was benzylidenated to give phenyl 3-O-benzyl-4,6-O-benzylidene-β-D-gluco-pyranoside (261), carrying a free hydroxyl group only at the site of development of the interglycosidic linkage in the following step.

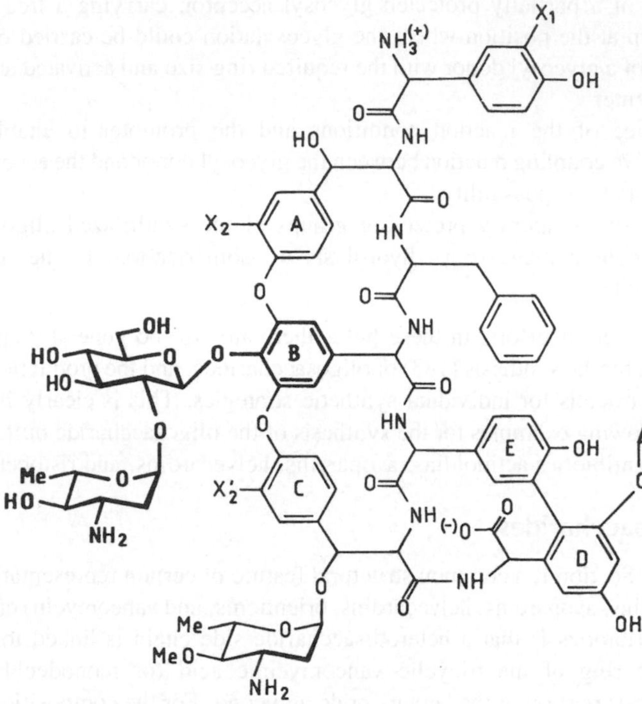

255 a   Actinoidin A : $X_1 = X_2 = H$, $X_2' = Cl$

255 b   Actinoidin B : $X_1 = X_2' = Cl$, $X_2 = H$

Fig. 57   Structural composition of the actinoidins.

The L-acosamine portion of the molecule of the disaccharide was synthesized according to the method of Gupta (90), and the methyl 3-azido-2,3,6-trideoxy-α-L-*arabino*-hexopyranoside (**60**) was 4-O-*para*-nitrobenzoylated to **262**. This glycoside was first converted into the free sugar **263** upon acid hydrolysis and then into 3-azido-1,4-di-O-(*p*-nitro-benzoyl)-2,3,6-trideoxy-L-*arabino*-hexopyranose (**264**), an intermediate for transforming (189) into the glycosyl chloride **265** by treatment with hydrochloric acid gas in dry dichloromethane.

Coupling of the glycosyl acceptor **261** with the glycosyl donor **265** was accomplished in the presence of silver triflate promoter according to the method of Hanessian and Banoub (187) to give a 4:1 mixture of **266a** and **266b**. At the same time, when the glycosylation was carried out with **264** and **261** and with tri-

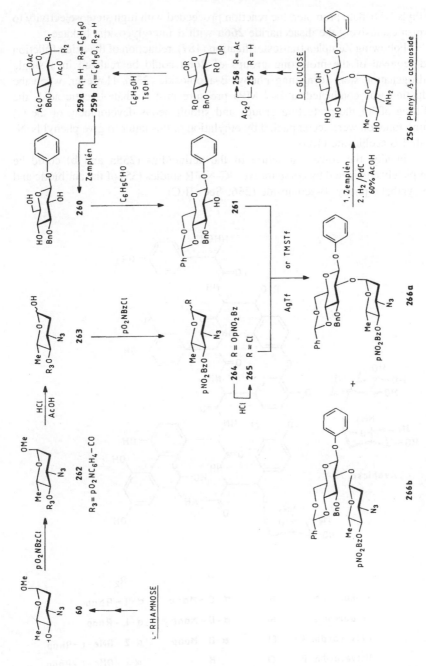

Fig. 58 Synthesis of phenyl β-acobioside.

methylsilyl'triflate promoter, the reaction proceeded with high stereoselectivity to furnish exclusively the disaccharide **266a** with α interglycosidic linkage.

Following Zemplén transesterification (187), reduction of the azido function and removal of the protecting groups of **266a** could be realized in a single hydrogenation step, furnishing phenyl β-acobioside (**256**). However, when the reduction was conducted at 2–7 MPa pressure in 4:1 ethanol-acetic acid, the splitting off of the protecting groups and simultaneous development of the C_3' amino function were accompanied by ethylation of this latter to give phenyl N,N-diethyl-β-acobioside (185).

Finally, the correct structure of the actinoidins (**255a** and **b**) could be completely elucidated by comparative ^{13}C-NMR studies (35) of the antibiotic and the synthetic phenyl β-acobioside (**256**, Sec. II.C).

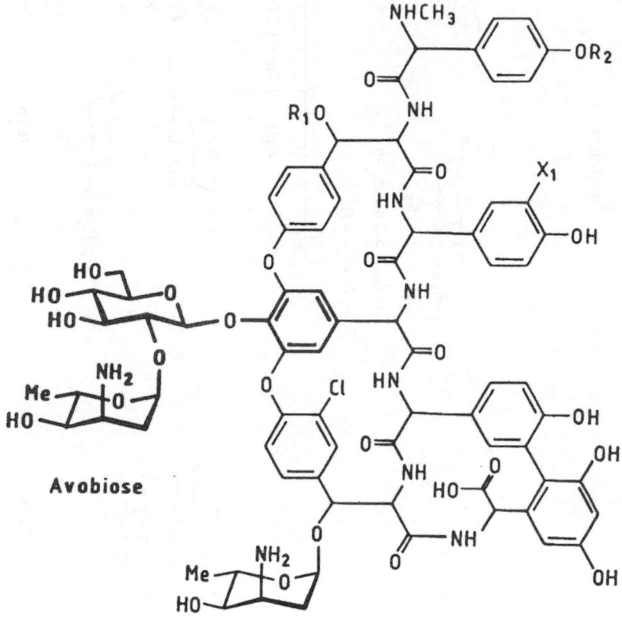

	X_1	R_1	R_2
Avoparcin α :	H	α-D-Manp	α-L-Rhap
Avoparcin β :	Cl	α-D-Manp	α-L-Rhap
Helvecardin A :	Cl	α-D-Manp	α-2-OMe-L-Rhap
Helvecardin B :	Cl	H	α-2-OMe-L-Rhap

Fig. 59   The structure of the avoparcins and helvecardins.

The composition of the disaccharide present in the avoparcins (44,45) and helvecardins (26) differs from that carried by the actinoidins only in that the former two antibiotics contain an L-ristosamine amino sugar component instead of L-acosamine (Fig. 59). Since the foregoing heterodisaccharide with the 2-O-(3-amino-2,3,6-trideoxy-α-L-*ribo*-hexopyranosyl)-β-D-glucopyranosyl structure was assigned, for the first time, from the avoparcins, it was named avobiose.

It appeared quite obvious that the synthetic strategy shown in Fig. 58 would be readily applicable also for the preparation (188) of phenyl β-avobioside (**277**, Fig. 61). However, the reaction of **261** (Fig. 60) with methyl 3-azido-4-O-(*p*-nitrobenzoyl)-2,3,6-trideoxy-α-L-*ribo*-hexopyranoside (**267**) (189) in dichloromethane in the presence of trimethylsilyl triflate gave exclusively phenyl 3-O-benzyl-4,6-O-benzylidene-2-O-(trimethylsilyl)-β-D-glucopyranoside (**268**). Zemplén transesterification of the latter led to the recovery of the glycosyl acceptor **261**. Glycosylation of **261** with the 1,4-di-O-(*p*-nitrobenzoyl)glycosyl donor **267** gave the protected disaccharide **269** with the undesired β-interglycosidic linkage. Zemplén O-deesterification of **269** then afforded **270**. Since the O-protecting groups of **270** could not be removed (upon catalytic hydrogenation or by treatment with trifluoroacetic acid or liquid ammonia) without extensive decomposition of the molecule, a new synthetic strategy (188) was elaborated by French and Hungarian researchers (Fig. 61). This novel procedure is completely different from the previously outlined approach and involves both another glycosyl donor and acceptor and a different glycosylation methodology.

D-Glucose was converted into the tri-O-acetyl-ortho-ester (**271**), and then the acetyl functions were replaced with benzyl-protecting groups. Refluxing of the resulting tri-O-benzyl-ortho ester **272** with phenol in chlorobenzene according to the method of Honma and Hamada (190) result in 91% phenyl 2-O-acetyl-3,4,6-tri-O-benzyl-β-D-glucopyranoside (**273**). O-Deacetylation of **273** afforded the desired glycosyl acceptor **274** having a free OH at C_2.

The glycosyl donor, 1,5-anhydro-2,3,6-trideoxy-4-O-(*p*-nitrobenzoyl)-3-trifluoroacetamido-L-*ribo*-hex-1-enitol (**275**), was synthesized (98) from 3,4-di-O-acetyl-L-rhamnal (**62**) in a four-step reaction sequence (including allylic azide addition, LAH reduction, N-trifluoroacetylation, and O-*p*-nitrobenzoylation).

Glycosylation of **274** with the L-*ribo* aminoglycal **275** in the presence of trimethylsilyl triflate promoter furnished exclusively the desired α-glycosidically linked protected heterodisaccharide (**276**) in 81% yield. After removal of the p-nitrobenzoyl and benzyl groups by successive treatment with 1 N aqueous sodium hydroxide and liquid ammonia, phenyl β-avobioside was isolated and identified as the tetra-O-acetyl derivative (**277**).

B. Heterotetrasaccharides

Among the known glycopeptide antibiotics only ristocetin (ristomycin) A has been shown (39,40) to contain a heterotetrasaccharide structural unit (Fig. 62), ristotetraose. The composition of this tetrasaccharide was elucidated according to the results of partial hydrolysis and acetolysis of the antibiotic, allowing the isolation and investigation of the sequence of attachment (Sec. II.C) of two reducing disaccharides (rutinose and ristobiose) and two reducing trisaccharides (ristotriose and ristriose). Since these compounds (except rutinose) were unknown in the literature, their structures were proved by syntheses elaborated by Sztaricskai and his coworkers (191).

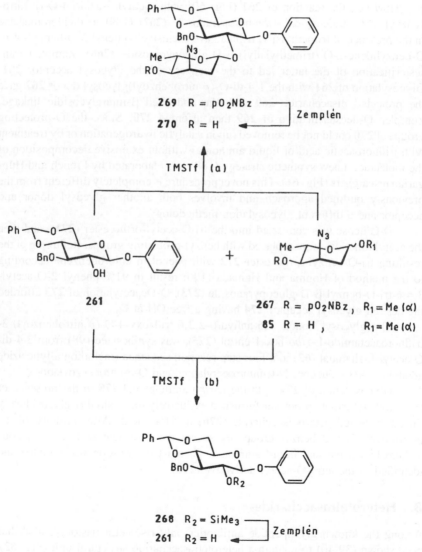

Fig. 60 Attempted synthesis of phenyl β-avobioside.

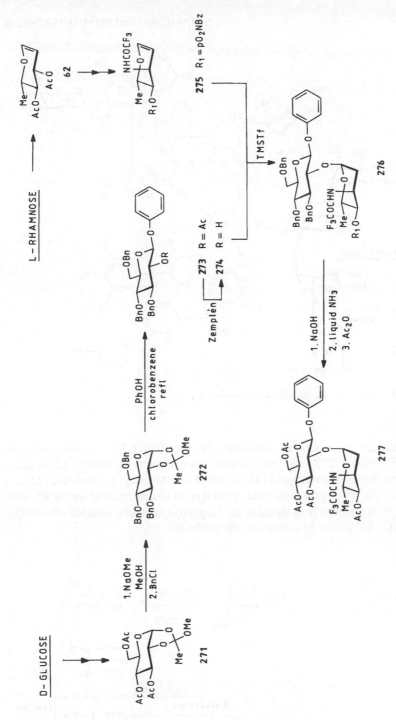

Fig. 61 Preparation of phenyl per-O-acetyl-N-trifluoroacetyl β-avobioside.

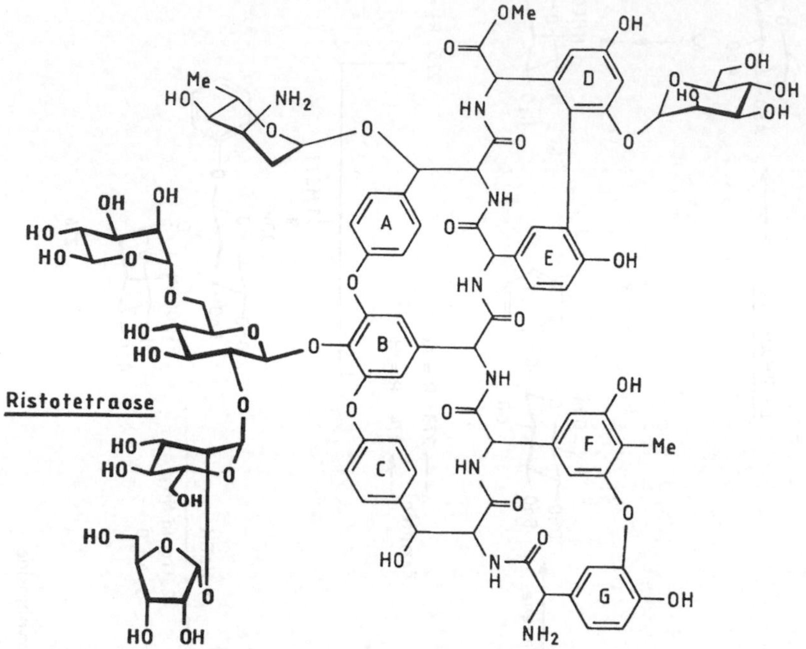

Fig. 62   Structure of ristocetin (ristomycin) A.

For the preparation of ristobiose (**39**, Fig. 63), 1,3,4,6-tetra-O-acetyl-α-D-glucopyranose (**278**) was reacted with α-acetobromo-D-mannose (**279**) under modified Helferich conditions (192) to obtain octa-O-acetyl-α-ristobiose (**42**) in a yield of 74%. This latter compound, as well as its O-deacetylated analog **39**, were shown to be completely identical to the respective authentic samples obtained by acetolysis and partial hydrolysis of the antibiotic.

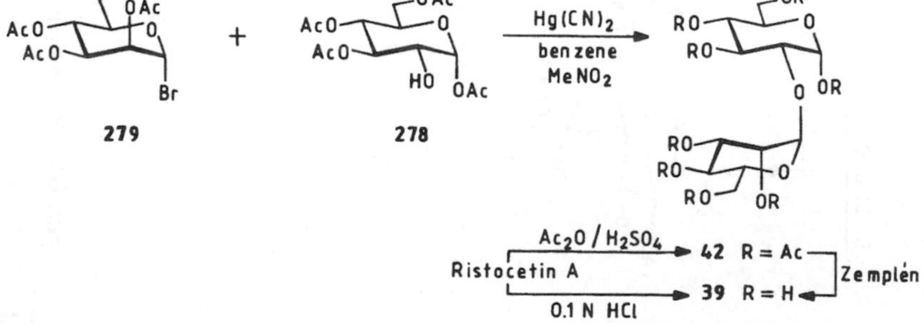

Fig. 63   Synthesis of ristobiose.

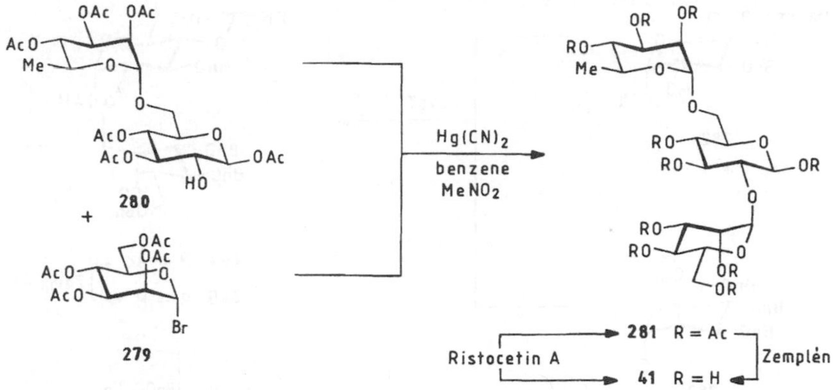

Fig. 64 Synthesis of ristotriose.

Hexa-O-acetyl-β-rutinose (**280**), the starting material for the synthesis (Fig. 64) of ristotriose, was prepared (191) from hexa-O-acetylrutinosyl chloride (193) according to the method of Helferich and Zirner (194). Mannosylation of **280** with **279** resulted in deca-O-acetyl ristotriose (**281**) (195) in 65% yield, whose Zemplén O-deacetylation (196) provided (191) ristotriose (**41**). The physico-chemical properties of **41** were in good agreement with those of an authentic sample derived from ristocetin (ristomycin) A. The structure and configuration of the interglycosidic linkages of both synthetic oligosaccharides (**39** and **41**) were unequivocally justified by ^{13}C-NMR spectroscopic measurements (197).

Since in the second trisaccharide (**43**) obtained by acetolysis (Fig. 11) and in the heterotetrasaccharide-containing fragment (**45**) produced upon reductive alkaline degradation (Fig. 12) D-arabinose appeared with different ring sizes (Sec. II.C), the definitive synthesis (198) of ristriose (Fig. 65) was deemed quite desirable.

Coupling of the glycosyl acceptor (**282**) prepared (199) from D-glucose with the glycosyl donor **283** (200) in the presence of silver triflate resulted in the protected disaccharide **284**, whose saponification with sodium methoxide afforded **285** in excellent yield. Glycosylation of the liberated C_2' hydroxyl group of **284** with 2,3,5-tri-O-benzoyl-D-arabinofuranosyl bromide (**286**) (201) under analogous conditions gave 91% of the fully protected ristriose (**287**). Following the removal of the benzylidene acetal function of **287** by hydrolysis with trifluoroacetic acid, the allyl glycosidic moiety was isomerized with tris(triphenylphospine)rhodium chloride (202) into prop-1-enyl, being easily removable upon treatment with mercuric chloride (203). Then O-debenzoylation and hydrogenolytic splitting off of the O-benzyl protecting groups furnished ristriose (**288**).

The isolation of **289** by selective removal of the benzylidene acetal function of **287** permitted the synthetic construction (Fig. 66) of ristotetraose, representing the entire heterotetrasaccharide side chain of the ristocetin (ristomycin) A molecule.

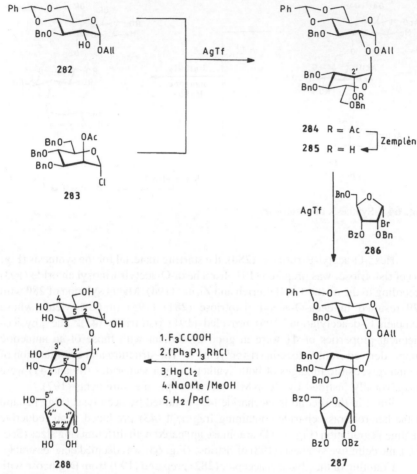

Fig. 65 Synthetic production of ristriose.

Knowing the difference between the reactivity of the primary and secondary hydroxyl groups of **289**, temporary protection of the C_4 hydroxyl group could be avoided. Thus, Koenigs-Knorr glycosylation of **289** with acetobromo-L-rhamnose (**290**) resulted in 50% heterotetrasaccharide **291**. Sequential removal of the acetyl and benzyl protecting groups by Zemplén transesterification (196) and catalytic hydrogenation at atmospheric pressure over a Pd catalyst then furnished the stable tetrasaccharide **292** with n-propyl-O-α-D-Araf-(1→2)-O-α-D-Manp-(1→2)-O-[α-L-Rhap]-(1→6)-O-α-D-Glcp structure (41,204). (Upon catalytic hydrogenation, the allyl function was reduced to the n-propyl moiety.)

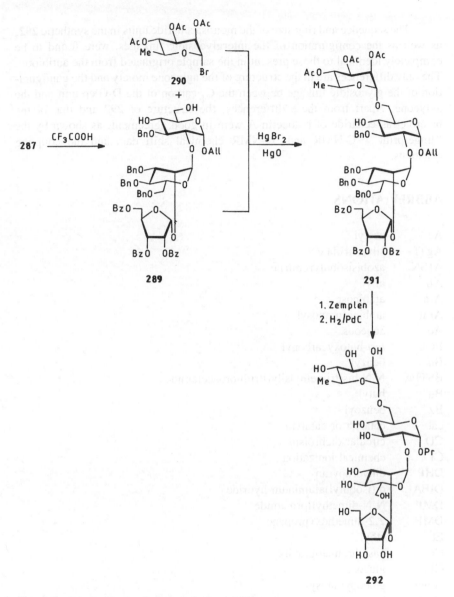

Fig. 66   Synthesis of *n*-propyl α-ristotetraoside.

The sequence and ring size of the monosaccharide units in the synthetic **292**, as well as the configuration of the interglycosidic linkages, were found to be completely identical to those present in the sample originated from the antibiotic. The only differences are in the structure of the aglycone moiety and the configuration of the glycosidic linkage between the C_1 carbon of the D-Glcp unit and the aglycone. Apart from these differences, the structure of **292** and that of the heterotetrasaccharide of ristocetin A were in good agreement, as shown by the "fingerprint" ^{13}C-NMR and ^{1}H-NMR chemical shift data and the coupling constants.

ABBREVIATIONS

Ac	acetyl
AgTf	silver triflate
AIBN	azobisisobutyronitrile
All	allyl
Ara	arabinose
Araf	arabinofuranosyl
Aq	aqueous
BOC	*tert*-butoxycarbonyl
Bn	benzyl
BSTFA	N,N-bis(trimethylsilyl)trifluoroacetamide
Bu	butyl
Bz	benzoyl
cat	catalyst or catalytic
CD	circular dichroism
CI	chemical ionization
DHP	dihydropyran
DIBAL	di(isobutyl)aluminum hydride
DMF	N,N-dimethylformamide
DMP	2,2-dimethoxypropane
Et	ethyl
GC	gas chromatography
Glu	glucose
Glucp	glucopyranosyl
HMPA	hexamethylphosphoric triamide
HPLC	high-performance liquid chromatography
LAH	lithium aluminum hydride
LDA	lithium dimethylamide
Man	mannose
Manp	mannopyranosyl

MCPBA	3-chloroperbenzoic acid
Me	methyl
Ms	methanesulfonyl
MS	mass spectrometry
NBS	N-bromosuccinimide
NIS	N-iodosuccinimide
NMO	N-methylmorpholine-N-oxide
NMR	nuclear magnetic resonance
pNO$_2$Bz	p-nitrobenzoyl
Ph	phenyl
PIV	pivaloyl
Pr	n-propyl
Pyr	pyridine or pyridinium
Refl	reflux
Rha	rhamnose
Rhap	rhamnopyranosyl
sec	secondary
$t/tert$	tertiary
THF	tetrahydrofuran
THP	tetrahydropyranyl
TMNO	trimethylamine-N-oxide
TMSCl	chlorotrimethylsilane
TMSTf	trimethylsilyl trifluoromethane sulfonate
Ts	p-toluenesulfonyl
TsOH	p-toluenesulfonic acid

ACKNOWLEDGMENTS

The authors are indebted to Miss Ibolya Kirják and Mr. Miklós Hornyák for the preparation of the manuscript and to Mrs. Erzsébet L. Magyar for the extensive artwork. Financial support for certain parts of the authors' research work on the scientific topics of this chapter was obtained from Grants TPB KKFA, OTKA 1728, and OTKA 298 from the Hungarian Academy of Sciences and the National Research Foundation (Hungary). The costs of preparation of the manuscript were covered by Grant OTKA 1181.

REFERENCES

1. Parenti F., Cavalleri B. (1990). Novel glycopeptide antibiotics of dalbaheptide group. Drugs Future 15:57.
2. Brandley B.K., Schnaar R.L. (1986). Cell-surface carbohydrates in cell recognition and response. J. Leukoc. Biol. 40:97.

3. Rao B.N., Moreland M., Brandley B.K. (1991). The potential of carbohydrates as cell adhesion inhibitors. Med. Chem. Res. 1:1.
4. Williams D.H., Rajanada V., Williamson M.P., Bojesen G. (1980). The vancomycin and ristocetin group of antibiotics. In (Sammes P. (Ed.), Topics in Antibiotics Chemistry. Ellis Horwood Limited, Chichester, p. 123.
5. Sztaricskai F., Bognár R. (1984). The chemistry of the vancomycin group of antibiotics. In Bognár R., Szántay C. (Eds.), Recent Developments in the Chemistry of Natural Carbon Compounds, Vol. 10. Akadémiai Kiadó, Budapest, p. 99.
6. Hauser F.M., Ellenberger S.R. (1986). Synthesis of 2,3,6-trideoxy-3-amino- and 2,3,6-trideoxy-3-nitrohexoses. Chem. Rev. 86:35.
7. Pelyvás I.F., Monneret C., Herczegh P. (1988). Synthetic Aspects of Aminodeoxy Sugars of Antibiotics. Springer-Verlag, Berlin.
8. Lomakina N.N., Spiridonova I.A., Sheinker Y.N., Vlasova T.F. (1973). Structure of the aminosugars from the antibiotic actinoidin (Russian). Khim. Prir. Soedein. 9:101.
9. Spiridonova I.A., Yurina M.S., Lomakina N.N., Sztaricskai F., Bognár R. (1976). Structure of the carbohydrate part of actinoidins A and B (Russian). Antibiotiki 21:304.
10. Bognár R., Sztaricskai F., Munk M.E., Tamás J. (1974). Structure and stereochemistry of ristosamine. J. Org. Chem. 39:2971.
11. Weringa W.D., Williams D.H., Feeney J., Borwn J.P., King R.W. (1972). The structure of an aminosugar from the antibiotic vancomycin. J. Chem. Soc. Perkin Trans. 1:443.
12. Smith R.M., Johnson A.W., Guthrie R.D. (1972). Vancosamine. A novel branched chain aminosugar from the antibiotic vancomycin. J. Chem. Soc. Chem. Commun. 361.
13. Johnson A.W., Smith R.M., Guthrie R.D. (1972). Vancosamine: The structure and configuration of a novel amino-sugar from vancomycin. J. Chem. Soc. Perkin Trans. 1:1972.
14. Debono M., Molloy R.M. (1980). Isolation and structure of the novel branched-chain aminosugar derived from antibiotic A-35512 B. J. Org. Chem. 45:4685.
15. Tsuji N., Kobayashi M., Kamigauchi T., Yoshimura Y., Terni Y.J. (1988). New glycopeptide antibiotics. I. The structures of orienticins. J. Antibiot. 41:819.
16. Borghi A., Coronelli C., Faninolo L., Allievi G., Pallanza R., Gallo G.G. (1984). Teichomycins, new antibiotics from Actinoplanes teichlomyceticus nov. sp. IV. Separation and characterization of the components of teichomycin (Teicoplanin). J. Antibiot. 37:615.
17. Barna J.C.J., Williams D.H., Stone D.J.M., Leung C.T-W., Doddrell D.M. (1984). Structure elucidation of the teicoplanin antibiotics. J. Am. Chem. Soc. 106:4895.
18. Sitrin R.D., Chan G.W., Dingerdissen J.J., Holl W., Hoover J.R.E., Valenta J.R., Webb L., Shader K.M. (1985). Aridicins, novel glycopeptide antibiotics. II. Isolation and characterization. J. Antibiot. 38:561.
19. Sitrin R.D., Chan G.W., Chapin F., Giovenella A.J., Grappel S.F., Jeffs P.W., Phillips L., Snader K.M., Nisbet L.J. (1986). Aridicins, novel glycopeptide antibiotics. III. Preparation, characterization and biological activities of aglycone derivatives. J. Antibiot. 39:68.

20. Jeffs P.W., Chan G., Sitrin R., Holder N., Roberts G.D., DeBrosse C. (1985). The structure of the glycolipid components of the aridicin antibiotic complex. J. Org. Chem. 50:1726.

21. Folena-Wasserman G., Poehland B.L., Yeung E.W-K., Staiger D., Killmer L.B., Snader K., Dingerdissen J.J., Jeffs P.W. (1986). Kibdelins (AAD-609), Novel glycopeptide antibiotics. II. Isolation, purification and structure. J. Antibiot. 19:1395.

22. Carr S.A., Mueller L., Shearer M.C., Christensen S.B. IV, Nisbet L.J., Sitrin R.D., Jeffs P.W., Roberts G.D. (1986). Antibiotics produced by *Actinomadura parvosata*. European Patent Appl. 0255258.

23. Waltho J.P., Williams D.H. (1987). Structure elucidation of the glycopeptide antibiotic complex A-40926. J. Chem. Soc. Perkin Trans. I:2103.

24. Williamson M.P., Williams D.H. (1981). Structure revision of the vancomycin. The use of nuclear Overhauser effect difference spectroscopy. J. Am. Chem. Soc. 103:6580.

25. Skelton N.J., Williams D.H. (1990). Structure elucidation of the novel glycopeptide antibiotic UK-68.597. J. Org. Chem. 55:3718.

26. Takeuchi M., Takahashi S., Inuka M., Nakamura T., Kinoshita T. (1991). Helvecardins A and B, novel glycopeptide antibiotics. II. Structural elucidation. J. Antibiot. 44:271.

27. Athalye M., Elson A., Gilpin M.L., Jeffries L.R. (1987). Antibiotic compounds. W.O. Patent 89/02441.

28. Skelton N.J., Williams D.H. (1990). Structure elucidation of UK-72.051, a novel member of vancomycin group of antibiotics. J. Chem. Soc. Perkin Trans. I:77.

29. Nagarajan R., Berry D.M., Hunt A.H., Occolowitz J.L., Schabel A.A. (1989). Conversion of antibiotic A-82846 B to orienticin A and structural relationships of related antibiotics. J. Org. Chem. 54:983.

30. Gause G.F., Brazhnikova M.G., Lomakina N.N., Berdnikova T.F., Fedorova G.B., Tokareva N.L., Borisova V.N., Batta G. (1989). Eremomycin—new glycopeptide antibiotic: Chemical properties and structure. J. Antibiot. 42:1790.

31. Chatterjee S., Vijakumar E.K.S., Chatterjee D.K., Ganguli B.N., Rupp R.H., Fehlhaber H.W., Kogler H., Seibert G., Teetz V. (1988). A novel glycopeptide antibiotic decaplanin a process for its production and its use as medicament. European Patent Appl. 0356894A.

32. Riva E., Gastaldo L., Beretta M.G., Ferrari P., Zerilli L.F., Cassani G., Selva E., Goldstein B.P., Berti M., Parenti F., Denaro M. (1989). A-42867, A novel glycopeptide antibiotic. J. Antibiot. 42:497.

33. Athalye M., Elson A.L., Gilpin M.L. (1988). Novel antibiotic compounds. W.O. Patent 89/07612.

34. Sztaricskai F., Harris C.M., Harris T.M. (1979). Structural investigation of the antibiotic actinoidin: Identification of the tris (aminoacid). Tetrahedron Lett. 2861.

35. Batta G., Sztaricskai F., Csanádi J., Komáromi I., Bognár R. (1986). ^{13}C-NMR study of actinoidins: Carbohydrate moieties and their glycosidic linkages. J. Antibiot. 39:910.

36. Athalye M., Coates N.J., Milner P.H. (1988). Antibiotic compounds. European Patent Appl. 0339982A$_1$.

37.   Dingerdissen J., Sitrin R.D., DePhillips P.A., Giorenella A.J., Grappel S.F., Mehta
      R.J., Oh Y.K., Pan C.H., Roberts G.D., Shearer M.C., Nisbet L.J. (1987). Acti-
      noidin A_2, a novel glycopeptide: Production, preparative HPLC separation and
      characterization. J. Antibiot. 40:165.
38.   Heald S.L., Mueller L., Jeffs P. (1987). Actinoidins A and A_2: Structure determina-
      tion using 2D NMR methods. J. Antibiot. 40:630.
39.   Sztaricskai F., Neszmélyi A., Bognár R. (1980). Structural investigation of the
      antibiotic ristomycin A: The ^{13}C-NMR evidence on the carbohydrate moieties and
      their linkages to the aglycone. Tetrahedron Lett. 21:2983.
40.   Sztaricskai F., Harris C.M., Neszmélyi A., Harris T.M. (1980). Structural studies of
      ristocetin A (ristomycin A): Carbohydrate-aglycone linkages. J. Am. Chem. Soc.
      702:7093.
41.   Takatsu T., Takahashi S., Nakajima M., Haneshi T., Nakamura T., Kuwaro H.,
      Kinoshita T. (1987). Chloropolysporins A, B and C, novel glycopeptide antibiotics
      from *Faenia interjecta* sp. nov. III. Structure elucidation of chloropolysporins. J.
      Antibiot. 40:933.
42.   Medakovic D., Batta G., Sztaricskai F., Harangi J. (1990). Proof of the structure of
      ristotetraose: Synthesis of propyl-α-ristotetraoside. Carbohydr. Res. 198:15.
43.   Box S.J., Coates N.J., Davis C.J., Gilpin M.L., Honge-Frydrych C.S.V., Milner P.H.
      (1991). MM-55266 and MM-55268, glycopeptide antibiotics produced by a new
      strain of *Amycolatopsis*. Isolation, purification and structure determination. J. Anti-
      biot. 44:807.
44.   Ellestad G.A., Leese R.A., Morton G.O., Barbatschi F., Gore W.E., McGahren W.J.
      (1981). Avoparcin and epiavoparcin. J. Am. Chem. Soc. 103:6522.
45.   Fesik S.W., Armitage I.M., Ellestad G.A., McGahren W.J. (1984). Nuclear mag-
      netic resonance studies on the antibiotic avoparcin. Conformation properties in
      relation to antibacterial activity. Mol. Pharmacol. 25:275.
46.   Sztaricskai F., Lomakina N.N., Spiridonova I.A., Yurina M.S., Puskás M.M.
      (1967). Quantitative determination of the carbohydrates of actinoidins (Russian).
      Antibiotiki 12:126.
47.   Philip J.E., Schenk J.R., Hargie M.P. (1956–1957). Ristocetins A and B two new
      antibiotics. Antibiot. Annu. 699.
48.   Philip J.E., Schenk J.R., Hargie M.P., Holper J.C., Grundy W.E. (1960). The
      increased activity of ristocetins A and B following acid hydrolysis. Antimicrob.
      Agents Annu. 10.
49.   Lomakina N.N., Muravyeva L.I., Puskás M.M. (1968). Ristomycin A and B.
      Carbohydrate moiety of the molecule (Russian). Antibiotiki 13:867.
50.   Higgins H.M., Harrison W.H., Wild G.M., Bungay M.R., McCormick M.H.
      (1957–1958). Vancomycin, a new antibiotic. Purification and properties of vanco-
      mycin. Antibiot. Annu. 906.
51.   Sztaricskai F. (1973). Unpublished results.
52.   Tsui N., Kobayashi M., Kamigauchi T., Yoshimura Y., Terni Y. (1988). New
      glycopeptide antibiotics. I. The structures of orienticins. J. Antibiot. 41:819.
53.   Cavalleri B., Pagani H., Volpe G., Selva E., Parenti F. (1984). A-16686, a new
      antibiotic from *Actinoplanes*, fermentation, isolation and preliminary physico-
      chemical characteristics. J. Antibiot. 37:309.

54. DeBono M., Molloy R.M., Barnhart M., Dorman D.E. (1980). A-35512, A complex of new antibacterial antibiotics produced by *Streptomyces candidus*. II. Chemical studies on A-35512 B. J. Antibiot. 33:1407.

55. Rosmus J., Deyl Z. (1971). Chromatographic methods in the analysis of protein structure. The methods for the identification of N-terminal amino acids in peptides and proteins. Chromatogr. Rev. 13:230.

56. Stoddart J.F. (1971). Stereochemistry of Carbohydrates. Wiley-Interscience, New York, p. 54.

57. Bognár R., Sztaricskai F., Munk M.E., Tamás J. (1974). Structure and stereochemistry of ristosamine (Hungarian). Magy. Kém. Foly. 80:385.

58. Reichstein R., Weiss E. (1962). The sugars of the cardiac glycosides. Adv. Carbohydr. Chem. 17:65.

59. Martin J.H., Ellestad G.A., Lovell F.M., Morton G.O., Hargreaves R.T., Perkinson N.A., Baker, J.L., Gamble J., McGahren W.J. (1981). A trimeric L-ristosamine-ammonia condensation product. Carbohydr. Res. 89:237.

60. Bukhari S.T.K., Guthrie R.D., Scott A.I., Wrixon A.D. (1970). Circular dichroism studies on cuprammonium complex of diols and amino-alcohols. Tetrahedron 26:3653.

61. Box S.J., Elson A.L., Gilpin M.L., Winstanley D.J. (1990). MM-47761 and MM-49721, glycopeptide antibiotics produced by a new strain of *Amycolatopsis orientalis*. Isolation, purification and structure determination. J. Antibiot. 43:931.

62. Lomakina N.N., Tokareva N.L., Potapova N.P. (1988). Structure of eremosamine, an amino sugar from antibiotic eremomycin (Russian). Antibiot. Chemother. (Moscow) 33:726.

63. Roberts P.J., Kennard O., Smith K.A., Williams D.H. (1973). Concerning the molecular weight and structure of the antibiotic vancomycin. J. Chem. Soc. Chem. Commun. 772.

64. Williams D.H., Kalman J.R. (1977). Structural and mode of action studies on the antibiotic vancomycin. Evidence from 270-MHz proton magnetic resonance. J. Am. Chem. Soc. 99:2768.

65. Takatsu T., Takahashi S., Takamatsu Y., Shioiri T., Iwado S., Haneishi T. (1987). Chloropolysporins A, B and C, novel glycopeptide antibiotics from *Faemia interjecta* sp. nov IV. Partially deglycosylated derivatives. J. Antibiot. 40:941.

66. Batta G., Sztaricskai F., Kövér K., Rüdel C., Berdnikova T.F. (1991). An NMR study of eremomycin and its derivatives. Full ^{1}H- and ^{13}C-NMR-assignment, motional behaviour, dimerization and complexation with Ac-D-Ala-D-Ala. J. Antibiot. 44: 1208.

67. Bochkov A.F., Zaikov G.E. (1979). Chemistry of the O-Glycosidic Bond: Formation and Cleavage. Pergamon Press, Oxford, p. 177.

68. Lomakina N.N., Bognár R., Brazhnikova M.G., Sztaricskai F., Muravyeva L.I. (1970). On the structure of ristomycin A (abstract). 7th International Symposium on the Chemistry of Natural Products, Zinate, Riga, p. 265.

69. Sztaricskai F., Bognár R., Puskás M.M. (1975). Structural investigation of the antibiotic ristomycin A. Chemical structure of the oligosaccharide part of ristomycin A. Acta Chim. Hung. (Budapest) 84:75.

70. Williamson M.P., Williams D.H. (1980). A ^{13}C-NMR study of the carbohydrate portion of ristocetin A. Tetrahedron Lett. 21:4187.

71. Williams D.H., Rajananda V., Bojesen G., Williamson M.P. (1979). Structure of the antibiotic ristocetin A. J. Chem. Soc. Chem. Commun. 906.

72. Hlavka J., Bitha P., Boothe J.H., Morton G. (1974). The partial structure of LL-AV290—a new antibiotic. Tetrahedron Lett. 175.

73. Kalman J.R., Williams D.H. (1980). An NMR study of the structure of the antibiotic ristocetin A. The negative nuclear Overhauser effect in structure elucidation. J. Am. Chem. Soc. 102:897.

74. Hunt A.H., Debono M., Merkel K.E., Barnhart M. (1984). Structure of pseudoaglycone of actaplanin. J. Org. Chem. 49:635.

75. Nielsen R.V., Hylding-Nielsen F., Jacobsen K. (1982). Biological properties of ristocetin Ψ-aglycone. J. Antibiot. 35:1561.

76. Pigman W., Horton D., Wander J.D. (Eds.) The Carbohydrates. Chemistry and Biochemistry (1980). Vol. 1.B. Academic Press, New York, pp. 727–728, 1023–1025.

77. Arcamone F. (1977). New antitumor anthracyclines. Lloydia 40:45.

78. Arcamone F. (1978). Daunomycin and related antibiotics. Top. Antibiot. Chem. 2:102.

79. Arcamone F. (1981). Doxorubicin. Anticancer Antibiotics. Academic Press, New York.

80. El Khadem H.S. (Ed.). Anthracycline Antibiotics. (1982). Academic Press, New York.

81. Yoshimura J. (1984). Synthesis of branched-chain sugars. Adv. Carbohydr. Chem. Biochem. 42:69.

82. McGarvey G.J., Kimura M., Oh T., Williams J.M. (1984). Acyclic stereoselective synthesis of carbohydrates. J. Carbohydr. Chem. 3:125.

83. Arcamone F., Penco S., Vigevani A., Redaelli S., Franchi G., DiMarco A., Casazza A.M., Dasdia T., Formelli F., Necco A., Soranzo C. (1975). Synthesis and antitumor properties of new glycosides of daunomycinone and adriamycinone. J. Med. Chem. 18:703.

84. Cassinelli G., Ruggieri D., Arcamone F. (1979). Synthesis and antitumor activity of 4'-O-methyldaunorubicin, 4'-O-methyladriamycin, and their 4'-epi analogues. J. Med. Chem. 22:121.

85. Sztaricskai F., Dinya Z., Batta G., Gergely L., Szabó B. (1992). Synthesis and anti-HIV activity of a new hexopyranoside analogue of AZT. J. Carbohydr. Nucleosides Nucleotides 11:11.

86. Lau J., Pedersen E.B., Nielsen C.M. (1991). Synthesis and evaluation of antiviral activity of L-acosamine and L-ristosamine nucleosides of furanose configuration. Acta Chem. Scand. 45:616.

87. Pelyvás I., Sztaricskai F., Bognár R. (1984). A novel approach to aminocyclitol analogues from azidotrideoxyhex-5-enopyranosides. J. Chem. Soc. Chem. Commun. 104.

88. László P., Pelyvás I.F., Sztaricskai F., Szilágyi L., Somogyi A. (1988). Novel aspects of the Ferrier carbocyclic ring-transformation reaction. Carbohydr. Res. 175:227.

89. Marsh J.P. Jr., Mosher C.W., Acton E.M., Goodman L. (1967). The synthesis of daunosamine. Chem. Commun. 973.

90. Gupta S.K. (1974). The synthesis of methyl 3-amino-2,3,6-trideoxy-α-L-*arabino*-hexopyranoside, a structural analog of daunosamine. Carbohydr. Res. 37:381.

91. Lee W.W., Wu H.Y., Christensen J.E., Goodman L., Henry D.W. (1975). Confirmation by synthesis of the structure of acosamine and methyl N-acetyl-actinosaminide. J. Med. Chem. 18:768.

92. Sztaricskai F., Pelyvás I., Szilágyi L., Bognár R., Tamás J., Neszmélyi A (1978). A synthesis of L-ristosamine and a derivative of its C-4 epimer. Carbohydr. Res. 65:193.

93. Monneret C., Gagnet R., Florent J-C. (1987). Regioselective alkylation and acylation of methyl 2,6-dideoxyhexopyranosides via their stannylene acetals: Key step for the synthesis of daunosamine and related amino-sugars. J. Carbohydr. Chem. 6:221.

94. Grethe G., Mitt T., Williams T.H., Uskokovic M.R. (1983). Synthesis of daunosamine. J. Org. Chem. 48:5309.

95. Grethe G., Sereno J., Williams T.H., Uskokovic M.R. (1983). Asymmetric synthesis of daunosamine. J. Org. Chem. 48:5315.

96. Martin A., Pais M., Monneret C. (1983). Synthese et réactivité vis-à-vis de réactifs nucléophiles des méthyl-3,4-anhydro-2,6-didésoxy-α- et β-L-*lyxo*- et -*ribo*-hexopyranosides. Carbohydr. Res. 113:189.

97. Heyns K., Lim M., Park J.I. (1976). Eine neue synthese von 3-amino-hexopyranosen. Tetrahedron Lett. 1477.

98. Boivin J., Pais M., Monneret C. (1980). Substitutions d'esters allyliques: Préparation de glycals aminés en C-3 et étude de leur glycosidation acido-catalysée. Application a l'hémisynthese de glycosides du groupe des anthracyclines. Carbohydr. Res. 79:193.

99. Boivin J., Montagnac A., Monneret C., Pais M. (1980). Hémisynthese de nonveaux glycosides analogues de la daunorubicine. Carbohydr. Res. 85:223.

100. Heyns K., Feldmann J., Hadamczyk D., Schwentner J., Thiem J. (1981). Ein verfahren zur synthese von α-glycosiden der 3-amino-2,3,6-tridesoxyhexopyranosen aus glycalen. Chem. Ber. 114:232.

101. Thiem J., Springer D. (1985). Synthesis and hydrogenation studies of 3-azidohex-2-eno-pyranosides, precursors of the sugar constituents of anthracycline glycosides. Carbohydr. Res. 136:325.

102. Springer D. (1985). Synthese von saccharideinheiten der anthrazyclin-antibiotika. Ph.D. Thesis University of Hamburg.

103. Boivin J. (1981). Synthesis and glycosylation of 2,6-dideoxyhexoses, application to semisynthesis of daunorubicin analogues. Ph.D. Thesis, University of Paris, Orsay.

104. Florent J-C., Monneret C. (1987). Stereocontrolled route to 3-amino-2,3,6-trideoxyhexopyranoses. K-10 Montmorillonite as a glycosidation reagent for acosaminide synthesis. J. Chem. Soc. Chem. Commun. 1171.

105. Abbaci B., Florent J-C., Monneret C. (1989). Addition of hydrazoic acid to pseudoglycals. Stereoselective synthesis of L-acosamine and L-daunosamine. Bull. Soc. Chim. Fr. 667.

106. Pelyvás I., Sztaricskai F., Bognár R. (1979). A convenient new route to methyl N,O-diacetyl-α-L-ristosaminide. Carbohydr. Res. 76:257.

107. Pelyvás I., Whistler R.L. (1980). Synthesis of L-acosamine and 1-thio-L-acosamine

derivatives by the stereoselective reduction of O-acetyloximes with borane. Carbohydr. Res. 84:C5.

108. Pelyvás I., Hasegawa A., Whistler R.L. (1986). Synthesis of N-trifluoroacetyl-L-acosamine, N-trifluoroacetyl-L-daunosamine, and their 1-thio analogs. Carbohydr. Res. 146:193.

109. Clode D.M., Horton D., Weckerle W. (1976). Reaction of derivatives of methyl 2,3-O-benzylidene-6-deoxy-α-L-mannopyranoside with butyllithium: Synthesis of methyl 2,6-dideoxy-4-O-methyl-α-L-erythro-hexopyranosid-3-ulose. Carbohydr. Res. 49:305.

110. Brimacombe J.S., Hanna R., Tucker L.C.N. (1985). Convenient synthesis of L-daunosamine and L-acosamine. Carbohydr. Res. 136:419.

111. Suarto A.S., Penco S., Vigevani A., Arcamone F. (1981). Anthracycline chemistry: Direct conversion of daunorubicin into the L-arabino, L-ribo, and L-xylo analogues, and selective deoxygenation at C-4'. Carbohydr. Res. 98:C1.

112. Sztaricskai F., Pelyvás I., Bognár R. (1975). The synthesis of N-benzoylristosamine. Tetrahedron Lett. 1111.

113. Lee W.W., Wu H.Y., Marsh J.J. Jr., Mosher C.W., Acton E.M., Goodman L., Henry DW (1975). Confirmation by synthesis of ristosamine as 3-amino-2,3,6-trideoxy-L-ribohexose. J. Med. Chem. 18:767.

114. Bartner P., Boxler D.L., Brambilla R., Mallams A.K., Morton J.B., Reichert P., Sancilio F.D., Surprenant H., Tomalesky G., Lukacs G., Olesker A., Thang T.T., Valente L. (1979). The megalomycins. Part 7. A structural revision by carbon-13 nuclear magnetic resonance and x-ray crystallography. Synthesis and conformational analysis of 3-dimethylamino- and 3-azido-D- and -L-hexopyranosides, and the crystal structure of 4"-O-(4-iodobenzoyl)megalomycin A. J. Chem. Soc. Perkin Trans. 1:1600.

115. Boivin J., Pais M., Monneret C. (1978). Synthese du méthyl-3-amino-2,3,6-tridesoxy-β-L-xylo-hexopyranoside. Carbohydr. Res. 64:271.

116. Richardson A.C. (1967). The synthesis of N-benzoyl-D-daunosamine. Carbohydr. Res. 4:422.

117. Arcamone F., Bargiotti A., Cassinelli G., Penco S., Hanessian H. (1976). Synthesis of a configurational analog of daunorubicin. Carbohydr. Res. 46:C3.

118. Brimacombe J.S., Hanna R., Saeed M.S., Tucker L.C.N. (1982). The reaction of derivatives of methyl 2,3-O-benzylidene-α-L-rhamnopyranoside with butyllithium. Carbohydr. Res. 100:C10.

119. Brimacombe J.S., Hanna R., Saeed M.S., Tucker L.C.N. (1982). Convenient syntheses of L-digitoxose, L-cymarose and L-ristosamine. J. Chem. Soc. Perkin Trans. 1:2583.

120. Bongini A., Cardillo G., Orena M., Sandri S., Tomasini C. (1983). A regio- and stereoselective synthesis of methyl α-L-ristosaminide hydrochloride. Tetrahedron 39:3801.

121. Pauls H.W., Fraser-Reid B. (1983). An efficient synthesis of ristosamine utilizing the allylic hydroxyl of an hex-2-enopyranoside. J. Org. Chem. 48:1392.

122. Pauls H.W., Fraser-Reid B. (1986). Stereocontrolled routes to cis-hydroxyamino sugars. Part VII. Synthesis of daunosamine and ristosamine. Carbohydr. Res. 150:111.

123. Sammes P.G., Thetford D. (1985). A new route to (±)-daunosamine and related amino-sugars. J. Chem. Soc. Perkin Trans. 1:352.

124. Sammes P.G., Thetford D. (1988). Synthesis of (L)-daunosamine and related amino sugars. J. Chem. Soc. Perkin Trans. I:111.

125. Pauls H.W., Fraser-Reid B. (1983). A short, efficient route to a protected daunosamine from L-rhamnose. J. Chem. Soc. Chem. Commun. 1031.

126. Cardillo G., Orena M., Sandri S., Tomasini C. (1984). Regio- and stereoselective synthesis of methyl α-L-daunosaminide hydrochloride. J. Org. Chem. 49:3951.

127. Giuliano R.M., Deisenroth T.W., Frank W.C. (1986). Synthesis of branched-chain nitro sugars. A stereoselective route to D-rubranitrose. J. Org. Chem. 51:2304.

128. Brimacombe J.S., Doner L.W. (1974). Branched-chain sugars. Part I. Characterization of methyl 2,4-di-O-acetyl-3,6-dideoxy-3-C-methyl-3-nitro-α-L-glucopyranoside and some of its chemical transformations. J. Chem. Soc. Perkin Trans. I:62.

129. Thang T.T., Winternitz F., Lagrange A., Olesker A., Lukacs G. (1980). Stereospecific access to branched-chain carbohydrate synthons. Tetrahedron Lett. 21:4495.

130. Brimacombe J.S., Miller J.A., Zakir U. (1975). The synthesis of branched-chain amino sugars from C-methylene sugars: A reassignment of structure. Carbohydr. Res. 44:C9.

131. Brimacombe J.S., Miller J.A., Zakir U. (1976). An approach to the synthesis of branched-chain amino sugars from C-methylene sugars. Carbohydr. Res. 49:233.

132. Thang T.T., Winternitz F., Olesker A., Lukacs G. (1979). Synthesis of a derivative of vancosamine, a component of the glycopeptide antibiotic vancomycin. J. Chem. Soc. Chem. Commun. 153.

133. Brimacombe J.S., Mengech A.S., Rahman K.M.M., Tucker L.C.N. (1982). An approach to branched-chain amino sugars, particularly derivatives of L-vancosamine (3-amino-2,3,6-trideoxy-3-C-methyl-L-lyxo-hexose) and its D enantiomer, via the cyanohydrin route. Carbohydr. Res. 110:207.

134. Brimacombe J.S., Rahman K.M.M. (1985). Branched-chain sugars. Part 17. A synthesis of L-rubranitrose (2,3,6-trideoxy-3-C-methyl-4-O-methyl-3-nitro-L-xylo-hexopyranose). J. Chem. Soc. Perkin Trans. I:1067.

135. Thang T.T., Imbach J.L., Fizames C., Lavelle F., Ponsinet G., Olesker A., Lukacs G. (1985). Synthesis and antitumor activity of 3'-C-methyl-daunorubicin. Carbohydr. Res. 135:241.

136. Klemer A., Wilbers H. (1987). Enolates of carbohydrates. 6. New syntheses of derivatives of 3-amino-2,3,6-trideoxy-3-C-methyl-L-xylo-hexopyranose, L-vancosamine, D-rubranitrose, and precursors of L-decilonitrose and D-kijanose. Liebigs Ann. Chem. 815.

137. Pelyvás I., Sztaricskai F., Szilágyi L., Bognár R., Tamás J. (1979). Stereoselective hydrogenation of methyl dideoxy- and trideoxy-β-D-hex-5-enopyranosides. Carbohydr. Res. 76:79.

138. Ahmad H.I., Brimacombe J.S., Mengech A.S., Tucker L.C.N. (1981). The synthesis of some derivatives of L-vancosamine (3-amino-2,3,6-trideoxy-3-C-methyl-L-lyxo-hexose). Carbohydr. Res. 93:288.

139. Brimacombe J.S., Mengech A.S. (1980). Branched-chain sugars. Part 10. Some

approaches to the synthesis of L-evernitrose (2,3,6-trideoxy-3-C,4-O-dimethyl-3-nitro-L-arabino-hexopyranose). J. Chem. Soc. Perkin Trans. 1:2054.

140. Yoshimura J., Yasumori T., Kondo T., Sato K. (1984). Branched-chain sugars. XXXV. The synthesis of L-rubranitrose (2,3,6-trideoxy-3-C-methyl-4-O-methyl-3-nitro-L-xylo-hexopyranose). Bull. Chem. Soc. Jpn. 57:2535.

141. Yasumori T., Sato K., Hashimoto H., Yoshimura J. (1984). Branched-chain sugars. XXXVI. A new synthesis of methyl 4-O-benzoyl-3-benzoyl-amino-2,3,6-trideoxy-3-C-methyl-α-L-xylo-hexo-pyranoside, a derivative of the branched-chain amino sugar of antibiotic A-35512 B. Bull. Chem. Soc. Jpn. 57:2538.

142. Brimacombe J.S., Hanna R., Tucker L.C.N. (1983). Branched-chain sugars. Part 16. The synthesis of a derivative of 3-amino-2,3,6-trideoxy-3-C-methyl-L-xylo-hexopyranose, the novel branched-chain amino sugar of antibiotic A-35512 B. J. Chem. Soc. Perkin Trans. 1:2277.

143. Brimacombe J.S., Mengech A.S., Saeed M.S. (1979). Some approaches to the synthesis of evernitrose and its enantiomer. Carbohydr. Res. 75:C5.

144. Jütten P., Scharf H-D. (1991). An improved synthesis of evernitrose. Carbohydr. Res. 212:93.

145. Yoshimura J., Matsuzawa M., Sato K., Nagasawa Y. (1979). The synthesis of evernitrose and 3-epi-evernitrose. Carbohydr. Res. 76:67.

146. Ishii K., Nishimura Y., Kondo S., Umezawa H. (1983). Decilonitrose and 4-O-succinyl-L-diginose, sugar components of decilorubicin. J. Antibiot. 36:454.

147. Dyong I., Bendlin H. (1978). Synthesis of N-acetyl-L-acosamine (3-acetylamino-2,3,6-trideoxy-L-arabino-hexose). Chem. Ber. 111:1677.

148. Welch J.T., Svahn B-M., Eswarakrishnan S., Hutchinson J.P., Zubieta J. (1984). The synthesis of 3-acetamido-2,3,5,6-tetradeoxy-5-fluoro-D,L-ribo-hexofuranose by the direct fluorination of methyl 3-acetamido-2,3,6-trideoxy-D,L-arabino-hexopyrano-side (methyl N-acetyl-D,L-acosaminide). Carbohydr. Res. 132:221.

149. Welch J.T., Svahn B-M. (1985). Synthesis and x-ray crystal structure of methyl D,L-desmethyl-holantosaminide. J. Carbohydr. Chem. 4:421.

150. Hirama M., Shigemoto T., Yamazaki Y., Ito S. (1985). Diastereoselective synthesis of N-acetyl-D,L-acosamine and N-benzoyl-D,L-ristosamine. Tetrahedron Lett. 26:4133.

151. Hirama M., Shigemoto T., Ito S. (1987). Stereodivergent total synthesis of N-acetyl-acosamine and N-benzoylristosamine. J. Org. Chem. 52:3342.

152. Hirama M., Ito S. (1989). Asymmetric induction in the intramolecular conjugate addition of γ- and δ-carbamoyl-α,β-unsaturated esters. A new method for diastereoselective amination and divergent syntheses of 3-amino-2,3,6-trideoxyhexoses. Heterocycles 28:1989.

153. Kametani T., Suzuki Y., Ban C., Kanada K., Honda T. (1987). A formal synthesis of N-acetyl-L-acosamine. Heterocycles 26:1789.

154. Trost B.M., Sudhakar A.R. (1987). A cis hydroxyamination equivalent: Application to the synthesis of (−)-acosamine. J. Am. Chem. Soc. 109:3792.

155. Fronza G., Fuganti C., Grasselli P. (1980). Synthesis of N-trifluoroacetyl-L-acosamine and -L-daunosamine. J. Chem. Soc. Chem. Commun. 442.

156. Fuganti C., Grasselli P., Pedrocchi-Fantoni G. (1981). Synthesis of N-benzoyl-L-daunosamine from D-threonine. Tetrahedron Lett. 22:4017.

157. Fuganti C., Grasselli P., Pedrocchi-Fantoni G. (1983). Stereospecific synthesis of N-benzoyl-L-daunosamine and L-ristosamine. J. Org. Chem. 48:909.

158. Dai L., Lon B., Zhang Y. (1988). A simple, divergent, asymmetric synthesis of all members of the 2,3,6-trideoxy-3-aminohexose family. J. Am. Chem. Soc. 110:5195.

159. Hiyama T., Nishide K., Kobayashi K. (1984). A new synthesis of N-benzoyl-L-acosamine. Tetrahedron Lett. 25:569.

160. Hiyama T., Kobayashi K., Nishide K. (1987). A new synthesis of 3-amino-2-alkenoates. Novel synthetic route to amino sugars N-benzoyl-L-daunosamine and -L-acosamine. Bull. Chem. Soc. Jpn. 60:2127.

161. DeShong P., Leginus J.M. (1983). Stereospecific synthesis of racemic daunosamine. Diastereofacial selectivity in a nitrone cycloaddition. J. Am. Chem. Soc. 105:1686.

162. Wovkulich P.M., Uskokovic M.R. (1981). A chiral synthesis of L-acosamine and L-daunosamine via an enantioselective [3 + 2] cycloaddition. J. Am. Chem. Soc. 103:3956.

163. Wovkulich P.M., Uskokovic M.R. (1985). Total synthesis of acosamine and daunosamine utilizing a diastereoselective intramolecular [3 + 2] cycloaddition. Tetrahedron 41:3455.

164. Hanessian S., Kloss J. (1985). Total synthesis of biologically important amino sugars via the nitroaldol reaction. Tetrahedron Lett. 26:1261.

165. Suami T., Tadano K., Suga A., Ueno Y. (1984). An alternative synthesis of acosamine and ristosamine. J. Carbohydr. Chem. 3:429.

166. Ha D-C., Hart D. (1987). Syntheses of methyl-N-benzylacosaminide and methyl-N-(benzoyloxalyl)-L-daunosaminide from (S)-ethyl-3-hydroxybutyrate. Tetrahedron Lett. 28:4489.

167. Galucci J.C., Ha D-C., Hart D.J. (1989). Preparation of aminosaccharides using ester-imine condensations: Syntheses of methyl N-benzoylacosaminide and methyl N-[oxo(phenylmethoxy)acetyl]daunosaminide from (S)-ethyl-3-hydroxybutyrate. Tetrahedron 45:1283.

168. Hatanaka M., Ueda I. (1991). A new approach to L-daunosamine and L-acosamine. Chem. Lett. 61.

169. Wade P.A., Appa Rao J., Bereznak J.F., Yuan C-K. (1989). A dihydroxyisoxazole-based route to 2,3,6-trideoxy-3-aminohexose derivatives. Tetrahedron Lett. 30:5969.

170. Kovács I., Herczegh P., Sztaricskai F.J. (1991). Tartraldehydes. III. Synthesis of N-benzoylristosamine and -L-acosamine. Tetrahedron 47:7837.

171. Fronza G., Fuganti C., Grasselli P., Marioni G. (1979). Synthesis of N-benzoyl-L- and -D-2,3,6-trideoxy-3-amino-xylo-hexose from noncarbohydrate precursors. Tetrahedron Lett. 20:3883.

172. Fronza G., Fuganti C., Grasselli P. (1980). Synthesis of N-benzoyl-L-ristosamine. Tetrahedron Lett. 21:2999.

173. Brandänge S., Lindqvist B. (1985). Synthesis of precursors to L-ristosamine and L-daunosamine from L-lactic acid. Acta Chem. Scand. 39B:589.

174. Heatcock C.H., Montgomery S.H. (1983). Acyclic stereoselection. 19. Total synthesis of (±)-ristosamine and (±) megalosamine. Tetrahedron Lett. 24:4637.

175. Hauser F.M., Ellenberger S.R. (1986). Stereoselective syntheses of (±)daunosamine, (±)vancosamine, and (±)ristosamine from acyclic precursors. J. Org. Chem. 51:50.

176.   Roush W.R., Straub J.A., Brown R.J. (1987). Stereochemistry of the epoxidations of acyclic allylic amides. Applications toward the synthesis of 2,3,6-trideoxy-3-aminohexoses. J. Org. Chem. 52:5127.

177.   Dyong I., Friege H. (1979). Synthesis of biologically important carbohydrates 20. N-Acetyl-1,4-di-O-acetyl-β-D,L-vancosamine. Chem. Ber. 112:3273.

178.   Dyong I., Friege H., Luftmann H., Merten H. (1981). Synthesis of biologically important carbohydrates. 27. Total synthesis and determination of the configuration of Me-C³-NHR-branched 2,3,6-trideoxyhexoses. Chem. Ber. 114:2669.

179.   Fronza G., Fuganti C., Grasselli P., Pedrocchi-Fantoni G. (1981). Synthesis of the N-benzoyl derivatives of L-arabino, L-xylo and L-lyxo (L-vancosamine) isomers of 2,3,6-trideoxy-3-C-methyl-3-aminohexoses from a non-carbohydrate precursor. Tetrahedron Lett. 22:5073.

180.   Fronza G., Fuganti C., Grasselli P., Pedrocchi-Fantoni G. (1983). Synthesis of the four configurational isomers of N-benzoyl-2,3,6-trideoxy-3-C-methyl-3-amino-L-hexose from the (2S,3R)-diol obtained from α-methylcinnamaldehyde by fermentation with baker's yeast. J. Carbohydr. Chem. 2:225.

181.   Hamada Y., Kawai A., Shioiri T. (1984). New methods and reagents in organic synthesis. 50. A stereoselective synthesis of L-vancosamine, a carbohydrate component of the antibiotics vancomycin and sporaviridin. Tetrahedron Lett. 25:5413.

182.   Sharon N. (1975). Complex Carbohydrates, Their Chemistry, Biosynthesis and Functions. Addison Westley, Reading.

183.   Paulsen H. (1982). Fortschritte bei der selectiver chemischen synthese komplexer oligosaccharide. Angew. Chem. 94:184.

184.   Sztaricskai F., Csanádi J., Batta G., Dinya Z., Bognár R. (1985). Synthesis of the disaccharide components of vancomycin-type antibiotics. Phenyl-β-acobioside and its N,N-diethyl derivative (Hungarian). Magy. Kém. Foly. 91:487.

185.   Csanádi J., Sztaricskai F., Batta G., Dinya Z., Bognár R. (1986). Synthesis of phenyl β-acobioside, a derivative of the disaccharide component of actinoidins. Carbohydr. Res. 147:211.

186.   Coleman G.H. (1963). 1,6-Anhydro-β-D-glucopyranose triacetate (levoglucosan triacetate). Methods Carbohydr. Chem. 2:397.

187.   Hanessian S., Banoub J. (1980). Preparation of 1,2-trans-glycosides in the presence of silver trifluoromethanesulfonate. Methods Carbohydr. Chem. 8:247.

188.   Dong X., Tillequin F., Monneret C., Horváth G., Sztaricskai F. (1992). Synthesis of a phenyl β-avobioside derivative of the disaccharide component of avoparcins. Carbohydr. Res. 232:107.

189.   Sztaricskai F., Menyhárt M., Bognár R. (1982). 7-O-(3-Azido-2,3,6-trideoxy α- and β-L-ribo-hexopyranosyl)carminomycinone: Novel analogue of anthracycline antibiotics. Carbohydr. Res. 100:C14.

190.   Honma K., Hamada A. (1976). Studies on glycosylation. III. A novel, stereospecific synthesis of 1-O-acyl- and 1-aryl-β-D-glucopyranose tetraacetates via the 1,2,-t-butylorthoacetate. Chem. Pharm. Bull. 24:1165.

191.   Sztaricskai F., Lipták A., Pelyvás I., Bognár R. (1976). Structural investigation of the antibiotic ristomycin A. Synthesis of ristobiose and ristotriose. J. Antibiot. 29:626.

192. Helferich B.. Weiss K. (1956). Concerning the synthesis of glucosides and non reducing disaccharides (German). Chem. Ber. 89:314.

193. Bognár R.. Szabó I.F.. Farkas I.. Gross H. (1967). Cleavage of oligosaccharide glycosides by means of dihalomethyl ethers: A novel preparation of rutinose. Carbohydr. Res. 5:241.

194. Helferich B.. Zirner J. (1963). 1,3,6,2',3',6'-Heptaacetyl-α-D-cellobiose. Chem. Ber. 96:385.

195. Lipták A.. Függedi P.. Szurmai Z. (1990). CRC Handbook of Oligosaccharides. Trisaccharides. Vol. 2. CRC Press. Boca Raton, FL, p. 88.

196. Zemplén G.. Gerecs A.. Hadácsi I. (1936). Saponification of acetylated carbohydrates (German). Chem. Ber. 69:1827.

197. Neszmélyi A., Sztaricskai F., Lipták A., Bognár R. (1978). Structural investigation of the antibiotic ristomycin A. ^{13}C-NMR spectral analysis of the interglycosidic linkages of the heterotetrasaccharide side-chain. J. Antibiot. 31:974.

198. Medakovic D.. Batta G.. Pelyvás I.F.. Sztaricskai F. (1990). Synthesis of ristriose. J. Carbohydr. Chem. 9:631.

199. Gent P.A.. Gigg R. (1976). Synthesis of benzyl and allyl ethers of D-glucopyranose. Carbohydr. Res. 49:325.

200. Garegg P.J.. Maron L. (1979). Improved synthesis of 3,4,6-tri-O-benzyl-α-D-mannopyranosides. Acta Chem. Scand. 33B:39.

201. Ness R.K.. Fletcher H.G. Jr. (1958). The anomeric 2,3,5-tri-O-benzoyl-D-arabinosyl bromides and other D-arabinofuranose derivatives. J. Am. Chem. Soc. 80:2007.

202. Corey P.A.. Suggs J.W. (1973). Selective cleavage of allyl ethers under mild conditions by transition metal reagents. J. Org. Chem. 38:3224.

203. Gigg R.. Warren C.D. (1968). The allyl ether as a protecting group in carbohydrate chemistry. Part II. J. Chem. Soc. (C) 1903.

204. Sztaricskai F.. Batta G.. Harangi J.. Medakovic D. (1989). Structure proof of ristotetraose by definitive synthesis. Preparation of n-propyl α-ristotetraoside (Hungarian). Magy. Kém. Foly. 95:382.

5
Structure–Activity Relationships of Vancomycin Antibiotics

RAMAKRISHNAN NAGARAJAN
Lilly Research Laboratories, Eli Lilly and Company, Indianapolis, Indiana

I. INTRODUCTION

A. General Considerations

Vancomycin (1) has been marketed for the past 35 years, as the hydrochloride, to treat deep-seated Gram-positive infections. It is the drug of choice for infections caused by *Staphylococcus aureus*, especially those caused by the methicillin-resistant and, more importantly the multiresistant strains. Chapter 9 covers the medical aspects of vancomycin in greater detail. Vancomycin is bactericidal to most Gram-positive organisms, but Gram-negative organisms are resistant. Vancomycin is not absorbed from the gastrointestinal tract, and the antibiotic is used to treat enterocolitis, especially that caused by *Clostridium difficile*. Vancomycin is in clinical use worldwide. In recent years there has been both a medical and scientific resurgence in vancomycin in particular and glycopeptide antibiotics in general. Teicoplanin (2,3) was recently launched in Italy and France and is undergoing clinical trials in Europe and the United States.

Vancomycin is produced by *Amycolatopsis orientalis* (previously designated *Norcardia orientalis* and *Streptomyces orientalis*) (4). There are 45 organisms that have been reported to produce this class of antibiotics (Chap. 1).

All glycopeptide antibiotics contain a core heptapeptide. This heptapeptide (Fig. 1), written in the conventional peptide structure, a high degree of homology is observed in the aromatic amino acids 2 and 4–7 for this class of antibiotics. In several glycopeptide antibiotics, the remaining amino acids, 1 and 3, are also aromatic. Vancomycin glycopeptide antibiotics are structurally unique in that amino acids 1 and 3 are aliphatic amino acids. The classification of glycopeptide

195

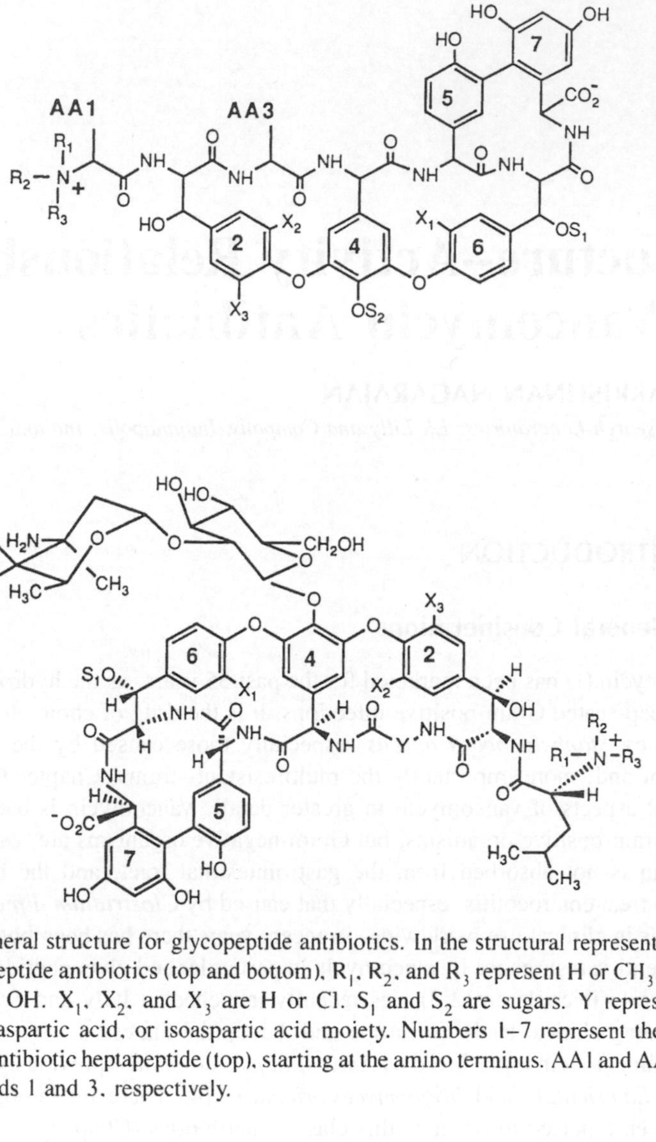

Fig. 1 General structure for glycopeptide antibiotics. In the structural representation of these glycopeptide antibiotics (top and bottom), R_1, R_2, and R_3 represent H or CH_3. R_4 and R_5 are H or OH. X_1, X_2, and X_3 are H or Cl. S_1 and S_2 are sugars. Y represents an asparagine, aspartic acid, or isoaspartic acid moiety. Numbers 1–7 represent the amino acids of the antibiotic heptapeptide (top), starting at the amino terminus. AA1 and AA3 refer to amino acids 1 and 3, respectively.

antibiotics into four classes is based on these and other structural differences (see Chap. 1).

As is common in microbial secondary metabolites, the glycopeptide-producing organisms elaborate a complex of antibiotics with minor chemical modifications. For example, the high-performance liquid chromatography (HPLC) of partially purified vancomycin shows no fewer than nine other structurally closely related factors (Fig. 2 and Table 1) (5). If we take such differences in the four classes of glycopeptides into consideration, well over 200 glycopeptides are known.

B. Scope of the SAR of Glycopeptides

Of the four classes of glycopeptide antibiotics, clearly the vancomycin type is clinically the most important group. Several close structural analogs of vancomycin have been reported recently, and these include the M43 group of antibiotics (6), A42867 (7), A51568 (8), A82846 (9), A83850 (10), chloroorienticin (11), decaplanin (12), eremomycin (13), MM45289 (14), MM47761 (15), OA7653 (16, 17), orienticin (18), and UK-72051 (19).

Second, several degradative and chemical interconversions of some of these antibiotics have been reported over the past 10 years (9,20,21,23–25).

Finally, recent reports have described a number of glycopeptides with long-chain acylamido side chains on a sugar residue (26–33). One of these, teicoplanin, has been claimed to have more favorable antibacterial and pharmacokinetic properties than vancomycin (34). Synthesis of several N-acyl and N-alkyl vancomycins and A82846 antibiotics (35–39) and their evaluation revealed that the N-alkyl derivatives have a substantial advantage over the parent antibiotics.

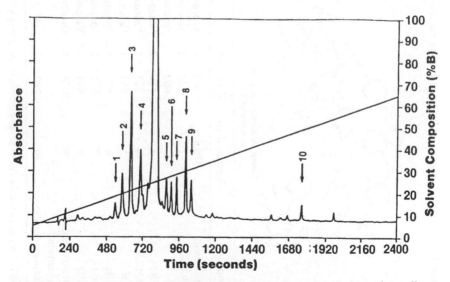

Fig. 2 HPLC profile of partially purified vancomycin hydrochloride using a linear gradient from 7 to 28% acetonitrile in triethylammonium phosphate buffer at pH 3.2 as described in Reference 22.

Table 1 Structural Assignments Proposed for Vancomycin and Related Compounds in the HPLC Chromatogram in Figure 2

General Vancomycin Structure (unrearranged)

Compounds 1 and 7 (rearranged)

Compound	Compound no.	T_f (s)	N	R_1	R_2	R_3	R_4	R_5	R_6
Vancomycin	Main	800	1	Vancosaminyl-O-glucosyl	—OH	—H	—CH₃	—CONH₂	—H
CDP-1	1[a]	570	1	Vancosaminyl-O-glucosyl	—OH	—H	—CH₃	—COOH	—H
	2	600		Structure unknown					
M43G	3	650	2	Vancosaminyl-O-glucosyl	—OH	—H	—CH₃	—CONH₂	—H
A51568A	4	723	1	Vancosaminyl-O-glucosyl	—OH	—H	—H	—CONH₂	—H
M43F	5	857	1	Vancosaminyl-O-glucosyl	—OH	—H	—CH₃	—COOH	—H
M43E	6	875	1	Vancosaminyl-O-glucosyl	—H	—H	—CH₃	—CONH₂	—H
CDP-1	7[a]	911	1	Vancosaminyl-O-glucosyl	—OH	—H	—CH₃	—COOH	—H
M43D	8	994	1	Vancosaminyl-O-glucosyl	—OH	—CH₃	—CH₃	—CONH₂	—H
Desvancosamine Vancomycin	9	1024	1	Glucosyl	—OH	—H	—CH₃	—CONH₂	—H
Aglucovancomycin	10	1767	1	—H	—OH	—H	—CH₃	—CONH₂	—H

[a]CDP-1 elutes as two peaks due to atropisomerism of the chlorophenyl residue of amino acid 2.

We (38,39) have examined the *in vitro* antimicrobial activities of these compounds against penicillin G-susceptible *S. aureus* XI.I (SAI), penicillin G-resistant *S. aureus* V4I (SA2), methicillin-resistant *S. aureus* X400 (SA3), methicillin-resistant *S. aureus* S13E (SA4), macrolide-resistant *Staphylococcus epidermidis* 270 (SEI), macrolide-susceptible *S. epidermidis* 222 (SE2), *Streptococcus pyogenes* C203 (SPy), *Streptococcus pneumoniae* Park (SPn), *Streptococcus (Enterococcus) faecium* X66 (SDI), and *Streptococcus (Enterococcus) faecalis* 2041 (SD2) (36).

In addition, all the compounds included in this chapter were evaluated against *S. aureus* XI.I, *Streptococcus pyogenes* C203, and *S. pneumoniae* Park infections in the *in vivo* mice model (36). Even though actual values for all compounds are not included in this chapter, the results are summarized for the SAR (structure-activity relationships).

With these data in hand, a SAR can now be delineated with reference to the antibacterial activity of the vancomycin classes of glycopeptide antibiotics. We examine how changes in each functional group of the vancomycin class of antibiotics affects the antibacterial activity.

II. SAR OF THE CORE HEPTAPEPTIDE AMINO ACIDS

A. Amino Acid 1

All the naturally occurring vancomycin glycopeptide antibiotics that have been reported contain leucine or its methylated analogs (6–15). The antibiotic activity of vancomycin, A51568A, and M43 factors A and D are similar, suggesting that the state of methylation of the leucine residue does not affect the antibacterial activity. Antibiotic OA-7563, factors A and B, have *N,N*-dimethylalanine (16,17) as the N-terminal amino acid in place of the *N*-methylleucine, and both factors A and B of OA-7563 are less active than vancomycin. However, the removal of this important terminal amino acid leucine (Edman degradation product) needed for the crucial binding of the N-acyl-D-alanyl-D-alanine (acyl-D-ala-D-Ala) carboxyl terminus of UDP-acetylmuramylpentapeptide to the N-terminal leucine of vancomycin, completely destroys the antibacterial activity of vancomycin (see Sec. IV). Acylation of the Edman degradation product of vancomycin with both aliphatic and aromatic amino acids afforded vancomycin modified at amino acid 1 (Table 2). None of these modified derivatives showed better activity than vancomycin (40).

B. Amino Acid 2

An analog of vancomycin in which the benzylic hydroxyl group of amino acid 2 is replaced by hydrogen (M43E) has been isolated from a strain of *A. orientalis* (39).

Table 2 Glycopeptide Antibiotics Modified at Amino Acid 1

Vancomycin: $R_1 = R_2 = H; R_3 = CH_3$
A51568A: $R_1 = R_2 = R_3 = H$
M3D: $R_1 = H; R_2 = R_3 = CH_3$
M43A: $R_1 = R_2 = R_3 = CH_3$
Edman degradation product = leucine removed

					MIC (μg/ml)					
Compound	SA1	SA2	SA3	SA4	SE1	SE2	SPy	SPn	SD1	SD2
Vancomycin	0.5	0.5	1	1	2	1	0.5	0.5	1	2
A51568A	0.5	1	1	1	2	1	2	0.5	1	4
M43D	1	1	2	1	4	1	0.5	0.25	2	4
M43A	1	1	2	1	4	2	1	0.5	2	4
Edman product	128⁺	128⁺	128⁺	128⁺	128⁺	128⁺	128⁺	128⁺	128⁺	128⁺

This naturally occurring deoxyvancomycin is half as active as vancomycin (Table 3).

Amino acid 2 contains a chlorine in the aromatic ring, and catalytic dechlorination of vancomycin (23) removes this chlorine to afford the monodechlorovancomycin. Vancomycin is twice as active as this monodechlorovancomycin (39).

To establish a similar SAR relationship for the A82846 antibiotics, A82846B was catalytically dechlorinated to its monodechloro A82846B analog, and this

Table 3 Glycopeptide Antibiotics Modified at Amino Acid 2

Vancomycin: $R = OH; R_1 = H; X = Cl$
M43E: $R = R_1 = H; X = Cl$
Monodechlorovancomycin: $R = OH; R_1 = X = H$

A82846B: $R = OH; X = Cl; R_1 =$

Orienticin A: $R = OH; X = H; R_1 =$

Compound	MIC (μg/ml)									
	SA1	SA2	SA3	SA4	SE1	SE2	SPy	SPn	SD1	SD2
Vancomycin	0.5	0.5	1	1	2	1	0.5	0.5	1	2
M43E	1	2	2	2	4	2	1	0.5	2	4
Monodechloro-vancomycin	1	2	2	2	2	2	1	1	2	8
A82846B	0.25	0.25	0.5	0.25	0.5	0.25	0.25	0.25	0.5	0.5
Orienticin A	0.5	0.5	1	0.5	0.5	0.5	0.5	0.5	2	2

Table 4 Glycopeptide Antibiotics Modified at Amino Acid 3

1. Vancomycin: $R_1 = R_2 = R_4 = H$; $R_3 = CH_3$; $R_5 = Cl$; $Y =$

2. A51568B: $R_1 = R_2 = R_3 = R_4 = H$; $R_5 = Cl$; Y

3. M43G: $R_1 = R_2 = R_4 = H$; $R_3 = CH_3$; $R_5 = Cl$; $Y =$

4. M43F: $R_2 = R_2 = R_4 = H$; $R_3 = CH_3$; $R_5 = Cl$; $Y =$

5. M43B: $R_1 = R_2 = R_3 = CH_3$; $R_4 = H$; $R_5 = Cl$; $Y =$

6. CDP-1 $R_1 = R_2 = R_5 = H$; $R_3 = CH_3$; $R_4 = Cl$; $Y =$

7. $R_1 = R_2 = R_3 = CH_3$; $R_4 = Cl$; $R_5 = H$; $Y =$

Table 4 Continued

Compound	MIC (μg/ml)									
	SA1	SA2	SA3	SA4	SE1	SE2	SPy	SPn	SD1	SD2
Vancomycin	0.5	0.5	1	1	2	1	0.5	0.5	1	2
A51568B	2	2	2	2	2	2	0.5	0.06	1	2
M43G	2	4	4	4	4	4	1	0.125	2	4
M43F	4	4	8	8	16	8	4	2	8	16
M43B	16	32	32	32	32	32	16	2	32	32
CDP-1	128[+]	128[+]	128[+]	128[+]	128[+]	128[+]	128[+]	128[+]	128[+]	128[+]
7	128[+]	128[+]	128[+]	128[+]	128[+]	128[+]	128[+]	128[+]	128[+]	128[+]

monodechloro A82846B was shown to be identical to orienticin A (9,18). The removal of chlorine in the aromatic amino acid 2 reduces the activity 10-fold (39). We discuss the effect of antibacterial activity on the removal of the second chlorine from vancomycin when we examine the SAR of amino acid 6.

C. Amino Acid 3

The third amino acid of the heptapeptide core of the vancomycin class of antibiotics is the aliphatic amino acid asparagine (Table 4). Several modifications, both naturally occurring analogs and the uniquely rearranged CDP analogs, are known.

The naturally occurring A51568B (8) and M43G (39) are the glutamine analogs of A51568A (8) and vancomycin, respectively. These two antibiotics have an additional methylene on the amino acid 3 (asparagine) of vancomycin. A51568B is fourfold less active than vancomycin, and M43G is half as active as vancomycin. Consequently, the increase in the chain length of asparagine to glutamine reduces the activity of vancomycin.

The naturally occurring antibiotics M43F and M43B are the aspartic acid analogs of vancomycin and M43A, respectively (6). These two desamido analogs are at least 10 times less active than vancomycin. The rearranged CDP-1 (41), the isoaspartic acid analog of vancomycin, and the corresponding isoaspartic analog of M43A (6) are devoid of antibacterial activity (39).

The negative charges of the aspartate and isoaspartate moieties in M43B, in M43F, and in the rearranged analogs near the binding site and the changed conformational geometry (24) in the rearranged derivatives seem to hinder the binding of the acyl-D-ala-D-Ala carboxyl terminus of UDP-N-acetylmuramyl-pentapeptide to the antibiotic and contributes dramatically to the diminution of the antibacterial activity (see Sec. IV).

D. Amino Acid 4

The only SAR of this amino acid pertains to the sugar moiety at the phenolic hydroxyl group. In several naturally occurring vancomycin glycopeptide antibiotics, the phenolic hydroxyl group carries a disaccharide unit. In many cases, the first sugar is D-glucose-linked β glycosidically, and the second sugar, usually an amino sugar, is either L-vancosamine or L-4-*epi*-vancosamine, linked to glucose α glycosidically. In A83850B (10), the amino sugar is α-L-4-keto-vancosamine. These two sugars were cleaved sequentially (20), and their antibacterial activity is shown in Table 5. In addition, similar analogs for A51568A, M43A, and A83850A (the dimethylleucine analog of A83850B) were prepared, and the SAR of these derivatives also support the conclusions discussed later.

Table 5   Glycopeptide Antibodies Modified at Amino Acid 4

1.	Vancomycin:	S_1 = α-O-L-vancosaminyl-β-O-D-glucosyl
2.	Desvancosamine vancomycin	S_1 = β-O-D-glucosyl
3.	Aglucovancomycin	S_1 = H
4.	A83850B	S_1 = α-O-L-4-ketovancosaminyl-β-O-D-glucosyl
5.	Reduced A83850B	S_1 = α-O-L-4-*epi*-vancosaminyl-β-O-D-glucosyl

	MIC (μg/ml)									
Compound	SA1	SA2	SA3	SA4	SE1	SE2	SPy	SPn	SD1	SD2
Vancomycin	0.5	0.5	1	1	2	1	0.5	0.5	1	2
2	8	8	8	8	16	8	4	4	8	32
3	1	1	1	1	2	2	0.5	1	2	4
4	0.5	1	1	0.5	1	1	0.5	0.125	1	4
5	1	1	1	1	2	2	0.5	0.125	1	4

The removal of the amino sugar, vancosamine, reduces the activity of the parent antibiotic by two- to fivefold. The removal of the second sugar, glucose, restores the activity, only insofar as *in vitro* activity, but the *in vivo* activity is reduced fivefold. We examined the *in vitro* and *in vivo* activities (in infections in mice), and they always parallel one another. The only exception seems to be the agluco derivatives. The sugars seem to play an important role in imparting enhanced pharmacokinetic properties to the vancomycin glycopeptide antibiotics. The keto group at the 4 position of L-vancosamine in A83850B does not seem to affect the antibacterial activity. Reduction of the ketone affords the 4-*epi*-vancomycin analog. This compound, the 4-*epi* analog of vancomycin, is less active than vancomycin.

E. Amino Acid 6

Recently, several glycopeptide antibiotics were isolated, which has made possible the delineation of the effect of chlorine on the benzene ring of the amino acid 6 on the antibacterial activity (Table 6). Accordingly, antibiotic A82846B (identical to chloroorienticin A) has two chlorines, one on the aromatic amino acid 2 and the other on aromatic amino acid 6. Antibiotic A82846A (identical to eremomycin, MM45289) has a chlorine on amino acid 2. Orienticin A (identical to UK72,051) possesses a chlorine on amino acid 6, and A82846C (identical to orienticin C, MM 4756) is devoid of a chlorine substituent. In all other aspects, these four antibiotics are identical.

The examination of both *in vitro* and *in vivo* (not included) antibacterial activities show that A82846A and A82846B are about 2–10 times more active than vancomycin; orienticin A and A82846C are 2-fold less active than vancomycin. (The differences in activities become clearer in the *in vivo* experiments.) These data suggest that the removal of chlorine in aromatic amino acid 6 has slight, if any, effect on the antibacterial activity. However, removal of chlorine on aromatic amino acid 2 diminishes the activity 10-fold.

Finally, let us consider the effect of the amino sugar on the benzylic hydroxyl group of aromatic amino acid 6. The selective and sequential removal of L-4-*epi*-vancosamine and then the second β-glucose from both A82846A and A82846B was accomplished and their antibacterial activity was evaluated. The SAR is identical in both cases, but we discuss here only the A82846B derivatives.

Removal of the L-4-*epi*-vancosamine on amino acid 4 from A82846B affords chloroorienticin B, and hydrolysis of the D-glucose affords chloroorienticin C. These compounds have also been isolated from natural sources (11). These two compounds are two to five times more active than vancomycin, as is A82846B itself. Consequently, the presence of this benzylic amino sugar, L-4-*epi*-

Table 6   Glycopeptide Antibiotics Modified at Amino Acid 6

1. A82846B S_1 = α-O-L-4-epi-vancosaminyl-β-O-D-glucose; X = Y = Cl
   (chloroorienticin A)
2. A82846A S_1 = α-O-L-4-epi-vancosaminyl-β-O-D-glucose; X = H; Y = Cl
3. Orienticin A S_1 = α-O-L-4-epi-vancosaminyl-β-O-D-glucose; X = Cl; Y = H
4. A82846C (orienticin C) S_1 = α-O-L-4-epi-vancosaminyl-β-O-D-glucose; X = Y = H
5. Chloroorienticin B S_1 = β-O-D-glucose; X = Y = Cl
6. Chloroorienticin C S_1 = H; X = Y = Cl

Compound	MIC (μg/ml)									
	SA1	SA2	SA3	SA4	SE1	SE2	SPy	SPn	SD1	SD2
Vancomycin	0.5	0.5	1	0.5	1	1	0.5	0.5	1	4
1	0.125	0.125	0.125	0.125	0.25	0.125	0.06	0.06	0.5	0.25
2	0.125	0.125	0.25	0.125	0.125	0.125	0.25	0.125	0.5	2
3	0.5	0.5	1	0.5	0.5	0.5	0.5	0.5	2	2
4	0.5	0.5	1	0.5	0.5	0.5	0.5	0.5	4	4
5	0.25	0.25	0.25	0.25	0.5	0.25	0.5	1	1	1
6	0.25	0.25	0.25	0.25	0.5	0.25	0.25	0.25	1	1

vancosamine, on aromatic amino acid 6 dramatically increases the activity of vancomycin by two- to fivefold.

F.   Amino Acids 5 and 7

No naturally occurring or chemically modified analogs of amino acids 5 and 7 are known. However, the carboxyl group on amino acid 7 affords an opportunity for chemical modification.

III. SAR OF SEMISYNTHETIC GLYCOPEPTIDE ANTIBIOTICS

A. N-acyl Derivatives

Several glycopeptides containing a long-chain aliphatic acyl residue on the amino sugar attached to the phenolic group of amino acid 4 have been reported (26–33). Teicoplanin, an antibiotic belonging to the long-chain acylamido glycopeptide class, has been claimed to have better antibacterial and more favorable pharmacokinetic properties than vancomycin (34). We undertook a systematic study of the SAR of N-acyl vancomycins and the more active A82846 antibiotics, especially derivatives acylated on the amino sugar of amino acid 4.

A total of 32 N-acyl vancomycins (35) and 7 N-acyl A82846 (39) derivatives were prepared and evaluated. For vancomycin, acylation yielded two mono-N-acyl derivatives substituted at the amino groups of the vancosamine and N-methylleucine moieties, respectively, and one di-N-acyl derivative. However, for the A82846A and A82846B antibiotics, three mono-N-acyl, three di-N-acyl, and one tri-N-acyl derivatives can be obtained. The ratio of three N-acyl derivatives for vancomycin and seven N-acyl derivatives for A82846A and A82846B derivatives depended on the reaction conditions (35,39).

Among the aliphatic N-acyl vancomycin derivatives, the mono-N-acyl derivatives functionalized on the vancosamine amino group are more active than the mono-N-acyl vancomycin substituted on the amino acid, N-methylleucine. Both mono-N-acyl vancomycins are more active than the di-N-acyl vancomycins. With the limited number of derivatives prepared, the SAR of N-acyl A82846 derivatives seems to be similar to that of the vancomycin series. Accordingly, the mono-N-acyl derivatives are more active than the di-N-acyl or tri-N-acyl A82846 derivatives.

A comparison of the more active aliphatic mono-N-acyl vancomycins substituted on the amino group of vancosamine reveals that increasing the length of the side chain increases activity. Optimum activity is found when the side chain is C_9–C_{11} straight-chain fatty acid residue. When the carbon chain length is greater than 11, the activity drops. Modification of the side chain with a branched chain, introduction of an amino, bromo, or carboxyl, or insertion of an oxygen or sulfur does not enhance the activity.

The SAR of the N-aracyl vancomycins follows a pattern similar to those of the aliphatic N-acyl vancomycins. Accordingly, the mono-N-aracyl vancomycins substituted on the vancosamine amino moiety are more active than the mono-N-acyl derivatives functionalized on the amino acid, N-methylleucine, and both the mono-N-aracyl vancomycins are more active than the corresponding di-N-aracyl vancomycins.

The aromatic mono-N-aracyl vancomycins substituted on the vancosamine sugar are more active than the corresponding aliphatic mono-N-acyl derivatives.

The most active compounds in the mono-N-aracyl series are the *p*-octylbenzoyl and *p*-octyloxybenzoyl vancomycin derivatives, with an aliphatic hydrocarbon chain attached to the aromatic ring.

Finally, a comparison of the antibacterial activities of the parent vancomycin with the N-acyl derivatives show that, even though in some cases there is an increase in the *in vitro* spectrum of the mono-N-acyl derivatives, the *in vivo* activities do not reveal any great increase over that of vancomycin.

B. N-Alkyl Derivatives

As an extension of the SAR of N-acyl glycopeptide derivatives, over 80 N-alkyl vancomycins (36) and 70 N-alkyl A82846 derivatives (38,39) were synthesized and evaluated.

Some derivatives in this category are the most active compounds known to date and exhibit interesting biological properties, such as longer half-lives even though they are not highly serum bound, and activity against vancomycin-resistant *Enterococcus* strains at clinically relevant mean inhibitory concentrations (MIC) (Table 7).

A comparison of the antibacterial activities of the N-decyl vancomycins with those of the N-decanoyl vancomycins show that the C_{10} alkyl analogs are more active than the corresponding alkanoyl series. Furthermore, the mono-N-decyl vancomycin substituted on the vancosamine sugar is more active *in vitro* than the parent vancomycin, is equivalent in activity to vancomycin *in vivo*, and shows a longer elimination half-life in rats. This was our first indication that the

Table 7 Comparison of the SAR of C_{10} N-Acyl and N-Alkyl Vancomycins

Compound	MIC (μg/ml)									
	SA1	SA2	SA3	SA4	SE1	SE2	SPy	SPn	SD1	SD2
Vancomycin	0.5	0.5	1	1	2	1	0.5	0.5	1	2
Mono-N-decanoyl (AA[a] 4)	0.5	0.5	0.5	0.5	2	1	0.5	1	0.5	1
Mono-N-decanoyl (AA 1)	0.5	0.5	0.5	1	4	2	4	4	2	4
Di-N-decanoyl	2	4	4	8	32	8	4	16	8	8
Mono-N-decyl (AA 4)	0.13	0.13	0.13	0.13	0.25	0.13	0.06	0.13	0.25	0.25
Mono-N-decyl (AA 1)	0.5	0.5	0.5	0.5	2	0.5	0.5	0.5	0.5	1
Di-N-decyl	2	4	4	4	16	4	2	4	4	4

[a]Amino acid.

N-alkyl vancomycin derivative showed interesting biological properties, and it triggered our interest in actively pursuing the SAR of N-alkyl vancomycin glycopeptide antibiotics.

As in the SAR of the N-acyl vancomycin series, the general SAR trend is that the N-alkyl derivatives substituted on the vancosamine sugar are more active than those substituted on the N-methylleucine, and both monosubstituted vancomycins are more active than the corresponding di-N-alkyl vancomycins.

In the N-alkyl A82846A and A82846B series, the SAR trend is similar to that for vancomycin (Table 8). Accordingly, in general, the mono-N-alkyl A82846A and A82846B derivatives substituted on either 4-*epi*-vancosamine sugar are more active than when substituted on the amino acid, N-methylleucine, and the mono-N-alkyl derivatives are more active than the di-N-alkyl A82846 derivatives.

As in the N-acyl vancomycins, increasing the aliphatic straight chain in the N-alkyl vancomycins enhanced activity (Table 9). The optimum length seems to be C_{10}. Branching the side chain or substituting with oxygen or sulfur did not increase the activity. In the N-aralkyl vancomycin series, the benzyl derivatives and the benzyl substituted at the 4 position with an aliphatic side chain were more active than the aliphatic series in the *in vivo* models. The most active vancomycin derivatives were the N-octylbenzyl and the N-octyloxybenzyl vancomycins. Substitutions at the *para* positions of the benzyl moiety with hetero atoms, such as oxygen, sulfur, nitrogen, or halogens, did not seem to enhance activity. Several compounds in the N-alkyl vancomycin series were more active than the parent vancomycin. Accordingly, benzyl, octylbenzyl, octyloxybenzyl, butylbenzyl, and butyloxybenzyl derivatives were up to five times more active than vancomycin.

The most active compounds in the N-alkyl A82846B series were the mono-N-octylbenzyl, mono-N-*p*-chlorobenzyl, and mono-N-*p*-bromobenzyl derivatives, substituted on the 4-*epi*-vancosamine linked to amino acid 4.

Table 8 SAR of N-Octyl Derivatives of A82846B

Compound	MIC (μg/ml)			ED$_{50}$ (mg/kg) × 2, SC		
	SA1	SP*y*	SP*n*	SA1	SP*y*	SP*n*
A82846B	.25	0.25	0.25	0.20	0.18	0.20
Mono-N-octyl (AA 4)	0.5	0.06	0.25	0.43	0.11	0.06
Mono-N-octyl (AA 1)	0.5	0.06	0.13	1.2	0.18	0.34
Di-N-octyl (AA 4, AA 6)	1	0.5	0.5	1.2	0.18	0.32
Di-N-octyl (AA 1, AA 6)	4	0.5	4	—	—	—

Table 9 SAR of N-Alkyl Vancomycins

Compound	MIC (μg/ml)			ED$_{50}$ (mg/kg) $\times$ 2, SC			$T_{1/2}$ (h)	Serum, IV Rat 5 Minute concentration (μg/ml)
	SA1	SPy	SPn	SA1	SPy	SPn		
Vancomycin	0.5	0.5	0.5	1.8	0.8	0.9	0.75	160
C$_6$H$_5$CH$_2$	0.06	0.06	0.015	0.8	1.0	0.9	1.8	89
p-C$_4$H$_9$C$_6$H$_4$CH$_2$	0.125	0.125	0.125	0.7	0.4	0.8	5.4	156
p-C$_8$H$_{17}$C$_6$H$_4$CH$_2$	0.5	0.25	0.25	0.19	0.62	0.23	—	—
p-C$_4$H$_9$OC$_6$H$_4$CH$_2$	0.125	0.06	0.06	0.7	0.5	0.6	2.4	205
p-C$_8$H$_{17}$OC$_6$H$_4$CH$_2$	0.25	0.25	0.25	0.2	0.2	0.2	—	—

C. Activity Against Resistant Enterococci

Since the recent reported isolation of clinical isolates of *E. facium* and *E. faecalis* resistant to vancomycin (42), several glycopeptides and their derivatives were tested against 34 susceptible and 26 resistant enterococcal strains (37,39). Some of the semisynthetic N-alkylvancomycin and A82846B derivatives exhibited excellent activity against these resistant strains (Table 10).

Whereas the geometric mean MIC of vancomycin were 1.0 and 263 μg/ml against 34 susceptible and 26 resistant enterococcal strains, respectively, the corresponding values for teicoplanin were 0.24 and 22 μg/ml. However, for the most active mono-N-alkyl vancomycins the values ranged between 0.5–0.75 μg/ml for susceptible strains and 4–6 μg/ml for resistant strains. For the most

Table 10 SAR of N-Acyl and N-Alkyl Vancomycins Against Resistant Enterococci

	MIC (μg/ml)		
	Resistant enterococci (N = 26)	Typical enterococci (N = 34)	Ratio
Vancomycin	263	1.0	261
Teicoplanin	22	0.24	92
A82846B	11	0.46	24
Mono-N-decyl	4.8	0.56	8.1
Mono-N-p-octylbenzyl	5.6	0.74	7.6
Mono-N-p-octyloxbenzyl	4.6	0.62	7.4

Table 11   SAR of Mono-N-Alkyl A82846B Derivatives

				Geometric mean MIC (μg/ml)		
	ED_{50} (mg/kg) $\times$ 2			Resistant strains[a]	Susceptible strains[b]	
Compound	SA1	SPy	SPn	(N = 25)	(N = 34)	Ratio
A82846B	0.2	0.18	0.2	10	0.46	21
Mono-N-octyl	0.43	0.11	0.06	2.1	0.73	2.1
Mono-N-p-chlorobenzyl	0.25	0.06	0.08	3.6	0.65	5.3
Mono-N-p-bromobenzyl	0.38	0.05	0.08	3.1	0.67	4.6

[a]Susceptibility of vancomycin-resistant and vancomycin-susceptible *Enterococcus faecium* and *E. faecalis*.
[b]All alkyl substituents are on *epi*-vancosamine in amino acid 4.

active mono-N-alkyl A82846B derivatives, the corresponding values were between 0.65 and 0.75 μg/ml for the susceptible strains and between 2.1 and 3.6 μg/ml for the resistant strains (Table 11). Thus, the N-alkyl vancomycin and A82846B derivatives are not only up to five times more active than vancomycin against Gram-positive bacteria, but more importantly, these semisynthetic N-alkyl derivatives are more active against the vancomycin-resistant enterococci strains at clinically relevant MIC.

IV. SAR AND MODES OF ACTION

The cell wall of Gram-positive bacteria appear uniform under the electron microscope and are made up of peptidoglycan, protein, and teichoic acids.

In the schematic representation of the peptidoglycan network of *S. aureus* (Fig. 3), the long horizontal chains are made up of N-acetylglucosamine (cross hatched) and muramic acid (white). Tetrapeptides (gray) are attached to the muramic acid units. Finally, the interpeptide pentaglycine (parallel lines) bonds complete the rigid peptidoglycan framework.

Biochemical studies suggest that the modes of action of vancomycin and other glycopeptides involve inhibition of the peptidoglycan synthesis. Vancomycin forms a stoichiometric 1:1 complex with the peptidoglycan precursor UDP-N-acetylmuramylpentapeptide (43) by forming hydrogen bonds (see Figs. 4 and 5).

The transglycosylase enzyme that transfers the disaccharide of the peptidoglycan precursor to the growing glycan polymer of the cell wall peptidoglycan is inhibited (44,45), presumably as a result of the steric bulkiness of the glyco-

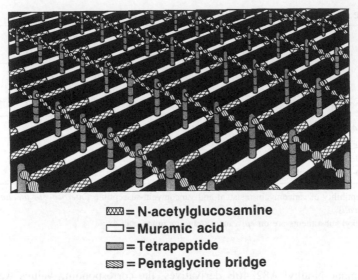

Fig. 3   A layer of peptidoglycan network in *S. aureus*.

peptide-peptidoglycan precursor (46). It seems that both the transglycosylase and transpeptidase enzyme reaction steps that enable the synthesis of the rigid cell wall peptidoglycan may be inhibited by the glycopeptides.

Computer and molecular modeling studies (47,48) reveal that the acyl-D-Ala-D-Ala carboxyl terminus of the cell wall subunit is held firmly by several hydrogen bonds to the N terminus of the vancomycin class of antibiotics (Fig. 5).

V.   CONCLUSIONS

The delineation of the SAR for the vancomycin class of antibiotics is entirely consistent with this mode of action for this class of antibiotics. There are two sites where chemical modification close to the binding area of the vancomycin molecule markedly affects the antibacterial activity of the glycopeptide antibiotic. First, the removal of the crucial N-terminal leucine, which is involved in the binding, completely destroys the antibacterial activity of the antibiotic. Second, the introduction of a new carboxylate moiety, by substituting the carboxamide moiety of asparagine of the vancomycin antibiotic with aspartic acid as in M43F or rearrangement of the asparagine to isoaspartic acid as in CDP-1, dramatically reduces the antibacterial activity. These two examples show clearly that changes near the vancomycin N-terminus that hinder the binding of the acyl-D-Ala-D-Ala carboxyl terminus of UDP-N-acetylmuramylpentapeptide drastically diminish the antibacterial activity of the vancomycin class of glycopeptide antibiotics.

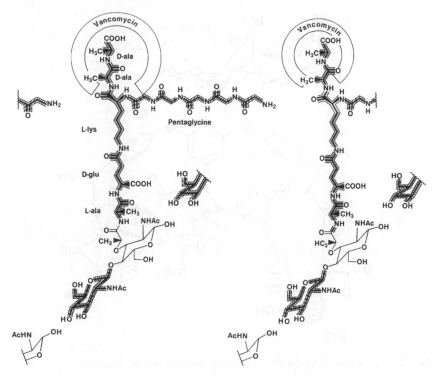

Fig. 4 Chemical representation of the mode of action of vancomycin.

Chemical modification of other functional groups of vancomycin antibiotics far removed from the binding site do not seem greatly to alter the antibacterial activity.

Two types of modifications decidedly seem to enhance the antibacterial activity and even impart better pharmacokinetic properties. First, the additional α-L-4-*epi*-vancosamine amino sugar on the benzylic hydroxyl group of the aromatic amino acid 6 enhances the antibacterial activity. Second, the alkylation of the amino group of amino sugars on aromatic amino acids 4 and 6, especially amino acid 4, enhances antibacterial activity and imparts better pharmacokinetic properties. More importantly, some N-alkyl derivatives show excellent activity against vancomycin-resistant enterococci. The most active compounds made to date are the three N-alkyl A82846B derivatives, mono-N-octyl, mono-N-*p*-chlorobenzyl, and mono-N-*p*-bromobenzyl derivatives. These three compounds incorporate the two modifications that enhance activity in the vancomycin antibiotics.

Structure-activity relationships of antibiotics are influenced by several

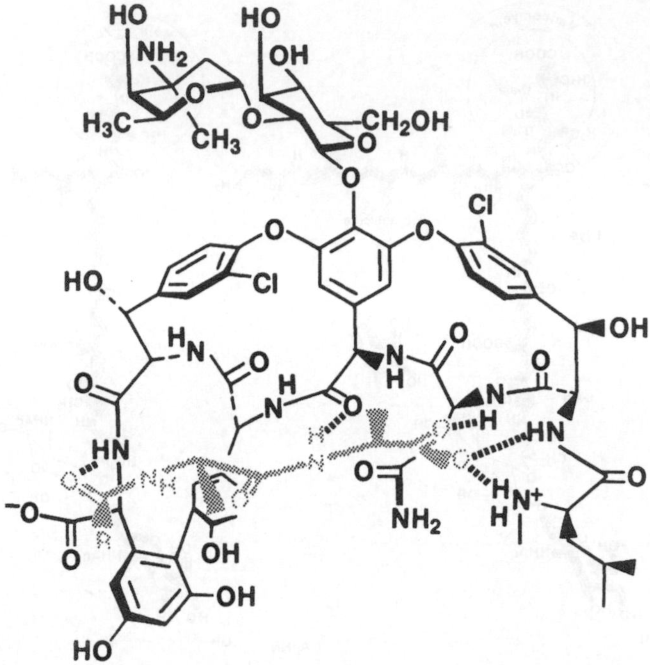

Fig. 5   Cell wall acyl-D-Ala-D-Ala terminus binding to vancomycin.

parameters, and vancomycin antibiotics are no exception. The SAR analysis of the vancomycin glycopeptides is an attempt to rationalize the SAR relationships based on one important parameter, that is, the mode of action.

Accordingly, it should be pointed out that although A82846A is 5- to 10-fold more active than vancomycin against Gram-positive bacteria, its affinity to the model peptidoglycan diacetyl-L-lysyl-D-alanine-D-alanine is 23-fold lower (49).

ACKNOWLEDGMENTS

I am pleased to acknowledge my colleagues N.E. Allen, D.R. Berry, F.T. Counter, M. Debono, A. Felty-Duckworth, R.L. Hamill, H. Higgins, Jr., M.M. Hoehn, A.H. Hunt, N.D. Jones, G.G. Marconi, K. Merkel, K. Michel, W. Nakatsukasa, T.I. Nicas, J.L. Occolowitz, J.L. Ott, A.A. Schabel, and R.C. Yao for collaboration on glycopeptide research and for helpful discussions.

REFERENCES

1. McCormick M.H., Stark W.M., Pittenger G.E., Pittenger R.C., McGuire J.R. (1956). Vancomycin, a new antibiotic. I. Chemical and biological properties. Antibiot. Annu. 1955–1956:601–611.

2. Somma S., Gastaldo L., Corti A. (1984). Teicoplanin, a new antibiotic from *Actinoplanes teichomyceticus* nov. sp. Antimicrob. Agents Chemother. 26:917–923.

3. Barna J.C., Williams D.H., Stone D.J.M., Leung T.W.C., Doddrel D.M. (1984). Structure elucidation of teicoplanin antibiotic. J. Am. Chem. Soc. 106:4895–4902.

4. Lechevalier M.D., Prauser H., Labeda D.P., Ruan J.-S. (1986). Two new genera of nocardioform actinomycetes: *Amycolata* gen. nov. and *Amycopatopsis* gen. nov. Int. J. Syst. Bacteriol. 36:29–37.

5. Nagarajan R., Inman E.L., Merkel K.E. unpublished results.

6. Nagarajan R., Merkel K.E., Michel K.H., Higgens H.M., Hoehn M.M., Hunt A.H., Jones N.D., Occolowitz J.L., Schabel A.A., Swartzendruber J.K. (1988). M43 antibiotics: Methylated vancomycins and unrearranged CDP-I analogs. J. Am. Chem. Soc. 110:7896–7897.

7. Riva E., Gastaldo L., Beretta M.G., Ferrari P., Zerilli L.F., Cassani, Selva E., Goldstein B.P., Berti M., Parenti F., Denaro M. (1989). A42867, a novel glycopeptide antibiotic. J. Antibiot. 42:497–505.

8. Hunt A.H., Marconi G.G., Elzey T.K., Hoehn M.M. (1984). A51568A: N-demethylvancomycin. J. Antibiot. 37:917–9.

9. Nagarajan R., Berry D.M., Hunt A.H., Occolowitz J.L., Schabel A.A. (1988). Conversion of antibiotic A82846B to orienticin A and structural relationships of related antibiotics. J. Org. Chem. 54:983–986.

10. Hamill R.L., Hunt A.H., Nagarajan R., Schabel A.A., Yao R.C. Process for producing A83850 antibiotics. U.S. Patent 07/568578, filed August 10, 1990.

11. Tsuji N., Kamigauchi T., Kobayashi M., Terui Y. (1988). New glycopeptide antibiotics. II. Isolation and structures of chloroorienticins. J. Antibiot. 41:1506–1510.

12. Franco C.M.M., Chatterjee S., Vijakumar E.K.S., Chatterjee D.K., Ganguli B.N., Rupp R.H., Fehlhaber H.W., Kogler H., Seibert G., Teetz V. (1990). Glycopeptide antibiotic decaplanin. European Patent Appl. 356894 (assigned to Hoechst A.-G.), March 7, 1990.

13. Lomakina N.N., Berdnikova T.F., Tokareva N.L., Abramova E.A., Dokshina N.Y. (1989). Structure of eremomycin, a novel polycyclic glycopeptide antibiotic. Antibiot. Khimioter. 34:254–258.

14. Athalye M., Elson A., Gilpin, M.L., Jeffries L.R. (1989). Production of antibiotics MM 45289 and MM 47756 by fermentation with *Amycolatopsis orientalis*. European Patent Appl. 309161 (assigned to Beecham Group plc), March 29, 1989.

15. Box S.J., Elson A.L., Gilpin M.L., Winstanley D.J. (1990). MM 47761 and MM 49721, glycopeptide antibiotics produced by a new strain of *Amycolatopsis orientalis*: Isolation, purification and structure determination. J. Antibiot. 43:931–937.

16. Jeffs P.W., Yellin B., Mueller L., Heald S.L. (1988). Structure of the antibiotic OA-7653. J. Org. Chem. 53:471–474.

17. Ang S.G., Williamson M.P., Williams S.H. (1988). Structure elucidation of a glycopeptide antibiotic, OA-7653. J. Org. Chem. Perkin Trans. 1:1949–1956.

18. Tsuji N., Kobayashi M., Kamigauchi T., Yoshimura Y., Terui Y. (1988). New glycopeptide antibiotics. I. The structure of orienticins. J. Antibiot. 41:819–822.

19. Skelton N.J., Williams D.H., Rance M.J., Ruddock J.C. (1990). Structure elucidation of UK-72051, a novel member of the vancomycin group of antibiotics. J. Chem. Soc. Perkin Trans. I:77–81.

20. Nagarajan R., Schabel A.A. (1988). Selective cleavage of vancosamine, glucose and N-methylleucine from vancomycin and related antibiotics. J. Chem. Soc. Chem. Commun. 1988:1306–1307.

21. Booth P.M., Stone D.J.M., Williams D.W. (1987). The Edman degradation of vancomycin: preparation of vancomycin hexapeptide. J. Chem. Soc. Chem. Commun. 1987:1694–1695.

22. Inman E.L. (1987). Determination of vancomycin related substances by gradient high pressure liquid chromatography. J. Chromatogr. 410:363–372.

23. Harris C.M., Kannan R., Kopecka H., Harris T.H. (1985). The role of chlorine substituents in the antibiotic vancomycin: Preparation and characterization of mono- and didechlorovancomycin. J. Am. Chem. Soc. 107:6652–6658.

24. Harris C.M., Kopecka H., Harris T.H. (1983). Vancomycin: Structure and transformation to CDP-I. J. Am. Chem. Soc. 105:6915–6922.

25. Nagarajan R., Berry D.M., Schabel A.A. (1989). The structural relationships of A82846B and its hydrolysis products with chloroorienticins A, B and C. J. Antibiot. 42:1438–1440.

26. Malabarba A., Strazzolini P., Depaoli A., Landi M., Berti M., Cavalleri B. (1984). Teicoplanin, antibiotics from *Actinoplanes teichomyceticus* nov. sp. VI. Chemical degradation: Physico-chemical and biological properties of acid hydrolysis products. J. Antibiot. 37:988–999.

27. Waltho J.P., Williams D.H., Selva E., Ferrari P. (1987). Structure elucidation of the glycopeptide antibiotic complex A40926. J. Chem. Soc. Perkin Trans. I:2103–21077.

28. Michel K.H., Yao R.C.F. (1991). New lipoglycopeptide antibiotic A84575. European Patent Appl. 424051 (assigned to Eli Lilly and Company), April 24, 1991.

29. Jeffs P.W., Muller L., DeBrosse C., Heald S.L., Fisher R. (1986). The structure of aridicin A. An integrated approach employing 2D NMR, energy minimization and distance constraints. J. Am. Chem. Soc. 108:3063–3075.

30. Sitrin R.D., Chan G.W., Chapin F., Giovenella A.J., Grappel S.F., Jeffs P.W., Phillips L., Snader K.M., Nisbet L.J. (1986). Aridicins, novel glycopeptide antibiotics. III. Preparation, characterization and biological activities of aglycone derivatives. J. Antibiot. 39:68–75.

31. Folena-Wasserman G., Poehland B.L., Yeung E.W.K., Staiger D., Killmer L.B., Snader K., Dingerdissen J.J., Jeffs P.W. (1986). Kibdelins (AAD-609), novel glycopeptide antibiotics. J. Antibiot. 39:1395–1406.

32. Box S.J., Coates N.J., Davis C.J., Gilpin M.L., Houge-Frydrych C.S., Milner P.H. (1991). MM 55266 and MM 55268, glycopeptide antibiotics produced by a new strain of *Amycolatopsis*. Isolation, purification and structure determination. J. Antibiot. 44:807–813.

33. Christensen S.B., Allaudeen H.S., Burke M.R., Carr S.A., Chung S.K., DePhillips P., Dingerdissen J.J., Dipaolo M., Giovenella A.J., Heald S.L., Killmer L.B., Mico B.A., Mueller L., Pan C.H., Poehland B.L., Rake J.B., Roberts G.D., Shearer M.C., Sitrin R.D., Nisbet L.J., Jeffs P.W. (1987). Parvodicin, a novel glycopeptide from a new species, *Actinomadura parvosata*: Discovery, taxonomy, activity and structure elucidation. J. Antibiot. 40:970–990.

34. Glupezynski Y., Lagast H., Van der Auwera P., Thys J.P., Crokaert F., Yourassowsky E., Meunier-Carpentier F., Klastersky J., Kains J.P., Serruys-Schoutens E., Legrand J.C. (1986). Clinical evaluation of teicoplanin for therapy of severe infections caused by gram-positive bacteria. Antimicrob. Agents Chemother. 29:52–57.

35. Nagarajan R., Schabel A.A., Occolowitz J.L., Counter F.T., Ott, J.L. (1988). Synthesis and antibacterial activity of N-acyl vancomycins. J. Antibiot. 41: 430–438.

36. Nagarajan R., Schabel A.A., Occolowitz J.L., Counter F.T., Ott J.L., Felty-Duckworth A.M. (1989). Synthesis and antibacterial evaluation of N-alkyl vancomycins. J. Antibiot. 42:62–73.

37. Nicas T.I., Cole C.T., Preston D.A., Schabel A.A., Nagarajan R. (1989). Activity of glycopeptides against vancomycin-resistant gram-positive bacteria. Antimicrob. Agents Chemother. 33:1477–1781.

38. Nagarajan R., Counter F.T., Nicas T.I., Schabel A.A. (1989). Synthesis and SAR of A82846 antibiotic derivatives. 1989 Int. Chem. Congr. Pacific Basin Soc., Honolulu, Hawaii, abstract 401.

39. Nagarajan R. (1988). Structure-activity relationships of vancomycin and related antibiotics. Session 8. Symposium of Recent Developments in Glycopeptide Antibiotics. 28th Intersci. Conf. Antimicrob. Agents Chemother.

40. Debono M. personal communication.

41. Marshall F.J. (1965). Structure studies on vancomycin. J. Med. Chem. 8:12–22.

42. Uttley A.H.C., Collins C.H., Naidoo J., George R.C. (1988). Vancomycin-resistant enterococci. Lancet i:57–58.

43. Chatterjee A.N., Perkins H.R. (1966). Compounds formed between nucleotides related to the biosynthesis of bacterial cell wall and vancomycin. Biochem. Biophys. Res. Commun. 24:489–494.

44. Anderson J.S., Matsuhashi M., Haskin M.A., Strominger J.L. (1965). Lipid-phosphoacetylmuramyl-pentapeptide and lipid-phosphodisaccharide-pentapeptide: Presumed membrane transport intermediates in cell wall synthesis. Proc. Natl. Acad. Sci. USA 53:881–889.

45. Anderson J.S., Matsuhashi M., Haskin M.A., Strominger J.L. (1967). Biosynthesis of the peptidoglycan of bacterial cell walls. Phospholipid carriers in the reaction sequence. J. Biol. Chem. 242:3180–3190.

46. Reynolds P.E. (1989). Structure, biochemistry and mode of action of glycopeptide antibiotics. Eur. J. Clin. Microb. Infect. Dis. 8:943–950.

47. Jeffs P.W., Nisbet L.J. (1988). Glycopeptide antibiotics.: A comprehensive approach to discovery, isolation and structure determination. In Actor P., Daneo-Moore L., Higgens M.L., Salton M.R.J., Stockman G.D. (Eds.). Antibiotic Inhibition of

Bacterial Cell Surface Assembly and Function. American Society of Microbiology, Washington, D.C., pp. 509–530.

48. Williams D.H., Rajananda V., Williamson M.P., Bojesen G. (1980). The vancomycin and ristocetin group of antibiotics. In Sammes P.G. (Ed.). Topics in Antibiotic Chemistry. Horwood, Chichester, England, pp. 119–158.

49. Good V.M., Gwynn M.N., Knowles D.J.C. (1990). MM 45289, a potent glycopeptide antibiotic which interacts weakly with diacetyl-L-lysyl-D-alanyl-D-alanine. J. Antibiot. 43:550–555.

6
Resistance and Mode of Action

THALIA I. NICAS and NORRIS E. ALLEN

Lilly Research Laboratories, Eli Lilly and Company, Indianapolis, Indiana

I. MECHANISM OF ACTION OF VANCOMYCIN AND OTHER GLYCOPEPTIDES

Glycopeptide antibacterial agents have the distinguishing feature of a rigid, sugar-substituted linear heptapeptide structure that binds to the terminal D-alanyl-D-alanine (D-Ala-D-Ala)-containing residues found in bacterial cell wall peptidoglycan and its precursors. The aglycone portions of glycopeptide antibiotics vary little among the compounds in this class: members differ primarily on the basis of their substituted sugar moieties (1). Some of the recently discovered glycopeptides contain fatty acid-substituted amino sugars (2). Although there are similarities, the glycopeptide antibiotics are structurally and mechanistically distinct from lipopeptide (e.g., daptomycin) (3,4) and lipoglycopeptide (e.g., ramoplanin) (5,6) antibiotics.

Molecular modeling studies show that the peptide backbone of the glycopeptide antibiotics can form a carboxylate binding pocket that can recognize and hold acyl-D-Ala-D-Ala residues by hydrogen bonding (7,8). Binding of the antibiotics to cell wall peptides blocks continued formation of cross-linked peptidoglycan. It is this action on the bacterial cell that accounts for the antibacterial activity of these compounds (9). It is noteworthy that the specificity of this binding reaction has been exploited by developing methodologies that have been used to screen for, isolate, and purify new glycopeptide antibiotics and their mixtures from fermentations (10,11).

A. Site of Action

Numerous studies since the discovery and clinical introduction of vancomycin have established that glycopeptide antibiotics act at the cell surface and interrupt cell wall peptidoglycan biosynthesis. Best and Durham (12) demonstrated a rapid and temperature-independent adsorption of vancomycin to both whole cells and cell walls isolated from *Bacillus subtilis*. Magnesium and other cations interfered with this adsorption. Subsequent reports (13,14) confirmed these findings and showed that the adsorption of vancomycin was noncovalent. Moreover, inhibitory effects of vancomycin in cell-free systems could be reversed by adding isolated cell wall preparations (15) or D-Ala-D-Ala-containing fragments of peptidoglycan (16). These studies all suggested an action of vancomycin at the cell surface.

Experiments using radiolabeled iodinated vancomycin to follow the distribution of antibiotic in intact bacteria (17) demonstrated that most of the antibiotic was associated with peptidoglycan. This observation, coupled with the finding that labeled vancomycin did not appear to associate with the membrane of protoplasts (9,17), indicated that vancomycin did not likely cross the cytoplasmic membrane and enter the cytoplasm.

Cell-free experiments demonstrating effects of vancomycin on peptidoglycan biosynthesis confirmed the site of antibacterial action of this antibiotic. A report by Chatterjee and Perkins (18) demonstrating that vancomycin formed a complex with the peptidoglycan precursor UDP-MurNAc-pentapeptide was the first evidence that vancomycin could bind to D-Ala-D-Ala-containing molecules. Further studies (19–23) confirmed complex formation and pointed to peptidoglycan biosynthesis as the lethal target of vancomycin. Studies investigating the membrane-associated reactions in peptidoglycan biosynthesis have shown that vancomycin inhibits the polymerization of UDP-GlcNAc and UDP-MurNAc-pentapeptide (24–28). Although secondary effects have been observed on the translocase reaction at higher concentrations of vancomycin (29), all the data indicate that the site of action of vancomycin is at the cell surface.

B. Effects on Peptidoglycan Biosynthesis

How do the glycopeptide antibiotics block formation of cell wall peptidoglycan? Figure 1 depicts the major stages and steps in bacterial peptidoglycan biosynthesis. Precursors providing the sugar backbone and the peptide side chains of the peptidoglycan polymer are synthesized in the cytoplasm. These nucleotide-linked molecules (UDP-GlcNAc and UDP-MurNAc-pentapeptide) are joined via the C_{55}-lipid carrier, undecaprenyl phosphate, to form a lipid pyrophosphate-linked disaccharide pentapeptide by membrane-associated enzymes. The disaccharide pentapeptide formed at this stage serves as the subunit by which the growing glycan chain is extended via transglycosylation. The lipid carrier is released and, following dephosphorylation, is made available to continue the cycle. The final stage of peptidoglycan biosynthesis involves cross-linking of growing nascent peptidoglycan to existing cell wall by transpeptidation. Details of peptidoglycan biosynthesis have been reviewed elsewhere (30,31).

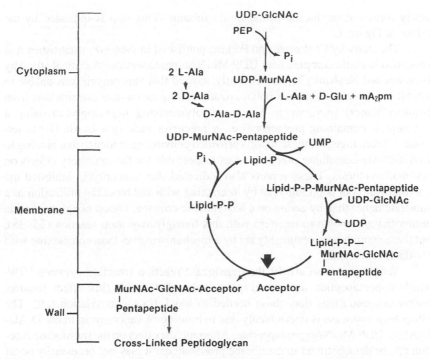

Fig. 1   Pathway for the biosynthesis of peptidoglycan in *Bacillus megaterium*. UDP-GlcNAc, uridine diphospho-*N*-acetylglucosamine; UDP-MurNAc-pentapeptide, uridine diphospho-*N*-acetylmuramyl-L-alanyl-D-glutamyl-*meso*-diaminopimelyl-D-alanyl-D-alanine.

Addition of vancomycin to whole cells of Gram-positive bacteria inhibits incorporation of cell wall amino acids into peptidoglycan and leads to a net accumulation of UDP-MurNAc-pentapeptide in the cytoplasm (32,33). Accumulation of this precursor is an indication of interference with a membrane-associated step, since blocking the cycle prevents regeneration of C_{55}-undecaprenyl phosphate needed to keep the cycle turning. The specific manner in which vancomycin blocked polymer formation required further studies using cell-free systems.

Studies employing particulate membranes from *Staphylococcus aureus* (24–26) and *Bacillus* spp. (27) to synthesize polymerized peptidoglycan starting with UDP-MurNAc-pentapeptide and UDP-GlcNac showed that addition of vancomycin induced an accumulation of lipid-linked sugar-peptide precursors. Accumulation of a lipid-bound disaccharide pentapeptide intermediate (26) indicated that vancomycin blocked the terminal transglycosylation reaction in which the acceptor end of the growing nascent peptidoglycan chain is transferred to the newly formed disaccharide pentapeptide subunit. This step is indicated by the arrow in Figure 1.

The study by Chatterjee and Perkins published in 1966 (18) established that vancomycin could complex with UDP-MurNAc-pentapeptide. A critical study by Hammes and Neuhaus (34) subsequently showed that vancomycin was unable to inhibit in vitro peptidoglycan polymerization using membrane preparations from *Gaffkya homari* (*Aerococcus viridans*) polymerizing peptidoglycan using a tetrapeptide-containing precursor (i.e., a precursor lacking a D-Ala-D-Ala terminus). Taken together, these reports provided convincing evidence that binding to D-Ala-D-Ala-containing precursors was responsible for the inhibitory effects on cell wall synthesis. These reports also indicated that vancomycin inhibited the polymerization of peptidoglycan by interacting with and blocking utilization of a substrate rather than by acting on a biosynthetic enzyme. Other, nonglycopeptide antibiotics are known to interfere with this transglycosylation reaction (35–38), but these compounds presumably act by a mechanism other than complexing with D-Ala-D-Ala termini.

Vancomycin also affects the translocase reaction (reaction between UDP-MurNAc-pentapeptide and undecaprenyl phosphate), but this effect requires higher concentrations than those needed to block transglycosylation (29). The effect on translocase is undoubtedly due to binding of vancomycin to the D-Ala-D-Ala of UDP-MurNAc-pentapeptide. Although the effect on the translocase reaction can be demonstrated in membrane preparations, it may not necessarily occur in vivo since vancomycin is not likely to have access to this reaction, which must occur at the interface between the cytoplasmic membrane and the cytoplasm (39).

As mentioned earlier and as described elsewhere in this monograph, vancomycin adopts a rigid configuration containing a cleft which an acetyl-D-Ala-D-Ala can bind via hydrogen bonding (39). Following the initial observation that vancomycin formed a complex with UDP-MurNAc-pentapeptide (18), Perkins (19) established that the minimal structure for forming a complex was acetyl-D-Ala-D-Ala. In a series of studies (19–23), synthetically prepared peptides were compared for complex-forming ability using ultraviolet difference spectroscopy. These studies established that although vancomycin could complex with acetyl-D-Ala-D-Ala, complex formation was largely determined by the terminal tripeptide moiety. In general, these early findings have been supported and extended by more recent reports (1,40). In addition, we now have a better understanding of the structural determinants of glycopeptides that facilitate interaction with the peptide substrates (41,42). For example, there are significant differences between the two glycopeptides ristocetin and vancomycin with respect to substrate binding specificities (8), and these differences may be related to the specific sugar moieties on the two antibiotics.

Taken together, the molecular and biochemical studies explain how glycopeptides inhibit peptidoglycan formation. By recognizing acyl-D-Ala-D-Ala residues in peptidoglycan and its precursors as a result of their three-dimensional structure, these compounds block the polymerization process. As pointed out by Reynolds (39), binding to the acyl-D-Ala-D-Ala termini could theoretically inhibit the transpeptidation reaction since this reaction also requires a presumably unshielded terminal D-Ala-D-Ala-containing pentapeptide. Although there remain unanswered questions regarding the exact mechanism by which glycopeptide antibiotics block the transglycosylation step (39,43), there is relative agreement about the basic molecular mechanism by which the glycopeptide antibiotics block peptidoglycan formation.

C. Other Glycopeptide Antibiotics

Much of our present understanding of the effects of glycopeptide antibiotics on cell wall peptidoglycan biosynthesis came from studies using vancomycin and, to a lesser extent, ristocetin. Although differences in detail are yet to be answered (44), the structural similarities shared by the glycopeptides dictate that all these antibiotics share very similar mechanisms of action. More recently discovered glycopeptides are aridicin (45), avoparcin (46), actaplanin (A4696) (47), A35512B (48,49), teicoplanin (50), A41030 (51), A82846B (LY264826) (52), and others (53). These compounds generally are thought to complex with acyl-D-Ala-D-Ala residues and block cell wall peptidoglycan biosynthesis.

A study by Hobbs and Allen (54) comparing the intrinsic activity of several glycopeptide antibiotics with vancomycin by measuring peptidoglycan formation in permeabilized cells revealed that although A41030A, A35512B, and actaplanin had similar activity against intact bacteria (Table 1), major differences in activity were detected against permeabilized cells (Fig. 2 and Table 1). Moreover, activity was demonstrated against both Gram-positive and Gram-negative bacteria as long as cells were permeabilized. These findings suggest that glycopeptide antibiotics

Table 1 Effect of Glycopeptide Antibiotics on Peptidoglycan Biosynthesis in Intact and Permeabilized Bacteria[a]

	Permeabilized cells[b]		Intact cells[c]	
Antibiotic	B. megaterium	E. coli	P. aeruginosa	B. megaterium
A41030A	5	50	10	0.5
A35512B	22	125	10	0.8
Actaplanin	17	370	50	1.0
Vancomycin	56	>400	200	0.6

[a]I_{50}, concentration (μg/ml) required to inhibit incorporation by 50%, is listed.
[b]Incorporation of UDP-N-acetyl-[^{14}C] glucosamine.
[c]Incorporation of [^{14}C] diaminopimelic acid.

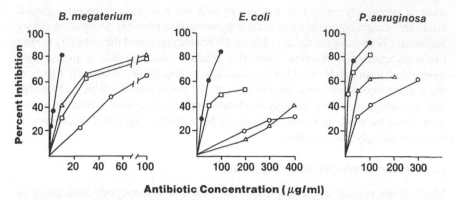

Fig. 2 Intrinsic activity of glycopeptide antibiotics: A41030A (filled circles); A35512B (squares); actaplanin (triangles); vancomycin (open circles). Intrinsic activity was measured as inhibition of incorporation of UDP-[14C]GlcNAc (uridine diphospho-N-acetyl [14C]glucosamine) in the presence of unlabeled UDP-MurNAc-pentapeptide into acid-insoluble peptidoglycan by permeabilized bacteria.

differ with respect to their intrinsic activity at the target site. These experiments also indicate that the apparent insensitivity of Gram-negative bacteria to glycopeptide antibiotics is due to the inability of these compounds to reach the target site. This is consistent with the report of an *Escherichia coli* mutant susceptible to vancomycin and other glycopeptide antibiotics (55).

Good et al. (56) reported that MM45289 (A82846A or eromomycin) was more active in vitro than vancomycin but had a 23-fold lower affinity for complexing with diacetyl-L-Lys-D-Ala-D-Ala. The findings with MM45289 raise the possibility that glycopeptide antibiotics may differ with respect to their mechanisms of action at the molecular level.

D. Implications for Resistance Development

Vancomycin has been used for more than 30 years to treat severe, life-threatening infections due to Gram-positive bacteria. Over the years, many of these bacteria developed resistance to β-lactam and other antibiotics but remained exquisitely susceptible to vancomycin. The unique ability of vancomycin and other glycopeptides to complex with a critical substrate involved in peptidoglycan formation was cited as an explanation of why resistance to these antibiotics had not appeared (9). The discovery of resistance to vancomycin and other glycopeptides in enterococci now obviates this point. Nevertheless, it is perhaps instructive that glycopeptide resistance involves a modification of the biosynthetic mechanism to survive the acyl-D-Ala-D-Ala-recognizing properties of these antibiotics (31,57–61).

II. GLYCOPEPTIDE RESISTANCE

A. Intrinsically Resistant Bacteria

Although the target site of vancomycin is clearly present in Gram-negative bacteria, resistance is readily explained by the presence of the permeability barrier imposed by the outer membrane, which effectively blocks access to the target.

A number of Gram-positive species also show intrinsic resistance to high concentrations of glycopeptides, for which no ready explanation is available. Intrinsically resistant Gram-positive species include members of the genera *Leuconostoc* (62,63) *Pediococcus* (64), and many species of lactobacilli (65). Typical vancomycin and teicoplanin minimum inhibitory concentrations (MIC) are ≥500–2000 μg/ml. Some of the actinomycete strains that produce glycopeptides also exhibit intrinsic resistance (66).

The basis for vancomycin resistance in the intrinsically resistant lactic acid bacteria (leuconostocs, pediococci, and lactobacilli) has not been systemically studied. Vancomycin inactivation by resistant leuconostocs has not been detected (63). Detailed studies of the peptidoglycan of *Leuconostoc* spp. and *Pediococcus* spp. have been reported, and the structure of the mature peptidoglycan reveals no clue to a resistance mechanism (67). The presence of L-alanyl-L-alanine in the interpeptide cross-bridges has been suggested by Isenberg et al. (68) as accounting for resistance in *Leuconostoc*, but this hypothesis is incompatible with the accepted mechanism of activity of these antibiotics, which involves the MurNAc-peptide precursor, not the interpeptide cross-bridge. Similar interpeptide bridges are present in vancomycin-susceptible species and are not present in vancomycin-resistant lactobacilli or in pediococci (67). Studies with the Gram-positive bacterium *G. homari* demonstrated that this organism can utilize either UDP-MurNAc-pentapeptide or UDP-MurNAc-tetrapeptide precursor in peptidoglycan synthesis, but only synthesis using pentapeptide precursor is sensitive to inhibition by vancomycin (34). This is a possible mechanism of intrinsic resistance.

With the increased use of vancomycin, clinical isolation of the intrinsically vancomycin-resistant lactic acid bacteria, particularly *Leuconostoc* spp. and *Pediococcus* spp., has been reported with increasing frequency (for examples see Refs. 68–74). Such isolates remain rare and are susceptible to other antibiotics, including penicillins, but their accurate identification poses a challenge to the clinical laboratory. Intrinsic resistance was recently recognized in the genus *Erysipelothrix*, an occasional cause of infections in humans (74).

The mechanism of resistance to teicoplanin by the teicoplanin-producing strain *Actinoplanes teichomyceticus* has been explored by cloning of resistance determinants into the related *Streptomyces lividans* (66). Acquisition of resistance correlated with increased binding of teicoplanin to cells, suggesting sequestration as a mechanism. Resistance to other glycopeptides in actinomycetes may occur by different mechanisms: many species, including the *S. lividans* host used in these cloning experiments, are intrinsically resistant to vancomycin (MIC > 100 μg/ml) (66).

B. Acquired and Newly Recognized Glycopeptide Resistance

Glycopeptide Resistance in Staphylococci

Despite over 30 years of experience with vancomycin, clinical isolates of *S. aureus* have remained uniformly susceptible. This contrasts with increasing occurrence of resistance to other agents (for a recent survey see Ref. 75). In vitro selection of resistant mutants is difficult (76). However, glycopeptide resistance in other staphylococci has now appeared: resistance to both vancomycin and teicoplanin has emerged in coagulase-negative staphylococci, most notably *Staphylococcus haemolyticus* (77–81), and teicoplanin-resistant vancomycin-sensitive isolates in coagulase-negative staphylococci (*S. haemolyticus* and *Staphylococcus epidermidis*) (77, 82–84) have appeared. More recently, an isolate of *S. aureus* resistant to teicoplanin but susceptible to vancomycin has been reported (85).

S. *haemolyticus* appears to be able to acquire resistance to vancomycin and teicoplanin fairly readily in the laboratory by passage on subinhibitory concentrations of vancomycin or teicoplanin (77,86,87). Selection of resistant mutants occurs more readily with teicoplanin than with vancomycin, and typically such strains retain some vancomycin susceptibility. Clinical isolates have followed a similar pattern (84,88,89). In one study of 23 isolates of glycopeptide-resistant strains of *S. haemolyticus* and *S. epidermidis*, MIC ranged from 2 to 4 μg/ml of vancomycin compared with 8 to 32 μg/ml of teicoplanin (88). A recent study of 362 isolates of coagulase-negative staphylococci found 23% to be "intermediate" in susceptibility to teicoplanin (MIC 4–32 μg/ml), with 1.7% resistant, but only a single isolate was intermediate to vancomycin (84).

Glycopeptide resistance in *S. aureus* emerging during therapy was reported recently by Kaatz et al. (85), who described a teicoplanin-resistant strain (MIC 8 μg/ml). The strain retained vancomycin susceptibility. Spontaneous mutants with the same phenotype could be isolated from a susceptible earlier isolate believed to be the parent strain.

The mechanism for resistance in staphylococci has not been elucidated, but alterations in wall structures have been observed and a 39 kD protein has been associated with resistance in both naturally occurring and laboratory-derived strains. Resistance is constitutively expressed and appears to be chromosomally mediated (77,85,89,90). This resistance can be clearly distinguished both genetically and phenotypically from the plasmid-mediated inducible resistance found in enterococci. However, it has been possible in the laboratory to transfer plasmid-mediated VanA glycopeptide resistance from *Enterococcus faecalis* (*Strepto-*

coccus faecalis) into *S. aureus* by conjugation (see later), raising the possibility of a new form of glycopeptide resistance in staphylococci in the future (91).

Clearly, the susceptibility of coagulase-negative staphylococci to teicoplanin cannot be inferred from results of tests of vancomycin susceptibility. Experience with resistance in staphylococci has led to the suggestion that there may be subtle differences in the mode of action of vancomycin and teicoplanin (86–89). This interesting concept requires further exploration.

Glycopeptide Resistance in Enterococci

Heterogeneity of Resistance.   Until recently, *Enterococcus faecium* (*Streptococcus faecium*) and *E. faecalis* were believed to be uniformly susceptible to vancomycin. Over the past few years several forms of glycopeptide resistance among the enterococci have been reported. Several reviews summarize recent developments (57,74,92). The designations VanA and VanB have been used to label the two best described phenotypes (57,93,94). VanA strains are highly resistant to both vancomycin and teicoplanin and share a common resistance gene, *vanA* (88). VanB strains are resistant to vancomycin but retain susceptibility to teicoplanin. Both forms of resistance are inducible, requiring exposure to glycopeptide for expression, and both are associated with the presence of novel membrane proteins with apparent molecular weights of 38,000–40,000 (95–98). This categorization will probably prove inadequate to describe all forms of resistance emerging in the enterococci. The originally described VanB strains showed only low-level resistance to vancomycin, with MIC of 8–32 μg/ml, but *E. faecium* resistant to high levels of vancomycin but susceptible to teicoplanin were recently reported (99,100). Constitutive expression of resistance to both teicoplanin and vancomycin can be selected in vitro in VanB strains (95), and such isolates are likely to appear in the clinic. In addition to these forms of emergent resistance, several enterococcal species appear to be intrinsically resistant to vancomycin. The designation VanC has been used for the type of resistance exhibited by *Enterococcus gallinarum* (*Streptococcus gallinarum*), intrinsic resistance to vancomycin but not teicoplanin (93,94). In vitro synergy between vancomycin and penicillins (94,101,102) may occur in enterococci of all three resistance phenotypes.

Noninducible Low-Level Vancomycin Resistance: VanC Phenotype. Recent experience has focused attention on vancomycin resistance among a number of enterococcal species, including *E. gallinarum*, *Enterococcus cassiflavus* (*Streptococcus cassiflavus*), *Enterococcus avium* (*Streptococcus avium*), and some strains of *E. faecalis* (74,94,103–106). These strains may exhibit vancomycin MIC of 8–32 μg/ml and are typically susceptible to teicoplanin (74,88). It has been noted that enterococci with moderate levels of vancomycin resistance (MIC 16–32 μg/ml) are rather difficult to detect, and their incidence may be underestimated (104,106,107).

The designation "VanC" has been applied to *E. gallinarum* (93). *E. gallinarum* exhibits intrinsic low-level vancomycin resistance that appears to be a characteristic of the species (74,93). Typical vancomycin MIC are 8–32 μg/ml (74,93,104,106). As in VanA resistance (see later), a protein related to D-alanine: D-alanine ligase has been found to be essential to expression of resistance in *E. gallinarum* (108). This gene has been named *vanC* and is found in all *E. gallinarum* strains (109). Studies by Shlaes and coworkers (110) also suggested a mechanism by which *E. gallinarum* may avoid vancomycin susceptibility and exhibit vancomycin-ampicillin synergy. They hypothesize that a low-molecular-weight penicillin binding protein functions as a carboxypeptidase that removes the terminal D-Ala from the wall precursor with a pentapeptide terminating in D-Ala-D-Ala, resulting in a tetrapeptide-containing precursor that is not susceptible to vancomycin inactivation. Vancomycin susceptibility in these strains increases at a concentration of ampicillin capable of inactivating the low-molecular-weight penicillin-binding protein (110).

Inducible Vancomycin Resistance with Teicoplanin Susceptibility: VanB Phenotype.   The resistance phenotype termed "VanB" was first characterized in a study of a French isolate, *E. faecium* D366 (95). This strain is resistant to vancomycin (MIC 32 μg/ml) but retains its susceptibility to teicoplanin (MIC 0.5 μg/ml). Strains with a similar resistance phenotype in both *E. faecium* and *E. faecalis* have since been reported (106,107,111). Although the original VanB phenotype described strains with moderate levels of vancomycin resistance, this designation is now being used to include strains with high-level vancomycin resistance and teicoplanin susceptibility in both *E. faecium* and *E. faecalis* (100,112). Resistance is induced by exposure to vancomycin but not by exposure to teicoplanin (95). Cells grown in the presence of vancomycin become teicoplanin resistant, however, indicating that the difference between the antibiotics results from differential ability to induce resistance, not differential activity against resistant cells. Resistance is associated with the presence of a membrane protein with an apparent molecular weight of 39,500. Recently a gene was identified in VanB isolates of both high-level and low-level vancomycin resistance that is homologous to *vanA*, the gene involved in synthesis of a substitute for D-Ala-D-Ala in the peptidoglycan precursor of VanA strains (see later) (112), suggesting a similar resistance mechanism.

High-Level Resistance to Vancomycin and Teicoplanin: VanA Phenotype.   The VanA resistance phenotype is the best characterized form of glycopeptide resistance. The first reports from Britain and France in 1988 and 1989 (96,113–115) have been joined by others from elsewhere in Europe and in the United States (57,74). Most isolates are *E. faecium*, but *E. faecalis* and possibly *E. avium* isolates are also found (88,113,115). Typical MIC are between 64 and 2000 μg/ml for vancomycin and 16–500 μg/ml for teicoplanin (88,115). Information is rapidly accumulating on both the genetics and the mechanism of resistance of

VanA strains. Characterization of resistant isolates has shown sufficient variety to rule out clonal dissemination of a single strain (57,88,113,115).

Genetics of Resistance

For VanA strains, resistance in the majority of strains has been shown to be transferable to other enterococci (96–98,114–117). Both self-transferable plasmids and a non-conjugative plasmid carrying vancomycin resistance have been reported (88,114). In one instance, resistance was found to be encoded on a pheromone-response plasmid that transferred at very high frequency (116). The resistance genes of VanA strains have been the subject of intensive study by Leclerq and Courvalin and coworkers, who cloned the major resistance determinants (see Ref. 57 for a review).

Characterization of plasmids for other antibiotic resistance markers, size, and restriction maps has shown that a number of different plasmids may carry vancomycin resistance, and hybridization studies have shown that the plasmids are related, sharing homology in the *vanA* gene (57,88). Recent data indicate that the resistance genes are on a transposon (118). Certain vancomycin plasmids can undergo conjugal transfer into some streptococcal species and to *Listeria monocytogenes*, as well as enterococci (57,114), and can be introduced by transformation into other streptococci (96,114). One group was able to introduce resistance into *Streptococcus pyogenes* and *Listeria* but not *S. aureus* (96,117). More recently, Noble et al. (91) were able to transfer vancomycin resistance into *S. aureus* by conjugation, selecting for a linked marker, erythromycin resistance. This work carries the ominous implication that VanA resistance may eventually spread into staphylococci.

Resistance in VanB and other enterococci have not been shown to be transferable or to be associated with plasmid DNA.

The inducibility of resistance is also an area of interest. Handwerger and Kolokathis (119) showed that moenomycin, an inhibitor of the transglycosylation reaction of peptidoglycan synthesis, acts as an inducer of vancomycin resistance in a VanA strain. Other inhibitors of wall biosynthesis did not. For the VanB strains, differences in the ability to induce resistance are among the determinants of relative activity of glycopeptides (95). A two-component regulatory system VanS-VanR has been shown to regulate expression of the vancomycin resistance genes in a VanA strain (120). Structural homologies evident from sequencing studies suggest the system may function as a signal-transducing system similar to the EnvZ-OmpR two-component system (120).

Studies of Resistance Mechanisms

Understanding of the biochemical mechanism of vancomycin resistance in enterococci has advanced rapidly. Detoxification by enzymatic alteration or antibiotic inactivation by complexation with excess peptidoglycan substrate were excluded by early studies (95,97,115). The appearance of a novel protein coincident with induction of resistance in both VanA and VanB strains suggested a major function of this protein in resistance. One suggestion was prevention of access of vancomycin to its target, MurNAc-pentapeptide, by some interaction with this substrate (58,59). Current evidence demonstrates that the actual function is enzymatic activity resulting in a modified substrate unable to interact with vancomycin.

Studies of the molecular biology of VanA resistance in the laboratories of Courvalin and Walsh have elucidated the mechanism of resistance in VanA strains and the specific role of the resistance-associated 39 kD protein (57,60,92,117,121–124). This group has shown three plasmid-encoded genes to be essential to the expression of vancomycin resistance. The VanA gene product is the membrane-associated resistance protein and has been shown to be a D-alanine:D-alanine ligase of altered substrate specificity, catalyzing the condensation of D-alanine to an alternative substrate, D-lactate. The *vanH* gene encodes an enzyme, VanH, required for generation of the substrate of VanA. The function of a third essential gene, *vanX*, is unknown at this time (120). A carboxypeptidase encoded by the *vanY* gene is also part of the resistance plasmid (125). The genes act in concert to produce an alternative peptidoglycan precursor in which the normal D-Ala-D-Ala terminus is replaced with D-Ala-D-lactate. This new precursor is unable to bind vancomycin and thus confers resistance (60,61,92,122).

Cloning of a VanA resistance determinant allowed the identification of the *vanA* gene encoding an inducible membrane protein, with a calculated molecular mass of 37,400. Expression of this protein is necessary but not sufficient for expression of resistance (57,117). Nucleic acid sequencing of the *vanA* gene revealed a high degree of homology with D-alanine:D-alanine ligase cell wall biosynthesis enzymes of *E. coli* and *Salmonella typhimurium*, and the gene was able to complement an *E. coli* mutant with defective D-alanine:D-alanine ligase activity (121). Subsequent studies of the purified VanA protein demonstrated activity as a D-alanine:D-alanine ligase of altered substrate specificity (60,122). The VanA protein preferentially catalyzed condensation of D-alanine to α-hydroxy acids, such as lactate and 2-hydroxybutyrate. The catalytic activity of the VanA protein differs from that of other D-alanine:D-alanine ligases not only in its substrate specificity but in that the bond formed is an ester linkage, resulting in a depsipeptide rather than a peptide product. The product formed by this reaction can be utilized by the next enzyme in peptidoglycan synthesis, the adding enzyme that adds D-Ala-D-Ala to UDP-MurNac-tripeptide to generate UDP-MurNac-pentapeptide (122). Thus it was predicted that the product of the VanA gene reaction would be incorporated as the terminus of the UDP-MurNac-pentapeptide precursor. VanA ligation products, such as D-Ala-D-lactate and D-Ala-D-hydroxybutryate, do not bind vancomycin (122) and thus could afford vancomycin resistance. The role of the VanH protein appears to be generation of substrate for the VanA ligation. Sequencing of the *vanH* gene demonstrated its homology to 2-hydroxycarboxylic acid dehydrogenases of *Leuconostoc* and *Lactobacillus* (123). Purification and characterization of the enzyme showed that it can synthesize the preferred substrates of VanA, catalyzing stereospecific reduction of 2-keto acids, such as 2-keto butyrate or pyruvate, to form the corresponding D-2-hydroxy acids, such as D-2-hydroxybutyrate or D-lactate (122). Both lactate and 2-keto-butyrate added to the growth medium of a *vanH*-deficient strain are able to restore vancomycin resistance (124). Additions of excess of other amino acids, such as glycine, that could compete as substrate can reduce vancomycin resistance in a VanA strain (126). Several subsequent studies demonstrated that the precursor found in resistant isolates is UDP-MurNac-alanyl-glutamyl-lysyl-alanyl-lactate, as predicted by biochemical studies (61,127,128), and that this product can be used in peptidoglycan synthesis by membrane systems of both resistant and sensitive cells (61). Tetrapeptide terminating in alanine has also been found, but it is uncertain if this is naturally present or an artifact of isolation, because the ester linkage in D-ala-D-lactate is labile (60).

Based on these studies, a model of resistance that highlights modification of the vancomycin target as the principal mechanism of resistance has been elaborated (60,92,122). The VanA gene product catalyzes the synthesis of a depsipeptide, D-Ala-D-lactate, rather than D-Ala-D-Ala. D-lactate is generated from pyruvate by the *vanH* gene product. Subsequent incorporation of D-Ala-D-lactate into peptidoglycan precursor leads to a substrate not recognized by vancomycin.

In the proposed model, incorporation of altered substrate into peptidoglycan in turn leads to a modified structure available for cross-linking by transpeptidation. An unresolved aspect of the model is the requirement for transpeptidation to occur using a modified pentapeptide structure. Studies of *G. homari*, which is able to synthesize peptidoglycan using UDP-MurNAc-pentapeptides containing modified amino acids at position 5 (129) and the activity of D-alanine carboxypeptidase (an enzyme related to the cross-linking transpeptidases) (130) on substrate containing D-Ala-D-lactate, have been cited in support (60,122).

This model does not exclude a role for the inducible carboxypeptidases observed in both VanA and VanB strains (60,131,132). Carboxypeptidase may contribute as an accessory enzyme by cleaving the terminal D-alanine residues of the standard D-Ala-D-Ala termini synthesized by the native chromosomally encoded peptidoglycan synthesis pathway. It has also been suggested that a reduction in the availability of the standard D-Ala-D-Ala termini may be achieved by downregulation of the normal D-alanine:D-alanine ligase (122).

Gutmann and coworkers proposed a model that emphasizes the role of carboxypeptidase activity in resistance. They demonstrated that induced vancomycin-resistant cells of both the VanA and VanB phenotypes have a reduction in vancomycin-inactivating material (presumably pentapeptide ending in D-Ala-D-Ala) and that membranes from resistant cells possess an activity able to abolish the ability of vancomycin to bind exogenous peptide (58). Subsequent experiments showed enhanced levels of D,D-carboxypeptidase activity in the cytoplasmic membrane of resistant cells (131). Gutmann's group proposed a mechanism of resistance involving induction of carboxypeptidase activity that would remove the terminal D-ala from MurNAc-pentapeptide, thus eliminating the vancomycin target. This model requires transpeptidation using a tetrapeptide rather than pentapeptide substrate. Some indirect evidence suggests that functional replacement of the activity of penicillin binding protein 5 by higher molecular weight penicillin binding proteins occurs after induction of resistance by vancomycin in an *E. faecium* strain with the VanB phenotype (132). Both the enhanced ampicillin sensitivity of induced, vancomycin-resistant isolates and altered morphology of VanB strains can be accounted for by this model (132). Transpeptidation using tetrapeptides in resistant enterococci is still a matter of speculation, but efficient peptidoglycan synthesis using UDP-MurNAc-tetrapeptide has been demonstrated in another Gram-positive bacterium, *G. homari* (34).

Diversity of Origin of Resistance

Despite the late emergence of glycopeptide resistance after over 30 years of clinical use, the current situation is one in which we see a wide variety of phenotypic and genotypic heterogeneity (57,81). The data available to date on the interrelationship between the various forms of vancomycin resistance in Gram-positive bacteria suggest diverse origins. Hybridization studies using the cloned *vanA* gene have shown that although this gene is homologous among VanA isolates regardless of source, no cross-hybridization is seen when this gene is used to probe VanB isolates, other resistant enterococci, staphylococci, or representatives of intrinsically resistant strains of *Leuconostoc, Pediococcus, Lactobacillus*, or glycopeptide-producing actinomycetes (57,74,121). Similarly, when antibody to the 39 kD protein of a VanA strain was used to probe intrinsically resistant strains, no cross-reactivity was found (133). Partial protein characterizations comparing the 39 kD protein of VanA strains and the 39.5 kD protein of VanB strains failed to demonstrate any relationship, although 8 of the first 11 amino acids of the N terminus are the same (134). Thus the origin of resistance remains unknown. The resistance mechanisms involved do not appear to originate from either the glycopeptide-producing actinomycetes or the highly resistant lactic acid genera *Leuconostoc, Pediococcus*, or *Lactobacillus*. It is intriguing, however, that the *vanH* gene is related to genes of D-lactate dehydrogenases of *Leuconostoc* and *Lactobacillus* (122). It remains to be established whether the emergent resistant bacteria share similar mechanisms of resistance. Probes based on the homologous regions of D-alanine:D-alanine ligase genes of *E. coli* and *Salmonella* and the *vanA* gene have been used to discover related genes, *vanC* and *vanB* in enterococci, and a similar role for these genes has been suggested (108,112). Target alteration in the form of modified pentapeptide is clearly part of the resistance mechanism for VanA strains, but other forms of target alteration, such as that generated by D,D-carboxypeptidase activity, may yet prove to be a major mechanism of resistance in other strains.

C. Clinical Implications of Glycopeptide Resistance

Glycopeptide resistance in the enterococci and staphylococci is apparently still uncommon, but uniform susceptibility can no longer be assumed. Problems with disk diffusion testing for the accurate identification of glycopeptide-resistant strains have been pointed out in several studies (57,74,104,106,107). Differential susceptibility to vancomycin and teicoplanin, especially among coagulase-negative staphylococci, suggests the need for separate determination of these two antibiotics.

The emergence of glycopeptide resistance in *Enterococcus* spp., especially in *E. faecium*, is of particular concern because of the few classes of antibiotics active against these bacteria, and resistance to other antibiotics, particularly gentamicin, β-lactams, and quinolones, is also increasing (74,135,136).

The emergence of glycopeptide resistance also highlights the need for ongoing development of new antibiotics. Glycopeptide-resistant strains are not cross-resistant to the lipopeptide antibiotic daptomycin or the glycolipodepsipeptide ramoplanin (88,93,96). Some experimental glycopeptides with significant activity against vancomycin-resistant enterococci have been reported (2,133). The insight gained with regard to the nature of glycopeptide resistance will help move toward a basis for the rational design of future agents.

REFERENCES

1. Barna J.C.J., Williams D.H. (1984). The structure and mode of action of glycopeptide antibiotics of the vancomycin group. Annu. Rev. Microbiol. 38:339.
2. Nagarajan R. (1991). Antibacterial activities and modes of action of vancomycin and related glycopeptides. Antimicrob. Agents Chemother. 35:605.
3. Allen N.E., Hobbs J.N., Jr., Alborn W.E., Jr. (1987). Inhibition of peptidoglycan biosynthesis in gram-positive bacteria by LY146032. Antimicrob. Agents Chemother. 31:1093.
4. Allen N.E., Alborn W.E. Jr., Hobbs J.N. Jr. (1991). Inhibition of membrane potential-dependent amino acid transport by daptomycin. Antimicrob. Agents Chemother. 35:2639.
5. Somner E.A., Reynolds P.E. (1990). Inhibition of peptidoglycan biosynthesis by ramoplanin. Antimicrob. Agents Chemother. 34:413.
6. Reynolds P.E., Somner E.A. (1990). Comparison of the target sites and mechanisms of action of glycopeptide and lipoglycodepsipeptide antibiotics. Drugs Exp. Clin. Res. 16:385.
7. Williams D.H., Waltho J.P. (1988). Molecular basis of the activity of antibiotics of the vancomycin group. Biochem. Pharmacol. 37:133.

8.   Williams D.H., Waltho J.P. 1989). Molecular basis of the activity of antibiotics of the vancomycin group. Pure Appl. Chem. 61:585.

9.   Gale E.F., Cundliffe E., Reynolds P.E., Richmond M.H., Waring M.J. (1981). The Molecular Basis of Antibiotic Action, 2nd ed. John Wiley & Sons, New York, p. 144.

10. Rake J.B., Gerber R., Mehta R.J., Newman D.J., Oh Y.K., Phelen C., Shearer M.C., Sitrin R.D., Nisbet L.J. (1986). Glycopeptide antibiotics: A mechanism-based screen employing a bacterial cell wall receptor mimetic. J. Antibiot. 39:58.

11. Folena-Wasserman G., Sitrin R.D., Chapin F., Snader K.M. (1987). Affinity chromatography of glycopeptide antibiotics. J. Chromatogr. 392:225.

12. Best G.K., Durham N.N. (1965). Vancomycin adsorption to *Bacillus subtilis* cell walls. Arch. Biochem. Biophys. 111:685.

13. Reynolds P.E. (1966). Antibiotics affecting cell wall biosynthesis. Symp. Soc. Gen. Microbiol. 16:47.

14. Best G.K., Best D.V., Ferguson D.V., Durham N.N. (1968). Adsorption of vancomycin by sensitive and resistant organisms. Biochim. Biophys. Acta 165:558.

15. Sinha R.K., Neuhaus F.C. (1968). Reversal of the vancomycin inhibition of peptidoglycan synthesis by cell wall. J. Bacteriol. 96:374.

16. Nieto M., Perkins R., Reynolds P.E. (1972). Reversal by a specific peptide (diacetyl-$\alpha\gamma$-L-diaminobutyryl-D-alanyl-D-alanine) of vancomycin inhibition in intact bacteria and cell-free preparations. Biochem. J. 126:139.

17. Perkins H.R., Nieto M. (1970). The preparation of iodinated vancomycin and its distribution in bacteria treated with antibiotic. Biochem. J. 116:83.

18. Chatterjee A.N., Perkins H.R. (1966). Compounds formed between nucleotides related to the biosynthesis of bacterial cell wall and vancomycin. Biochem. Biophys. Res. Commun. 24:489.

19. Perkins H.R. (1969). Specificity of combination between mucopeptide precursors and vancomycin or ristocetin. Biochem. J. 111:195.

20. Nieto M., Perkins H.R. (1971). Physicochemical properties of vancomycin and iodovancomycin and their complexes with diacetyl-L-lysyl-D-alanyl-D-alanine. Biochem. J. 123:773.

21. Nieto M., Perkins H.R. (1971). Modifications of the acyl-D-alanyl-D-alanine terminus affecting complex formation with vancomycin. Biochem. J. 123:789.

22. Nieto M., Perkins H.R. (1971). The specificity of combination between ristocetins and peptides related to bacterial cell wall mucopeptide precursors. Biochem. J. 124:845.

23. Perkins H.R., Nieto M. (1972). The molecular basis for the antibiotic action of vancomycin, ristocetin and related drugs. Proceedings of the Symposium on Molecular Mechanisms of Antibiotic Action on Protein Biosynthesis and Membranes, Elsevier, Amsterdam, p. 363.

24. Anderson J.S., Matsuhashi M., Haskin M.A., Strominger J.L. (1965). Lipid-phosphoacetylmuramylpentapeptide and lipid-phosphodisaccharide-pentapeptide: Presumed membrane transport intermediates in cell wall synthesis. Proc. Natl. Acad. Sci. USA 53:881.

25. Anderson J.S., Meadow P.M., Haskin M.A., Strominger J.L. (1966). Biosynthesis of the peptidoglycan of bacterial cell walls. I. Utilization of uridine diphosphate

acetylmuramyl pentapeptide and uridine diphosphate acetylglucosamine for peptidoglycan synthesis by particulate enzymes from *Staphylococcus aureus* and
Micrococcus lysodeikticus. Arch. Biochem. Biophys. 116:487.

26. Anderson J.S., Matsuhashi M., Haskin M.A., Strominger J.L. (1967). Biosynthesis
of the peptidoglycan of bacterial cell walls. II. Phospholipid carriers in the reaction
sequence. J. Biol. Chem. 242:3180.

27. Reynolds P.E. (1971). Peptidoglycan synthesis in bacilli. I. Effect of temperature on
the in vitro system from *Bacillus megaterium* and *Bacillus stearothermophilus*.
Biochim. Biophys. Acta 237:239.

28. Reynolds P.E. (1971). Peptidoglycan synthesis in bacilli. II. Characteristics of
protoplast membrane preparations. Biochim. Biophys. Acta 237:255.

29. Struve W.G., Sinha R.K., Neuhaus F.C. (1966). On the initial stage in peptidoglycan
synthesis. Phospho-*N*-acetylmuramyl-pentapeptide translocase (uridine monophos-
phate). Biochemistry 5:82.

30. Pellon G. (1990). Biosynthesis of peptidoglycans from eubacterial cell envelopes.
Bull. Inst. Pasteur 88:203.

31. Bugg T.D.H., Walsh C.T. (1992). Intracellular steps of bacterial cell wall peptidoglycan biosynthesis: Enzymology, antibiotics, and antibiotic resistance. Nat. Prod.
Rep. 1992:199.

32. Reynolds P.E. (1961). Studies on the mode of action of vancomycin. Biochim.
Biophys. Acta 52:403.

33. Jordan D.C. (1961). Effect of vancomycin on the synthesis of the cell wall mucopeptide of *Staphylococcus aureus*. Biochem. Biophys. Res. Commun. 6:167.

34. Hammes W.P., Neuhaus F.C. (1974). On the mechanism of action of vancomycin:
Inhibition of peptidoglycan synthesis in *Gaffkya homari*. Antimicrob. Agents
Chemother. 6:722.

35. Lugtenberg E.J.J., van Schijndel-van Dam A., van Bellegem T.H.M. (1971). In vivo
and in vitro action of new antibiotics interfering with the utilization of *N*-acetyl-
glucosamine-*N*-acetyl-muramyl-pentapeptide. J. Bacteriol. 108:20–29.

36. Somma S., Merati W., Parenti F. (1977). Gardimycin, a new antibiotic inhibiting
peptidoglycan synthesis. Antimicrob. Agents Chemother. 11:396.

37. Linnett P.E., Strominger J.L. (1973). Additional antibiotic inhibitors of peptidoglycan synthesis. Antimicrob. Agents Chemother. 4:231.

38. Spiri-Nakagawa P., Oiwa R., Tanaka Y., Tanaka H., Omura S. (1980). The site of
inhibition of bacterial cell wall peptidoglycan synthesis by azureomycin B, a new
antibiotic. J. Biochem. 2:565.

39. Reynolds P.E. (1989). Structure, biochemistry and mechanism of action of glycopeptide antibiotics. Eur. J. Clin. Microbiol. Infect. Dis. 8:943.

40. Kim K.-H., Martin Y., Otis E., Mao J. (1989). Inhibition of iodine-125 labeled
ristocetin binding to *Micrococcus luteus* cells by the peptides related to bacterial
cell wall mucopeptide precursors: Quantitative structure-activity relationships. J.
Med. Chem. 32:84.

41. Molanari H., Pastore A., Lian L., Hawkes G.E., Sales K. (1990). Structure of
vancomycin and a vancomycin/D-Ala-D-Ala complex in solution. Biochemistry
29:2271.

42. Arriaga P., Laynez J., Menendez M., Canada J., Garcia-Bianco F. (1990). Thermo-
     dynamic analysis of the interaction of the antibiotic teicoplanin and its aglycone with
     cell-wall peptides. Biochem. J. 265:69.
43. Jeffs P.W., Nisbet L.J. (1988). Glycopeptide antibiotics: A comprehensive approach
     to discovery, isolation, and structure determination. In P Actor, L Daneo-Moore,
     ML Higgins, MRJ Salton, GD Shockman (eds.), Antibiotic Inhibition of Bacterial
     Cell Surface Assembly and Function. American Society for Microbiology, Washing-
     ton, D.C., p. 509.
44. Felice G., Lanzarini P., Orsolini P., Soldini L., Perversi L., Carnevale G., Comolli
     G., Castelli F. (1990). Effect of teicoplanin and vancomycin on *Staphylococcus
     aureus* ultrastructure. Microbiologica 13:231.
45. Shearer M.C., Actor P., Bowie B.A., Grappel S.F., Nash C.H., Newdman D.J., Oh
     Y.K., Pan S.H., Nisbet L.J. (1985). Aridicins, novel glycopeptide antibiotics. I.
     Taxonomy, production and biological activities. J. Antibiot. 38:555.
46. Ellsstad G.A., Leese R.A., Morton G.O., Barbatschi F., Gore W.E., McGahren
     W.R., Armitage I.M. (1981). Avoparcin and epiavoparcin. J. Am. Chem. Soc.
     103:622.
47. Hunt A.H., Elzey T.K., Merkel K.E., Debono M. (1984). Structures of the acta-
     planins. J. Org. Chem. 49:641.
48. Michel K.H., Shah R.M., Hamill R.L. (1980). A35512, a complex of new anti-
     bacterial antibiotics produced by *Streptomyces candidus*. J. Antibiot. 33:1397.
49. Hunt A.H., Vernon P.D. (1981). Complexation of acetyl-D-alanyl-D-alanine by
     antibiotic A35512B. J. Antibiot. 34:469.
50. Somma S., Gastaldo L., Corti A. (1984). Teicoplanin, a new antibiotic from
     Actinoplanes teichomyceticus nov.sp. Antimicrob. Agents Chemother. 26:917.
51. Hunt A.H., Dorman D.E., Debono M., Molloy R.M. (1985). Structure of antibiotic
     A41030A. J. Org. Chem. 50:2031.
52. Nagarajan R., Berry D.M., Hunt A.H., Occolowitz J.L., Schabel A.A. (1989).
     Conversion of antibiotic A82846B to orienticin A and structural relationships of
     related antibiotics. J. Org. Chem. 58:983.
53. Gadebusch H.H., Stapley E.O., Zimmerman S.B. (1992). The discovery of cell wall
     active antibacterial antibiotics. Crit. Rev. Biotechnol. 12:225.
54. Hobbs J.N. Jr., Allen N.E. (1983). A41030: A glycopeptide-aglycone antibiotic that
     interferes with peptidoglycan biosynthesis. Abstracts of 23rd Interscience Confer-
     ence on Antimicrobial Agents and Chemotherapy, p. 442.
55. Shlaes D.M., Shlaes J.H., Davies J., Williamson R. (1989). *Escherichia coli* suscep-
     tible to glycopeptide antibiotics. Antimicrob. Agents Chemother. 33:192.
56. Good V.M., Gwynn M.N., Knowles D.J.C. (1990). MM45289, a potent glycopep-
     tide antibiotic which interacts weakly with diacetyl-L-lysyl-D-alanyl-D-alanine. J.
     Antibiot. 43:550.
57. Courvalin P. (1990). Resistance of enterococci to glycopeptides. Antimicrob.
     Agents Chemother. 34:2291.
58. Al-Obeid S., Collatz E., Gutmann L. (1990). Mechanism of resistance to vanco-
     mycin in *Enterococcus faecium* D399 and *Enterococcus faecalis* A256. Antimicrob.
     Agents Chemother. 34:252.
59. Knox J.R., Pratt R.F. (1990). Different modes of vancomycin and D-alanyl-

D-alanine peptidase binding to cell wall peptide and a possible role for the vancomycin resistance protein. Antimicrob. Agents Chemother. 34:1342.

60. Bugg T.T.H., Dutka-Malen S., Arthur M., Courvalin P., Walsh C.T. (1991). Identification of vancomycin resistance protein VanA as a D-alanine:D-alanine ligase of altered substrate specificity. Biochemistry 30:2017.

61. Allen N.E., Hobbs J.N. Jr., Richardson J.M., Riggin R.M. (1992). Biosynthesis of modified peptidoglycan precursors by vancomycin-resistant *Enterococcus faecium*. FEMS Microbiol. Lett. 98:109.

62. Buu-Hoï A., Branger C., Acar J.F. (1985). Vancomycin-resistant streptococci or *Leuconostoc* sp. Antimicrob. Agents Chemother. 128:458.

63. Orberg P.K., Sandine W.E. (1984). Common occurrence of plasmid DNA and vancomycin resistance in *Leuconostoc* spp. Appl. Environ. Microbiol. 48:1129.

64. Colman G., Efstratiou A. (1987). Vancomycin-resistant leuconostocs, lactobacilli and now pediococci. J. Hosp. Infect. 10:1

65. Vescovo M., Morelli L., Bottazzi V. (1982). Drug resistance plasmids in *Lactobacillus acidophilus* and *Lactobacillus reuteri*. Appl. Environ. Microbiol. 43:50.

66. Moroni M.C., Granozzi C., Lorenzzi R., Parenti F., Sosio M., Denaro M. (1989). Cloning of a DNA region of *Actinoplanes teichomyceticus* conferring teicoplanin resistance. FEBS Lett. 253:108.

67. Schleifer H.K., Kandler O. (1972). Peptidoglycan types of bacterial cell walls and their taxonomic implications. Bacteriol. Rev. 36:407.

68. Isenberg H.D., Vellozi E.M., Shapiro J., Rubin L.G. (1988). Clinical laboratory challenges in the recognition of *Leuconostoc* spp. J. Clin. Microbiol. 26:479.

69. Coovadia Y.M., Sowa A., van den Ende J. (1987). Meningitis cased by vancomycin-resistant *Leuconostoc* sp. J. Clin. Microbiol. 25:1784.

70. Horowitz H.W., Handwerger S., van Horn K.G., Wormser G.P. (1987). *Leuconostoc*, an emerging vancomycin-resistant pathogen. Lancet 2:1829.

71. Ruoff K.L., Kuritzkes D.R., Wolfson J.S., Ferraro M.J. (1988). Vancomycin-resistant Gram-positive bacteria isolated from human sources. J. Clin. Microbiol. 26:2064.

72. Facklam R., Hollis D., Collins M.D. (1989). Identification of Gram-positive coccal and coccobacilly vancomycin resistant bacteria. J. Clin. Microbiol. 27:724.

73. Golledge C.L., Stingemore N., Aravena M., Joske D. (1990). Septicemia caused by vancomycin-resistant *Pediococcus acidilactici*. J. Clin. Microbiol. 28:1678.

74. Johnson A.P., Uttley A.H.C., Woodford N., George R.C. (1990). Resistance to vancomycin and teicoplanin: An emerging clinical problem. Clin. Microbiol. Rev. 3:280.

75. Maple P.A.C., Hamilton-Miller J.M.T., Brumfitt W. (1989). World-wide antibiotic resistance in methicillin-resistant *Staphylococcus aureus*. Lancet 1:537.

76. Griffith R.D. (1984). Vancomycin in use—an historical review. J. Antimicrob. Chemother. 14(Suppl. D):1.

77. Schwalbe R.S., Stapleton J.I., Gilligan P.H. (1987). Emergence of vancomycin resistance in coagulase-negative staphylococci. N. Engl. J. Med. 216:927.

78. Arioli V., Pallanza R. (1987). Teicoplanin-resistant coagulase-negative staphylococci. Lancet 1:39.

79. Froggatt J.W., Johnston J.L., Galetto D.W., Archer G.L. (1989). Antimicrobial

resistance in nosocomial isolates of *Staphylococcus haemolyticus*. Antimicrob. Agents Chemother. 33:460.

80. Aubert G., Passot S., Lucht F., Dorche E. (1991). Selection of vancomycin- and teicoplanin-resistant *Staphylococcus haemolyticus* during teicoplanin treatment of *S. epidermidis* infection. J. Antimicrob. Chemother. 25:491.

81. Sanyal D., Johnson A.P., George R.C., Cookson B.D., Williams A.J. (1991). Peritonitis due to vancomycin-resistant *Staphylococcus epidermidis*. Lancet 1 (337):54.

82. Grant A.C., Lacey R.W., Brownjohn A.M., Turney J.H. (1986). Teicoplanin-resistant coagulase-negative *Staphylococcus*. Lancet 2:1166.

83. Wilson A.P.R., O'Hare M.D., Felmingham D., Grünberg R.N. (1986). Teicoplanin-resistant coagulase-negative staphylococci. Lancet 2:973.

84. Goldstein F., Coutrot A., Seiffer A., Acar J.F. (1990). Percentages and distributions of teicoplanin- and vancomycin-resistant strains among coagulase-negative staphylococci. Antimicrob. Agents Chemother. 34:899.

85. Kaatz G.W., Seo S.M., Dorman N.J., Lerner S.A. (1990). Emergence of teicoplanin resistance during therapy of *Staphylococcus aureus* endocarditis. J. Infect. Dis. 162: 103–108.

86. Biavasco F., Giovanetti E., Montanari M.P., Varaldo P.E. (1991). Development of in-vitro resistance to glycopeptide antibiotics: Assessment in staphylococci of different species. J. Antimicrob. Chemother. 27:71.

87. Watanakunakorn C. (1988). In-vitro induction of resistance in coagulase-negative staphylococci to vancomycin and teicoplanin. J. Antimicrob. Agents Chemother. 22:321.

88. Dutka-Malen S., Leclercq R., Coutant V., Duval J., Courvalin P. (1990). Phenotypic and genotypic heterogeneity of glycopeptide resistance determinants in Gram-positive bacteria. Antimicrob. Agents Chemother. 34:1875.

89. O'Hare M.D., Reynolds P.E. (1992). Novel membrane proteins present in teicoplanin-resistant, vancomycin-sensitive, coagulase-negative *Staphylococcus* spp. J. Antimicrob. Chemother. 30:735.

90. Daum R.S., Gupta S., Sabbagh R., Milewski W.M. (1992). Characterization of *Staphylococcus aureus* isolates with decreased susceptibility to vancomycin and teicoplanin: Isolation and purification of a constitutively produced protein associated with decreased susceptibility. J. Infect. Dis. 166:1066.

91. Noble W.C., Virani Z., Cree R.G. (1992). Co-transfer of vancomycin and other resistance genes from *Enterococcus faecalis* NCTC 12201 to *Staphylococcus aureus*. FEMS Microbiol. Lett. 72:195.

92. Wright G.D., Walsh C.T. (1992). D-alanyl-D-alanine ligases and the molecular mechanism of vancomycin resistance. Acc. Chem. Res. 25:468.

93. Shlaes D.M., Al-Obeid S., Gutmann L. (1989). Vancomycin resistant enterococci. APUA Newslett. 7:1.

94. Shlaes D.M., Etter L., Gutmann L. (1991). Synergistic killing of vancomycin-resistant enterococci of classes A, B, and C by combinations of vancomycin, penicillin and gentamicin. J. Antimicrob. Chemother. 35:776.

95. Williamson R., Al-Obeid S., Shlaes J.H., Goldstein F.W., Shlaes D.M. (1989). Inducible resistance to vancomycin in *Enterococcus faecium* D366. J. Infect. Dis. 159:1095.

96. Leclercq R., Derlot E., Duval J., Courvalin P. (1988). Plasmid-mediated resistance to vancomycin and teicoplanin in *Enterococcus faecium*. N. Engl. J. Med. 319:157.

97. Shlaes D.M., Al-Obeid S., Shlaes J.H., Boisivon A., Williamson R. (1989). Inducible, transferable resistance to vancomycin in *Enterococcus faecium* D399. J. Antimicrob. Chemother. 23:503.

98. Nicas T.I., Wu C.Y.E., Hobbs J.N. Jr., Preston D.A., Allen N.E. (1989). Characterization of vancomycin resistance in *Enterococcus faecium* and *Enterococcus faecalis*. Antimicrob. Agents Chemother. 33:1112.

99. Fowler C., Scully G., Willey M., Gill V., Standiford H. (1991). Vancomycin resistant enterococci: Successful treatments of infection in two lymphoma patients using teicoplanin. Abstracts of 91st General Meeting of the American Society for Microbiology, Dallas, Texas, Abstract A-57, p. 10.

100. Handwerger S., Perlman D.C., Altarac D., McAuliffe V. (1992). Concomitant high-level vancomycin and penicillin resistance in clinical isolates of enterococci. Clin. Infect. Dis. 14:655.

101. Bingen E., Lambert-Zechovsky N., Leclercq R., Doit C., Mariani-Kurkdjian P. (1990). Bactericidal activity of vancomycin, daptomycin, ampicillin and aminoglycosides against vancomycin-resistant *Enterococcus faecium*. J. Antimicrob. Chemother. 26:619–626.

102. Leclercq R., Bingen E., Su Q.H., Lambert-Zechovsky N., Courvalin P., Duval J. (1991). Effects of combinations of β-lactams, daptomycin, gentamicin, and glycopeptides against glycopeptide-resistant enterococci. J. Antimicrob. Chemother. 35:92.

103. Kaplan A.H., Gilligan P.H., Facklam R.R. (1988). Recovery of resistant enterococci during vancomycin prophylaxis. J. Clin. Microbiol. 26:1216.

104. Swenson J.M., Hill B.C., Thornsberry C. (1989). Problems with the disk diffusion test for detection of vancomycin resistance in enterococci. J. Clin. Microbiol. 23:503.

105. MacLaughlin J., Harrington S., Dick J., Froggatt V. (1991). Vancomycin resistant enterococci: Species distribution and antibiotic susceptibility. Abstracts of 91st General Meeting of the American Society for Microbiology, Dallas, Texas, Abstract A-118, p. 20.

106. Sahm D.F., Olsen L. (1990). In vitro detection of enterococcal vancomycin resistance. Antimicrob. Agents Chemother. 34:1846.

107. Sahm D.F., Kissinger J., Gilmore M.S., Murray P.R., Mulder R., Solliday J., Clarke B. (1989). In vitro susceptibility studies of vancomycin-resistant *Enterococcus faecalis*. Antimicrob. Agents Chemother. 33:1588.

108. Dutka-Malen S., Molinas C., Arthur M., Courvalin P. (1992). Sequence of the *vanC* gene of *Enterococcus gallinarum* BM4174 encoding a D-alanine-D-alanine ligase-related protein necessary for vancomycin resistance. Gene 112:53.

109. Leclercq R., Dutka-Malen S., Duval J., Courvalin P. (1992). Vancomycin resistance gene *vanC* is specific to *Enterococcus gallinarum*. Antimicrob. Agents Chemother. 36:2005.

110. Vincent S., Minkler P., Bibcziewski B., Etter L., Shlaes D.M. (1992). Vancomycin resistance in *Enterococcus gallinarum*. Antimicrob. Agents Chemother. 37:1392.

111. Woodford N., Johnson A.P., Morrison D., Chin A.T.L., Stephenson J.R., George (1990). Two distinct forms of vancomycin resistance among enterococci in the UK. Lancet 1:226.

112. Evers S., Sahm D.F., Courvalin P. (1993). The *vanB* gene of vancomycin-resistant *Enterococcus faecalis* V583 is structurally related to genes encoding D-ala:D-ala ligases and gypcopeptide-resistance proteins VanA and Vanc. Gene 124:143.

113. Uttley, A.H.C., Collins C.H., Naidoo J., George R.C. (1988). Vancomycin-resistant enterococci. Lancet 1:57–58.

114. Leclercq R., Derlot E., Weber M., Duval J., Courvalin P. (1989). Transferable vancomycin and teicoplanin resistance in *Enterococcus faecium*. Antimicrob. Agents Chemother. 33:10.

115. Shlaes D.M., Bouvet A., Devine C., Shlaes J.H., Al-Obeid S., Williamson R. (1989). Inducible, transferable resistance to vancomycin in *Enterococcus faecalis* A256. Antimicrob. Agents Chemother. 33:198.

116. Handwerger S., Pucci M., Kolokathis A. (1990). Vancomycin resistance is encoded on a pheromone response plasmid in *Enterococcus faecium* 228. Antimicrob. Agents Chemother. 34:358.

117. Brisson-Noël A., Dutka-Malen S., Molinas C., Leclercq R., Courvalin P. (1990). Cloning and heterospecific expression of the resistance determinant *vanA* encoding high-level resistance to glycopeptides in *Enterococcus faecium* BM4147. Antimicrob. Agents Chemother. 34:924.

118. Arthur M., Molinas C., Depardieu F., Courvalin P. (1993). Characterization of Tn1546, a Tn3-related transposon conferring glycopeptide resistance by synthesis of depsipeptide peptidoglycan precursors in *Enterococcus faecium* BM4147. J. Bacteriol. 175:117.

119. Handwerger S., Kolokathis A. (1990). Induction of vancomycin resistance in *Enterococcus faecium* by inhibition of transglycosylation. FEMS Microbiol. Lett. 90:167.

120. Arthur M., Molinas C., Courvalin P. (1992). The VanS-VanR two-component regulatory system controls synthesis of depsidpeptide peptidoglycan precursors in *Enterococcus faecium* BM4147. J. Bacteriol. 174:2582.

121. Dutka-Malen S., Molinas C., Arthur M., Courvalin P. (1990). The VANA glycopeptide resistance protein is related to D-alanyl-D-alanyl ligase cell wall biosynthesis enzymes. Mol. Gen. Genet. 224:364.

122. Bugg T.D.H., Wright G.D., Dutka-Malen S., Arthur M., Courvalin P., Walsh C.T. (1991). Molecular basis for vancomycin resistance in *Enterococcus faecium* BM4147: Biosynthesis of a depsipeptide peptidoglycan precursor by vancomycin-resistance proteins VanH and VanA. Biochemistry 30:10408.

123. Arthur M., Molinas C., Dutka-Malen S., Courvalin P. (1991). Structural relationship between the vancomycin resistance protein VanH and 2-hydroxycarboxylic acid dehydrogenases. Gene 103:133.

124. Arthur M., Molinas C., Bugg T.D.H., Wright G.D., Walsh C.T., Courvalin P. (1992). Evidence for in vivo incorporation of D-lactate into peptidoglycan precursors of the vancomycin-resistanct enterococci. Antimicrob. Agents Chemother. 36:867.

125. Wright G.D., Molinas C., Arthur M., Courvalin P., Walsh C.T. (1992). Characterization of vanY, a DD-carboxypeptidase from vancomycin-resistant *Enterococcus faecium* BM4147. Antimicrob. Agents Chemother. 36:1514.

126. Zarlenga L.J., Gilmore M.S., Sahm D.F. (1992). Effects of amino acids on expression of enterococcal vancomycin resistance. Antimicrob. Agents Chemother. 36:902.

127. Messer J., Reynolds P.E. (1992). Modified peptidoglycan precursors produced by glycopeptide-resistant enterococci. FEMS Microbiol. Lett. 94:195.

128. Handwerger S, Pucci MJ, Volk KJ, Liu J, Lee MS (1992). The cytoplasmic precursor of rancomycin-resistant *Enterococcus faecalis* terminates in lactate. J. Bacteriol. 174:5982.

129. Carpenter C.V., Goyer S., Neuhaus F.C. (1976). Steric effects on penicillin sensitive peptidoglycan synthesis in a membrane-wall system from *Gaffkya homari*. Biochemistry 15:3146.

130. Rasmussen J.R., Strominger J.L. (1978). Utilization of a depsipeptide substrate for trapping acyl-enzyme intermediates of penicillin-sensitive D-alanine carboxypeptidases. Proc. Natl. Acad. Sci. USA 75:84.

131. Gutmann L., Billot-Klein D., Al-Obeid S., Kare I., Francoual S., Collatz E., van Heijenoort J. (1992). Inducible carboxypeptidase activity in vancomycin-resistant enterococci. Antimicrob. Agents Chemother. 36:77.

132. Gutmann L., Billot-Klein D., Collatz E., van Heijenoort J. (1992). Replacement of the essential penicillin-binding protein 5 by high-molecular mass PBPs may explain vancomycin-β-lactam synergy in low-level vancomycin-resistant *Enterococcus faecium*. FEMS Microbiol. Lett. 91:79.

133. Nicas T.I., Cole C.T., Preston D.A., Schabel A.A., Nagarajan R. (1989). Activity of glycopeptides against vancomycin-resistant gram-positive bacteria. Antimicrob. Agents Chemother. 33:1477.

134. Al-Obeid S., Gutmann L., Shlaes D.M., Williamson R., Collatz E. (1990). Comparison of vancomycin-inducible proteins from four strains of enterococci. FEMS Microbiol. Lett. 70:101.

135. George R.C., Uttley A.H.C. (1989). Susceptibility of enterococci and epidemiology of enterococcal infections in the 1980s. Epidemiol. Infect. 103:403.

136. Hoffman S.A., Moellering R.O. (1987). The enterococcus: Putting the bug in our ear. Ann. Intern. Med. 106:757.

7
Analytical Quantitation and Characterization

EUGENE L. INMAN and C. L. WINELY
Lilly Research Laboratories, Eli Lilly and Company, Indianapolis, Indiana

Glycopeptides have demonstrated their utility as antibiotics effective against many Gram-positive infections. As a class of compounds, they offer a wide variety of structural diversity, resulting in the development of a number of potential therapeutic products. The chemistry involved in the production, metabolism, and degradation of these compounds has been the focus of decades of study.

Vancomycin is the most widely studied member of this family. It offers the analytical chemist significant challenge because it has a relatively large molecular weight compared to synthetic antibiotics and its peptide and carbohydrate moieties possess characteristic chemical properties. As a result, vancomycin is a model compound to study, aiding the understanding of more complex proteins and glycoproteins. The availability of commercial vancomycin products adds to the significance of these studies. These products include vancomycin hydrochloride for injection, oral solution, and capsules, necessitating the contributions of analytical chemists to study the biological profile of vancomycin and to study the quality of vancomycin products.

This chapter, which describes the analytical quantitation and characterization of vancomycin and related glycopeptide antibiotics, has been divided into three sections. The first section details the analytical methodologies used to quantitate vancomycin in biological matrices in support of clinical applications. This review is not intended to include a comprehensive summary of every reported clinical study that conducted analytical monitoring of vancomycin concentrations but has been limited to those references in which advances in methodology have been provided. The next section details the analytical methodologies used to characterize the quality and stability of vancomycin products on the shelf and during clinical use. A review of the regulatory implications of product quality is also included. Finally, a review of other analytical methodologies for the characterization of related glycopeptide antibiotics is provided.

I. VANCOMYCIN BIOANALYTICAL METHODOLOGIES

The therapeutic range for vancomycin has been suggested to be from 5 to 50 μg/ml. Therapeutic monitoring is required because nephrotoxicity and permanent damage to the auditory nerve have been reported for serum levels above 50 μg/ml (1). This damage has been observed to be especially significant for patients with impaired renal function resulting in reduced creatinine clearance.

The literature includes a wide array of analytical techniques that have been applied to vancomycin serum monitoring. These include microbiological assays that monitor total antibiotic activity; chromatographic methods that monitor vancomycin, the largest component of vancomycin products; and immunoassays (Table 1). The complex profile of vancomycin products has caused some difficulty in comparing data generated by these different techniques because of the difference in chemistries employed. Therefore, each technique that has been applied to vancomycin measurements is described here, and a discussion comparing results from these techniques is included. The analytical method of choice is dependent upon the information needed by the analyst.

A. Microbiological Assay

The microbiological assay has historically been the method of choice for the measurement of vancomycin in biological matrices (2). This method measures the total activity of antibiotic in serum. Vancomycin is a mixture of similarly structured compounds, with vancomycin the compound of greatest abundance. The vancomycin factors vary widely in microbiological activity, as described in Chapter 5. Therefore, the microbiological assay yields a result representing a concentration-weighted summation of the individual factor activities. This provides the advantage of measuring antibacterial activity without a detailed understanding of the individual vancomycin factors. However, specificity is also lost because of the method's inability to differentiate from other antibiotics unrelated to vancomycin that may also demonstrate activity against the test organism. The advantages and disadvantages of the microbiological assay must be carefully considered when applied to clinical samples.

The microbiological assays used for vancomycin determinations are agar diffusion methods primarily employing strains of *Bacillus subtilis* or *Bacillus globigii* as assay organisms. For the most part, standard agar media have been employed with assay sensitivities or dose-response concentrations in the 5–40 μg/ml range. Minor sample preparation is required, with multiple dilutions tested for each sample to ensure that they fall within the concentration range spanned by

Table 1 Clinical Methods for Vancomycin

Method	Concentration range (μg/ml)	Sample preparation	Assay time	Detection limit	Interferences
Microbiological assay	0.8–25	None	Hours	0.8 μg/ml	Other antibiotics
Radioimmunoassay	2–32	None	Minutes	0.04 ng/ml	Vancomycin degradation products
High-performance liquid chromatography	1–100	Column extraction. Protein precipitation	Minutes	0.1 μg/ml	Dependent upon separation system
Fluorescence polarization immunoassay	0.6–100	None	Minutes	0.6 μg/ml	Vancomycin degradation products
Enzyme immunoassay	5–50	None	Minutes	5 μg/ml	None known

the standards. Total activity is calculated from a measure of the size of the zone of inhibited bacterial growth observed.

A number of improvements have been proposed in the original microbiological assay to improve selectivity and sensitivity. Sabath et al. (3) introduced a reduction in interference from penicillins and cephalosporins for the measurement of vancomycin, gentamicin, kanamycin, neomycin, and streptomycin. Samples were treated with β-lactamase II to eliminate the activity of these interferences. Walker and Kopp (4) reported an increase in assay sensitivity by describing a disk diffusion technique employing buffered glucose minimal salts agar. A detection limit of 0.8 μg/ml was reported compared with a 5–10 μg/ml detection limit previously reported. The changes in agar produced larger zones of inhibition, with good accuracy found from 0.8 to 25 μg/ml of vancomycin concentrations.

Walker (5) proposed another assay modification to increase the selectivity of the microbiological assay for vancomycin in the presence of other antibiotics, specifically rifampin and aminoglycosides. A new strain of *B. subtilis* from the American Type Culture Collection, ATCC 6633, that is resistant to rifampin was selected from colonies of *B. subtilis* ATCC 6633 that grew in the presence of rifampin at 200 μg/ml. In addition, the sodium chloride concentration in the minimal salts-glucose agar was increased from 0.2 to 6.0% to eliminate the aminoglycoside activity that may be present in biological samples. The detection limit was found to match the level of 0.8 μg/ml reported earlier (4).

Torres-R. et al. (6) described the application of the microbiological assay to human tissue samples, including kidney, liver, aorta, lung, and heart, collected during a postmortem examination of a patient who had had *Staphylococcus aureus* sepsis. The modified assay of Walker and Kopp (4) was used for adequate sensitivity. The vancomycin concentration found in the kidney was significantly higher than in the other tissues.

B. Immunoassays

The selectivity limitations of the microbiological assay have prompted the development of analytical methods designed specifically to measure vancomycin in the presence of other compounds with antibacterial activity. These methods include radioimmunoassays (RIA), fluorescence polarization immunoassays (FPIA), and an enzyme immunoassay. These methods offer different advantages and have found widespread application in the clinical laboratory.

Crossley et al. (7) described their evaluation of a commercial RIA for vancomycin developed to overcome the limitations of the microbiological assay. In this method, rabbit antiserum was produced to vancomycin conjugated to bovine serum albumin. Sample vancomycin concentrations were determined through competitive binding against ^{125}I-labeled vancomycin. Although the RIA results for

84 samples correlated well with data obtained by microbiological analysis, in general the RIA results were higher.

Fong et al. (8) reported a similar RIA using vancomycin labeled with ^{3}H or [125]I. Detection limits of 4 or 0.04 ng/ml were obtained depending on the tracer used, offering significant sensitivity improvements over the microbiological assay. An incubation time of 1 h was proposed for routine applications. No cross-reactivity was found for a number of other antibiotics or chemotherapeutic agents. The RIA provides a means for the measurement of vancomycin at low concentrations that are undetected by the microbiological assay.

An automated FPIA using a homogeneous competitive binding assay format has been suggested as an alternative to the RIA for a number of common antibiotics (9). A commercial FPIA analyzer, the TDx, was developed by Abbott Laboratories and applied to the clinical determination of several pharmaceuticals, including gentamicin, tobramycin, and amikacin (10). The application of FPIA to vancomycin determinations based on the TDx format was described by Schwenzer et al. (11). Sample vancomycin competes with fluorescein-labeled vancomycin by binding to rabbit antiserum, resulting in reduced fluorescence polarization with increased vancomycin concentrations. A detection limit of 0.6 μg/ml and coefficients of variation that were less than 5% were reported. Filburn et al. (12) evaluated the TDx system for vancomycin and compared their results with those obtained by microbiological assay and high-performance liquid chromatography (HPLC). The TDx system provides shorter assay times because of greatly reduced sample preparation and incubation times. Ackerman et al. (13) compared the FPIA to RIA for 123 serum samples from 34 patients. Comparable results were obtained by the two techniques ($R = 0.99$). More recently, no interferences were found for mezlocillin or piperacillin (14).

A homogeneous enzyme immunoassay has been reported for the measurement of vancomycin in serum, based on the EMIT (enzyme multiplied immunoassay technique) format, available through Syva Company (15). The assay is based on competition between sample vancomycin and enzyme-labeled vancomycin for antibody binding sites. Yeo et al. (16) reported their evaluation of this technique for vancomycin. Vancomycin is labeled with glucose-6-phosphate dehydrogenase, with glucose-6-phosphate as the substrate. Acceptable precision was found at concentrations below 30 μg/ml; however, marginal precision was found at concentrations above 30 μg/ml because of the relative flatness of the calibration curve in this region. Comparison of results generated by the EMIT system with results generated by FPIA suggest that further study is needed.

C. High-Performance Liquid Chromatography

Many chromatographic techniques have been described for the isolation of vancomycin since the first report in 1968 (17). Although Kirchmeier and Upton

(18) reported the chromatographic analysis of vancomycin in antibiotic formulations, Uhl and Anhalt (19) first described an analytical HPLC method for the determination of vancomycin in serum, which included sample extraction over a Sephadex column followed by reversed-phase chromatographic analysis using ultraviolet (UV) detection. Ristocetin was used as an internal standard. A 30 cm C_{18} column was used along with isocratic elution with 9% acetonitrile and 91% 0.05 M phosphate buffer at pH 6.0 and a detection wavelength of 210 nm. A working concentration range of 2–128 μg/ml and a between-day coefficient of variation of less than 4% were found. A run time of about 20 minutes was used for each injection.

McClain et al. (20) developed a modified HPLC method. The sample preparation scheme was changed to include the use of cold trichloroacetic acid for the precipitation of serum proteins rather than ion-exchange column chromatography. Again, a 30 cm C_{18} column with 10 μm packing was employed with a mobile phase of 12% acetonitrile and 88% 0.01 M one-heptane sulfonic acid. Detection was achieved at 280 nm, but the detection limit of 5 μg/ml was higher than that reported by Uhl and Anhalt. A run time of 20 minutes matched that previously described.

Other modifications to the original HPLC method of Uhl and Anhalt have been reported in the literature. Bever et al. (21) maintained the ion-exchange chromatographic sample preparation scheme but noted that most of the sample vancomycin was found in the first borate buffer eluate rather than the second, as Uhl and Anhalt suggested. In addition, an alternative Sephadex supplier was recommended. The mobile phase for chromatographic analysis was modified to include 1% methanol, resulting in narrower peak shapes. While testing for interferences from other common pharmaceuticals, they found that theophylline and acetaminophen at high concentrations overlapped with the ristocetin peak. Hoagland et al. (22) described an HPLC method with several modifications to the Uhl and Anhalt procedure. They used trichloroacetic acid during sample preparation for protein separation instead of the ion-exchange chromatographic cleanup. Analysis was performed on a 25 cm, 5 μm cyano column in tandem with a 5 cm guard column. The mobile phase consisted of 10% acetonitrile and 90% triethylammonium acetate buffer at pH 4.6. An elevated temperature of 60°C was used during the separation. A detection wavelength of 240 nm was utilized, resulting in a detection limit of 2 μg/ml. The run time was reduced to less than 10 minutes because of the shorter elution times for vancomycin and ristocetin. Jehl et al. (23) proposed other changes to existing HPLC methodology. In addition to a cold trichloroacetic acid in acetonitrile extraction, a methylene chloride extraction step was introduced to remove acetonitrile and lipids from the aqueous phase. A 15 cm C_{18} 5 μm particle column was used with a mobile phase of 9% acetonitrile and 91% ammonium acetate buffer. With detection at 214 nm, a detection limit of 0.1 μg/ml was reported.

Rosenthal et al. (24) reported limitations to these sample preparation schemes. Instead of trichloroacetic acid extraction, methanol was used to remove serum proteins. The internal standard was changed to 3,4,5-trimethoxyphenylacetonitrile to provide longer retention times, which resulted in fewer interferences. The mobile phase consisted of 5% methanol, 25% acetonitrile, and 70% aqueous octane sulfonic acid at pH 2.6. EDTA (0.35 mM) was introduced to reduce variability in vancomycin retention time.

Sample preparation was further simplified with the introduction of a solid-phase extraction cleanup step. Bauchet et al. (25) described their work with a Sep-Pak C_{18} cartridge. Other modifications included the use of trimethoprim as an internal standard, a mobile phase of 22% methanol and 78% phosphate buffer at pH 2.8, and detection at 229 nm. This method was applied to the determination of vancomycin in serum and heart valves during cardiac surgery. Greene et al. (26) introduced a solid-phase extraction step for the determination of vancomycin in plasma, bone, atrial appendage, and pericardial fluid. This work describes the use of Bond-Elut C_8 cartridges for sample preparation. Analysis was performed on a 25 cm C_8 5 μm particle column using a mobile phase of 10% acetonitrile and 90% phosphate buffer at pH 5.0. With detection at 254 nm, a detection limit of 3 μg/ml was obtained.

A normal-phase chromatographic method for vancomycin was described by Hosotsubo (27). The method includes protein precipitation with acetonitrile, followed by analysis on an aminopropyl column. A mobile phase of 36% phosphate buffer and 64% acetonitrile was employed with detection at 240 nm. A detection limit of 0.01 μg/ml of vancomycin in plasma was reported.

Finally, an HPLC method has been described for the quantitation of vancomycin and crystalline degradation product 1 (CDP-1), its major solution degradation product (28). An isopropanol-acetonitrile sample extraction followed by methylene chloride extraction was employed for sample preparation. A multisegmented gradient elution profile was used, ranging in solvent strength of 1–50% acetonitrile. Vitamin B_{12} was used as an internal standard, with dual-wavelength detection at 235 and 360 nm. This method was used to study the impact of CDP-1 on vancomycin concentrations measured by HPLC, FPIA, and enzyme immunoassay.

D. Data Correlations Among Techniques

The complex nature of vancomycin makes data correlation among the techniques just described especially challenging. As technology advances, it is important to be able to trace results obtained by a new method to the generally accepted method. This has been difficult for vancomycin because of its multifactorial nature and the varied relative activities of the vancomycin-related factors. As a result, many scientists have compared assay methods for vancomycin and reported their observations in the literature. It is important to understand data correlations among techniques for effective utility in the clinical laboratory.

The historical base for vancomycin measurements for clinical samples has been the microbiological assay. As new methods of analysis have been developed, direct comparison of patient samples using the microbiological assay has been routinely reported. Generally, these results have been deemed comparable because of the relatively poor precision of the microbiological assay. A more comprehensive evaluation was needed to assess the comparability of these assay methods.

Pfaller et al. (29) reported their results from the evaluation of five assay methods for vancomycin: microbiological assay, HPLC, fluorescence polarization immunoassay, radioimmunoassay, and fluorescence immunoassay (FIA). The microbiological assay employed was designed to inactivate aminoglycosides and β-lactams. The HPLC method was a modified version of that described by Uhl and Anhalt (19) using an ion-exchange column sample cleanup step followed by a second column extraction. A C_{18}, 30 cm column was utilized with a 10% acetonitrile and 90% phosphate buffer at pH 3.8–4.0, mobile phase. The FPIA method utilized the TDx system previously described. The RIA method was purchased from American Diagnostics and was unmodified. The FIA method was purchased from American Diagnostics. The immunochemistry resembles that of the FPIA method, but this assay is run as a heterogeneous assay because bound vancomycin is removed from solution. The fluorescence of the supernatant is measured using a fluorometer. The FIA assay yielded the worst precision of all methods tested. A total of 106 patient samples were assayed by these five methods. From their study, the investigators concluded that the FPIA vancomycin assay is an excellent alternative to HPLC and the microbiological assay, the FIA lacks acceptable precision for clinical use, the need to dilute samples containing more than 32 μg/ml for the RIA assay compromises its practical utility, and although the microbiological assay and HPLC assay are accurate and precise, their turnaround times make them less attractive than the FPIA method.

Ristuccia et al. (30) reported their comparison of the microbiological assay, HPLC assay, and FPIA for the determination of vancomycin in serum. They utilized the microbiological assay described by Walker and Kopp (4) and the HPLC method described by Uhl and Anhalt (19). These methods were applied to the analysis of 100 clinical samples from patients receiving vancomycin. Their results matched those reported by other researchers. The microbiological assay was accurate and precise, but it lacked specificity. The HPLC method was accurate and precise, but it was labor intensive as a result of the sample preparation scheme. The FPIA method was reported to be fast, easy, and practical. Good correlation was observed between HPLC and FPIA ($R = 0.9996$), with poor correlation found between the HPLC and microbiological assay ($R = 0.7779$) and between the FPIA and microbiological assay ($R = 0.7773$). The poor correlations were suggested to result from the greater variability of the microbiological assay.

In 1987, Morse et al. (31) reported their results of a comparison of HPLC and FPIA for vancomycin determinations for patients on peritoneal dialysis. They suggested that vancomycin concentrations were overestimated by FPIA when vancomycin degradation products accumulated in CAPD patients. Both assay methods yielded comparable results during early stages of treatment. Significant differences between methods were found at times greater than 48 h, with differences of more than 200% found after 15 days. Because of poor renal function and the slow removal of vancomycin by dialysis, significant vancomycin degradation was expected among these patients. It was proposed that the antivancomycin antibody cross-reacted with these degradation products. The nature of these degradation products was unknown at the time of the study.

White et al. (32) studied the degradation of vancomycin in serum, CAPD fluid, and phosphate-buffered saline to understand better the differences observed by Morse et al. The degradation of vancomycin at 37°C in these matrices was monitored by microbiological assay, HPLC, and FPIA. The major solution degradation products, CDP-1M and CDP-1m (the two atropisomers of CDP-1, see pp. 198, 263, and 265), were monitored by HPLC. Significant degradation was observed over 10 days, with a loss of about 50% vancomycin observed for serum and buffer saline and with concomitant increases in the two degradation peak areas. The FPIA assay showed only a 20% decrease. Vancomycin acid hydrolysis yielded four peaks, which were isolated and tested for cross-reactivity. The cross-reactivity of these peaks ranged from 10 to 42%. It was concluded that the FPIA assay becomes nonspecific when samples contain vancomycin degradation products, yielding overestimated vancomycin concentrations.

In 1989, Anne et al. (33) reported differences they observed during the comparison of the enzyme multiplied immunoassay technique with the FPIA for patient samples. White et al. were able to generate CDP-1 in vitro in serum samples; Anne et al. attempted to determine if CDP-1 could be found in patient sera. Liquid chromatography–mass spectrometry was utilized to confirm the identity of CDP-1. CDP-1 is formed in vivo as a result of the prolonged half-lives for vancomycin in renally impaired patients. Its cross-reactivity was measured for EMIT, FPIA, and RIA. High cross-reactivity was found for FPIA (40%) and RIA (70%), with less than 3% cross-reactivity measured for EMIT. It was proposed that this was due to the polyclonal antibodies used in the FPIA and RIA; the EMIT assay employs a monoclonal antibody. Hu et al. (28) continued their study of CDP-1 to understand the differences between FPIA and EMIT. An improved gradient HPLC method was described for the quantitation of CDP-1 along with vancomycin. At vancomycin concentrations of 15–35 μg/ml, good correlation was observed between HPLC and EMIT. Data from FPIA yielded high results compared with EMIT at concentrations greater than 10 μg/ml. It was suggested that the accumulation of CDP-1 in renally impaired patients may yield a 10% difference between FPIA and HPLC and EMIT. This limitation should be considered when applying FPIA to vancomycin.

E.   Summary of Clinical Assays

The complex nature of vancomycin and its degradation products makes direct comparison of clinical results from different techniques difficult for some patient populations. Each analytical technique offers advantages to the clinical laboratory but may introduce a limitation to data interpretation. The disadvantages of these techniques for vancomycin determinations may limit their practical application. These advantages and disadvantages are summarized here.

The microbiological assay is inexpensive and utilizes small sample sizes. It provides a true assessment of antimicrobial activity, providing a concentration-weighted summation of the individual factor activities even if the identity or relative activity of a specific factor is unknown. Disadvantages include interferences from other antibiotics, although selected interferences can be removed. This can be a serious limitation: a large percentage of vancomycin patients also receive other antimicrobial agents. A minimum sample turnaround time of 6 h is required, with overnight analysis usually employed. The microbiological assay provides a reference method for the assessment of other techniques for vancomycin determinations and is the method of choice when total activity is desired.

HPLC is specific, precise, and accurate and provides individual factor information. It has been shown to be useful for the study of degradation products and can contribute to the understanding of vancomycin as a mixture of factors. The technique suffers from extensive sample preparation and relatively long analysis time (about 15 minutes per sample). This is partially offset by the widespread availability of HPLC instrumentation in the clinical laboratory and automated sample introduction allowing around-the-clock analysis. HPLC is most appropriate when more information is needed, such as for the individual measurement of several components in a sample. A summary of the chromatographic conditions proposed for the determination of vancomycin found in the literature is shown in Table 2.

RIA provides significant sensitivity enhancement over other techniques. It has been found to be accurate and precise. Disadvantages include the need for unique instrumentation that is not universally available and the growing concern for radioactive waste disposal. In addition, the ^{125}I label has a short half-life, resulting in reduced shelf life for assay reagents. RIA offers unique advantages when ultimate sensitivity is a requirement.

FPIA has been found to be rapid and available in a commercial automated instrument, resulting in an analysis time of 5 minutes per sample. The commercial instrument stores the calibration curve for up to 1 month. For patients with normal renal function, results from this method correlate well with those obtained for other methods. However, for those patients with impaired renal function, FPIA tends to overestimate vancomycin concentrations because of cross-reactivity of vancomycin degradation products found in vivo. Thus FPIA is selective, reliable,

Table 2 HPLC Methods for Vancomycin

Column	Mobile phase	Detection (nm)	Sample preparation	Internal standard	Run time (minutes)	Reference
C_{18}, 10 μm	9% Acetonitrile (ACN) 91% Phosphate buffer	210	Sephadex column	Ristocetin	20	19
C_{18}, 10 μm	12% ACN 88% 1-Heptanesulfonic acid	280	Cold trichloroacetic acid Methylene chloride	None	20	20
C_{18}, 10 μm	1% Methanol 9% ACN 90% Phosphate buffer	210	Sephadex column	Ristocetin	10	21
CN, 5 μm 60°C	10% ACN 90% Triethylammonium acetate	240	Cold trichloroacetic acid	Ristocetin	8	22
C_{18}, 5 μm	9% ACN 91% Ammonium acetate	214	Cold trichloroacetic acid Methylene chloride	None	15	23
C_{18}, 5 μm	5% Methanol 25% ACN 70% 1-Octane sulfonic acid EDTA	230	Methanol	3,4,5-Trimethoxy-phenylacetonitrile	10	24
C_{18}, 5 μm	22% Methanol 78% Phosphate buffer	229	Sep-Pak C_{18} cartridge	Trimethoprin	6	25
C_8, 5 μm	10% ACN 90% Phosphate buffer	254	Bond-Elut C_8 cartridge	Cephazolin	12	26
Aminopropyl	64% ACN 36% Phosphate buffer	240	Acetonitrile	None	12	27
C_{18}, 5 μm	Gradient 1–50% ACN 50–99% Phosphate buffer	235 360	Isopropanol-acetonitrile Methylene chloride	Vitamin B_{12}	33	28
C_{18}, 5 μm	10% ACN 90% Phosphate buffer	210	Sephadex column Polypropylene column	Ristocetin	20	29

and inexpensive for normal patient populations. Special care is recommended for patients with impaired renal function.

EMIT overcomes the selectivity limitations of the FPIA by utilizing a monoclonal antibody rather than a polyclonal antibody used in FPIA and RIA. Precision limitations have been suggested for vancomycin concentrations greater than 30 μg/ml. Further evaluation of this method is needed to establish its widespread application to clinical determinations of vancomycin.

II. VANCOMYCIN CHARACTERIZATION METHODOLOGIES

Vancomycin is administered as the hydrochloride salt of the glycopeptide produced by the growth of *Amycolatopsis orientalis*. It has a molecular weight of 1485 and consists of seven amino acids fused together to form the aglycone and two hexose rings. The wide array of functional groups collectively give vancomycin a unique set of chemical properties. These properties are used to distinguish vancomycin from other organic molecules as well as from other structurally related glycopeptides. The analytical chemist can exploit these chemical properties to characterize vancomycin used in chemical and clinical applications.

Like all other pharmaceutical products, manufacturers of vancomycin-containing products must assure the customer that each dose meets established requirements for strength, identity, purity, and quality. These requirements have evolved as new technologies are introduced into the quality control arsenal for vancomycin. This section provides a summary of current analytical methodologies used in the characterization of vancomycin products, the chemical and biochemical properties measured by these methodologies, and a historical perspective of their development. These methods are described within the context of marketed vancomycin products, including sterile vancomycin hydrochloride for injection, vancomycin hydrochloride for oral administration, and bulk vancomycin hydrochloride.

This discussion includes the regulatory requirements for vancomycin hydrochloride, a detailed discussion of the analytical tests used for vancomycin quality control, a literature review of other techniques used to characterize vancomycin, a review of published stability properties of vancomycin products, and a summary of reference standards available for vancomycin testing. Representative compendial monographs provide the source for established regulatory requirements for vancomycin products. Potency as determined by microbiological assay and chromatographic purity as determined by HPLC are presented in detail because of their historical significance. An understanding of the solution and solid-phase stability properties of vancomycin are critical to understanding the in vitro and in vivo properties of vancomycin, as illustrated in Section I. The scientific literature illustrates the ongoing evolution of analytical methodologies applied to vancomycin characterization. This section summarizes this evolution.

A. Compendia

Representative monographs for vancomycin products can be found in the *United States Pharmacopeia* (USP) (34), the *British Pharmacopoeia* (BP) (35), and the *Minimum Requirements for Antibiotic Products of Japan* (36). These monographs contain vancomycin bulk and product requirements, the testing methodologies, and corresponding test limits. These monographs overlap extensively, but there are significant differences among them. These differences include the product forms covered by monographs and specific test requirements. The significance of these differences is also reviewed here.

The vancomycin monographs found in USP XXII are summarized in Table 3. Monographs are included for bulk vancomycin hydrochloride and three formulations, sterile vancomycin hydrochloride, vancomycin hydrochloride for oral solution, and vancomycin hydrochloride capsules. Table 3 includes the chemical properties, analytical tests, and test limits. Together, they probe a variety of vancomycin characteristics that control product quality.

The vancomycin monographs found in the 1988 *British Pharmacopoeia* are summarized in Table 4. These monographs include vancomycin hydrochloride and vancomycin injection. Similarities to the USP include the use of the microbiological assay for potency and other tests, such as pH, water, pyrogens, and sterility. Significant differences include the chemical tests used for identification and the control of related impurities by chromatographic purity determinations, found only in the USP. The extent of testing and test limits are generally comparable between the two compendia.

The vancomycin monographs found in the *Minimum Requirements for Antibiotic Products of Japan* are summarized in Table 5. These monographs include vancomycin hydrochloride and vancomycin hydrochloride powder. These monographs provide additional spectroscopic measurements, such as UV absorbance and specific rotation, and extensively overlap with the tests found in the USP and BP. Again, potency is controlled by microbiological assay, and no chromatographic purity test is included. Other differences can be found.

B. Potency

Potency is the single most important property of a pharmaceutical product. As an antibiotic, vancomycin potency is controlled with the microbiological assay. The basic principles of the microbiological assay have remained unchanged through the years, but the experimental details of the assays included in the regulatory monographs differ. These differences and their significance are described here.

The experimental conditions for the microbiological assays included in the

Table 3 *United States Pharmacopeia XXII* Vancomycin Monographs

Vancomycin hydrochloride

Assay	Microbial assay	NLT 900 μg/mg anhydrous[a]
Identification	Infrared spectrum	Compares to reference standard
pH	50 mg/ml	2.5–4.5
Water	Titration	NMT 5.0%[b]
Chromatographic purity	HPLC	NLT 80.0% vancomycin B
		NMT 9.0% any other peak

Sterile vancomycin hydrochloride

Assay	Microbial assay	NLT 900 μg/mg anhydrous
		NLT 90.0% label
		NMT 115.0% label
Constituted solution		Meets test
Pyrogen	5 mg/ml	Meets test
Sterility		Meets test
Particulate matter		Meets test
Heavy metals		NMT 0.003%
Uniformity of dosage units		Meets test

Vancomycin hydrochloride for oral solution

Assay	Microbial assay	NLT 90.0% label
		NMT 115.0% label
pH	Constituted per label	2.5–4.5
Water	Titration	NMT 5.0%

Vancomycin hydrochloride capsules

Assay	Microbial assay	NLT 90.0% label
		NMT 115.0% label
Identification	Paper chromatography Bioautography	Compares to reference standard
Dissolution		NLT 85% (Q) label in 45 minutes
Uniformity of dosage units		Meets test
Water	Titration	NMT 8.0%

Vancomycin hydrochloride for injection

Potency	Microbial assay	NLT 925 μg/mg anhydrous
		NLT 90.0% label
		NMT 115.0% label
Constituted solution		Meets test
Pyrogen	5 mg/ml	Meets test
Sterility		Meets test
Particulate matter		Meets test
Heavy metals		NMT 0.003%
Chromatographic purity	HPLC	NLT 88.0% vancomycin B
		NMT 4.0% any other peak
Uniformity of dosage units		Meets test

[a]NLT, not less than.
[b]NMT, not more than.

Table 4   *British Pharmacopoeia* Vancomycin Monographs

Vancomycin hydrochloride		
Identification	Descending paper chromatography	Compares to reference standard
	Cupric tartrate test	Produces red precipitate
	Chloride reactions test	Meets test
pH	5%	2.8–4.5
Completeness of solution	0.10 g/1 ml water	Dissolves completely
Sulfated ash		NMT 1.0%
Water		NMT 4.5%
Assay	Biological assay	NLT 900 units/mg anhydrous
		NLT 95% estimated potency
		NMT 105% estimated potency
Abnormal toxicity		Meets test
Pyrogens		Meets test
Sterility		Meets test
Vancomycin injection		
Identification	Descending paper chromatography	Compares to reference standard
	Cupric tartrate test	Produces red precipitate
	Chlorides reactions test	Meets test
pH	5%	2.8–4.5
Completeness of solution	0.10 g/1 ml water	Dissolves completely
Water		NMT 4.5%
Pyrogens		Meets test
Assay	Biological assay	NLT 95% estimated potency
		NMT 105% estimated potency

three representative compendia are summarized in Table 6. All three methods are agar diffusion-based assays with *B. subtilis* as the assay organism. Since the three methods differ in the media employed and, in particular, the pH of the media, the test sensitivities differ. The methods also differ in the number of standard and sample concentrations used. The individual compedia should be consulted for complete experimental details for this assay.

C.   Chromatographic Characterization

Potency as determined by the microbiological assay provides a measure of biological activity of vancomycin and related factors, and chromatographic characterization provides chemical purity information, distinguishing vancomycin factors on a molecular basis. A number of separation schemes have been proposed, differing in the purpose of the analysis. These include paper chromatography, thin-

Table 5 *Minimum Requirements for Antibiotic Products of Japan* Vancomycin Monographs

Vancomycin hydrochloride		
Identification	Basic cupric sulfate test	Forms purple color
	Anthrone test	Forms brown color
	Silver nitrate test	Forms white turbidity
	Absorption spectrum	Maximum 279–284 nm
Potency	Microbiological assay	NLT 900 µg/mg anhydrous
pH	50 mg/ml	2.5–4.5
Absorbance	281 nm	45–55 AU/cm%
Specific rotation	Sodium D, 20°C	−30 to −40°C
Sterile		Meets test
Toxicant		Negative
Moisture	Karl Fischer titration	NMT 5.0%
Heavy metal		NMT 20 ppm
Vancomycin factor A content	TLC	Conformable
Vancomycin hydrochloride powder		
Potency	Microbiological assay	NLT 90% label
		NMT 120% label
pH	50 mg/ml	2.5–4.5
Sterile		Meets test
Toxicant		Negative
Moisture	Karl Fischer titration	NMT 5.0%

layer chromatography, and high-performance liquid chromatography. These have been used for component identification through relative retention times, for vancomycin quantitation in mixed formulations, and for product profiling. A review of these techniques and their applications is provided here.

Paper chromatography has provided a means for the separation of antibiotics, allowing the classification of groups of compounds into structurally related families (37). Descending paper chromatography is included in the BP monograph for vancomycin hydrochloride as an identification test (35). A solvent mixture of 2-methylbutan-2-ol, water, and acetone is used for chromatogram development, Bioautography with *B. subtilis* is used to confirm antimicrobial activity. This method is simple and inexpensive and can provide classification information. Its separating power is limited, however, and offers minimal specificity.

Thin-layer chromatography (TLC) provides separation improvements over paper chromatography and can be coupled to chemical detection methods for enhanced sensitivity. Thomas and Newland (38) have described four TLC systems for the separation of vancomycin, related antibiotics, and degradation products.

Table 6 Compendial Microbiological Assay Methods for Vancomycin

	USP	BP	Japan

Assay organism
Medium, pH
	B. subtilis ATCC 6633
	Antibiotic medium no. 8.
	pH 5.8–6.0
		B. subtilis NCTC[a] 8236
		Antibiotic medium no. 11, pH 7.8
			B. subtilis ATCC 6633
			Medium composition
			  Peptone 5.0 g
			  Meat extract 3.0 g
			  Agar 13.0–20.0 g
			  Distilled water, sufficient
			    quantity to 1000 ml
			  pH 6.2–6.4 after sterilization

Agar volume	20 ml base layer 4 ml inoculated layer	Single inoculated layer 3–4 mm thickness	20 ml Base layer 4 ml Inoculated layer
Assay design	Cylinder, three plates per standard solution. three plates per sample	Cylinder, well, or disk, three-point assay	Cylinder, two-point assay
Diluent	0.1 M phosphate buffer, pH 4.5	0.1 M phosphate buffer, pH 8.0	0.1 M phosphate buffer, pH 4.5
Standard concentrations	Five concentrations with 10 µg/ml concentrations as median	At least three concentrations 20–200 µg/ml range in logarithmic progression	25 and 100 µg/ml
Sample concentrations	~10 µg/ml	At least three concentrations presumed equal to standards	~25 and ~100 µg/ml
Incubation	36–37.5°C 16–18 h	37–39°C 16–18 h	32–37°C 16–20 h

[a]NCTC. National Collection of Type Cultures.

These systems include stationary phases of silica gel, carboxymethylcellulose, and reversed-phase bonded silica and mobile phases of various aqueous and organic solvents. Detection included spraying the developed plate with p-nitrobenzene diazonium tetrafluoroborate or bioautography. Analytes of interest included the potential degradation products demethylvancomycin, aglycovancomycin, and CDP-1 and other glycopeptides, including ristocetin A and B, avoparcin, teicoplanin, actinoidin A and B, carboxyristomycin, and A35512B. These compounds can be reasonably separated by one or more of these systems. However, further characterization of minor vancomycin components requires the specificity offered by HPLC.

The initial application of HPLC to the analytical characterization of vancomycin was reported by Kirchmeier and Upton (18). A method was developed for the simultaneous determination of vancomycin, anisomycin, and trimethoprim lactate in antibiotic mixtures. A 30 cm C_{18} column with 10 μm packing was utilized with a mobile phase of 12.5% acetonitrile and 87.5% phosphate buffer at pH 6. Detection was achieved at 225 nm. Calibration curves spanned a concentration range of 0.01–0.07 mg/ml. The precision of this method was found to be less than 2% for each major compound, suggesting the potential use of HPLC as a quality control tool for vancomycin.

Sztaricskai et al. (39) reported their use of HPLC for the examination of vancomycin-related antibiotics. The analytes of interest included ristocetin A, ristomycin A, A35512B, carboxyristomycin A, vancomycin HCl, avoparcin, and actinoidin A and B. Two methods were compared using C_8 and C_{18} bonded phases with 10 μm particles. Mobile phases included 12% methyl cellosolve in citrate buffer at pH 6.40 and 10% acetonitrile in ammonium formate buffer at pH 7.30, respectively. With detection at 254 nm, at least five minor components were separated from the vancomycin B peak. The general purity of the commercially available vancomycin sample was significantly higher than for the other materials studied. A detection limit of 140 ng vancomycin was reported.

In 1987, Inman (40) reported the development of an HPLC method specifically designed for the separation and quantitation of vancomycin cofermentation factors and degradation products. For this application, emphasis was placed on detecting a wide range of factors with accuracy and specificity. A 25 cm C_{18} column with 5 μm packing was selected for this separation. A linear gradient program spanning a solvent strength from 5 to 90% acetonitrile was initially used to demonstrate that vancomycin-related factors elute within a relatively narrow window. A narrower linear gradient spanning a range of 7–28% acetonitrile was found to define adequately the elution window for vancomycin-related factors, as shown in Figure 1. This method was proposed for inclusion in the USP (41) as a test to measure chromatographic purity for vancomycin hydrochloride and sterile vancomycin hydrochloride. This method has limited effectiveness as a quality

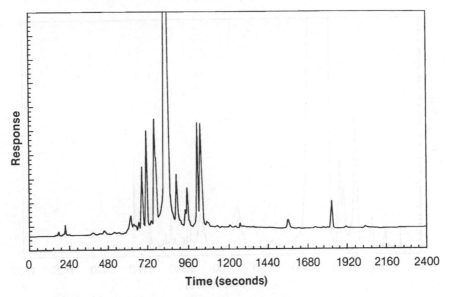

Fig. 1 HPLC profile of vancomycin hydrochloride using a linear gradient from 7 to 28% acetonitrile in triethylammonium phosphate buffer at pH 3.2 as described in Reference 40.

control tool, however, because of variability demonstrated between instruments as a result of mixing and gradient program differences.

The chromatographic properties of vancomycin are characteristic of those observed for other peptides, with retention times that are extremely sensitive to mobile-phase changes (42). A multisegmented gradient program was developed to include an initial isocratic region in which the majority of the separation takes place and a gradient ramp to higher solvent strength to ensure complete elution (see Fig. 2). By premixing the solvents, instrumental differences in mobile-phase composition are reduced. The mobile-phase compositions included 7% aceto-nitrile, 1% tetrahydrofuran, and 92% triethylammonium phosphate (TEAP) buffer at pH 3.2 during the isocratic portion of the chromatogram ramped to a mobile phase of 29% acetonitrile, 1% tetrahydrofuran, and 70% triethylammonium phosphate buffer. Quantitation of vancomycin-related factors was based on rela-tive peak area comparisons because the chromophore generally remains un-changed. This method, with minor changes, has been added to the USP (43) to measure the chromatographic purity of vancomycin hydrochloride. A resolution solution was developed as a system suitability check for both the isocratic and gradient portions of the separation. A solution of vancomycin hydrochloride is heated at 65°C for 24 h to produce significant quantities of CDP-1. One CDP-1 peak

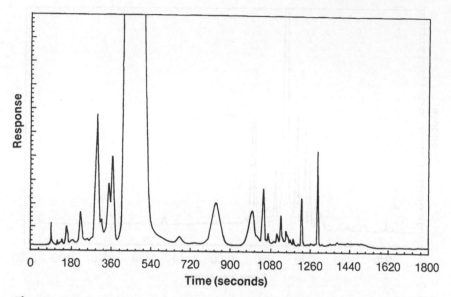

Fig. 2 HPLC profile of vancomycin hydrochloride using a multisegment gradient mobile-phase program. The program includes an initial isocratic region of 7% acetonitrile, 1% tetrahydrofuran, and 92% TEAP buffer for 12 minutes, a linear gradient ramp to 29% acetonitrile, 1% tetrahydrofuran, and 70% TEAP buffer over 8 minutes, and an isocratic region at 29% acetonitrile, 1% tetrahydrofuran, and 70% TEAP buffer for 2 minutes. Complete chromatographic conditions are described in Reference 43.

elutes before vancomycin during isocratic elution, and the other CDP-1 peak elutes during the gradient ramp (see Fig. 3). Test limits include a main peak purity of no less than 80.0% and no more than 9% for any peak other than the main peak. The chromatographic purity assay has been added to the monograph for vancomycin hydrochloride for injection (44) with test limits of no less than 88.0% vancomycin B and no more than 4.0% for any peak other than the main peak. This method provides adequate measurement of chromatographic purity of vancomycin products both at the time of manufacture and throughout its shelf life.

Thomas and Newland (38) also described an HPLC method for the examination of vancomycin and other glycopeptide antibiotics. A polymeric column with reversed-phase properties and 8 μm particles was used with a mobile phase of acetonitrile in borate buffer, pH 8.6. A gradient elution profile from 8 to 16% acetonitrile was used for these separations. Commercially available vancomycin was found to contain at least 7 minor components, and artificially degraded vancomycin contained at least 13 peaks in addition to vancomycin B. This work demonstrated the utility of HPLC to monitor vancomycin stability.

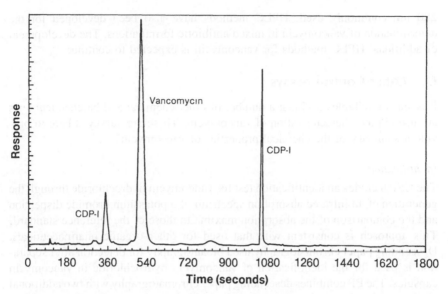

Fig. 3 HPLC profile of the vancomycin resolution solution using the conditions described in Figure 2. The CDP-1 peaks demonstrate adequate resolution during the isocratic and gradient regions of the separation.

Vancomycin has been observed to possess other chromatographic properties. Inman et al. (45) found that inaccuracies in HPLC quantitation can be generated as a result of sample-solvent interactions. Erroneously high vancomycin peak areas were observed when water was used as the sample solvent without matching the ionic strength of the comparator solutions. Additional work suggested that this was limited to specific injection valves. Recently, Dolan (46) explained this effect as adsorption of the analyte onto the polymeric material used in the injector rotor. Vancomycin detection was studied to demonstrate the utility of absorbance ratioing to detect changes in the spectral characteristics of the related factors of a pharmaceutical (47). This method has been used to ensure the accuracy of the normalized peak area calculation for related factor quantitation. This is especially valuable when many of the related factors have not been identified. An HPLC method has also been reported for the determination of EDTA in vancomycin formulations (48).

Thus, HPLC has proven to be an effective tool to measure the chemical purity of vancomycin-containing products. It complements potency measurements obtained by the microbiological assay. A wide array of vancomycin-related factors have been separated from vancomycin using reversed-phase conditions. UV detection provides an accurate means for peak quantitation, with 210, 240, and

280 nm commonly used. HPLC methods have also been developed for the determination of vancomycin in mixed antibiotic formulations. The development of additional HPLC methods for vancomycin is expected to continue.

D. Other Control Assays

It is shown in Tables 3–5 that a number of other chemical and biochemical tests are used in the characterization of vancomycin. These are surveyed here to provide a summary of the chemical properties of vancomycin.

Identification

The USP includes an identification test for vancomycin hydrochloride through the generation of an infrared absorption spectrum of a potassium bromide dispersion and the comparison of the absorption maxima to those of the reference standard. This approach is consistent with that used for other organic pharmaceuticals. Descending paper chromatography used bioautography for detection after separation is used for the identification of vancomycin hydrochloride in vancomycin capsules. The BP combines descending paper chromatography with two additional chemical tests. A cupric tartrate method is added for the detection of the two sugar moieties, and a chloride reactions test is included to confirm the presence of hydrochloride salt. In Japan, vancomycin UV absorbance measurement at 281 nm and specific rotation measurement confirm the presence of vancomycin. These spectroscopic tests complement the chemical tests, which include the basic cupric sulfate test to confirm the presence of phenol groups, the anthrone test to confirm the presence of sugar moieties, and the silver nitrate test to confirm the presence of chloride.

pH

The pH of a 50 mg/ml solution of vancomycin falls between 2.5 and 4.5. All current compendial monographs include this test.

Water Content

Vancomycin hydrochloride is hygroscopic and does not equilibrate to a stoichiometric level of water. The Karl Fischer titration effectively measures water content, with compendial limits ranging from 4.5 to 8.0%.

"Vancomycin Factor A"

"Vancomycin factor A" is a term used in the vancomycin compendium to describe a spot that separates in paper chromatography from vancomycin, when impure vancomycin is used and is composed of one or more vancomycin-related factors that can be separated from vancomycin by TLC (49). This measurement falls within the scope of the chromatographic purity test found in the USP, and a separate test is not included. The Japanese requirements include the TLC test method.

Sterility, Pyrogens, Particulate Matter, Sulfate Ash, and Heavy Metals

These tests and limits are consistent with those included for other parenterally administered antibiotics.

E. Stability

The stability of vancomycin hydrochloride directly impacts the process for vancomycin manufacture and the storage and use conditions for vancomycin products. The analytical tests and limits that are included in the regulatory compendia apply to vancomycin products throughout their shelf lives. The stability characteristics for vancomycin include both solid-state and solution stability and are impacted by the chemical properties of vancomycin formulations and the diluents used to reconstitute these formulations. The following is a review of published stability studies for vancomycin products.

Marshall (50) reported his findings of the chemical properties of vancomycin hydrochloride as they impact its chemical structure. This provided a solid foundation for later structural studies, but a complete chemical structure was not elucidated. This report included discussions of several related factors, including aglucovancomycin, a crystalline degradation product, CDP-1, and a second crystalline degradation product, CDP-2. The latter two compounds are formed as a result of mild acid hydrolysis of vancomycin. A general scheme for the conversion of vancomycin to aglycovancomycin and to CDP-1 followed by conversion to CDP-2 was proposed by Smith et al. (51). Harris et al. (52) proposed a complete chemical structure for vancomycin, along with a mechanism for the conversion of vancomycin to the two forms of CDP-1, CDP-1m (minor) and CDP-1M (major). The rearrangement involves the formation of a succinimide intermediate that is converted to CDP-1m. The identity of each of these compounds was confirmed by nuclear magnetic resonance; however, only CDP-1m and CDP-1M are found in equilibrium and in the HPLC chromatogram.

The kinetics of solid-state degradation of vancomycin have not been reported in the scientific literature. A number of vancomycin-related factors have been identified and included in chromatographic method development to demonstrate resolution, but mechanisms for formation or rates of formation studies have not been reported. The recent introduction of HPLC techniques for chromatographic profiling has not led to differentiation between cofermentation products and solid-state degradation products. Thomas and Newland (38) included an evaluation of a sample of the international vancomycin standard from the World Health Organization that was degraded by storage at 56°C, but this provided qualitative information only. This is clearly an area for further study.

Solution stability studies for vancomycin complement the CDP-1 mechanism previously described. Das Gupta et al. (53) investigated the stability of vancomycin in 5% dextrose and 0.9% sodium chloride injections. Solutions of 5 mg/ml of vancomycin were stored at −10, 5, and 24°C and evaluated by HPLC. At 24°C, about 5% degradation was observed after 17 days for both dextrose and saline solutions. No degradation was found at −10 or 5°C for up to 63 days. Considerable degradation was found at other pH values. At pH 1.4, nearly 20% degradation was observed within 5 days at 24°C. At pH 5.6, more than 10% degradation was observed at the end of 17 days. At pH 7.1, more than 10% degradation was observed within 5 days. This study confirmed that the pH range of 2.5–4.5 for vancomycin formulations provides optimal solution stability. Greenberg et al. (54) studied the stability of vancomycin hydrochloride solutions at 37°C for up to 4 weeks. Using a microbiological assay and HPLC, they found that vancomycin degraded very rapidly under these conditions, with a 20% loss observed in 4 days. Extensive turbidity was observed, probably because of the precipitation of the CDP-1 components.

Nahata et al. (55) reported their study of the stability of vancomycin hydrochloride in plastic syringes containing high concentrations of dextrose injection. Solution storage modeled normal use, with 24 h storage at 4°C and 2 h storage at room temperature. The solutions were monitored by an isocratic HPLC method capable of detecting CDP-1. All changes in vancomycin concentration fell within the variability of the method. Nahata (56) also studied the stability of vancomycin hydrochloride in total parenteral nutrient solutions. Two concentrations of two commonly used total parenteral nutrient solutions were used. Vancomycin concentrations were monitored by an HPLC method. Solutions were stored for up to 4 h at 22°C. Vancomycin concentrations did not change significantly during this study.

Thus the stability characteristics of vancomycin have received moderate study. The formation of CDP-1 has been reported, and a mechanism has been proposed. The solid-phase stability of vancomycin has not been studied extensively. Perhaps this is because of the limited availability of effective analytical tools for this study. HPLC has been proven to provide good resolution of vancomycin-related factors, and CDP-1 formation is readily monitored by existing HPLC methods. Further study of the stability characteristics of vancomycin by HPLC is needed, especially for vancomycin formulations before reconstitution.

F.   Reference Standards

Vancomycin reference standards are available from the *United States Pharmacopeia* and the World Health Organization. These are used for comparative assays, such as potency by microbiological assay and identification by infrared absorption. The USP reference standard is available as a preweighed vial in which the entire contents are dissolved without drying. It has a potency of 1100 μg/mg and a chemical purity of 94.0%. The potency unit of measure, the microgram, is a unit of activity compared to previously characterized vancomycin reference standards, referenced to the original vancomycin master standard (34). No correlation is defined between activity units and chemical purity. Thus a potency of 1000 μg/mg does not have a chemical purity of 100%. The international vancomycin standard available from the World Health Organization has a potency of 1007 μg/mg, also defined in terms of units of activity. These reference standards are important for maintenance of a consistent measure of antimicrobial activity as product changes evolve over time.

III. CHARACTERIZATION OF OTHER GLYCOPEPTIDES

Other glycopeptides that are structurally related to vancomycin have been reported in the literature. These have not been studied as extensively by analytical chemists, but the reports that are available provide additional understanding of this family of compounds. This review includes teichomycin, aricidins, ristomycins, avoparcin, and actinoidin.

The chromatographic characterization of vancomycin as described in Section II has included several other glycopeptides for comparison purposes. For example, Thomas and Newland (38) included demethylvancomycin, actinoidin A and B, carboxyristomycin, avoparcin, ristocetin A and B, and teicoplanin in their evaluation of an HPLC method for vancomycin. They found that these compounds were effectively separated by this system. The TLC systems that were described also were effective for these glycopeptides. Also, Sztaricskai et al. (39) included ristomycin, A35512B, carboxyristomycin A, avoparcin, and actinoidin A and B in their evaluation of two HPLC systems. The relatively low purity of these materials made extensive evaluation impossible.

Two additional reports provided analytical characterization of vancomycin-related glycopeptide antibiotics through exploiting their binding properties to acyl-D-alanyl-D-alanine. Folena-Wasserman et al. (57) developed an affinity chromatographic process for the isolation and purification of glycopeptide antibiotics through the immobilization of L-Lys-D-Ala-D-Ala, known to bind to glycopeptides from the vancomycin family. An HPLC method is described for separation efficiency measurements using a linear gradient ramp of 7–30% acetonitrile in phosphate buffer at pH 3.2 on a 15 cm C_{18} column with 5 μm particles. The method proved to be effective for the development of appropriate elution conditions for the affinity column. Corti et al. (58) reported their development of a solid-phase enzyme receptor assay for glycopeptides of the vancomycin class. The assay relies on their interaction with acyl-D-Ala-D-Ala. The antibiotics and enzyme-labeled teicoplanin compete for a synthetic analog of the biological receptor. Various glycopeptides were tested, including teicoplanin, vancomycin, ristocetin, and avoparcin. The antibiotics produced 50% displacement at concentrations ranging from 0.04 to 4 µg/ml, with vancomycin the weakest competitor. This method is effective for those antibiotics that rely on this binding mechanism for their antimicrobial activity.

A. Teichoplanin

Teichoplanin is a glycopeptide produced by *Actinoplanes teichomyceticus*. Characterization of the teichoplanin glycopeptides relies on many of the techniques applied to vancomycin. Bardone et al. (59) used paper and thin-layer chromatography with a variety of elution systems to separate teichoplanin A_1 and A_2. Vancomycin and teichoplanin A_2 are resolved by TLC. Spectroscopic analysis by UV and infrared absorption yielded spectra typical of glycopeptides. Borghi et al. (60) utilized reversed-phase HPLC to characterize teichoplanin and related factors. The method included a linear gradient program from 10 to 40% acetonitrile in phosphate buffer at pH 6.0. They found teichoplanin to be a mixture of five closely related compounds, differing in their side aliphatic chain.

Several methods have been described for the determination of teicoplanin in biological matrices. A solid-phase enzyme receptor assay has been described (58) and compared to the microbiological assay for teicoplanin (61). A receptor-antibody sandwich assay provides greater selectivity for teicoplanin and enhanced sensitivity over these other methods (62). This assay involves a sandwich reaction between bovine serum albumin (BSA)-ε-Aca-D-Ala-D-Ala, teicoplanin, and antiteicoplanin antibodies. Detection results from a chromogenic reaction with *o*-phenylenediamine. When other antibiotics were tested, no response was obtained at concentrations between 0.001 and 100 µg/ml, demonstrating excellent specificity. However, at high vancomycin concentrations, teicoplanin recoveries were reduced because of the limited availability of binding sites on the BSA conjugate. Thus, complementary techniques are needed to assess potential interferences from vancomycin-related antibiotics.

B. Aridicin

The aridicins are glycopeptides produced by the fermentation of *Kibdelosporangium aridum*. A reversed-phase HPLC method was developed to characterize the aridicin complex (63). Three major components were separated using gradient elution from 27 to 39% acetonitrile in phosphate buffer at pH 3.2. Similar spectroscopic characteristics were found for these compounds, matching those of other vancomycin-related glycopeptides. Paper and thin-layer chromatography were again proven effective for general separation of these compounds from other class members.

REFERENCES

1. Moellering R.C., Krogstad D.J., Greenblatt D.J. (1981). Vancomycin therapy in patients with impaired renal function: A nomogram for dosage. Ann. Intern. Med. 94:343.

2. Wenk M., Vozeh S., Follath F. (1984). Serum level monitoring of antibacterial drugs: A review. Clin. Pharmacokinet. 9:475.

3. Sabath L.D., Casey J.I., Ruch P.A., Stumpf L.L., Finland M. (1971). Rapid microassay of gentamicin, kanamycin, neomycin, streptomycin, and vancomycin in serum or plasma. J. Lab. Clin. Med. 78:457.

4. Walker C.A., Kopp B. (1978). Sensitive bioassay for vancomycin. Antimicrob. Agents Chemother. 13:30.

5. Walker C.N. (1980). Bioassay for determination of vancomycin in the presence of rifampin or aminoglycosides. Antimicrob. Agents Chemother. 17:730.

6. Torres-R J.R., Sanders C.V., Lewis A.C. (1979). Vancomycin concentration in human tissues—preliminary report. J. Antimicrob. Chemother. 5:475.

7. Crossley K.B., Rostchafer J.C., Chern M.M., Mead K.E., Zaske D.E. (1980). Comparison of a radioimmunoassay and a microbiological assay for measurement of serum vancomycin concentrations. Antimicrob. Agents Chemother. 17:654.

8. Fong K.L.L., Ho D.H.W., Bogerd L., Pan T., Brown N.S., Gentry L., Bodey G.P. (1981). Sensitive radioimmunoassay for vancomycin. Antimicrob. Agents Chemother. 19:139.

9. Jolley M.E. (1981). Fluorescence polarization immunoassay for the determination of therapeutic drug levels in human plasma. J. Anal. Toxicol. 5:236.

10. Jolley M.E., Stroupe S.D., Schwenzer K.S., Wang C.J., Lu-Steffes M., Hill H.D., Popekla S.R., Holen J.T., Kelso D.M. (1981). Fluorescence polarization immunoassay. III. An automated system for therapeutic drug determination. Clin. Chem. 27:1575.

11. Schwenzer K.S., Wang C.H.J., Anhalt J.P. (1983). Automated fluorescence polarization immunoassay for monitoring vancomycin. Ther. Drug Monit. 5:341.

12. Filburn B.H., Shull V.H., Tempera Y.M., Dick J.D. (1983). Evaluation of an automated fluorescence polarization immunoassay for vancomycin. Antimicrob. Agents Chemother. 24:216.

13. Ackerman B.H., Berg H.G., Strate R.G., Rotschafer J.C. (1983). Comparison of radioimmunoassay and fluorescent polarization immunoassay for quantitative determination of vancomycin concentrations in serum. J. Clin. Microbiol. 18:994.

14. Merritt G.J., Hunter B.H., Hall W.C. (1987). Lack of mezlocillin and piperacillin interference in measurement of vancomycin in the Abbott TDx. Clin. Chem. 33:2304.

15. Anne L., Hammad N., Chang C.C., Laungani D., Gottwald K., Alexander S., Centofanti J. (1988). Development of EMIT assay for the measurement of vancomycin in serum. Clin. Chem. 34:1256.

16. Yeo K.T., Traverse W., Horowitz G.L. (1989). Clinical performance of the EMIT vancomycin assay. Clin. Chem. 35:1504.

17. Best G.K., Best N.H., Durham N.N. (1968). Chromatographic separation of the vancomycin complex. Antimicrob. Agents Chemother. 4:115.

18. Kirchmeier R.L., Upton R.P. (1978). Simultaneous determination of vancomycin, anisomycin, and trimethoprim lactate by high pressure liquid chromatography. Anal. Chem. 50:349.

19. Uhl J.R., Anhalt J.P. (1979). High performance liquid chromatographic assay of vancomycin in serum. Ther. Drug Monit. 1:75.

20. McClain J.B.L., Bongiovanni R., Brown S. (1982). Vancomycin quantitation by high-performance liquid chromatography in human serum. J. Chromatogr. 231:463.

21. Bever F.N., Finley P.R., Fletcher C., Williams J. (1984). Liquid-chromatographic determination of vancomycin evaluated and improved. Clin. Chem. 30:1586.

22. Hoagland R.J., Sherwin J.E., Phillips J.M. (1984). Vancomycin: A rapid HPLC assay for a potent antibiotic. J. Anal. Toxicol. 8:75.

23. Jehl F., Gallion C., Theirry R.C., Monteil H. (1985). Determination of vancomycin in human serum by high-pressure liquid chromatography. Antimicrob. Agents Chemother. 27:503.

24. Rosenthal A.F., Sarfati I., A'Zary E. (1986). Simplified liquid-chromatographic determination of vancomycin. Clin. Chem. 32:1016.

25. Bauchet J., Pussard E., Garaud J.J. (1987). Determination of vancomycin in serum and tissues by column liquid chromatography using solid-phase extraction. J. Chromatogr. 414:472.

26. Greene S.V., Abdalla T., Morgan S.L., Bryan C.S. (1987). High performance liquid chromatographic analysis of vancomycin in plasma, bone, atrial appendage tissue, and pericardial fluid. J. Chromatogr. 417:121.

27. Hosotsubo H. (1989). Rapid and specific method for the determination of vancomycin in plasma by high-performance liquid chromatography on an aminopropyl column. J. Chromatogr. 487:421.

28. Hu M.W., Anne L., Forni T., Gottwald K. (1990). Measurement of vancomycin in renally impaired patient samples using a new high-performance liquid chromatography method with vitamin B_{12} internal standard: Comparison of high-performance liquid chromatography, EMIT, and fluorescence polarization immunoassay methods. Ther. Drug Monit. 12:562.

29. Pfaller M.A., Krogstad D.J., Granich G.G., Murray P.R. (1984). Laboratory evaluation of five assay methods for vancomycin: Bioassay, high-pressure liquid chromatography, fluorescence polarization immunoassay, radioimmunoassay, and fluorescence immunoassay. J. Clin. Microbiol. 20:311.

30. Ristuccia P.A., Ristuccia A.M., Bidanset J.H., Cunha B.A. (1984). Comparison of bioassay, high-performance liquid chromatography, and fluorescence polarization immunoassay for quantitative determination of vancomycin in serum. Ther. Drug Monit. 6:238.

31. Morse G.D., Nairn D.K., Bertino J.S., Walshe J.J. (1987). Overestimation of vancomycin concentrations utilizing fluorescence polarization immunoassay in patients of peritoneal dialysis. Ther. Drug Monit. 9:212.

32. White L.O., Edwards R., Holt H.A., Lovering A.M., Finch R.G., Reeves D.S. (1988). The in vitro degradation at 37°C of vancomycin in serum, CAPD fluid, and phosphate-buffered saline. J. Antimicrob. Chemother. 22:739.

33. Anne L., Hu M., Chan K., Colin L., Gottwald K. (1989). Potential problem with

fluorescence polarization immunoassay cross-reactivity to vancomycin degradation product CDP-I: Its detection in sera of renally impaired patients. Ther. Drug Monit. 11:585.

34. *United States Pharmacopeia XXII* (1990). p. 1441.
35. *British Pharmacopoeia* (1988). p. 593.
36. Japan Antibiotics Research Association Minimum Requirements for Antibiotic Products of Japan (1986). Yakugyo Jiho Co., p. 896.
37. Betina V. (1964). A systematic analysis of antibiotics using paper chromatography. J. Chromatogr. 15:379.
38. Thomas A.H., Newland P. (1987). Chromatographic methods for the analysis of vancomycin. J. Chromatogr. 410:373.
39. Sztaricskai F., Borda J., Puskas M.M., Bognar R. (1983). High performance liquid chromatography (HPLC) of antibiotics of vancomycin type. J. Antibiot. 36:1691.
40. Inman E.L. (1987). Determination of vancomycin related substances by gradient high-performance liquid chromatography. J. Chromatogr. 410:363.
41. United States Pharmacopeial Convention (1985). Pharmacopeial Forum. Rockville, MD, p. 1058.
42. Snyder L.R., Stadalius M.A. (1986). In Horvath C.S. (Ed.), High Performance Liquid Chromatography. Academic Press, New York, p. 195.
43. *United States Pharmacopeia XXII*, First Supplement (1989). p. 2179.
44. *United States Pharmacopeia XXII*, Third Supplement (1990). p. 2389.
45. Inman E.L., Maloney A.M., Rickard E.C. (1989). Inaccuracies due to sample-solvent interactions in high performance liquid chromatography. J. Chromatogr. 465:201.
46. Dolan J.D. (1991). Sample adsorption in liquid chromatography injection valves. LC-GC 9:22.
47. Inman E.L., Lantz M.D., Strohl M.M. (1990). Absorbance ratioing as a screen for related substances of pharmaceutical products. J. Chromatogr. Sci. 28:578.
48. Inman E.L., Clemens R.L., Olsen B.A. (1990). Determination of EDTA in vancomycin by liquid chromatography with absorbance ratioing for peak identification. J. Pharm. Biomed. Anal. 8:513.
49. Fooks J.R., McGilveray I.J., Strickland R.D. (1968). Colorimetric assay and improved method for identification of vancomycin hydrochloride. J. Pharm. Sci. 57:314.
50. Marshall F.J. (1965). Structural studies on vancomycin. J. Med. Chem. 8:18.
51. Smith K.A., Williams D.H., Smith G.A. (1974). Structural studies on the antibiotic vancomycin: the nature of the aromatic rings. J. Chem. Soc. Perkin Trans. 120:2369.
52. Harris C.M., Kopecka H., Harris T.M. (1983). Vancomycin: Structure and transformation to CDP-I. J. Am. Chem. Soc. 105:6915.
53. Das Gupta V., Stewart K.R., Nohria S. (1986). Stability of vancomycin hydrochloride in 5% dextrose and 0.9% sodium chloride injections. Am. J. Hosp. Pharm. 43:1729.
54. Greenberg R.N., Saeed A.M.K., Kennedy D.J., McMillian R. (1987). Instability of vancomycin in Infusaid drug pump model 100. Antimicrob. Agents Chemother. 31:610.
55. Nahata M.C., Miller M.A., Durrell D.E. (1987). Stability of vancomycin hydrochloride in various concentrations of dextrose injection. Am. J. Hosp. Pharm. 44:802.

56. Nahata M.C. (1989). Stability of vancomycin hydrochloride in total parenteral nutrient solutions. Am. J. Hosp. Pharm. 46:2055.
57. Folena-Wasserman G., Sitrin R.D., Chapin F., Snader K.M. (1987). Affinity chromatography of glycopeptide antibiotics. J. Chromatogr. 392:225.
58. Corti A., Rurali C., Borghi A., Cassini G. (1985). Solid-phase enzyme receptor assay (SPERA): A competitive-binding assay for glycopeptide antibiotics of the vancomycin class. Clin. Chem. 31:1606.
59. Bardone M.R., Paternoster M., Coronelli C. (1978). Teichomycins, new antibiotics from *Actinoplanes teichomyceticus* nov. sp. II. Extraction and chemical characterization. J. Antibiot. 31:170.
60. Borghi A., Coronelli C., Faniuolo L., Allievi G., Pallanza R., Gallo G.G. (1984). Teichomycins, new antibiotics from *Actinoplanes techomyceticus* nov. sp. IV. Separation and characterization of the components of teichomycin. J. Antibiot. 37:615.
61. Cavenaghi L., Corti A., Cassani G. (1986). Comparison of the solid-phase enzyme receptor assay (SPERA) and the microbiological assay for teicoplanin. J. Hosp. Infect. 7:85.
62. Corti A., Cavenaghi L., Giani E., Cassani G. (1987). A receptor-antibody sandwich assay for teicoplanin. Clin. Chem. 33:1615.
63. Sitrin R.D., Chan G.W., Dingerdissen J.J., Holl W., Hoover J.R.E., Valenta J.R., Webb L., Snader K.M. (1985). Aridicins, novel glycopeptide antibiotics. II. Isolation and characterization. J. Antibiot. 38:561.

8
Teicoplanin

BETH P. GOLDSTEIN
Lepetit Research Center, Marion Merrell Dow Research Institute, Gerenzano, Italy

RITA ROSINA and FRANCESCO PARENTI
Marion Merrell Dow Europe AG, Thalwil, Switzerland

Teicoplanin and vancomycin are currently the only glycopeptide antibiotics registered for human use. Teicoplanin was introduced into clinical practice in Europe starting in 1988, decades after the first introduction of vancomycin. Registration has been applied for in Canada and the United States. Teicoplanin is in early phase III clinical trials in Japan.

Although some features of teicoplanin have been touched on in the preceding chapters, our aim is to give an overview of the chemistry, antimicrobial activity, pharmacokinetics, and toxicology of this new antibiotic.

I. DISCOVERY

Teicoplanin was discovered in the course of a program aimed at detecting novel molecules of microbial origin that inhibit bacterial cell wall synthesis (2). This target was selected for several reasons: because the cell wall has no counterpart in mammalian cells, agents that inhibit its formation are likely to be relatively nontoxic; known inhibitors of cell wall formation were usually bactericidal; and only a few classes of cell wall inhibitors had been described.

Our source of microbial metabolites was fermentation broths of a large collection of rare actinomycetes of the genus *Actinoplanes*. These microorganisms had been shown in our laboratory to be versatile producers of a diverse array of antimicrobial agents (3). Detection of cell wall inhibitors was based on an antimicrobial assay using *Staphylococcus aureus* and its L form (cells lacking the wall, which can grow when cultured in a hypertonic medium). Any fermentation broth showing activity against the normal cells but not against the L form was tentatively considered to possess a specific cell wall inhibiting activity, which was then confirmed with crude extracts of the broth using appropriate biochemical assays.

Teicoplanin was rediscovered several times in later years using a different assay specific for agents that can form a complex with peptides terminating in D-alanyl-D-alanine. The original producer of teicoplanin is a strain of *Actinoplanes teichomyceticus* nov. sp. isolated from an Indian soil sample (2).

II. CHEMISTRY

Teicoplanin is a glycopeptide antibiotic of the dalbaheptide group (4) belonging to the "ristocetin type" (5). It contains a modified linear peptide of seven amino acids, five of which are common to all members of the group. The remaining two (in positions 1 and 3), *p*-hydroxyphenylglycine and *m,m'*-hydroxyphenylglycine, are linked together by an ether bond; this generates an extra cycle characteristic of the ristocetin type. A chlorine atom is found on each of the tyrosine residues (amino acids 2 and 6). Three sugar moieties are attached to the aryl groups: α-D-mannose in position 42 of amino acid 7, N-acetyl-β-D-glucosamine in position 34 of amino acid 6, and N-acyl-β-D-glucosamine in position 54 of amino acid 4. The acyl moiety is a fatty acid residue containing 10 or 11 carbon atoms.

Teicoplanin is a complex of five closely related molecules, denoted T-A_2-1 to T-A_2-5, which differ only in the nature of the acyl residue (see Fig. 1). The complex of these components is called teicoplanin A_2. In addition to A_2, there is a minor component called A_3 that lacks the N-acetyl-β-D-glucosamine in position 54 of amino acid 4.

Teicoplanin is an off-white amorphous powder that is freely soluble in water. The apparent pK_a for the terminal carboxyl and amino groups are 5.0 and 7.1, respectively. The four free phenolic groups have pK_a between 9 and 12.5. These values were determined in a mixture of methylcellosolve and water (4:1); when extrapolated to a totally aqueous system the pK_a of the carboxyl group is calculated as 3.1, explaining the isoelectric point of 5.1 found with electrofocusing.

In mixtures of *n*-octanol and aqueous buffer at pH 1.0, 5.2, 7.4, and 9.0, the partition coefficients (log P) of teicoplanin were -1.9, -1.1, -1.3, and -2.1, respectively. Teicoplanin is about five times as lipophilic as vancomycin.

III. FERMENTATION

Teicoplanin is a complex of closely related molecules, differing only in the fatty acid residue. We have studied parameters of the fermentation process that influence the relative amounts of the various components synthesized by the producer

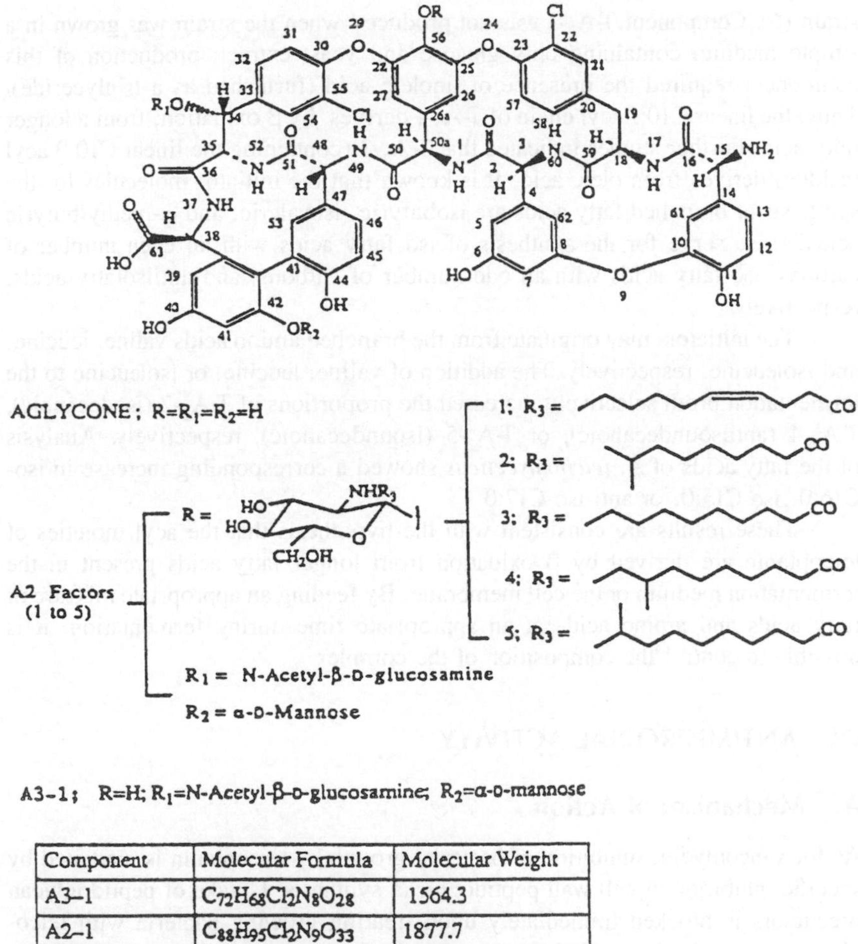

AGLYCONE; R=R₁=R₂=H

A2 Factors
(1 to 5)

$R =$ (sugar structure with NHR₃, HO, HO, CH₂OH)

1; $R_3 =$ (acyl chain)

2; $R_3 =$ (acyl chain)

3; $R_3 =$ (acyl chain)

4; $R_3 =$ (acyl chain)

5; $R_3 =$ (acyl chain)

$R_1 =$ N-Acetyl-β-D-glucosamine

$R_2 =$ α-D-Mannose

A3–1; R=H; R_1=N-Acetyl-β-D-glucosamine; R_2=α-D-mannose

Component	Molecular Formula	Molecular Weight
A3–1	$C_{72}H_{68}Cl_2N_8O_{28}$	1564.3
A2–1	$C_{88}H_{95}Cl_2N_9O_{33}$	1877.7
A2–2	$C_{88}H_{97}Cl_2N_9O_{33}$	1879.7
A2–3	$C_{88}H_{97}Cl_2N_9O_{33}$	1879.7
A2–4	$C_{89}H_{99}Cl_2N_9O_{33}$	1893.7
A2–5	$C_{89}H_{99}Cl_2N_9O_{33}$	1893.7

Fig. 1 Structure of the individual components of teicoplanin.

strain (6). Component T-A$_2$-1 was not produced when the strain was grown in a simple medium containing only glucose and yeast extract; production of this component required the presence of linoleic acid (furnished as a triglyceride). Thus, the linear C10:1 acyl chain of T-A$_2$-1 derives, by β oxidation, from a longer fatty acid. Further studies indicated that T-A$_2$-3 (containing the linear C10:0 acyl residue) derives from oleic acid. It is known that the initiator molecules for the synthesis of branched fatty acids are isobutyric, isovaleric, and α-methylbutyric acids; these serve for the synthesis of iso fatty acids with an even number of carbons, iso fatty acids with an odd number of carbons, and antiisofatty acids, respectively.

The initiators may originate from the branched amino acids valine, leucine, and isoleucine, respectively. The addition of valine, leucine, or isoleucine to the fermentation broth selectively increased the proportions of T-A$_2$-2 (isodecanoic), T-A$_2$-4 (antiisoundecanoic), or T-A$_2$-5 (isoundecanoic), respectively. Analysis of the fatty acids of A. teichomyceticus showed a corresponding increase in iso-C16:0, iso-C15:0, or anti-iso-C17:0.

These results are consistent with the hypothesis that the acyl moieties of teicoplanin are derived by β oxidation from longer fatty acids present in the fermentation medium or the cell membrane. By feeding an appropriate mixture of fatty acids and amino acids at an appropriate time during fermentation, it is possible to control the composition of the complex.

IV. ANTIMICROBIAL ACTIVITY

A. Mechanism of Action

As for vancomycin, inhibition of bacterial growth by teicoplanin is mediated by specific inhibition of cell wall peptidoglycan synthesis. Uptake of peptidoglycan precursors is blocked immediately upon treating growing bacteria with teicoplanin; the synthesis of other macromolecules is not directly affected (7). Teicoplanin causes accumulation of UDP-MurNAc-pentapeptide, indicating that it blocks later steps in peptidoglycan synthesis (7). Like most other glycopeptides teicoplanin binds strongly to peptides terminating in D-alanyl-D-alanine (8), which simulate the substrate for subsequent transglycosylation and transpeptidation steps (9,10).

B. Antibacterial Activity In Vitro

The antibacterial spectrum of teicoplanin is similar to that of vancomycin and encompasses most species of important Gram-positive pathogens. Teicoplanin does not penetrate the outer membrane of most Gram-negative bacteria.

Minimum inhibitory concentrations (MIC) of teicoplanin were usually significantly lower than those of vancomycin against isolates of *Enterococcus* and *Streptococcus* spp. (11–39). *Micrococcus* spp. had similar sensitivity to teicoplanin and vancomycin (33).

Most investigators have reported teicoplanin to be equivalent to or slightly more active than vancomycin against *S. aureus*, including strains resistant to methicillin, aminoglycosides, or quinolones (12–17,19–21,24,25,29–33,36–39, 40–53). In these studies, teicoplanin MIC tended to be somewhat higher when agar dilution methodology was used.

In many studies, teicoplanin and vancomycin also appeared to have similar activity against coagulase-negative staphylococci (CNS) (12–15,17,19,25,31, 32,38,39,45,50). In the last several years, CNS (including methicillin-resistant strains) have emerged as an important cause of infection, particularly of foreign bodies (54–56). With the investigation of larger numbers of CNS, it has become apparent that the teicoplanin MIC distribution for these isolates is broader than that of vancomycin (16,20,24,27,29,30,33,36,37,42,43,49,53,57–60). Thus, although a significant proportion (sometimes more than 50%) of strains may be at least as susceptible to teicoplanin as to vancomycin, many strains have somewhat higher teicoplanin MIC, and some vancomycin-susceptible strains are moderately susceptible or resistant to teicoplanin. The latter isolates are most often methicillin-resistant *Staphylococcus haemolyticus*; among this group, vancomycin resistance has also been occasionally reported (61–63). Reduced susceptibility to teicoplanin or vancomycin has also been documented among *Staphylococcus epidermidis* isolates (60,64–69).

In the studies of CNS just cited, the frequency of teicoplanin intermediate or resistant strains varied from none to greater than 10%. This may reflect repeated isolation of endemic strains in some hospitals or true regional or temporal differences in occurrence. However, methodology also affects the apparent sensitivity of CNS to teicoplanin to a greater extent than that to vancomycin. Several studies have shown that teicoplanin MIC or inhibition zones, particularly for CNS, are influenced by such factors as inoculum size, medium and/or supplements, and agar dilution versus broth (70–75). The highest frequencies of reduced susceptibility to teicoplanin have been reported in France, where agar dilution is the preferred method (58,59). The clinical significance of condition-dependent MIC has not been established.

Teicoplanin is also active against *Corynebacterium* spp., including *Corynebacterium jeikeium*, with MIC similar to those for vancomycin (15,20,27,33,37, 39,50,76–79). It is somewhat more active than vancomycin against *Listeria monocytocytogenes* (20,27,29,33,37,39) and *Bacillus cereus* (37).

Among anaerobic bacteria, *Clostridium*, *Peptococcus*, and *Peptostreptococcus* spp. were usually reported as somewhat more susceptible to teicoplanin than to vancomycin, and the MIC of the two glycopeptides for *Propionibacterium acnes* were similar (12,14,15,20,27,33,36,39,70–84).

A relatively weak activity of teicoplanin (MIC_{90} 16 mg/L; range 1–32 mg/L) has been reported against a limited number of isolates of *Erysipelothrix rhusiopathiae*; most of these strains were resistant to vancomycin (85). Teicoplanin is also active against *Eubacterium* spp. and *Actinomyces israelii* (86).

Teicoplanin has been reported to have potentially useful activity against *Gardnerella vaginalis* and some species of *Bacteroides* (33). Vancomycin had similar activity against *G. vaginalis* but not against *Bacteroides*.

Bactericidal Activity

Teicoplanin has usually been reported as bactericidal for sensitive staphylococci and most species of streptococci and weakly bactericidal for enterococci and viridans streptococci (for review, see Ref. 87). Growth conditions and interpretive criteria may be responsible for some of the reports of the bacteriostatic activity of teicoplanin.

As is generally the case for cell wall inhibitors, glycopeptides are more bactericidal when added to cultures of actively growing bacteria (12,88). The bactericidal action of glycopeptides is usually slow (89). For this reason, the use of MBC to define "glycopeptide-tolerant" strains may be misleading. Our own experience, and that of others (89), has been that minimal bactericidal concentration (MBC) end points of glycopeptides may not be reproducible for staphylococci and enterococci, although a significant bactericidal activity of teicoplanin is generally observed at concentrations near the MIC.

Resistance to Teicoplanin

Certain Gram-positive species (*Leuconostoc*, *Pediococcus*, and some species of *Lactobacillus*) are intrinsically resistant to high concentrations of teicoplanin and vancomycin (for review, see Ref. 90). Some species of *Lactobacillus* may be resistant to vancomycin but sensitive to teicoplanin (91). The physiological basis of this resistance is unknown.

As discussed, resistance in CNS (particularly *S. haemolyticus*)—that is, MIC above the therapeutic breakpoint—has been observed more frequently for teicoplanin than for vancomycin. Teicoplanin MIC are often condition dependent for these isolates. The teicoplanin MIC distribution is broad, continuous, and apparently unimodal. It may therefore be difficult to establish the physiological and genetic basis of teicoplanin resistance in CNS.

Acquired resistance to vancomycin has been described in enterococci, where it may be associated with plasmids or chromosomal transposons (92–100). Among the phenotypes so far described, one class of highly vancomycin resistant (VanA) isolates shows partial cross-resistance with teicoplanin (MIC 8 to >32 mg/L). For this class, which includes isolates of *Enterococcus faecalis*, *Enterococcus faecium*, and *Enterococcus avium*, the somewhat lower teicoplanin MIC may simply reflect the greater sensitivity of enterococci to teicoplanin and may have no therapeutic significance. A second group (VanB) has been described that has a lower level of vancomycin resistance (MIC 16–32 mg/L) and is susceptible to <1 mg/L of teicoplanin. Susceptibility to teicoplanin seems to be due to the inability of this antibiotic to induce the resistant phenotype; pregrowth with vancomycin renders these strains teicoplanin resistant. A third phenotype (VanC), which also involves moderate resistance to vancomycin and susceptibility to teicoplanin, is distinguished by lack of inducibility; it occurs mainly in such species as *Enterococcus gallinarum* and *Enterococcus casseliflavus* (91,101). It is not clear why VanC isolates are susceptible to teicoplanin.

A fourth phenotype recently described involves high-level resistance to vancomycin coupled with normal susceptibility to teicoplanin (102). In this case, vancomycin must be present in the medium for teicoplanin resistance to be expressed. The potential usefulness of teicoplanin in combating infections due to such isolates and to those of the a VanB class depends on the ease with which constitutively resistant mutants may arise in vivo. The results of experimental endocarditis experiments in rabbits with a VanB strain of *E. faecium* suggest that teicoplanin may be efficacious in these cases (103).

Despite intensive study over the last few years (discussed in Chap. 6), we do not yet have a complete understanding of the resistance mechanisms that govern the diverse resistance phenotypes.

Susceptibility Testing

Teicoplanin diffuses poorly in agar. A recent study showed that only 20% of teicoplanin diffused out of a 30 μg disk in 6 h (compared with 70% of vancomycin); at the critical time for inhibition zone formation, relatively low concentrations of teicoplanin were found at distances from the disk, where an inhibition zone should form (104). This behavior is probably related to the relatively lipophilic nature of teicoplanin and explains why investigators have observed smaller inhibition zones (relative to MIC) for teicoplanin than for vancomycin (71,105–108).

Recently, Jones et al. (109) reported that the statistical correlation between inhibition zone size and MIC was poor. The number of major errors (sensitive with one method and resistant with the other) has been very low in well-controlled studies (71,106,107). However, our experience has been that reports of false resistance to teicoplanin are rather frequent in routine disk susceptibility testing. This has been borne out by a study in which significant proportions of fully susceptible staphylococcal isolates appeared to be only moderately susceptible with the disk susceptibility test (60). Because of the poor diffusibility of teicoplanin, zone sizes may be inordinately affected by relatively small increases in agar depth or inoculum density.

There have been occasional reports of the reverse phenomenon (high teicoplanin MIC with large inhibition zones) for *S. haemolyticus* isolates (71,110). One may speculate that this is caused by relatively slow growth of these isolates combined with inoculum-dependent MIC.

C. Animal Infection Studies

Septicemia

Early studies in acute mouse septicemia models demonstrated that teicoplanin was efficacious in vivo against staphylococci and streptococci (12). Subcutaneous (SC) ED_{50} ranged from 0.1 mg/kg/day for *Streptococcus pyogenes* to 0.7 mg/kg/day for *S. aureus* and were almost always lower than those of the comparison agents, which included vancomycin, β-lactams, erythromycin, and lincomycin. Because these experiments employed once daily treatment for 3 days, the relatively long half-life of teicoplanin undoubtedly played a role in the outcome. In a subacute septicemia model in mice, 33 mg/kg of vancomycin or teicoplanin (once daily intraperitoneally, IP, for 10 days) gave similar survival rates (111).

Endocarditis

There have been several studies of teicoplanin in endocarditis in rats or rabbits. Arioli et al. (112) found that once daily treatment with 25 or 50 mg/kg SC of teicoplanin was more efficacious than the same dose of vancomycin in reducing heart loads of *S. aureus* in rats. Against *S. aureus* endocarditis in rabbits, Chambers and Sande (113) found that 4.5 mg/kg/12 h intramuscularly (IM) of teicoplanin (first dose 6 mg/kg) was as efficacious as 200 mg/kg/8 h IM of nafcillin or 25 mg/kg/12 h intravenously (IV) of vancomycin. In this model, using higher doses of teicoplanin (36 mg/kg loading dose followed by 18 mg/kg/12 h), IM treatment produced a greater reduction in vegetation titer of a methicillin-resistant *S. aureus* MRSA strain than IV treatment, suggesting that higher trough levels may be important (114). However, Contrepois et al. (115) found that, upon IV administration to rabbits, 4.5 mg/kg/16 h of teicoplanin was as efficacious as 9 mg/kg/12 h of vancomycin in reducing *S. aureus* in vegetations. Kaatz et al. (116) found that, in rabbits, 12.5 mg/kg/12 h IV of teicoplanin was as efficacious as 17.5 mg/kg/6 h IV of vancomycin against one *S. aureus* strain but that higher doses of teicoplanin were needed to obtain comparable results against an isolate that was less sensitive to teicoplanin. In the rat endocarditis model, an intermediate dosage of teicoplanin was efficacious against this same strain (117).

Against *S. epidermidis* endocarditis in rabbits, Galetto et al. (118) observed similar efficacies with 12.5 mg/kg/12 h IM of teicoplanin (25 mg/kg loading dose) and 25 mg/kg/12 h IV of vancomycin. Gentamicin and rifampin potentiated the efficacy of both glycopeptides. Tuazon and Washburn (119) also observed potentiation of teicoplanin efficacy by rifampin in *S. epidermidis* endocarditis in rabbits.

In the rat model, teicoplanin (20 mg/kg/12 h IV, 40 mg/kg loading dose) and vancomycin (40 mg/kg/12 h) reduced heart loads of *S. epidermidis* strains to a similar extent (117).

In *E. faecalis* endocarditis in rats, 3.1 mg/kg/24 h SC of teicoplanin was more efficacious than once daily 25 mg/kg of vancomycin (112). In rabbits, 12.5 mg/kg/12 h IV of teicoplanin reduced *E. faecalis* in vegetations to a similar extent as ampicillin (100 mg/kg/6 h IM), and the efficacy of teicoplanin was potentiated by gentamicin (120). When antibiotics were administered to rabbits by continuous IV infusion, 30 mg/kg/day of teicoplanin and 150 mg/kg/day of vancomycin produced similar reductions in vegetation titers of *E. faecalis* (121); relapses were less frequent in the teicoplanin group. Bush et al. (122) observed potentiation by gentamicin of the efficacy of teicoplanin and vancomycin in *E. faecium* endocarditis in rats. Recently, Fantin et al. showed that 20 mg/kg IM twice daily of teicoplanin (loading dose 40 mg/kg) was efficacious in reducing heart bacterial loads in rabbits infected with a VanB isolate of *E. faecium* (103). Neither vancomycin (30 mg/kg twice daily) nor gentamicin (6 mg/kg twice daily) alone had a significant effect. In these experiments, addition of gentamicin potentiated the efficacy both of vancomycin and of a lower dose of teicoplanin (10 mg/kg twice daily; loading dose 20 mg/kg).

In *Streptococcus sanguis* endocarditis in rats (112) a significant reduction in heart titer was seen at teicoplanin dosages as low as 6.3 mg/kg/day SC. In a rabbit model, efficacy against *S. sanguis* appeared to be somewhat dose dependent over the range 4.5–18 mg/kg/12 h IM (with doubled loading doses) (114).

Other Infections

In a rat granuloma pouch model, 6.3 mg/kg/day SC of teicoplanin significantly reduced the titer of *S. aureus* (112); the better efficacy of teicoplanin relative to once daily vancomycin was attributed to longer persistence of teicoplanin levels above the MIC in the exudate fluid. In a subcutaneous abscess model in mice, 6.3 or 12.5 mg/kg SC of teicoplanin (at 1 and 24 h after infection) was more efficacious than the same dosages of vancomycin (123). Teicoplanin and vancomycin were equally efficacious in a thigh infection model in neutropenic mice (124). Teicoplanin at 6.6 mg/kg/12 h IP was not active in a foreign body (tissue cage) model of *S. epidermidis* infection in guinea pigs (125).

Neither teicoplanin nor vancomycin was efficacious in an experimental osteomyelitis model in rabbits (126). The authors attributed this to reduced activity of the antibiotics under anaerobic conditions, but we have been unable to confirm this observation in our laboratory (unpublished data). In fact, teicoplanin is very active against Gram-positive anaerobes, and a dose of 5 mg orally every 8 h prolonged the survival of hamsters with *Clostridium difficile* colitis (127).

Teicoplanin (20 mg/kg/12 h SC) prevented neonatal bacteremia and meningitis due to group B streptococci in rats (128). It was efficacious against MRSA meningitis in rabbits when given intracisternally (1 mg/kg 18 h after infection) (129) but not when given systemically (130).

Both teicoplanin and vancomycin had poor activity against experimental listeriosis in normal mice and were inactive in nude mice (131).

V. TOXICOLOGY

A. Acute Toxicity

The acute toxicity of teicoplanin administered by various routes was determined in mice, rats, rabbits, and dogs. The approximate LD_{50} values (in mg/kg) were 1600 IM, 700–900 IV, 2860 SC, 1000–1360 IP, and >10,000 orally in the mouse; >500 IV in the rabbit; and >900 IM and 750 IV in the dog.

B. Subchronic and Chronic Toxicity

Multiple-dose toxicity studies were performed in rats with dosages of up to 80 mg/ kg per day IV for 1 month and up to 150 mg/kg per day SC for 6 months. In the dog, dosages were up to 40 mg/kg per day IM for 1 or 6 months and IV for 1 month. Similar results were obtained in the various studies. No deaths were attributed to a direct effect of teicoplanin. Body weight gain and food consumption were only slightly affected, indicating low general toxicity.

The target organ in all studies was the kidney. At the higher doses, the kidneys appeared pale at postmortem examination and showed slight focal degeneration of the epithelium of the cortical tubules with signs of regeneration. Blood urea nitrogen was not increased except in rats treated for 1 month at the highest dosage (150 mg/kg/day). After the drug-free recovery period of 9 weeks in rats and 10 weeks in dogs (after the end of the 6 month studies), signs of functional or histological changes were absent or there was a marked trend to recovery.

Among the various studies, there were no consistent treatment-related changes in nonrenal clinical pathology. Slight modifications were seen in some other organs, however, differing somewhat with the species, sex, dose, route of administration, and duration of treatment. These included an inflammatory reaction at the site of injection, the severity of which increased with the dosage and duration of treatment. Leukocytosis and histiocytic proliferation, a consequence of this subcutaneous inflammation, were seen in the liver, kidneys, spleen, lungs, and adrenals. Hypertrophy of the follicular epithelium of the thyroid and enlargement of the adrenals secondary to injection site irritation were noted in rats.

In dogs treated for 6 months, the cytoplasm of hepatocytes had a more granular appearance than in the controls. Vacuolization of the pancreatic acini was noted in rats treated IV or SC. Both the hepatocytic granules and the pancreatic vacuoles were found upon ultrastructural examination to be autolysosomes.

The doses producing minimal or no toxic signs were considered 10–20 mg/kg/day IV in 1 month studies in rats and dogs; 20 mg/kg/day SC for 1 month and 5–10 mg/kg/day SC for 6 months in the rat; and 20–40 mg/kg/day IM for 1 month and 10 mg/kg/day IM for 6 months in the dog.

C. Reproduction Studies

Potential teratogenic effects were studied in rats and rabbits treated during the period of organogenesis. In a peri- and postnatal study, rats were treated from day 15 of gestation to weaning. In a male fertility study, male rats were treated for 70 days prebreeding and 14 days during mating. In a female fertility study, female rats were treated for 15 days before mating and through breeding, gestation, and lactation. Rabbits were treated from day 6 to 18 of gestation.

No teratogenic effects attributable to treatment were found in either rats or rabbits. There were indications of maternal toxicity in rats, manifested as an increased incidence of stillbirths at doses of 100 and 200 mg/kg/day and of mortality of sucklings at 200 mg/kg/day, which were attributed to reduced maternal care of the offspring by some dams that were excessively stressed as a result of severe inflammation at the site of SC injection. In the peri- and postnatal and fertility studies (10, 20, and 40 mg/kg/day SC) there were no effects of treatment on fertility or gestational parameters of F_0 rats, on fertility of the F_1 generation, or on survival and development of F_2 offspring.

D. Mutagenicity Studies

Teicoplanin showed no evidence of mutagenic potential in a series of tests, including the Ames test with *Salmonella typhimurium* (at concentrations of up to 5000 μg/plate); a point mutation test in the yeast *Schizosaccharomyces pombe* (at up to 2000 μg/ml); a gene conversion test in *Saccharomyces cerevisiae* (at up to 4000 μg/ml); a chromosome aberration study in rat lymphocytes at concentrations up to 300 μg/ml; and a mouse bone marrow micronucleus test at doses up to 620 mg/kg IV. All the in vitro tests were performed both with and without metabolic activation.

E. Other Studies

In vitro and in vivo pharmacological studies were performed to detect any possible effects of teicoplanin on the central nervous system, cardiovascular, or respiratory systems. Teicoplanin had no relevant central nervous system effects. In rats, doses of 15 mg/kg or more caused reduction in blood pressure when given IV but not SC or IP; prior administration of lower doses protected the rats from this effect. Cardiovascular effects were not seen in conscious dogs treated IV with 10 mg/kg of teicoplanin.

VI. PHARMACOKINETICS

The pharmacokinetics of teicoplanin has been studied in various animal species (mice, rats, and dogs) and in humans, including healthy volunteers, patients with different degrees of renal function, children, and the elderly. The penetration of teicoplanin in tissues and the effect of various disease states on teicoplanin pharmacokinetics were also investigated (132).

A. Analytical Methods

Techniques currently available for the determination of teicoplanin in various biological matrices include microbiological assay, high-performance liquid chromatography (HPLC), solid-phase enzyme receptor assay (SPERA), receptor-antibody sandwich assay (RASA), and the fluorescence polarization immunoassay (FPIA). In addition, the radiochemical method was used after the administration of radiolabeled teicoplanin.

Teicoplanin in biological fluids has most often been determined using the microbiological assay. This method assays teicoplanin complex and uses *Bacillus subtilis* American Type Culture Collection (ATCC) 6633 as the indicator organism (133–137). The limit of detection is 0.05–0.15 mg/L; upon repeated assay, within- and between-day coefficients of variation are less than 10%. With appropriate modifications (such as the use of resistant derivatives of *B. subtilis* ATCC 6633), the microbiological method can be used to assay teicoplanin in the presence of many of the other commonly administered antibiotics (135,138).

A number of HPLC methods have been developed, which resolve the teicoplanin complex into its six components (Fig. 2) (139–144). A selective extraction of teicoplanin from the biological matrices using affinity chromatography is usually needed. The sensitivity of the method, considering the volume of sample to be processed, is usually lower than that of the microbiological assay; however, good sensitivity can be achieved (143,144).

The SPERA method is based on competition for a synthetic analog of the biological receptor (albumin-ε-amino-caproyl-D-alanyl-D-alanine) between teicoplanin in a biological sample and enzyme-labeled teicoplanin (133). The results of this assay are not significantly affected by the presence of other commonly used antibacterial drugs, with the exception of vancomycin, which is unlikely to be coadministered with teicoplanin. There is a good correlation between the microbiological assay and both HPLC and SPERA.

The RASA method extends SPERA by adding a "sandwich" reaction between teicoplanin, bound to the synthetic analog of the biological receptor, and antiteicoplanin antibodies (145). It is a highly sensitive method whose limit of detection is 0.03 mg/L; it is particularly useful when sample size is limited and for teicoplanin determination in tissues.

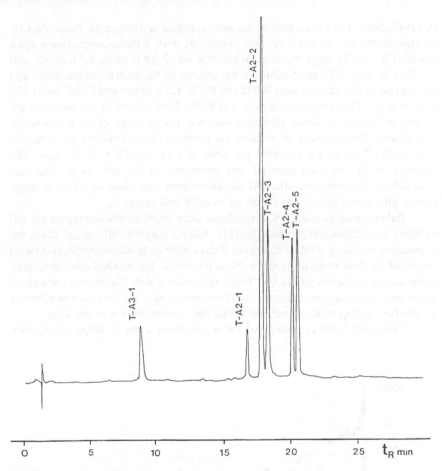

Fig. 2   HPLC chromatogram of teicoplanin complex.

The FPIA method involves the use of teicoplanin antiserum and a fluoresceinated teicoplanin tracer (146). Over the range 5–100 mg/L, the intra- and interassay variability is less than 10%. The limit of detection is 1 mg/L. Analytical results obtained using this method correlated well with those obtained using the microbiological assay ($r = 0.985$).

B.   Animal Studies

Single-dose pharmacokinetic studies on teicoplanin were carried out in mice (1, 5, and 10 mg/kg SC), rats (10 mg/kg IV), and dogs (5 mg/kg IV and IM and 40 mg/kg

IV) (147–149). The disposition of the antimicrobial activity in the body after IV administration to rats and dogs was consistent with a three-compartment open model (Fig. 3). The apparent terminal half-life was 2.5 h in mice, 8.5 h in rats, and 27–30 h in dogs. The mean values of the volume of the central compartment and the volume of distribution were 0.095 and 0.724 L/kg in rats and 0.080 and 0.323 L/kg in dogs. The systemic clearance was 6.19 ml/minute-m^2 in rats and 3.46 ml/minute-m^2 in dogs. Renal excretion was the major route of elimination for teicoplanin. The recovery of microbiologically active teicoplanin in urine accounted for 47% of the administered dose in rats and 53–67% in dogs. The majority of the excreted material was recovered in the first 24 h. The bioavailability of teicoplanin after IM administration was close to 100% in dogs. Linear pharmacokinetics was observed in mice and dogs.

Balance studies using [^{14}C]teicoplanin were conducted following single and multiple administrations in rats (150,151). After a single (~10 mg/kg) dose, the cumulative recovery of total ^{14}C activity 5 days after drug administration averaged 76.3% of the dose in the urine and 8.7% in the feces. The residual dose remaining in the animal carcasses was 11.1%. Teicoplanin was widely distributed throughout the body. In most organs, the maximum concentration of teicoplanin was achieved by the first killing time (15 minutes) after the administration of the drug.

The liver, kidneys, skin, and fat represented a deep compartment: they

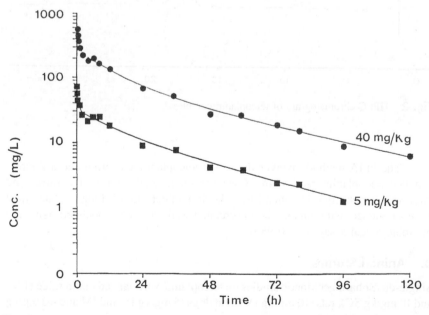

Fig. 3 Plasma concentrations in the dog after IV administration of teicoplanin.

contained most of the residual dose found in the rat carcasses 120 h after administration. The apparent elimination half-life, calculated from the radiochemical data over the interval 0–72 h postdose, was about 25 h, which is much higher than the half-life calculated using a microbiological assay over the 36 h postdose. Comparison of the tissue distribution pattern following a single dose with that found after the last of seven 10 mg/kg daily doses showed no major differences.

In a study designed to investigate the metabolism of teicoplanin in rats, urine samples were collected over 24 h following a 7.6 mg/kg ^{14}C-labeled dose (152). HPLC analysis of urine indicated that teicoplanin undergoes minimal metabolism and/or chemical transformation in the range of 3–5%.

C. Human Studies

The basic pharmacokinetic properties of teicoplanin have been derived from single-dose studies in healthy volunteers.

Absorption

Teicoplanin is not significantly absorbed from the gastrointestinal tract, probably because of its hydrophilic nature, and is consequently commonly administered via the intravenous or intramuscular route (153). Absorption occurs rapidly from the intramuscular site, resulting in a peak plasma concentration of 5 to 7 mg/L 2–4 h after a 3 mg/kg dose (134,154,155) and a proportional concentration of 12.3 mg/L 4 h after a single 6 mg/kg dose (156). The extent of systemic absorption following IM administration was reported as 90–100% (134,154–156).

In continuous ambulatory peritoneal dialysis (CAPD) patients, teicoplanin may be administered into the peritoneal cavity; there was 77% absorption of a 3 mg/kg dose in a 5 h dwell time (157). Intraventricular administration has been used in shunt-associated ventriculitis; absorption from the intraventricular site was very low (158). The endobronchial route of administration produced no detectable levels in plasma after multiple doses (159).

Distribution

The mean values of some pharmacokinetic variables obtained in healthy volunteers after intravenous administration are reported in Table 1 (134,137,155,156, 160–163). In most cases, the pharmacokinetics of teicoplanin are best described by a triexponential equation, however two- or four-compartment and model-independent analyses have also been used. The values reported for the first (20–30 minutes) and second (1.6–15.4 h) exponential terms indicate that distribution equilibrium is not reached for at least 24 h following drug administration, possibly longer. The terminal half-life of teicoplanin appears to vary to a great extent, from 32 to 130 h, in different studies. This variability is mostly due to methodological factors, however, since half-life values increase with the duration of sampling, which varied from 4 to 25 days after drug administration among the different

Table 1 Mean ± SD Pharmacokinetics of Teicoplanin After Intravenous Administration to Subjects with Normal Renal and Hepatic Function[a]

Dose (mg/kg)	No.	Last sampling time (days)	Model[b]	CL (L/h/kg)	CL_R (L/h/kg)	Phase 1 (minutes)	Phase 2 (h)	Terminal phase[c] (h)	V_{ss} (L/kg)	V_c (L/kg)	Reference
								Half-life			
3	6	4.2	3	0.012 ± 0.00124[d]	0.0062 ± 0.00084[e]	19.2 ± 3.6	3.60 ± 0.21	47.4 ± 8.5	0.61 ± 0.090	0.057 ± 0.012	134
6	6	4.2	3	0.0118 ± 0.00246[d]	0.00613 ± 0.0011[e]	20.4 ± 2.4	3.26 ± 0.37	44.1 +5.6	0.56 ± 0.083	0.057 ± 0.020	159
2	5	4	3	0.0162 ± 0.0008[d]	0.0095 ± 0.00135[e]	9.9 ± 3.7	1.62 ± 0.65	31.9 ± 6.8	0.68 ± 0.100	0.078 ± 0.026	
3	5	4	3	0.0159 ± 0.00121[d]	0.0103 ± 0.00116[e]	30.7 ± 13.1	4.14 ± 2.00	49.1 ± 21.7	0.84 ± 0.23	0.108 ± 0.014	160
2.9	6	4	2	0.0056 ± 0.00032[d]	NR	NR	NR	33.2 ± 5.1	0.28 ± 0.02	0.097 ± 0.023	
5.3	8	7	NR	0.0094 ± 0.00118	0.0059 ± 0.0012[e]	NR	NR	70	0.71 ± 0.10	NR	161
5.1	5	10	NR	0.0098 ± 0.00101[d]	0.0078 ± 0.00129[e]	NR	NR	77[f]	0.76[f] ± 0.11	NR	

2.8	10	4	3	0.0129 ± 0.0025[d]	0.0076 ± 0.00219[g]	NR	NR	44.5 ± 3.4	0.58 ± 0.096	0.05 ± 0.007	155
3	6	25	3	0.0085 ± 0.00124[h]	0.0085	NR	NR	130 ± 14.9	1.10 ± 0.028	0.076 ± 0.010	137
15	5	13	3	0.0109 ± 0.00163	0.0081 ± 0.00248	29 ± 9.6	9.7 ± 3.5	88.0 ± 8.6	0.80 ± 0.075	0.070 ± 0.20	162
19	4	13	3	0.011 ± 0.0011	0.0089 ± 0.00069	34 ± 6.7	9.4 ± 2.5	83.0 ± 4.4	0.87 ± 0.14	0.087	
26	5	13	3	0.0113 ± 0.0014	0.0088 ± 0.0013	61 ± 19	15.4 ± 2.7	91.8 ± 6.95	0.86 ± 0.21	0.11 ± 0.025	156
6	23	21	NR	0.0102[i]	0.010[i]	NR	NR	168	1.62[i]	NR[i]	

[a] All studies used microbiological assay.
[b] Number of exponential terms that best describe the data.
[c] Half-life of terminal log-linear phase.
[d] Calculated as dose/AUC, where AUC was estimated numerically from the data.
[e] Calculated as $A_e(0, t)$/AUC(0, t), where AUC(0, t) is the area to the last observation and A_e is the corresponding cumulative amount excreted unchanged.
[f] Calculated as MRT × CL, where MRT is the mean residence time.
[g] Calculated as f_e/CL, where f_e is the fraction of dose excreted in urine unchanged.
[h] Estimated from dose/AUC, where AUC is calculated from the sum of the ratios of coefficients to exponential coefficients.
[i] No standard deviation (SD) reported.

Abbreviations: No. = number of subjects; CL = total clearance; CL_R = renal clearance; V_i = Initial volume of distribution; V_{ss} = volume of distribution at steady-state; V_t = volume of distribution during the terminal phase; NR = not reported; AUC = total area under the concentration-time curve.

Source: Adapted from Ref. 132. with permission.

studies. The initial volume of distribution V_c is slightly in excess of the plasma volume, and the steady-state volume of distribution V_{ss} ranges from 0.56 to 1.1 L/kg; the higher values were reported in the studies that used later sampling times. The pharmacokinetics of teicoplanin appears to be linear over a wide dose range from 2 to 26 mg/kg.

Teicoplanin is highly bound to plasma proteins, mainly to albumin, with an unbound fraction f_u of 0.064–0.124, irrespective of drug concentration over the range of 4–280 mg/L (164,165).

The concentrations of teicoplanin in various body tissues and fluids after single intravenous doses are reported in Table 2 (87,166–175). Teicoplanin does not distribute into erythrocytes (176) or penetrate effectively into cerebrospinal fluid (CSF) even in patients with bacterial meningitis (177). Therapeutic CSF concentrations, however, are obtained upon intraventricular administration (158,178).

Elimination

Teicoplanin is eliminated almost entirely by renal mechanisms. In a study with the ^{14}C-labeled drug, 83% of the dose was recovered over 16 days: 80% in urine and about 3% in feces (176). The urinary excretion of a single 3 mg/kg dose increased with time from 25% after 1 day to 80% at 16 days and to 100% at 35 days. Comparison of radioactive drug recovery with microbiological assay indicated that there was little or no metabolism of teicoplanin into inactive compounds (137,176).

Teicoplanin has a low total clearance (CL of the order of 0.006–0.0162 L/h/kg); lower values have often been reported for renal clearance (Table 1). This difference possibly reflects, in part, the failure to collect urine for a sufficient period of time after administration to produce a good estimate of the fraction of the administered dose that was excreted unchanged. Considering the high protein binding of teicoplanin, the unbound renal clearance of teicoplanin has been calculated as of the order of 120 ml/minute/70 kg, a value close to the glomerular filtration rate, and suggests that teicoplanin is excreted almost entirely by glomerular filtration (132).

Multiple Dosing

The pharmacokinetics of teicoplanin appears to be predictable from single-dose data. In general, there is reasonable agreement between the findings of Carver et al. (137) and those of earlier studies (154,162,179). In all the studies, teicoplanin had to be given for at least a week to attain the steady state. The calculated values of degree of accumulation for once daily dosing of teicoplanin given over 5 minutes are 1.2, 2.3, and 3.4 for the maximum, average, and minimum concentrations, respectively (137).

Disposition Kinetics of Individual Components

The disposition kinetics of the major components of teicoplanin complex have been studied using HPLC to resolve the individual components (180,181). Gener-

Table 2 Concentrations of Teicoplanin Achieved in Various Body Tissues and Fluids After Single Intravenous Doses

Tissue/fluid	No. patients	Dose (mg)	Time (h)	Mean tissue or fluid concentration (mg/kg or mg/L)[a]	Tissue or fluid/serum concentration (%)	Reference
Blister fluid	6[b]	440	2	13.2	62.9[c]	87
						166
						167
	7	600	2.1	8.7	42.9[c]	168
Peritoneal fluid	34	400	≈1	>20	94.6	167
Gallbladder wall	24	400	3	3.2	57.1	169
	19	400	11	15.7	231	170
Gallbladder bile	24	400	1	1.9	10.3	169
Liver	15	400	13	10.3	229	170
Pancreas	3	400	13	4.41	158	170
Heart	32	6 mg/kg	≈1	706	318	171
	17	12 mg/kg	0.8	140	247	171
Fat	10	400	0.5	1.3	4.47	172
Fat	10	400	3.25	1.2	16.2	172
Fat (presternal)	10	12 mg/kg	—[d]	6.1	35	173
Skin (sternal)	10	12 mg/kg	—[d]	9.0	52	173
Tonsils	3	400	24	19	339	174
Cartilage	3	400	24	6.4	114	174
Mucosa	3	400	24	6.9	123	174
Bone	3	400	24	6.8	121	174
	10	400	0.5	6.6	22.3	172
	10	400	3.25	6.3	85.1	172
Synovial fluid	7	400 (× 3)	25	11	42.3	175

[a]Highest concentration reported per study.
[b]Healthy volunteers.
[c]Based on ratio of AUC.
[d]Sampled at the end of cardiac bypass surgery (teicoplanin administered during induction of anesthesia).
Source: Adapted from Reference 87.

ally, the composition of the complex appears to change only slightly with time; within the major A_2 group there was a slight trend toward enrichment with the more lipophilic components $A_2$4 and $A_2$5. Since the A_2 components (A_2-2 to A_2-5), which accounted for up to 75% of the teicoplanin complex, differ only marginally in their pharmacokinetics, teicoplanin can be regarded for practical purposes as a single compound whose pharmacokinetics can be adequately characterized using the microbiological assay.

Effects of Age on Pharmacokinetics

The pharmacokinetics of teicoplanin in elderly patients has been studied after single- and multiple-dose IV administration (182,183). In a group of 12 elderly patients with a moderate degree of renal impairment, the pharmacokinetics of teicoplanin after a single dose generally correlated well with that of younger patients with similar renal functions; the average elimination half-life was 107 h and the average CL was 0.0106 L/h/kg, with CL_R accounting for about 40% of the total, possibly because of incomplete urine collection (182). In a multiple-dose study conducted in 5 elderly patients with normal renal function given 7 consecutive IV doses of 400 mg, the elimination half-life was reported to be about 52 h, but in a group of similar patients receiving a single dose, the elimination half-life was 19.5 h. However, these results were obtained with sampling times of 72 and 24 h, respectively, after the last administration. Consequently, as discussed earlier, the sampling time was not long enough for precise determination of the terminal half-life.

The pharmacokinetics of teicoplanin in pediatric patients aged 2–12 years given a single IV 3 mg/kg dose were reported to be similar to those in normal volunteers (184). The values of the main pharmacokinetic parameters were V_c = 0.15 L/kg, V_{ss} = 0.8 L/kg, CL = 0.0148 L/h/kg, CL_R = 0.009 L/h/kg, and terminal half-life = 58 h. The low values for renal clearance reported in this study, however, may reflect the difficulties encountered in collecting complete urine samples for a long period of time in this patient population. At variance with this study are results obtained with another group of subjects of similar age (185). These authors, who used the HPLC method, reported a CL of 0.028 L/h/kg, which is twice the greatest value reported in adults, a proportionally shorter half-life of 20 h, and a renal clearance accounting for 40% of the total. In the same study, the pharmacokinetics of teicoplanin was studied in four neonatal patients (mean age 8.5 days) who received 6 mg/kg intravenously over 20 minutes and were followed for 6 days. In these patients, clearance was greater than in the group of older children; consequently, the terminal half-life was also greater in the neonates (30 h). The poorly developed renal and extrarenal functions of neonates accounted for the difference. However, the number of subjects studied was probably insufficient to draw any firm conclusions. These data prompted the authors to recommend a daily dosage of 10 mg/kg for pediatric patients, a recommendation that was also supported by the pharmacokinetics results obtained in a clinical study (186).

Altered Pathophysiological States

In subjects with impaired renal function both the CL and CL_R of teicoplanin are decreased; there is a linear correlation with creatinine clearance (180,187–192). As a consequence, the elimination half-life is increased and may reach 124–237 h in severe renal impairment; the half-lives reported vary somewhat according to the length of the sampling period after administration. Renal dysfunction did not affect teicoplanin distribution, as judged by the values of V_c and V_{ss} and by the time taken to achieve distribution equilibrium. In most of these studies nonrenal clearance CL_{NR} was found to be unaffected by impairment of renal function. It has been suggested (132) that there may have been a major error in the calculation of total area under the curve (AUC), and hence in the estimates of CL and CL_{NR}, in these studies. In a study of the comparative pharmacokinetics of the six major components of teicoplanin, no effect of renal impairment was demonstrated (180). Teicoplanin is not effectively removed from the circulation by hemodialysis since only traces could be detected in dialysates obtained from anuric patients on intermittent 4 h dialysis (180,192,193). These results are supported by the estimates, based on theoretical grounds, of a teicoplanin dialysis clearance CL_D of 0.12–0.24 L/kg and suggest that hemodialysis is unlikely to have a significant effect on the disposition of teicoplanin (132). Teicoplanin dosage normograms for various degrees of renal function have been reported (192).

In patients on CAPD, intravenous injection of 3 mg/kg of teicoplanin resulted in concentrations in the peritoneal effluents that were likely to be ineffective in the treatment of peritonitis (194–197). In a study in which continuous collection of the dialysate was performed for 15 days, only 5% of the dose was recovered in the dialysate, with a net peritoneal clearance of 0.006–0.012 L/h/kg. In these patients, values of V_{ss} are similar to those reported in healthy subjects (1.2–1.3 L/kg); however, CL values appear to be much reduced (0.0024–0.0026 L/h/kg, corresponding to 0.168 L/h/70 kg) and the terminal half-life was substantially prolonged (range 364–386 h) (197,198).

Intraperitoneal administration of a single 3 mg/kg dose in the dialysate during a 6 h dwell time resulted in approximately 70% absorption, with peak serum concentrations of 2.8–5.5 mg/L. The calculated absorption half-life from the peritoneal cavity was 2.2 h; reverse exchange, back into the peritoneal effluent, was negligible (157,197). As expected, no significant differences were observed in these patients in the systemic pharmacokinetics of teicoplanin when intravenous and intraperitoneal administration were compared (157,197,198).

VII. PLACE OF TEICOPLANIN IN THERAPY

The place of teicoplanin in therapy has been reviewed by Williams and Grüneberg (198) and Campoli-Richards et al. (87). Their conclusions may be summarized as follows. Teicoplanin is indicated for the treatment of infections for which vancomycin is also indicated, including Gram-positive septicemia, endocarditis, skin and soft tissue infections, and infections associated with venous catheters. Teicoplanin has a longer half-life than vancomycin, allowing once daily administration. In general, clinical efficacy has been associated with dosages of 200–400 mg daily IV or IM, except for staphylococcal endocarditis, for which doses of 800–1200 mg IV daily may be required.

The adverse effects most frequently associated with teicoplanin include local reactions, such as pain upon IM injection. Hypersensitivity reactions occur less frequently and are less severe than with vancomycin.

According to Campoli-Richards et al. (87),

Teicoplanin is a well tolerated and highly effective anti-bacterial for the treatment of Gram-positive infections. It is likely to prove particularly useful in methicillin-resistant staphylococcal infections and for patients hypersensitive to or unable to tolerate injection of vancomycin. Providing equivalent efficacy and improved tolerability are confirmed, the convenience and potential economic benefits of teicoplanin's once daily administration, and the availability of an intramuscular formulation, should make it preferable to vancomycin.

REFERENCES

1. Parenti F., Ciabatti R., Cavalleri B., Kettenring J. (1990). Ramoplanin a review of its discovery and its chemistry. Drugs Exp. Clin. Res. XVI:451.
2. Parenti F., Beretta G., Berti M., Arioli V. (1978). Teichomycins, new antibiotic from *Actinoplanes teichomyceticus*, nov. sp. I. Description of the producer strain, fermentation studies and biological properties. J. Antibiot. 31:276.
3. Parenti F., Coronelli L. (1979). Members of the genus *Actinoplanes* and their antibiotics. Annu. Rev. Microbiol. 33:389.
4. Parenti F., Cavalleri B. (1989). Proposal to name the vancomycin-ristocetin like glycopeptides as dalbaheptides. J. Antibiot. 42:1882.
5. Parenti F. (1988). Glycopeptide antibiotics. J. Clin. Pharmacol. 38:136.
6. Borghi A., Duncan E., Zerilli L.F., Lancini G.C. (1991). Factors affecting the normal and branched-chain acyl moieties of teicoplanin components produced by *Actinoplanes teichomyceticus*. J. Gen. Microbiol. 137:587.
7. Somma S., Gastaldo L., Corti A. (1984). Teicoplanin, a new antibiotic from *Actinoplanes teichomyceticus* nov. sp. Antimicrob. Agents Chemother. 26:917.
8. Corti A., Soffientini A., Cassani G. (1985). Binding of the glycopeptide antibiotic teicoplanin to D-alanyl-D-alanine-agarose: The effect of micellar aggregates. J. Appl. Biochem. 7:133.
9. Reynolds P.E. (1989). Structure, biochemistry and mechanism of action of glycopeptide antibiotics. Eur. J. Clin. Microbiol. Infect. Dis. 8:943.
10. Lancini G.C., Cavalleri B. (1990). Glycopeptide antibiotics of the vancomycin group. In Kleinhauf H., Van Dohren H. (Eds.), Biochemistry of Peptide Antibiotics. Walter de Gruyter, Berlin, p. 159.

11. Bauernfeind A., Petermüller C. (1982). In vitro activity of teichomycin A_2 in comparison with penicillin and vancomycin against Gram-positive cocci. Eur. J. Clin. Microbiol. 1:278.

12. Pallanza R., Berti M., Goldstein B.P., Mapelli E., Randisi E., Scotti R., Arioli V. (1983). Teichomycin in-vitro and in-vivo evaluation in comparison with other antibiotics. J. Antimicrob. Chemother. 11:419.

13. Fietta A., Bersani C., Gialdroni Grassi G. (1983). In vitro activity of teichomycin against isolates of Gram-positive bacteria. Chemotherapy 29:275.

14. Grüneberg R.N., Ridgway G.L., Cremer A.W.F., Felmingham D. (1983). The sensitivity of Gram-positive pathogens to teichomycin and vancomycin. Drugs Exp. Clin. Res. IX:139.

15. Varaldo P.E., Debbia E., Schito G.C. (1983). In vitro activity of teichomycin and vancomycin alone and in combination with rifampin. Antimicrob. Agents Chemother. 23:402.

16. Fainstein V., LeBlanc B., Bodey G.P. (1983). Comparative in vitro study of teichomycin A_2. Antimicrob. Agents Chemother. 23:497.

17. Tuazon C.U., Miller H. (1984). Comparative in vitro activities of teichomycin and vancomycin alone and in combination with rifampin and aminoglycosides against staphylococci and enterococci. Antimicrob. Agents Chemother. 25:411.

18. Chandrasekar P.H., Price S., Levine D.P. (1985). In-vitro evaluation of cefpirome (HR 810), teicoplanin and four other antimicrobials against enterococci. J. Antimicrob. Chemother. 16:179.

19. Pohlod D.J., Saravolatz L.D., Somerville M.M. (1987). In-vitro susceptibility of Gram-positive cocci to LY146032, teicoplanin, sodium fusidate, vancomycin and rifampicin. J. Antimicrob. Chemother. 20:197.

20. Hodinka R.L., Jack-Wait K., Wannamaker N., Walden T.P., Gilligan P.H. (1987). Comparative in vitro activity of LY146032 (daptomycin), a new lipopeptide antimicrobial. Eur. J. Clin. Microbiol. 6:100.

21. Wright D.N., Saxon B., Matsen J.M. (1987). In vitro activity in daptomycin (LY-146032) compared with other antimicrobial agents against Gram-positive cocci. Diagn. Microbiol. Infect. Dis. 7:283.

22. Breyer S., Georgopoulos A., Graninger W. (1987). In vitro evaluation of teicoplanin and other antimicrobials against streptococci group D bacteria. Drugs Exp. Clin. Res. 13:479.

23. Baiocchi P., Venditti M., Santini C., Gelfusa V., Brandimarte C., Raccah R., Santilli S., Serra P. (1989). In vitro antibiotic susceptibilities of *Streptococcus viridans* blood isolates from neutropenic patients. J. Chemother. 1(Suppl. 4):18.

24. Georgopoulous A., Breyer S., Czejka M.J., Georgopoulos M., Graninger W. (1989). Activity of teicoplanin alone and in combination. J. Chemother. 1(Suppl. 4):204.

25. Maskell J.P., Tang T., Asad S., Williams J.D. (1989). Comparative inhibitory and bactericidal activities of FCE 22101 against Gram-positive cocci and anaerobes in vitro. J. Antimicrob. Chemother. 23(Suppl. C):65.

26. LePennec M.P., Berardi-Grassias L. (1989). In-vitro activity of 13 antibiotics against clinical isolates of *Streptococcus milleri*. J. Antimicrob. Chemother. 24:618.

27. Jorgensen J.H., Redding J.S., Maher L.A. (1989). Antibacterial activity of the new glyco-peptide antibiotic SKF104662. Antimicrob. Agents Chemother. 33:560.

28. Venditti M., Baiocchi P., Santini C., Brandimarte C., Serra P., Gentile G., Girmenia C., Martino P. (1989). Antimicrobial susceptibilities of *Streptococcus* species that cause septicemia in neutropenic patients. Antimicrob. Agents Chemother. 33:580.

29. Yao J.D., Eliopoulos G.M., Moellering R.C. (1989). In vitro activity of SK&F 104662, a new glycopeptide antibiotic. Antimicrob. Agents Chemother. 33:965.

30. Gorzynski E.A., Amsterdam D., Beam T.R., Rotstein C. (1989). Comparative in vitro activities of teicoplanin, vancomycin, oxacillin, and other antimicrobial agents against bacteremic isolates of Gram-positive cocci. Antimicrob. Agents Chemother. 33:2019.

31. Mini E., Novelli A., Mazzei T., Periti P. (1989). Comparative in vitro activity of the new oxazolidinones DuP 721 and DuP 105 against staphylococci and streptococci. Eur. J. Clin. Microbiol. Infect. Dis. 8:256.

32. Ravizzola G., Pinsi G., Gonzales R., Colombrita D., Pirali F., Turano A. (1989). Antibacterial activity of the new carbapenem meropenem (SM-7338) against clinical isolates. Eur. J. Clin. Microbiol. Infect. Dis. 8:1053.

33. O'Hare M.D., Ghosh G., Felmingham D., Grüneberg R.N. (1990). In-vitro studies with ramoplanin (MDL 62,198): A novel lipoglycopeptide antimicrobial. J. Antimicrob. Chemother. 25:217.

34. Phillips I., King A., Grandsden W.R., Eykyn S.J. (1990). The antibiotic sensitivity of bacteria isolated from the blood of patients in St. Thomas' Hospital, 1969–1988. J. Antimicrob. Chemother. 25(Suppl. C):59.

35. Edwards R., Greenwood D. (1990). Antibiotic resistance in enterococci in Nottingham. J. Antimicrob. Chemother. 26:155.

36. Bartoloni A., Colao M.G., Orsi A., Deo R., Giganti E., Parenti F. (1990). In-vitro activity of vancomycin, teicoplanin, daptomycin, ramoplanin, MDL 62873 and other agents against staphylococci, enterococci and *Clostridium difficile*. J. Antimicrob. Chemother. 26:627.

37. Rolston K.V.I., Nguyen H., Messer M. (1990). In vitro activity of LY264826, a new glycopeptide antibiotic against Gram-positive bacteria isolated from patients with cancer. Antimicrob. Agents Chemother. 34:2137.

38. Kenny M.T., Dulworth J.K., Brackman M.A. (1991). Comparative in vitro activity of teicoplanin and vancomycin against United States teicoplanin clinical trial isolates of Gram-positive cocci. Diagn. Microbiol. Infect. Dis. 14:29.

39. Niu W.-W., Neu H.C. (1991). Activity of mersacidin, a novel peptide, compared with that of vancomycin, teicoplanin, and daptomycin. Antimicrob. Agents Chemother. 35:998.

40. Thabaut A., Meyran M. (1983). Comparative in vitro study of the activity of teichomycin, vancomycin, and *N*-formimidoyl-thienamycin on *Staphylococcus aureus*. Chemioterapia 2(Suppl. 5):12.

41. Aldridge K.E., Janney A., Sanders C.V. (1985). Comparison of the activities of coumermycin, ciprofloxacin, teicoplanin, and other non-β-lactam antibiotics against clinical isolates of methicillin-resistant *Staphylococcus aureus* from various geographical locations. Antimicrob. Agents Chemother. 28:634.

42. Van der Auwera P., Klastersky J. (1986). In vitro activity of coumermycin alone or in combination against *Staphylococcus aureus* and *Staphylococcus epidermidis*. Drugs Exp. Clin. Res. XII:307.

43. DelBene V.E., John J.F., Twitty J.A., Lewis J.W. (1986). Anti-staphylococcal activity of teicoplanin, vancomycin, and other antimicrobial agents: The significance of methicillin resistance. J. Infect. Dis. 154:349.

44. Traub W.H., Spohr M., Bauer D. (1987). Gentamicin- and methicillin-resistant *Staphylococcus aureus*: In vitro susceptibility to antimicrobial drugs. Chemotherapy 33:361.

45. Ravizzola G., Pirali F., Foresti I., Turano A. (1987). Comparison of the in vitro antibacterial activity of teicoplanin and vancomycin against Gram-positive cocci. Drugs Exp. Clin. Res. XIII:225.

46. French G.L., Ling J., Ling T., Hui Y.W. (1988). Susceptibility of Hong Kong isolates of methicillin-resistant *Staphylococcus aureus* to antimicrobial agents. J. Antimicrob. Chemother. 21:581.

47. Maple P.A.C., Hamilton-Miller J.M.T., Brumfitt W. (1989). World-wide antibiotic resistance in methicillin-resistant *Staphylococcus aureus*. Lancet 1:537.

48. Maple P.A.C., Hamilton-Miller J.M.T., Brumfitt W. (1989). Comparative in-vitro activity of vancomycin, teicoplanin, ramoplanin (formerly A16686), paldimycin, DuP 721 and DuP 105 against methicillin and gentamicin resistant *Staphylococcus aureus*. J. Antimicrob. Chemother. 23:517.

49. Jones R.N., Barry A.L., Gardiner R.V., Packer R.R. (1989). The prevalence of staphylococcal resistance to penicillinase-resistant penicillins, a retrospective and prospective national surveillance trial of isolates from 40 medical centers. Diagn. Microbiol. Infect. Dis. 12:385.

50. Fernández-Roblas R., Jiménez-Arriero M., Romero M., Soriano F. (1989). In vitro activity of quinolones and other agents against problematic Gram-positive pathogens. Rev. Infect. Dis. 11(Suppl. 5):S931.

51. Bauernfeind A., Przyklenk B., Matthias C., Jungwirth R., Bertele R.M., Harms K. (1990). Staphylococcal aspects of cystic fibrosis. Infection 18:68.

52. Maple P.A.C., Hamilton-Miller J.M.T., Brumfitt W. (1991). Differing activities of quinolones against ciprofloxacin-susceptible and ciprofloxacin-resistant, methicillin-resistant *Staphylococcus aureus*. Antimicrob. Agents Chemother. 35:345.

53. Jones R.N., Goldstein F.W., Zhou X.Y. (1991). Activities of two new teicoplanin amide derivatives (MDL 62211 and MDL 62873) compared with activities of teicoplanin and vancomycin against 800 recent staphylococcal isolates from France and the United States. Antimicrob. Agents Chemother. 35:584.

54. Karchmer A.W., Archer G.L., Dismukes W.E. (1983). *Staphylococcus epidermidis* causing prosthetic valve endocarditis: Microbiologic and clinical observations as guides to therapy. Ann. Intern. Med. 98:447.

55. Gruer L.D., Bartlett R., Ayliffe G.A.J. (1984). Species identification and antibiotic sensitivity of coagulase-negative staphylococci from CAPD peritonitis. J. Antimicrob. Chemother. 13:577.

56. Bailey E.M., Constance T.D., Albrecht L.M., Rybak M.J. (1990). Coagulase-negative staphylococci: Incidence pathogenicity, and treatment in the 1990s. DICP Ann. Pharmacother. 24:714.

57. Kropec A., Daschner F. (1989). In vitro activity of fleroxacin and 14 other antimicrobials against slime- and non-slime-producing *Staphylococcus epidermidis*. Chemotherapy 35:351.

58. Goldstein F.W., Coutrot A., Sieffer A., Acar J.F. (1990). Percentages and distributions of teicoplanin- and vancomycin-resistant strains among coagulase-negative staphylococci. Antimicrob. Agents Chemother. 34:899.

59. Maugein J., Pellegrin J.L., Brossard G., Fourche J., Leng B., Reiffers J. (1990). In vitro activities of vancomycin and teicoplanin against coagulase-negative staphylococci isolated from neutropenic patients. Antimicrob. Agents Chemother. 34:901.

60. Bannerman T.L., Wadiak D.L., Kloos W.E. (1991). Susceptibility of *Staphylococcus* species and subspecies to teicoplanin. Antimicrob. Agents Chemother. 35:1919.

61. Schwalbe R.S., Stapleton J.T., Gilligan P.H. (1987). Emergence of vancomycin resistance in coagulase-negative staphylococci. N. Engl. J. Med. 316:927.

62. Aubert G., Passot S., Lucht F., Dorche G. (1990). Selection of vancomycin- and teicoplanin-resistant *Staphylococcus haemolyticus* during teicoplanin treatment of *S. epidermidis* infection. J. Antimicrob. Chemother. 25:491.

63. Veach L.A., Pfaller M.A., Barrett M., Koontz F.P., Wenzel R.P. (1990). Vancomycin resistance in *Staphylococcus haemolyticus* causing colonization and bloodstream infection. J. Clin. Microbiol. 28:2064.

64. Cherubin C.E., Corrado M.L., Sierra M.F., Gombert M.E., Shulman M. (1981). Susceptibility of gram-positive cocci to various antibiotics, including cefotaxime, moxalactam, and *N*-formimidoyl thienamycin. Antimicrob. Agents Chemother. 20:553.

65. Tuazon C.U., Miller H. (1983). Clinical and microbiologic aspects of serious infections caused by *Staphylococcus epidermidis*. Scand. J. Infect. Dis. 15:347.

66. Wilson A.P.R., O'Hare M.D., Felmingham D., Grüneberg R.N. (1986). Teicoplanin-resistant coagulase-negative *Staphylococcus*. Lancet 2:973.

67. Grant A.C., Lacey R.W., Brownjohn A.M., Turney J.H. (1986). Teicoplanin-resistant coagulase-negative *Staphylococcus*. Lancet 2:1166.

68. Moore E.P., Speller D.C.E. (1988). In-vitro teicoplanin-resistance in coagulase-negative staphylococci from patients with endocarditis and from a cardiac surgery unit. J. Antimicrob. Chemother. 21:417.

69. Sanyal D., Johnson A.P., George R.C., Cookson B.D., Williams A.J. (1991). Peritonitis due to vancomycin-resistant *Staphylococcus epidermidis*. Lancet 337:54.

70. Greenwood D. (1988). Microbiological properties of teicoplanin. J. Antimicrob. Chemother. 21(Suppl. 4):1.

71. Jones R.N. (1989). Antimicrobial activity of teicoplanin: A review of in vitro activity, including susceptibility testing guidelines and a critical evaluation of potency against *Staphylococcus haemolyticus* isolates. In Phillips I., (Ed.), Focus on Coagulase-Negative Staphylococci. Royal Society of Medicine Services International Congress and Symposium Series, No. 151, Royal Society of Medicine Services, Ltd., London, p. 55.

72. Chomarat M., Espinouse D., Flandrois J.-P. (1991). Coagulase-negative staphylococci emerging during teicoplanin therapy and problems in the determination of their sensitivity. J. Antimicrob. Chemother. 27:475.

73. Greenwood D., Bidgood K., Turner M. (1987). A comparison of the responses of staphylococci and streptococci to teicoplanin and vancomycin. J. Antimicrob. Chemother. 20:155.

74. Felmingham D., Solomonides K., O'Hare M.D., Wilson A.P.R., Grüneberg R.N.

(1987). The effect of medium and inoculum on the activity of vancomycin and teicoplanin against coagulase-negative staphylococci. J. Antimicrob. Chemother. 20:609.

75. O'Hare M.D., Felmingham D., Grüneberg R.N. (1989). The bactericidal activity of vancomycin and teicoplanin against methicillin-resistant strains of coagulase negative *Staphylococcus* spp. J. Antimicrob. Chemother. 23:800.

76. Jadeja L., Fainstein V., LeBlanc B., Bodey G.P. (1983). Comparative in vitro activities of teicomycin and other antibiotics against JK diphtheroids. Antimicrob. Agents Chemother. 24:145.

77. Spitzer P.G., Eliopoulos G.M., Karchmer A.W., Moellering R.C. (1987). Comparative in vitro activity of the new cyclic lipopeptide LY146032 against *Corynebacterium* species. Eur. J. Clin. Microbiol. 6:183.

78. Fernández-Roblas R., Prieto S., Santamaria M., Ponte C., Soriano F. (1987). Activity of nine antimicrobial agents against *Corynebacterium* group D₂ strains isolated from clinical specimens and skin. Antimicrob. Agents Chemother. 31:821.

79. Dealler S.F. (1988). In-vitro activity of LY146032 daptomycin and other agents against JK diphtheroids. J. Antimicrob. Chemother. 21:807.

80. Newsom S.W.B., Matthews J., Rampling A.M. (1985). Susceptibility of *Clostridium difficile* strains to new antibiotics: Quinolones, efrotomycin, teicoplanin and imipenem. J. Antimicrob. Chemother. 15:648.

81. Traub W.H., Karthein J., Spohr M. (1986). Susceptibility of *Clostridium perfringens* type A to 23 antimicrobial drugs. Chemotherapy 32:439.

82. Faruki H., Niles A.C., Heeren R.L., Murray P.R. (1987). Effect of calcium on in vitro activity of LY146032 against *Clostridium difficile*. Antimicrob. Agents Chemother. 31:461.

83. Robbins M.J., Marais R., Felmingham D., Ridgway G.L., Grüneberg R.N. (1987). In vitro activity of vancomycin and teicoplanin against anaerobic bacteria. Drugs Exp. Clin. Res. XIII:551.

84. Biavasco F., Manso E., Varaldo P.E. (1991). In vitro activities of ramoplanin and four glycopeptide antibiotics against clinical isolates of *Clostridium difficile*. Antimicrob. Agents Chemother. 35:195.

85. Venditti M., Gelfusa V., Tarasi A., Brandimarte C., Serra P. (1990). Antimicrobial susceptibilities of *Erysipelothrix rhusiopathiae*. Antimicrob. Agents Chemother. 34:2038.

86. Anzivino D., Nani E., Gulletta E., Lembo M., Covelli B. (1989). Evaluation of in vitro activity of teicoplanin against Gram-positive anaerobic bacteria. J. Chemother. 1(Suppl. 4):201.

87. Campoli-Richards D.M., Brogden R.N., Faulds D. (1990). Teicoplanin A review of its antibacterial activity, pharmacokinetic properties and therapeutic potential. Drugs 40:449.

88. Pasticci M.B., Moretti A., Frongillo R.F., Baldelli F., Moretti M.V., Pauluzzi S. (1989). Technical factors influencing the bactericidal activity of teicoplanin against staphylococci. Boll. Ist. Sieroter. Milan. 68:101.

89. Brumfitt W., Maple P.A.C., Hamilton-Miller J.M.T. (1990). Ramoplanin versus methicillin-resistant *Staphylococcus aureus*: In vitro experience. Drugs Exp. Clin. Res. XVI:377.

90. Johnson A.P., Uttley A.H.C., Woodford N., George R.C. (1990). Resistance to vancomycin and teicoplanin: An emerging clinical problem. Clin. Microbiol. Rev. 3:280.

91. Green M., Barbadora K., Michaels M. (1991). Recovery of vancomycin-resistant Gram-positive cocci from pediatric liver transplant recipients. J. Clin. Microbiol. 29:2503.

92. Kaplan A.H., Gilligan P.H., Facklam R.R. (1988). Recovery of resistant enterococci during vancomycin prophylaxis. J. Clin. Microbiol. 26:1216.

93. Leclercq R., Derlot E., Duval J., Courvalin P. (1988). Plasmid-mediated resistance to vancomycin and teicoplanin in *Enterococcus faecium*. N. Engl. J. Med. 319:157.

94. Uttley A.H.C., George R.C., Naidoo J., Woodford N., Johnson A.P., Collins C.H., Morrison D., Gilfillan A.J., Fitch L.E., Heptonstall J. (1989). High-level vancomycin resistant enterococci causing hospital infections. Epidemiol. Infect. 103:173.

95. Sahm D.F., Kissinger J., Gilmore M.S., Murray P.R., Mulder R., Solliday J., Clarke B. (1989). In vitro susceptibility studies of vancomycin-resistant *Enterococcus faecalis*. Antimicrob. Agents Chemother. 33:1588.

96. Shlaes D.M., Al-Obeid S., Shlaes H.H., Boisivon A., Williamson R. (1989). Inducible, transferable resistance to vancomycin in *Enterococcus faecium* D399. J. Antimicrob. Chemother. 23:503.

97. Shlaes D.M., Al-Obeid S., Shlaes J.H., Williamson R. (1989). Activity of various glycopeptides against an inducibly vancomycin-resistant strain of *Enterococcus faecium* (D366). J. Infect. Dis. 159:1132.

98. Shlaes D.M., Bouvet A., Devine C., Shlaes J.H., Al-Obeid S., Williamson R. (1989). Inducible, transferable resistance to vancomycin in *Enterococcus faecalis* A256. Antimicrob. Agents Chemother. 33:198.

99. Williamson R., Al-Obeid S., Shlaes J.H., Goldstein F.W., Shlaes D.M. (1989). Inducible resistance to vancomycin in *Enterococcus faecium* D366. J. Infect. Dis. 159:1095.

100. Woodford N., Johnson A.P., Morrison D., Chin A.T.L., Stephenson J.R., George R.C. (1990). Two distinct forms of vancomycin resistance amongst enterococci in the UK. Lancet 1:226.

101. Shlaes D.M., Etter L., Guttmann L. (1991). Synergistic killing of vancomycin-resistant enterococci of classes A, B, and C by combinations of vancomycin, penicillin, and gentamicin. Antimicrob. Agents Chemother. 35:776.

102. Shwalbe R., Capacio E., Margaret B., Drusano G., Fowler C., Verma P., Walsh T., Standiford H. (1991). Characterization of enterococcal strains exhibiting a novel glycopeptide susceptibility pattern. Abstracts of the 91st General Meeting of the American Society for Microbiology, Abstract A-117, American Society for Microbiology, Washington, D.C., p. 20.

103. Fantin B., Leclercq R., Arthur M., Duval J., Carbon C. (1991). Influence of low-level resistance to vancomycin on efficacy of teicoplanin and vancomycin for treatment of experimental endocarditis due to *Enterococcus faecium*. Antimicrob. Agents Chemother. 35:1570.

104. Cavenaghi L.A., Biganzoli E., Danese A., Parenti F. (1992). Diffusion of teicoplanin and vancomycin in agar. Diagn. Microbiol. Infect. Dis. 15:253.

105. Gialluca G., Vaiani R., Pallanza R. (1983). Study of the correlation between MIC

values and disc inhibition zones for teicoplanin and vancomycin. In Spitzy K.H., Karrer K. (Eds.), Proceedings of the 13th International Congress of Chemotherapy, Verlag H. Egermann, Vienna, 112:34.

106. Barry A.L., Thornsberry C., Jones R.N. (1986). Evaluation of teicoplanin and vancomycin disk susceptibility tests. J. Clin. Microbiol. 23:100.

107. Barry A.L., Jones R.N., Gavan T.L., Thornsberry C., Collaborative Antimicrobial Susceptibility Testing Group (1987). Quality control limits for teicoplanin susceptibility tests and confirmation of disk diffusion interpretive criteria. J. Clin. Microbiol. 25:1812.

108. Vedel G., Leruez M., Lémann F., Hraoui E., Ratovohery D. (1990). Prevalence of Staphylococcus aureus and coagulase-negative staphylococci with decreased sensitivity to glycopeptides as assessed by determination of MICs. Eur. J. Clin. Microbiol. Infect. Dis. 9:820.

109. Jones R.N., Goldstein F.W., Biavasco F., Grimm H., Acar J.F., Varaldo P., Goldstein B.P., Cavenaghi L., Parenti F. (1990). International studies to determine optimal teicoplanin (TPN) disk diffusion and dilution susceptibility tests for staphylococci. 10th Interdisciplinary Meeting on Anti-infectious Chemotherapy, Paris, December 6–7, 1990.

110. Low D.E., McGeer A., Poon R. (1989). Activities of daptomycin and teicoplanin against Staphylococcus haemolyticus and Staphylococcus epidermidis, including evaluation and susceptibility testing recommendations. Antimicrob. Agents Chemother. 33:585.

111. Carper H.T., Sullivan G.W., Mandell G.L. (1987). Teicoplanin, vancomycin, rifampicin: In-vivo and in-vitro studies with Staphylococcus aureus. J. Antimicrob. Chemother. 19:659.

112. Arioli V., Berti M., Candiani G. (1986). Activity of teicoplanin in localized experimental infections in rats. J. Hosp. Infect. 7(Suppl. A):91.

113. Chambers H.F., Sande M.A. (1984). Teicoplanin versus nafcillin and vancomycin in the treatment of experimental endocarditis caused by methicillin-susceptible or -resistant Staphylococcus aureus. Antimicrob. Agents Chemother. 26:61.

114. Chambers H.F., Kennedy S. (1990). Effects of dosage, peak and trough concentrations in serum, protein binding, and bactericidal rate on efficacy of teicoplanin in a rabbit model of endocarditis. Antimicrob. Agents Chemother. 34:510.

115. Contrepois A., Joly V., Abel L., Pangon B., Vallois J.-M., Carbon C. (1988). The pharmacokinetics and extravascular diffusion of teicoplanin in rabbits and comparative efficacy with vancomycin in an experimental endocarditis model. J. Antimicrob. Chemother. 21:621.

116. Kaatz G.W., Seo S.M., Reddy V.N., Bailey E.M., Rybak M.J. (1990). Daptomycin compared with teicoplanin and vancomycin for therapy of experimental Staphylococcus aureus endocarditis. Antimicrob. Agents Chemother. 34:2081.

117. Berti M., Candiani G.P. (1989). Activity of teicoplanin (T) in rat endocarditis caused by staphylococci with reduced in vitro susceptibility to T. Program and Abstracts of the 29th Interscience Conference on Antimicrobial Agents and Chemotherapy, September 17–20, Houston Texas, Abstract 976, American Society for Microbiology, Washington, D.C.

118. Galetto D.W., Boscia J.A., Kobasa W.D., Kaye D. (1986). Teicoplanin compared

with vancomycin for treatment of experimental endocarditis due to methicillin-resistant *Staphylococcus epidermidis*. J. Infect. Dis. 154:69.

119. Tuazon C.U., Washburn D. (1987). Teicoplanin and rifampicin singly and in combination in the treatment of experimental *Staphylococcus epidermidis* endocarditis in the rabbit model. J. Antimicrob. Chemother. 20:233.

120. Sullam P.M., Täuber M.G., Hackbarth C.J., Sande M.A. (1985). Therapeutic efficacy of teicoplanin in experimental enterococcal endocarditis. Antimicrob. Agents Chemother. 27:135.

121. Yao J.D.C., Thauvin-Eliopoulos C., Eliopoulos G.M., Moellering R.C. (1990). Efficacy of teicoplanin in two dosage regimens for experimental endocarditis caused by a β-lactamase-producing strain of *Enterococcus faecalis* with high-level resistance to gentamicin. Antimicrob. Agents Chemother. 34:827.

122. Bush L.M., Calmon J., Cherney C.L., Wendeler M., Pitsakis P., Poupard J., Levison M.E., Johnson C.C. (1989). High-level penicillin resistance among isolates of enterococci. Implications for treatment of enterococcal infections. Ann. Intern. Med. 110:515.

123. Berti M., Romanò G., Pallanza R., Arioli V. (1985). Antistaphylococcal activity of teicoplanin. Chemioterapia 4(Suppl. 2):688.

124. Peetermans W.E., Hoogeterp J.J., Hazekamp-van Dokkum A.-M., van den Broek P., Mattie H. (1990). Antistaphylococcal activities of teicoplanin and vancomycin in vitro and in an experimental infection. Antimicrob. Agents Chemother. 34:1869.

125. Widmer A.F., Frei R., Rajacic Z., Zimmerli W. (1990). Correlation between in vivo and in vitro efficacy of antimicrobial agents against foreign body infections. J. Infect. Dis. 162:96.

126. Norden C.W., Niederreiter K., Shinners E.M. (1986). Treatment of experimental chronic osteomyelitis due to *Staphylococcus aureus* with teicoplanin. Infection 14:136.

127. Nord C.E., Lindmark A., Persson I. (1989). Efficacy of daptomycin compared to teicoplanin in the treatment of *Clostridium difficile* colitis in hamsters. J. Chemother. 1(Suppl. 4):209.

128. Kim K.S., Kang J.H., Bayer A.S. (1987). Efficacy of teicoplanin in experimental group B streptococcal bacteremia and meningitis. Chemotherapy 33:177.

129. Manquat G., Stahl J.P., Pelloux I., Morand P., Garaut J.J., Micoud M. (1990). Efficacy of intracisternally administered teicoplanin in experimental methicillin-resistant *Staphylococcus aureus* meningitis of rabbits. In Grüneberg R.N. (Ed.), Teicoplanin: Further European Experience. Royal Society of Medicine Services International Congress and Symposium Series 156, Royal Society of Medicine Services, Ltd., London, p. 59.

130. Manquat G., Stahl J.P., Pelloux I., Micoud M. (1990). Influence of mannitol on the penetration of teicoplanin into infected CSF of experimental *Staphylococcus aureus* meningitis of rabbits. Infection 18:113.

131. Berner R., Hof H. (1988). Therapeutic activity of teicoplanin on experimental listeriosis compared with that of vancomycin and ampicillin. Zentralbl. Bakteriol. Hyg. |A| 268:50.

132. Rowland M. (1990). Clinical pharmacokinetics of teicoplanin. Clin. Pharmacokinet. 18:184.

133. Cavenaghi L., Corti A., Cassani G. (1986). Comparison of the solid phase enzyme receptor assay (SPERA) and the microbiological assay for teicoplanin. J. Hosp. Infect. 7(Suppl. A):85.

134. Verbist L., Tjandramaga B., Hendrickx B., Van Hecken A., Van Melle P., Verbesselt R., Verhaegen J., De Schepper P.J. (1984). In vitro activity and human pharmacokinetics of teicoplanin. Antimicrob. Agents Chemother. 26:881.

135. Erickson R.C., Hildebrand A.R., Hoffman P.F., Gibson C.B. (1989). A sensitive bioassay for teicoplanin in serum in the presence or absence of other antibiotics. Diagn. Microbiol. Infect. Dis. 12:235.

136. Patton K.R., Beg A., Felmingham D., Ridgway G.L., Grüneberg R.N. (1987). Determination of teicoplanin in serum using a bioassay technique. Drugs Exp. Clin. Res. 13:547.

137. Carver P.L., Nightingale C.H., Quintiliani R., Sweeney K., Stevens R.C., Maderazo E. (1989). Pharmacokinetics of single- and multiple-dose teicoplanin in healthy volunteers. Antimicrob. Agents Chemother. 33:82.

138. Kenny M.T., Dulworth J.K., Brackman M.A., Torney H.L., Gibson C.B., Hildebrand A.R., Weckbach L.S., Staneck J.L. (1989). Bioassay of teicoplanin in serum containing rifampin or a beta-lactam antibiotic. Diagn. Microbiol. Infect. Dis. 12:449.

139. Riva E., Ferry N., Cometti A., Cuisinaud G., Gallo G.G., Sassard J. (1987). Determination of teicoplanin in human plasma and urine by affinity and reversed-phase high-performance liquid chromatography. J. Chromatogr. 421:99.

140. Jehl F., Monteil H., Tarral A. (1988). HPLC quantisation of the six main components of teicoplanin in biological fluids. J. Antimicrob. Chemother. 21(Suppl. A):53.

141. Joos B., Luthy R. (1987). Determination of teicoplanin concentrations in serum by high-pressure liquid chromatography. Antimicrob. Agents Chemother. 31:1222.

142. Levy J., Truong B.L., Goignau H., Van Laethem Y., Butzler J.P., Bourdoux P. (1987). High pressure liquid chromatographic quantitation of teicoplanin in human serum. J. Antimicrob. Chemother. 19:533.

143. Georgopoulos A., Czejka M.J., Starzengruber N., Jager W., Lackner H. (1989). High-performance liquid chromatographic determination of teicoplanin in plasma: Comparison with a microbiologic assay. J. Chromatogr. 494:340.

144. Taylor R.B., Reid R.G., Gould I.M. (1991). Determination of teicoplanin in plasma using microbore high-performance liquid chromatography and injection-generated gradients. J. Chromatogr. 563:451.

145. Corti A., Cavenaghi L., Giani E., Cassani G. (1987). A receptor-antibody sandwich assay for teicoplanin. Clin. Chem. 33:1615.

146. Rybak M.J., Bailey E.M., Reddy V.N. (1991). Clinical evaluation of teicoplanin fluorescence polarization immunoassay. Antimicrob. Agents Chemother. 35:1586.

147. Verbist L. (1985). Pharmacokinetics of teicoplanin in mice, rats, dogs and humans (abstract). In Ishigami J. (Ed.), Abstracts of the 14th International Congress of Chemotherapy (WS-34-3), Kyoto, June 23–28, 1985, University of Tokyo Press, p. 91.

148. Bernareggi A., Cavenaghi L., Assandri A. (1985). Kinetics of ¹⁴C-teicoplanin in rats, In: Deuxièmes journées Méditerranéennes de Pharmacocinetique. Actualité

Pharmacocinetique 1985 et Horizon 1990. La Grande Motte, March 14–16, 1985, preprint.

149. Bernareggi A., Rosina R., Zenatti C. (1984). Pharmacokinetics of teicoplanin in dogs. In Aiache J.M., Hirtz J. (Eds.), Proceedings of the 2nd European Congress of Biopharmaceutics and Pharmacokinetics, Vol. 2, Experimental Pharmacokinetics. Salamanca, April 24–27, 1984, University of Clermont-Ferrand Press, p. 553.

150. Bernareggi A., Cavenaghi L., Assandri A. (1986). Pharmacokinetics of [^{14}C] teicoplanin in male rats after single intravenous dose. Antimicrob. Agents Chemother. 30:733.

151. Zanolo G., Bernareggi A., Cavenaghi L., Giachetti C., Tognolo C. (1990). Distribution and excretion of teicoplanin in rats after single and repeated intravenous administration (abstract). Eur. J. Drug. Metab. Pharmacokinet. 15(Suppl.):33.

152. Zerilli L.F., Cavenaghi L., Bernareggi A., Assandri A. (1989). Teicoplanin metabolism in rats. Antimicrob. Agents Chemother. 33:1791.

153. Buniva G., Cavenaghi L., Frigo G.M., Rosina R., Lewis P. (1986). Microbiological activity of teicoplanin in feces following oral administration. Comparison with vancomycin (abstract). In Abstracts of the 7th International Symposium on Future Trends in Chemotherapy, Tirrenia, May 26–28, 1986, p. 172.

154. Buniva G., Cavenaghi L. (1985). Pharmacokinetics of teicoplanin after repeated i.m. injections. In Ishigami J. (Ed.), Recent Advances in Chemotherapy. Antimicrobial Section 3. University of Tokyo Press, Tokyo, 1985, p. 1897.

155. Ripa S., Ferrante L., Mignini F., Falcioni E. (1988). Pharmacokinetics of teicoplanin. Chemotherapy 34:178.

156. Anthony K.K., Lewis E.W., Kenny M.T., Dulworth J.K., Brackman M.B., Kuzma R., Yuh L., Eller M.G., Thompson G.A. (1991). Pharmacokinetics and bioavailability of a new formulation of teicoplanin following intravenous and intramuscular administration to humans. J. Pharm. Sci. 60:605.

157. Bonati M., Traina G.L., Gentile M.G., Fellini G., Rosina R., Cavenaghi L., Buniva G. (1988). Pharmacokinetics of intraperitoneal teicoplanin in patients with chronic renal failure on continuous ambulatory peritoneal dialysis. Br. J. Clin. Pharmacol. 25:761.

158. Maserati R., Cruciani M., Azzini M., Carnevale C., Suter F., Concia E. (1987). Teicoplanin in the therapy of staphylococcal neuroshunt infections. Int. J. Clin. Pharmacol. Res. 7:207.

159. Amaducci S., Barbieri D., Pollice P., Altieri A., Samori G., Rosina R. (1991). Endobronchial local teicoplanin therapy (abstract). In Adam D., Lode H., Rubinstein E. (Eds.), Abstracts of the 17th International Congress of Chemotherapy, Berlin, June 23–28, 1991, Abstract 1831, Futuramed Publishers, Munich, 1992.

160. Traina G.L., Bonati M. (1984). Pharmacokinetics of teicoplanin in man after intravenous administration. J. Pharmacokinet. Biopharm. 12:119.

161. Lagast H., Dodion P., Klastersky J. (1986). Comparison of pharmacokinetics and bactericidal activity of teicoplanin and vancomycin. J. Antimicrob. Chemother. 18:513.

162. Buniva G., Cavenaghi L., Taglietti M., Frigo G.M. (1988). Pharmacokinetics of teicoplanin after single and repeated i.v. injections. In: Proceedings of the Interna-

tional Symposium on the Control of Hospital Infections. Rome, April 27–29, 1987. Istituto Superiore di Sanità, Rome, Edizioni Riviste Scientifiche, Florence, p. 192.

163. Del Favero A., Patoia L., Rosina R., Buniva G., Danese A., Bernareggi A., Molini E., Cavenaghi L. (1991). Pharmacokinetics and tolerability of teicoplanin in healthy volunteers after single increasing doses. Antimicrob. Agents Chemother. 35:2551.

164. Assandri A., Bernareggi A. (1987). Binding of teicoplanin to human serum albumin. Eur. J. Clin. Pharmacol. 33:191.

165. Bernareggi A., Borgonovi M., Del Favero A., Rosina R., Cavenaghi L. (1991). Teicoplanin binding in plasma following administration of increasing intravenous doses to healthy volunteers. Eur. J. Drug Metab. Pharmacokinet. 16(Suppl. III):236.

166. McNulty C.A.M., Garden G.M.F., Wise R., Andrews J.M. (1985). The pharmacokinetics and tissue penetration of teicoplanin. J. Antimicrob. Chemother. 16:743.

167. Wise R., Donovan I.A., McNulty C.A.M., Waldron R., Andrews J.M. (1986). Teicoplanin, its pharmacokinetics, blister and peritoneal fluid penetration. J. Hosp. Infect. 7(Suppl. A):47.

168. Novelli A., Mazzei T., Reali E.F., Mini E., Periti P. (1989). Clinical pharmacokinetics and tissue penetration of teicoplanin. Int. J. Clin. Pharmacol. Res. 9:233.

169. Antrum R.M., Bibby S.R., Ramsden C.H., Kester R.C. (1989). Teicoplanin. Part 2. Evaluation of its use in the biliary system. Drugs Exp. Clin. Res. 15:25.

170. Pederzoli P., Falconi M., Girelli R., Martini N., Guaglianone H. (1989). In Rubinstein E., Adam D. (Eds.), Penetration of teicoplanin into cholecystic, hepatic and pancreatic tissue. 16th ICC Proceedings, Jerusalem, Israel, June 11–16, 1989, E. Lewin-Epstein, LTD., p. 7.

171. Bergeron M.G., Saginur R., Desaulniers D., Trottier S., Goldstein W., Foucault P., Lessard C. (1990). Concentrations of teicoplanin in serum and atrial appendages of patients undergoing cardiac surgery. Antimicrob. Agents Chemother. 34:1699.

172. Wilson A.P.R., Taylor B., Treasure T., Grüneberg R.N., Patton K., Felmingham D., Sturridge M.F. (1988). Antibiotic prophylaxis in cardiac surgery: Serum and tissue levels of teicoplanin, flucloxacillin and tobramycin. J. Antimicrob. Chemother. 21:201.

173. Wilson A.P.R., Shankar S., Felmingham D., Treasure T., Grüneberg R.N. (1989). Serum and tissue levels of teicoplanin during cardiac surgery: The effect of a high dose regimen. J. Antimicrob. Chemother. 23:613.

174. Lenders H.E., Walliser D., Schumann K., Dieterich H.A., Fell J.J., Cavenaghi L.A. (1987). Teicoplanin pharmacokinetics and tissue penetration in patients undergoing ENT-surgery (abstract). In Paul-Ehrlich-Society for Chemotherapy e.V. (Ed.), Program and Abstract of the Biennial Conference on Chemotherapy of Infectious Diseases and Malignancies, Munich, April 26–29, 1987, No. 124, Futuramed Publications, Munich.

175. Morgan J.R., Williams B.D., Williams K. (1989). In Rubinstein E., Adam D. (Eds.), Teicoplanin concentrations during multiple dosage, in blood and synovial fluid of rheumatoid arthritis patients. 16th ICC Proceedings, Jerusalem, Israel, June 11–16, E. Lewin-Epstein, LTD., p. 357.1.

176. Buniva G., Del Favero A., Bernareggi A., Patoia L., Palumbo R. (1988). Pharmacokinetics of ^{14}C-teicoplanin in healthy volunteers. J. Antimicrob. Chemother. 21(Suppl. A):23.

177. Stahl J.P., Croize J., Wolff M., Garaud J.J., Leclercq P., Vachon F., Micoud M. (1987). Poor penetration of teicoplanin into cerebrospinal fluid in patients with bacterial meningitis (letter). J. Antimicrob. Chemother. 20:141.
178. Venditti M., Micozzi A., Serra P., Buniva G., Palma L., Martino P. (1988). Intraventricular administration of teicoplanin in shunt associated ventriculitis caused by methicillin resistant *Staphylococcus aureus* (letter). J. Antimicrob. Chemother. 21:513.
179. Verbist L. (1985). Teicoplanin pharmacokinetics: Multiple dose study. In Gialdroni-Grassi G., Mitsuhashi S., Williams J.D. (Eds.), Proceedings of a Workshop, 14th International Congress of Chemotherapy, Kyoto, 1985. University of Tokyo Press, Tokyo, p. 15.
180. Falcoz C., Ferry N., Pozet N., Cuisinaud G., Zech P.Y., Sassard J. (1987). Pharmacokinetics of teicoplanin in renal failure. Antimicrob. Agents Chemother. 31:1255.
181. Bernareggi A., Danese A., Cometti A., Buniva G., Rowland M. (1990). Pharmacokinetics of individual components of teicoplanin in man. J. Pharmacokinet. Biopharm. 18:525.
182. Rosina R., Villa G., Danese A., Cavenaghi L., Picardi L., Salvadeo A. (1988). Pharmacokinetics of teicoplanin in the elderly. J. Antimicrob. Chemother. 21(Suppl. A):39.
183. Quin J.D., Mackay I., Thompson A., Paterson K.R. (1990). Multiple dose teicoplanin pharmacokinetics in the elderly. R. Soc. Med. Int. Cong. Symp. Ser. 156:29.
184. Terragna A., Ferrea G., Loy A., Danese A., Bernareggi A., Cavenaghi A., Rosina R. (1988). Pharmacokinetics of teicoplanin in pediatric patients. Antimicrob. Agents Chemother. 32:1223.
185. Tarral E., Jehl F., Tarral A., Simeoni V., Monteil H., Willard D., Geisert J. (1988). Pharmacokinetics of teicoplanin in children. J. Antimicrob. Chemother. 21(Suppl. A):47.
186. Lemerle S., De La Roque F., Lamy R., Fremaux A., Bernaudin F., Labut J.B., Reinert P. (1988). Teicoplanin in combination therapy for febrile episodes in neutropenic and non-neutropenic paediatric patients. J. Antimicrob. Chemother. 21(Suppl. A):113.
187. Bonati M., Traina G.L., Villa G., Salvadeo A., Gentile M.C., Fellin G., Rosina R., Cavenaghi L., Buniva G. (1987). Teicoplanin pharmacokinetics in patients with chronic renal failure. Clin. Pharmacokinet. 12:292.
188. Bonati M., Traina G.L., Rosina R., Buniva G. (1988). Pharmacokinetics of a single intravenous dose of teicoplanin in subjects with various degrees of renal impairment. J. Antimicrob. Chemother. 21(Suppl. A):29.
189. Domart Y., Pierre C., Clair B., Garaud J.J., Regnier B., Gibert C. (1987). Pharmacokinetics of teicoplanin in critically ill patients with various degrees of renal impairment. Antimicrob. Agents Chemother. 31:1600.
190. Derbyshire N., Webb D.B., Roberts D., Glew D., Williams J.D. (1989). Pharmacokinetics of teicoplanin in subjects with varying degrees of renal function. J. Antimicrob. Chemother. 23:869.
191. Roberts D.E., Webb D.B., Williams J.D. (1988). Teicoplanin pharmacokinetics in

subjects with normal and impaired renal function. J. Pharm. Pharmacol. 40 (Suppl.):154.

192. Lam Y.W.F., Kapusnik-Uner J.E., Sachdeva M., Hackbarth C., Gambertoglio J.G., Sande M.A. (1990). The pharmacokinetics of teicoplanin in varying degrees of renal function. Clin. Pharmacol. Ther. 47:655.

193. Brumfitt W., Baillod R., Smith G.W., Grady D., Hamilton-Miller J.M.T., Chuah P. (1985). Teicoplanin for patients undergoing dialysis: Microbiological and pharmacokinetic aspects. In Gialdroni-Grassi G., Mitsuhashi S., Williams J.D. (Eds.), Proceedings of a Workshop, 14th International Congress of Chemotherapy, Kyoto, 1985. University of Tokyo Press, Tokyo, p. 23.

194. Neville L.O., Baillod R., Grady D., Brumfitt W., Hamilton-Miller J.M.T. (1987). Teicoplanin in patients with chronic renal failure on dialysis: Microbiological and pharmacokinetic aspects. Int. J. Clin. Pharm. Res. VII:485.

195. Traina G.L., Gentile M.G., Fellin G., Rosina R., Cavenaghi L., Buniva G., Bonati M. (1986). Pharmacokinetics of teicoplanin in patients on continuous ambulatory peritoneal dialysis. Eur. J. Clin. Pharmacol. 31:501.

196. Brouard R., Kapusnik J.E., Sachdeva M., Schoenfeld P., Gambertoglio J.G., Tozer D.T.N. (1988). Teicoplanin pharmacokinetics in continuous ambulatory peritoneal dialysis (abstract). Miner. Electrolyte Metab. 14:178.

197. Guay D.R.P., Awni W.M., Fant B., Kenny M., Matzke G.R. (1988). Pharmacokinetics of teicoplanin in continuous ambulatory peritoneal dialysis (abstract). In Program and Abstracts of the 28th Interscience Conference on Antimicrobial Agents and Chemotherapy, Los Angeles, October 23–26, 1988, American Society for Microbiology, Washington, D.C., p. 134.

198. Williams H.H., Grüneberg R.N. (1988). Teicoplanin revisited. Antimicrob. Chemother. 22:397.

9
Vancomycin
A Clinical Overview

MICHAEL L. ZECKEL and JAMES R. WOODWORTH
Lilly Research Laboratories, Eli Lilly and Company, Indianapolis, Indiana

Although vancomycin has been available for 35 years, its use has dramatically increased only over the last decade. This resurgence is the result of several interacting factors, including (1) the increasing incidence of methicillin-resistant *Staphylococcus aureus*, (2) a growing awareness that coagulase-negative staphylococci are potential pathogens, (3) the increasing use of prosthetic devices, (4) antibiotic-selective pressures, and (5) advances in patient support requiring prolonged vascular access.

The renewed interest in vancomycin over the last decade has produced several excellent and comprehensive summaries (1–4; see also Ref. 59). This chapter seeks to provide an overview of clinically useful information regarding the pharmacokinetics, clinical uses, and safety profile of vancomycin.

I. PHARMACOKINETICS

Several extensive reviews have provided a comprehensive overview of vancomycin pharmacokinetic behavior (1,5,6; see also Ref. 33).

A. Absorption

Oral Administration

Vancomycin exhibits poor oral bioavailability. Following oral administration, only trace amounts of vancomycin have been detected in the urine of patients with normal renal function (7,8; see also Ref. 674). Among five anephric patients receiving 2 g oral vancomycin daily for 16 days, serum concentration was ≤0.66 μg/ml (see Ref. 775). Since orally administered vancomycin is minimally absorbed, stool concentrations reach high levels. Following oral administration of 500 mg vancomycin every 6 h for 7 days, mean stool concentrations in eight patients reached 3100 ± 400 mg/kg wet weight, and serum concentrations ranged from <1 to 3.9 μg/ml (9). Vancomycin administered as a semisolid matrix produced fecal concentrations similar to those achieved using a solution; however, the average urinary recovery was <0.2% of the administered dose (8).

Despite this lack of oral absorption, measurable serum concentrations and potential systemic toxicity have been documented in patients with renal insufficiency and/or colitis following oral administration (9–13; see also Ref. 776). Therefore, measurement of serum vancomycin concentrations may be advisable for patients with concomitant renal impairment and inflammatory gastrointestinal disorders receiving multiple doses of oral vancomycin.

Intravenous Administration

For systemic infection in patients with normal renal function, vancomycin is generally administered intravenously at 6 or 12 h intervals to provide a total daily dosage of 30–40 mg/kg. Under these circumstances, vancomycin should be administered intravenously at a concentration of no more than 5 mg/ml and at a rate not to exceed 10 mg/minute. In addition, the infusion should be administered over at least 60 minutes (14). Following administration of 15 mg/kg of vancomycin over 60 minutes, mean plasma concentrations reach 63 μg/ml immediately after completion of the infusion, decreasing to 23 μg/ml 2 h later. Mean plasma concentrations 12 h postinfusion are approximately 8 μg/ml in patients with normal renal function. Following infusion of 500 mg over 30 minutes at steady state, mean plasma concentrations reach 49 μg/ml at the end of infusion, 19 μg/ml 2 h after infusion, and 10 μg/ml at 6 h after infusion. Vancomycin should not be administered intramuscularly since it may result in severe pain (7) and tissue injury.

Intraperitoneal Administration

Vancomycin is readily absorbed from the peritoneal cavity in the absence of peritoneal inflammation, although absorption may be increased in the presence of peritonitis (15). Rogge et al. (16) found mean peak plasma vancomycin concentrations of 9.1 μg/ml 5 h following intraperitoneal administration of a 1 g vancomycin dose. Bastani et al. (17) noted that 74% of an intraperitoneally administered vancomycin dose was absorbed after 6 h in the presence of peritoneal inflammation compared with 51% during the same time interval in the absence of peritoneal inflammation. Following intraperitoneal vancomycin administration to six patients receiving continuous ambulatory peritoneal dialysis (CAPD), Neal and Bailie (18) noted serum concentrations at the end of a 6 h dwell to be 14–18 μg/ml following a 15 mg/kg dose and 6.75–24 μg/ml following a 1 g dose. Dialysis clearance was 0.70–0.73 L/h. Following intraperitoneal loading doses of 30 mg/kg, mean plasma vancomycin concentrations reached 26.5 μg/ml at 4 h and 37 μg/ml at 6 h.

B. Distribution

Theoretical Considerations

Tissue distribution is generally influenced by three factors: (1) plasma and tissue protein binding, (2) molecular physicochemical properties (size, spatial configuration, molecular weight, and polarity), and (3) tissue blood flow. Only the first two factors are issues related to the drug itself. The low plasma protein binding of vancomycin (see Protein Binding, p. 312) suggests that it would be available for wide tissue distribution; however, the relatively small volume of distribution measured with vancomycin (V_{ss} approximately 0.65 L/kg) suggests otherwise, implying only limited distribution outside the circulating plasma. The large molecular size (1449 D) and polarity of vancomycin may be the factors limiting wider tissue distribution. Nonetheless, vancomycin achieves therapeutic concentrations in most tissues (6; see also Ref. 39), with the notable exception of cerebrospinal fluid and the eye.

Intraocular Concentrations

Vancomycin does not achieve high intraocular fluid concentrations following intravenous administration. MacIlwaine et al. (19) found mean aqueous humor vancomycin concentrations to be less than 0.78 µg/ml in five subjects having concomitant mean serum vancomycin concentrations of 14 µg/ml. Because vancomycin achieves only limited intraocular concentrations, local administration has been advocated (see Ocular Use, p. 343)

Cerebrospinal Fluid Concentrations

Cerebrospinal fluid (CSF) vancomycin concentrations following intravenous administration have generally been low in the absence of cerebrospinal fluid inflammation. Nolan et al. (20) noted marginal cerebrospinal fluid vancomycin concentrations following therapeutic intravenous doses in hemodialysis patients. LeRoux et al. (see Ref. 406) reported intraventricular concentrations ranging from 0.1 to 1.5 µg/ml 1 h after completion of a 60 minute intravenous infusion of 1 g vancomycin to hydrocephalic patients undergoing cerebrospinal fluid shunt surgery. These concentrations could not be correlated with ventricular fluid volume, cerebrospinal fluid protein concentrations, or serum vancomycin concentrations.

Vancomycin cerebrospinal fluid penetration following intravenous administration may be enhanced in the presence of inflammation. In a study of 16 patients, Redfield et al. (21) noted mean CSF vancomycin concentrations of 2.0 µg/ml if the CSF white blood cell count was under 200/mm³ compared with 8.0 µg/ml if the CSF white blood cell count was over 200/mm³. Schaad et al. (see Ref. 59) noted cerebrospinal fluid concentrations from 1.0 to 12.3 µg/ml (mean 3.9 µg/ml) in children with meningitis. Moellering et al. (36) reported cerebrospinal fluid concentrations of 0–8.0 µg/ml (mean 2.5 µg/ml) in 5 adults with meningitis. Gump (see Ref. 279) noted vancomycin concentrations of 4.3 and 7.6 µg/ml in a cerebral cyst 48 and 96 h, respectively, after initiation of vancomycin therapy in an infant with ventriculoperitoneal shunt infection. McGee et al. (see Ref. 558) reported ventricular fluid concentrations ranging from <1.0 to 17.3 μg/ml in a group of pediatric patients receiving between 30 and 80 mg/kg/day of intravenous vancomycin. In that study, the ventricular fluid vancomycin concentration following intravenous administration correlated with ventricular fluid protein concentration, the logarithm of the white blood cell count, and the glucose concentration. Barois et al. (22) noted mean cerebrospinal fluid concentrations of 2 μg/ml in 13 patients with meningitis receiving intravenous vancomycin. One report (23) documented vancomycin concentrations of 15–18 μg/ml in an intracerebral abscess during intravenous vancomycin therapy.

Marginal cerebrospinal fluid concentrations following intravenous administration have prompted some investigators to recommend intrathecal or intraventricular routes of administration when adequate cerebrospinal fluid concentrations are believed to be critical (see Intraventricular and Intrathecal, p. 344).

Protein Binding

Although early studies suggested that less than 10% of vancomycin was bound to plasma proteins (24), subsequent studies using ultrafiltration techniques have documented 55% protein binding (see Ref. 55). The unbound fraction of vancomycin may be correlated with serum albumin concentration and renal clearance (25). Disease may also affect the serum protein binding of vancomycin. In a study of six patients with end-stage renal insufficiency, Tan et al. (26) noted mean serum protein binding to be 18.5 ± 12.0% compared with 46% in pooled serum from normal controls. In another report, vancomycin concentrations as high as 464 μg/ml were noted in a patient with IgA myeloma, presumably due to abnormal binding to myeloma protein (see Ref. 774).

C. Elimination

Renal Elimination

Vancomycin is primarily eliminated unchanged by glomerular filtration (27); however, tubular secretion may play a role (see Ref. 47). Golper et al. (27) found that up to 30% of vancomycin clearance was effected through concentration-dependent nonrenal mechanisms. Patients with acute renal failure continue to display significant nonrenal clearance; however, as the renal failure persists, nonrenal clearance declines substantially (28). Although there is a good correlation between creatinine clearance and vancomycin clearance (see Refs. 35, 36, and 39), other factors, such as age, weight, sex, and underlying disease (see Ref. 34), may have an effect. Notable interindividual differences in pharmacokinetic behavior have been noted among patients with apparently normal creatinine clearances. The terminal serum half-life in adult patients with normal renal function has varied from 2.9 to 9.1 h, and the volume of distribution has varied from 0.39 to 0.92 L/kg (5). These variations encountered in patients with normal and impaired renal function suggest that serum vancomycin concentrations should be monitored in selected patients.

Hepatic Metabolism

Hepatic metabolism may play a minor role in vancomycin elimination (see Ref. 34). Vancomycin is present in bile and stool following intravenous dosing (29), suggesting at least some hepatic excretion. Schaad et al. (see Ref. 63) found fecal concentrations ranging from 4.1 to 35.8 µg/ml following intravenous vancomycin administration, suggesting that biliary excretion may be present. Although Matzke et al. (5) suggested that nonrenal mechanisms account for <5% of total vancomycin clearance, hepatic clearance may be clinically relevant in some clinical settings. Brown et al. (30) studied the effect of hepatic impairment on the pharmacokinetics of vancomycin in 15 cancer patients. The mean serum half-life in the 6 patients with normal hepatic function was 2.6 h compared with 37 h in 9 patients with hepatic impairment. One cannot exclude the possibility of coexistent renal insufficiency among the patients with hepatic impairment, however, since estimates of creatinine clearance may be unreliable in such patients (31). One report (32) described a patient with both renal and hepatic dysfunction who had a terminal serum vancomycin half-life of 244 h, suggesting that hepatic excretion may be important in patients with concurrent renal insufficiency.

D. Dosing Considerations

Adults

Various vancomycin dosage guidelines for adults have been published to address the problem of interpatient pharmacokinetic variability and the changes in dosing required for various disease states. As comprehensively reviewed by Pryka et al. (33), dosing methods can be classified into three categories: predictive algorithms, pharmacokinetics-based dosing systems, and Bayesian methods.

Predictive Nomograms. Predictive algorithms or nomograms attempt to define the appropriate initial dose of vancomycin before serum concentrations are available. Most nomograms depend on the relationship between creatinine clearance and vancomycin clearance. However, only 50% of the variance in vancomycin serum terminal half-life can be explained by renal function, volume of distribution, age, sex, and race (34). Therefore, calculations of vancomycin dosage based on a knowledge of renal function do not uniformly assure that target vancomycin serum concentrations will be achieved.

Several nomograms have been devised to guide vancomycin dosage adjustment in patients with various levels of renal function. Nielsen et al. (35) demonstrated a linear relationship between vancomycin clearance and creatinine clearance, suggesting that vancomycin dosage could be calculated on the basis of creatinine clearance. Moellering et al. (36) subsequently derived a widely used nomogram for calculating initial vancomycin doses in patients with impaired renal function. As illustrated in Figure 1, the total daily dosage of vancomycin may be determined once the creatinine clearance has been estimated or measured. This method seeks to achieve mean serum vancomycin concentrations of 15 μg/ml. Although the total daily dose may be determined with the aid of the nomogram, specific guidance on the appropriate dosage intervals is not provided. Furthermore, calculations of daily dosage cannot be applied reliably to patients with severe renal insufficiency. Subsequently, Brown and Mauro (37) extended the dosage nomogram of Moellering et al. by suggesting appropriate dosing intervals.

Lake and Peterson (38) developed a simplified dosing formula using a constant dose (8 mg times lean body weight in kg) administered at intervals dependent upon creatinine clearance (calculated using the modified Cockcroft and Gault method). Their dosage interval recommendations were 6, 8, 12, 18, and 24 h for creatinine clearances of >90, 70–89, 46–69, 30–45, and 15–29 ml/minute, respectively.

Matzke et al. (39) formulated a dosing nomogram based on an analysis of 56

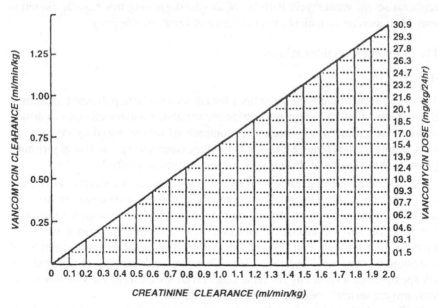

Fig. 1 Dosage nomogram for vancomycin in patients with impaired renal function. The nomogram is not valid for functionally anephric patients on dialysis. For such patients, the dose is 1.9 mg/kg-24 h. (Reproduced with permission from Ref. 36.)

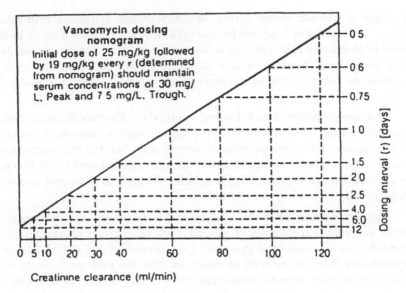

Fig. 2 Dosage nomogram for vancomycin in patients with various degrees of renal dysfunction. The nomogram is not valid for peritoneal dialysis patients. The therapeutic goal is peak and trough serum concentrations of 30 and 7.5 mg/L. (Reproduced with permission from Ref. 39.)

patients with various levels of renal function. Following an initial loading dose of 25 mg/kg, subsequent doses of 19 mg/kg are administered at intervals derived from Figure 2.

Rodvold et al. (40) developed a dosing nomogram after a study of 37 adult patients. According to this nomogram, the dose (mg/kg/day) is calculated by multiplying the creatinine clearance by 0.227 and adding 5.67. The creatinine clearance estimate used in the calculation is standardized to milliliters per minute per 70 kg. The dosage intervals used in this dosing method are 8, 12, 24, and 48 h for standardized creatinine clearances of >65, 40–65, 20–39, and 10–19 ml/ minute per 70 kg, respectively.

Although all these dosing nomograms depend on the relationship between vancomycin clearance and creatinine clearance, such relationships are not always highly correlated. The nomograms derived from one population may not be reliably applied to another patient population, such as critically ill patients (41–44), burn patients (vancomycin clearance increased) (45–47), and possibly intravenous drug users (47).

As reviewed by Pryka et al. (33), the various predictive nomograms have been compared in several studies (48–51). The dosing nomogram of Matzke et al.

(39) tends to provide higher serum vancomycin concentrations than other methods. The method of Lake and Peterson (38) appears to be comparable to the method of Moellering et al. (36). No nomogram, however, is consistently predictive in all patients. Therefore, dosage adjustments based on serum vancomycin concentrations are advisable in the majority of patients requiring vancomycin therapy.

Pharmacokinetics-Based Dosing Methods.   Pharmacokinetics-based dosing methods appear to provide more precise estimates of vancomycin concentrations compared with nomogram-based methods (52–54). The pharmacokinetic behavior of vancomycin has been described using monoexponential (39), biexponential (34,40), and triexponential (55) models, providing an additional source of variability (56).

Bayesian Dosing Methods.   Bayesian methods have been developed to address the inter- and intrapatient variability of vancomycin pharmacokinetic parameters. As reviewed by Pryka et al. (33), several studies (54,57,58) have suggested that Bayesian methods of dosing provide further improvements over results obtained using pharmacokinetically based dosing methods or nomograms.

Neonates, Infants, and Children

Several reviews have summarized the pharmacokinetic behavior of vancomycin in children (59–61). Because neonates generally have larger volumes of distribution and incompletely developed renal function, vancomycin dosing guidelines differ from those advocated for adults. A review of pharmacokinetic parameters derived from various studies involving neonates (60) suggests a wide variation in the estimates of volume of distribution and a positive correlation between vancomycin clearance and postconceptional or postnatal age. Recent studies stressed the importance of postconceptional age rather than postnatal age as determinant of vancomycin clearance. Schaible et al. (62), using stepwise multiple regression analysis, found postconceptional age to be the best single predictor of vancomycin clearance in 11 neonates.

In a study of 55 children, Schaad et al. (63) found that the half-life decreased and the vancomycin clearance increased with increasing postnatal age. Serum half-lives for newborns, older infants, and children were 5.9–9.8, 4.1, and 2.2–3.0 h, respectively. Based on these data, Schaad et al. (63) recommended that vancomycin, 15 mg/kg, be infused over 30–60 minutes at 12 h intervals for newborns up to 1 week of age and at 8 h intervals for newborns from 8 to 30 days of age. Doses of 10 mg/kg every 6 h were recommended for older infants and children.

The results of Schaad et al. may not be universally applicable to other pediatric populations. Alpert et al. (64) reported that the recommendations of Schaad et al. (63) resulted in peak concentrations greater than 30 µg/ml 30–60 minutes after the end of the infusion in 70, 39, and 30% of determinations in patients in the first month of life, second month of life, and older ages, respectively. Mercier et al. (65) also noted relatively high peak concentrations following vancomycin administration to neonates. Among 33 children greater than 1 month of age, administration of 10 mg/kg of vancomycin over 30 minutes every 6 h resulted in peak serum concentrations of 33 µg/ml and trough concentrations of 6 µg/ml. Among 5 neonates less than 8 days of age, 10 mg/kg over 30 minutes every 8 h resulted in peak concentrations of 54 µg/ml and trough concentrations of 22 µg/ml. Among neonates between 8 days and 1 month of age, 15 mg/kg every 8 h resulted in peak concentrations of 42 µg/ml and trough concentrations of 10 µg/ml.

Gross et al. (66) studied vancomycin pharmacokinetics in nine premature infants (29–35 weeks gestation) at 26–62 days of age. A dosing nomogram derived from pharmacokinetic measurements suggested that a dose of 15 mg/kg every 12 h would provide peak concentrations of 19 and 25–40 µg/ml for infants weighing over or under 1000 g, respectively.

Naqvi et al. (67) administered vancomycin in doses of 10 or 15 mg/kg to 17 infants every 8 h (age < 4 weeks) or every 6 h (age > 4 weeks). Mean peak and trough concentrations were 30.4 and 11.6 µg/ml, respectively, following a 10 mg/kg vancomycin dose. For patients receiving 15 mg/kg every 6 h the peak and trough vancomycin concentrations were 57.8 and 15.1 µg/ml, respectively. Naqvi et al. therefore recommended an initial loading dose of 15 mg/kg followed by 10 mg/kg at 8 h intervals for infants under 42 weeks postconceptional age. Reed et al. (68), however, subsequently reported accumulation of vancomycin when dosage was administered on an 8 h schedule in premature infants under 36 weeks postconceptional age.

James et al. (69) devised a dosage nomogram designed to achieve steady-state peak concentrations of 30 µg/ml and trough concentrations of 6 µg/ml. For neonates under 27 weeks (<800 g), 27–30 weeks (800–1200 g), 31–36 weeks (1200–2000 g), and greater than 37 weeks (>2000 g) postconceptional age, dosage recommendations were 27 mg/kg every 36 h, 24 mg/kg every 24 h, 18 mg/kg every 12 h or 27 mg/kg every 18 h, and 22.5 mg/kg every 12 h, respectively. Koren and James (70) found these recommendations to provide reasonable serum concentrations, although use of the nomogram did not obviate the need to monitor serum concentrations in this patient population. Although these dosage recommendations have provided good serum concentrations in a prospective evaluation (70), Kildoo et al. (71) suggested that serum creatinine should be considered along with postconceptional age in determining vancomycin dosage in premature infants. According to these investigators, neonates over 2 weeks postnatal age and 30 weeks postconceptional age should receive doses of 10 mg/kg at intervals of 8 h if serum creatinine is less than or equal to 0.6 mg/dl or at intervals of 12 h if serum creatinine is 0.7–1.2 mg/dl. In a subsequent prospective study of 37 patients (72), this regimen achieved target serum vancomycin concentrations (20–40 μg/ml peak and <10 μg/ml trough) in 79% of neonates.

Lisby-Sutch and Nahata (73) developed a dosage nomogram for infants based on a study of 13 premature infants. Vancomycin clearance was correlated with both postconceptional age (PCA) and actual body weight. Based on their pharmacokinetic analysis, a dose of 10 mg/kg was recommended at intervals of 12, 8, and 6 h for infants 30–34 weeks PCA (<1200 g), 30–42 weeks PCA (>1200 g), and >42 weeks PCA (>2000 g), respectively.

Leonard et al. (74), after studying vancomycin pharmacokinetics in 12 very low birth weight infants (mean gestational age 25.9 weeks and mean body weight 769.2 g), found a statistically significant positive correlation between postconceptional age and vancomycin clearance. Based on these findings, a dose of 15 mg/kg every 24 h was recommended for these very low birth weight neonates.

Extracorporeal membrane oxygenation has been used to sustain neonates through critical illness. Vancomycin pharmacokinetics in neonates undergoing extracorporeal membrane oxygenation are similar to those reported for neonates not receiving such therapy (75).

Specific dosing recommendations, based on the studies just outlined, have provided widely divergent advice. Therefore, routine monitoring of serum vancomycin concentrations appears to be justified in virtually all neonates receiving parenteral vancomycin.

Elderly Patients

Age appears to have an important effect on vancomycin clearance independent of the known inverse relationship between age and creatinine clearance. Cutler et al. (76) compared vancomycin pharmacokinetics following a single infusion of 6 mg/kg of vancomycin in 6 elderly (mean age 68.5 years) and 6 young adult patients (mean age 23 years). The mean vancomycin half-life was longer in the elderly patients (12.1 h) than in the young adult patients (7.2 h), reflecting the lower vancomycin clearance and larger volume of distribution documented in elderly patients. Vance-Bryan et al. (77) compared the vancomycin pharmacokinetic profile of 122 young patients and 108 elderly patients with normal serum creatinine levels using a Bayesian forecasting technique. This retrospective study confirmed a longer mean half-life (17.8 ± 11.8 versus 7.9 ± 8.0 h), a larger mean volume of distribution (0.98 ± 0.28 versus 0.83 ± 0.29 L/kg), a lower mean total body vancomycin clearance (50 ± 24 versus 90 ± 35 ml/h/kg), and a smaller mean daily dose requirement (18 ± 6 versus 25 ± 8 mg/kg total body weight/day) for older patients compared with younger patients. Dosage adjustments based on serum vancomycin concentrations are advisable in elderly patients.

Obesity

Blouin et al. (78) compared the vancomycin pharmacokinetics of four normal and six morbidly obese patients following single-dose administration of 1 g vancomycin. The groups were not comparable, since the morbidly obese patients were studied 3–4 h following gastric bypass surgery. The mean half-life and volume of distribution for the normal and obese patients were 4.8 h and 0.39 L/kg versus 3.2 h and 0.26 L/kg, respectively. These differences were statistically significant. In view of the strong correlation between total body weight and both volume of distribution and total body clearance, as well as the reduced volume of distribution in obese patients, the authors suggested that obese patients receive daily dosage based on total body weight divided into more frequent intervals. Rotschafer et al. (79) noted significant increases in half-life in patients greater than 60% above ideal body weight, suggesting that dosing should be based on total body weight unless the patient is greater than 50% above ideal body weight. Although initial dosing may be based on total body weight in morbidly obese patients less than 50% over ideal body weight, serum concentrations are advisable to assure appropriate dosage modifications.

Severe Renal Impairment

The serum half-life is considerably prolonged in patients with renal failure, varying from 160 to 200 h (39,80,81). Therapeutic serum concentrations may be achieved by administration of a 1 g dose every 7–10 days in patients with severe renal insufficiency (82,83); however, dosage modification may be necessary depending upon the method of dialysis and type of dialysis membranes used.

Hemodialysis. Vancomycin is not significantly removed by the cellulose acetate or cuprophane membranes usually used in hemodialysis (24,83). Therefore, vancomycin may generally be administered in a dose of 1 g administered every 7–10 days in patients undergoing hemodialysis using these membranes.

Several reports have suggested that vancomycin clearance may be increased during hemodialysis using high-flux dialysis membranes. Bastani et al. (84), in a study of five hemodialysis patients, noted mean vancomycin clearance rates of 8.9, 21.1, and 31.1 ml/minute for cuprophane, cellulose acetate, and polyacrylonitrile membranes, respectively. Barth et al. (85) noted a 34% reduction in serum vancomycin concentrations following dialysis with a polyacrylonitrile (AN69) membrane. The reductions in serum vancomycin concentrations during dialysis using polyacrylonitrile membranes may be due to ionic binding (86). Suboptimal serum vancomycin concentrations were encountered 100 h following a vancomycin loading dose in eight patients undergoing hemodialysis using polyacrylonitrile membranes (87). Lanese et al. (88) noted a marked increase in vancomycin clearance during hemodialysis using polysulfone dialyzers (clearance 44.7–85.2 ml/minute) compared with cuprophane (clearance 9.6 ml/minute). These authors suggested that a supplemental vancomycin dose (50% of the loading dose) may be needed following a 4 h dialysis using a polysulfone F-80 dialyzer. Patients receiving hemodialysis using highly permeable membranes may require individualized dosing based on serum concentrations; however, serum levels may rebound several hours after hemodialysis is completed (89), thereby complicating dosage calculations.

Hemoperfusion.   Unlike hemodialysis, vancomycin adheres well to charcoal and the primary resin (Amberlite XAD-4) used in hemoperfusion. Ahmad et al. (90) reported vancomycin clearance of 23.6 and 85.2 ml/minute using charcoal and Amberlite resin XAD-4 hemoperfusion with resulting marked decreases in serum vancomycin concentrations. Supplemental doses of vancomycin may be needed in patients undergoing hemoperfusion.

Hemofiltration.   Hemofiltration is a method that is gaining greater acceptance in treating patients with renal disease, especially those with hemodynamic instability and acute renal insufficiency. Matzke et al. (91), using polyamide hemofilters in a series of five patients, measured a mean vancomycin clearance of 152.6 ml/minute, and a shortening of mean half-life from 136 to 4.1 h during the hemofiltration procedure. An average of 398.5 mg vancomycin was removed with each treatment, resulting in a 32% loss of total-body vancomycin stores. Following completion of hemofiltration, the authors noted a 52% rebound increase in serum vancomycin concentration, probably due to the redistribution of vancomycin into the intravascular space. Such a marked increase in vancomycin clearance during hemofiltration may require careful monitoring of vancomycin concentrations and subsequent dosage adjustment. Golper (92) noted that hemofiltration generates ultrafiltrate, which contains vancomycin in a concentration 69 ± 6% that of serum.

Rawer and Seim (93), in a study of vancomycin clearance during hemofiltration in 27 treatments in four patients, noted a vancomycin clearance of 61–157 ml/minute and a strong correlation between the vancomycin concentration at the start of hemofiltration and the total amount of vancomycin removed during the procedure. Based on daily hemofiltration using a HF 88 hemofilter (Gambro, Germany), a blood flow rate of 200–400 ml/minute, and a filtrate flow rate of 70–180 ml/minute, the authors recommended maintenance doses of 250 mg every 24 h or 500 mg every 48 h following a 1 g loading vancomycin dose. Dupuis et al. (94) also reported markedly increased vancomycin clearance in a patient receiving hemofiltration.

The amount of vancomycin removed by hemofiltration may be estimated by one of two methods (95). The first method consists of collecting a timed volume of ultrafiltrate, determining its vancomycin concentration, and multiplying the concentration by the total volume of the ultrafiltrate. The second method assumes that the concentration in the ultrafiltrate equals the free drug concentration in plasma, a result derived by multiplying the measured serum concentration by the free fraction of vancomycin (1 − fraction bound to protein). Although either method may provide a rationale for vancomycin dosage calculations, serum concentrations of vancomycin should be closely monitored in these patients.

Continuous Arteriovenous Hemodialysis. Continuous arteriovenous hemodialysis or hemodiafiltration combines the convective clearance of hemofiltration with the diffusive effect of dialysis. Bellomo et al. (96) found vancomycin clearance to be greater during continuous arteriovenous hemodialysis than during hemofiltration. The clearance of vancomycin was correlated with the ultrafiltration rate. Kox and Davies noted vancomycin clearance of 15.58 ± 2.07 ml/minute at a flow rate of 2 L/h (97). The large amounts of vancomycin removed during continuous arteriovenous hemodialysis may require supplemental doses of vancomycin [average 500 mg daily (96) to 1000 mg every 48 h (97)] to maintain adequate concentrations in serum; however, such doses should be based upon measurement of serum vancomycin concentrations.

Continuous Ambulatory Peritoneal Dialysis. As reviewed by Matzke et al. (5), several studies have confirmed the shortened half-life of vancomycin during CAPD. Estimates of vancomycin half-life during CAPD range from 61.5 h (98) to 93.5 h (99). In three detailed pharmacokinetic studies (100–102) peritoneal clearance of vancomycin was estimated as 1.35–1.48 ml/minute.

Maintenance of vancomycin serum concentrations may be achieved during CAPD by intermittent intravenous administration regimens. Intravenous vancomycin, 23 mg/kg initially followed by 17 mg/kg every 7 days (39) or 1 g intravenously every 7 days (see Refs. 364 and 365), should provide adequate serum and peritoneal fluid concentrations throughout the course of therapy.

Vancomycin may also be administered intraperitoneally as (a) a continuous regimen of 30 mg/kg intraperitoneally initially followed by 1.5 mg/kg in each dialysis bag every 6 h (102) or (b) intermittently at a dose of 30 mg/kg intraperitoneally every 7–10 days.

Intermittent Peritoneal Dialysis. As reviewed by Matzke et al. (5), the clearance of vancomycin in patients on peritoneal dialysis is variable, yielding clearances from 1.0 ml/minute (103) to 14.2 ml/minute (104). Therefore, serum concentrations should be monitored and supplemental dosages considered in patients undergoing prolonged 24–48 h courses of peritoneal dialysis (5).

E. Monitoring Serum Vancomycin Concentrations

Rationale

The role of serum vancomycin concentration measurements in patient management is controversial (105,106). Data suggesting an association between serum concentrations and therapeutic outcome (63,73,107,108,155) are limited. No system of dosing can uniformly achieve target peak and trough concentrations. Moreover, there is only anecdotal evidence suggesting an association between high serum vancomycin concentrations and ototoxicity (see Ref. 151), nephrotoxicity (see Ref. 720), or other adverse events (see Refs. 577 and 724). However, until the safety and efficacy of dosing without measurement of serum concentrations is demonstrated, it may be advisable to measure vancomycin concentrations in patients who exhibit at least one of the following conditions: extremes of age, renal impairment, impaired or changing renal function, hepatic impairment, hearing impairment, neutropenia, critical illness, septicemia, concomitant treatment with aminoglycosides, or treatment with other ototoxic or nephrotoxic agents. Since most patients requiring vancomycin therapy have at least one of these factors, it is prudent to consider measuring vancomycin concentrations in almost all patients receiving more than a few days therapy.

Rodvold et al. (106) suggest that vancomycin serum concentrations be measured once a week after steady state is achieved. More frequent monitoring may be considered in patients with changing renal function, critical illness, and concomitant oto- or nephrotoxic drug therapy (i.e., aminoglycosides).

Therapeutic Serum Concentrations

Widely differing recommendations on the appropriate vancomycin therapeutic peak and trough concentrations have been reviewed by Fitzsimmons et al. (109). These differences arise from uncertainty regarding the definition of the term "peak" (the highest measured versus the highest extrapolated serum concentration), the relationship between serum concentrations and efficacy, and the relationship between serum concentrations and toxicity. Any recommendation regarding target concentrations should consider patient factors (age, concomitant illness, risk of toxicity, and severity of infection), microbiological factors (minimum inhibitory concentration, MIC, and minimum bactericidal concentration, MBC), the timing of serum samples, and the rate of vancomycin infusion. Therefore, no standardized guideline for target concentrations will meet the needs of all patients.

In the absence of definitive recommendations, the following may serve as reasonable guidelines for most patients. Serum sampled 1 h following the end of a 1 h infusion should achieve concentrations of 25–40 μg/ml following a 1000 mg infusion and 15–30 μg/ml following a 500 mg infusion (110). Serum sampled 2 h after the end of a 1 h infusion of 1000 mg vancomycin should achieve concentrations of 18–26 μg/ml (1). In neonates, serum sampled immediately after the end of a 1 h infusion should achieve a peak concentration of 25–40 μg/ml (63). Serum drawn immediately before the subsequent vancomycin dose should achieve concentrations in the 8–10 μg/ml range (see Ref. 245). Serum concentrations are most representative if obtained after steady state is achieved, usually 24–36 h following initiation of therapy in patients with normal renal function (106).

Assay Methods

Several assay methods are available for quantitating vancomycin concentrations in serum, including microbiological (111), radioimmunoassay (RIA) (112,113), fluorescence polarization immunoassay (FPIA) (114,115), high-performance liquid chromatography (HPLC) (116–121), and monoclonal enzyme multiplied immuno-assay (EMIT) (122). FPIA is widely available as a commercial test kit.

Studies comparing these various assay methods have demonstrated good correlation between immunoassay and bioassay (123,124), between immunoassay and FPIA (125), and between HPLC and FPIA (126). In a comparison of five assay methods (bioassay, HPLC, FPIA, RIA, and fluorescence immunoassay), FPIA was judged the most precise method (127). Among patients with renal insufficiency, the FPIA method may overestimate vancomycin concentrations as measured by HPLC (128,129) or EMIT (129,130). In one study of three patients undergoing peritoneal dialysis, peak vancomycin concentrations were 13–52% higher by FPIA than by HPLC. The discrepancy between FPIA and either HPLC or EMIT in patients with renal insufficiency may be due to the accumulation of vancomycin crystalline degradation products (131). Such products react in the polyclonal antibody-based FPIA but not in the HPLC or monoclonal EMIT assays (132,133). Therefore, patients with renal insufficiency and unexpectedly high serum concentrations may require confirmation using a non-FPIA methodology, such as EMIT, HPLC, or bioassay.

II. CLINICAL EFFICACY

A. Specific Pathogens

Vancomycin has in vitro activity against a wide range of Gram-positive bacteria, many of which have only recently been recognized as pathogens. Although vancomycin has activity against many of the Gram-positive cocci susceptible to β-lactam agents, it also has in vitro activity against methicillin-resistant S. aureus (see Ref. 143), methicillin-resistant coagulase-negative staphylococci (see Ref. 168), and some strains of ampicillin-resistant enterococci (134). Unfortunately, as vancomycin use has grown, increasingly frequent reports of glycopeptide resistance have emerged (135,136), with potentially grave clinical implications.

S. aureus

Soon after its discovery in the 1950s, vancomycin was found to be effective in the treatment of serious staphylococcal infections (137–141), including septicemia (142) and endocarditis (see Ref. 212), providing cure or improvement in the majority of patients (1). Further information relevant to specific clinical conditions is reviewed in the appropriate sections.

Methicillin-Resistant S. aureus

Vancomycin is effective in the treatment of methicillin-resistant S. *aureus* (MRSA) infection (143; see also Refs. 157 and 153). Other agents reported to be effective include rifampin combined with fusidic acid (144), trimethoprim-sulfamethoxazole (145), minocycline (see Ref. 175), and teicoplanin. Clinical failures have been documented with cefamandole (see Ref. 149) and the combination of cephalothin and gentamicin (146). In a randomized, double-blind comparative trial of 101 intravenous drug users with *S. aureus* infection, 57 (98.3%) of 58 vancomycin recipients were cured, compared with 37 (86.0%) of 43 trimethoprim-sulfamethoxazole recipients; however, most of the failures occurred in patients with methicillin-susceptible isolates (147). Following the initial description of vancomycin efficacy in MRSA infection by Benner and Morthland (148), several clinical series (149–153; see also Ref. 633) documented cure rates from 59 to 88% in MRSA bacteremia. In one series of 48 patients with MRSA bacteremia, 14 of 18 episodes treated with vancomycin responded fully compared with only 1 of 15 episodes treated with antibiotics possessing no in vitro activity (154). Of 17 drug abusers with MRSA endocarditis 14 (88%) were cured with vancomycin alone administered for 4 weeks (150). Craven et al. (151) noted an overall 1 month survival rate of 74% in 29 patients with MRSA bacteremia, a rate similar to that achieved in 29 patients with oxacillin-susceptible *S. aureus* bacteremia treated with appropriate β-lactam agents. Myers and Linnemann (152) reported a response rate of 86% in patients receiving vancomycin for the treatment of MRSA bacteremia.

MRSA infections may be resistant to treatment, however. Klastersky et al. (153) noted favorable clinical outcomes in only 16 of 27 patients (59%) with MRSA infection. Vancomycin-tolerant strains of MRSA have been associated with therapeutic failure (155,156). Levine et al. (see Ref. 251) noted a response to vancomycin or vancomycin combined with rifampin in 36 of 42 consecutive patients (80%) with MRSA endocarditis; however, the duration of bacteremia and duration of fever were prolonged, suggesting a slow response to vancomycin. Occasional patients who fail to respond to vancomycin alone may benefit from the addition of gentamicin or rifampin (157).

Coagulase-Negative Staphylococci

Coagulase-negative staphylococci are increasingly recognized as pathogens in medical and surgical patients (158). Morrison et al. (159) noted an increase in the proportion of nosocomial coagulase-negative staphylococcal bacteremias from 1.4 to 4.5% over a 7 year period. Martin et al. (160) reported that coagulase-negative staphylococci accounted for 26% of nosocomial bacteremias, resulting in an overall rate of 38.6 episodes per 10,000 admissions. This increased incidence is probably related to the growing use of intravascular devices and prosthetic materials, changes in patient characteristics (161), and antibiotic selection pressures (162). Coagulase-negative staphylococci have become prominent pathogens in newborn infants with implanted devices and intravenous lines (163–165). Archer and Armstrong (166) noted that the incidence of methicillin-resistant *Staphylococcus epidermidis* colonization among cardiac surgery patients increased from 6% on admission to 68% 7 days postoperatively.

The majority of coagulase-negative staphylococci are methicillin resistant. Although cephalosporines may display in vitro activity against methicillin-resistant staphylococci, routine susceptibility testing may underestimate cephalo-sporine resistance (167). Karchmer et al. (see Ref. 176), using high inoculum techniques to detect heteroresistance, found that 83% of S. epidermidis isolates were resistant to methicillin. Glycocalyx (slime), a factor important in staphylo-coccal adherence to polymer surfaces and a marker of pathogenicity (168), inhibits the antibacterial effect of vancomycin and teicoplanin but not that of the lipopep-tide antibiotic daptomycin (169).

Although vancomycin is effective against S. epidermidis, resistance among coagulase-negative strains has been reported. Most of these cases have been reported in patients who received prior vancomycin. Schwalbe et al. (170) reported a progressive increase in resistance of Staphylococcus haemolyticus strain during an 88 day course of vancomycin therapy for dialysis-associated peritonitis; the MIC reached 8 μg/ml, with subpopulations demonstrating greater resistance. This strain was susceptible to daptomycin but not teicoplanin. Bacte-remia caused by a vancomycin-resistant strain (MIC 8.0–16 μg/ml) (171) was reported in a leukemic patient receiving prolonged vancomycin therapy. A vancomycin-resistant strain of S. epidermidis was implicated as a cause of peritonitis in a patient who received prior vancomycin (172). S. haemolyticus strains, in contrast to other coagulase-negative staphylococcal species, may contain subpopulations stably resistant to vancomycin, a characteristic that may be detected by demonstration of a double zone of growth around imipenem agar diffusion disks (173). Glycopeptide resistance among coagulase-negative staphy-lococci may portend resistance in S. aureus, since resistance may be transmitted between species (174).

Treatment options for vancomycin-resistant coagulase-negative staphylo-cocci are limited; potential alternatives, based on in vitro data, may include imipenem, teicoplanin, ciprofloxicin, minocycline (175), and daptomycin; how-ever, experience with these agents is limited. In prosthetic valve endocarditis due to methicillin-resistant S. epidermidis, Karchmer et al. noted improved cure rates with vancomycin combined with either rifampin or an aminoglycoside (176). Although in vitro synergy has been demonstrated between vancomycin and gentamicin or rifampin (177,178), the combination of rifampin and vancomycin may be antagonistic for some strains (179). The clinical significance of in vitro synergy or antagonism in S. epidermidis remains unresolved.

Enterococcus sp.

Vancomycin has a role in the treatment of enterococcal infection in patients allergic to penicillin (see Ref. 259) and in patients infected with ampicillin-resistant strains (180). Although vancomycin-aminoglycoside combinations are synergistic when the enterococcal isolate is susceptible to each component (181), such synergy does not exist in *Enterococcus faecalis* (*Streptococcus faecalis*) (182–185) and *Enterococcus faecium* (*Streptococcus faecium*) (186) strains resistant to streptomycin or gentamicin.

Vancomycin resistance among enterococci is being recognized with increasing frequency (187). Concomitant high-level resistance to penicillin and vancomycin has also been described among clinical enterococcal isolates (188). Infection with vancomycin-resistant enterococci has been associated with prolonged treatment with broad-spectrum antibiotics, prolonged intensive care unit treatment, and exposure to contaminated fomites (189). Vancomycin prophylaxis or treatment may be associated with an increased incidence of colonization or infection due to resistant enterococci (190,191); however, some patients may carry vancomycin-resistant enterococci even without prior vancomycin exposure (192). In one study conducted on a hematology ward, 9 of 70 adult but none of 25 children were found to have resistant enterococci in stool samples in the absence of prior vancomycin administration (193). In a prospective study among 49 pediatric liver transplant recipients, resistant enterococcal species were isolated from stool samples of 31 patients (194). In that study, stool colonization increased steadily during the first month following transplantation and persisted in 22 of the 31 patients. Only 3 of the 31 patients developed infection with one of these resistant strains. Vancomycin resistance among enterococci may be underestimated using the usual in vitro methods (195–197). Treatment options for infections due to ampicillin-resistant and vancomycin-resistant strains are undefined. Some investigators (198,199) have suggested that synergistic killing may be achieved in aminoglycoside-susceptible strains by combining vancomycin, penicillin, and gentamicin; however, such in vitro findings are not consistently found (200,201). Although animal models suggest that such combinations may prove efficacious (202), clinical evidence to support these in vitro findings is not available. Although not yet encountered in clinical isolates, Nobel et al. demonstrated transfer of vancomycin resistance from *E. faecalis* to *S. aureus* in the laboratory (203).

Other Streptococci

Vancomycin has been recommended as treatment for streptococcal infections occurring in patients allergic to β-lactam agents or those infected with resistant strains. Although viridans streptococci are generally susceptible to vancomycin, occasional cases of clinical failure have been observed apparently associated with antibiotic resistance (204). Vancomycin was successful in the treatment of patients with penicillin-resistant *Streptococcus pneumoniae* meningitis (205), nosocomial pneumonia (206), and endocarditis (207); however, 7.5 mg/kg of vancomycin every 6 h failed to cure 4 of 10 consecutively treated patients with uncomplicated *S. pneumoniae* meningitis (208). Although penicillin-resistant *S. pneumoniae* infections may be treated with third-generation cephalosporins, resistance to these agents has been reported (209).

Corynebacterium

Vancomycin is effective in clinical infections due to *Corynebacterium* JK, including pneumonia (210), bacteremia (211), and endocarditis (212,213). Vancomycin has also been effective in infections caused by *Corynebacterium pseudodiphtheriticum* (214). *Corynebacterium* group D_2, a slow-growing organism implicated as a cause of urinary tract infection among hospitalized, elderly, and immunocompromised patients, responds well to vancomycin; all nine patients who received vancomycin were cured in one report (215). Endocarditis due to this organism has been successfully treated with vancomycin (216). Vancomycin-resistant *Corynebacterium* JK [MIC 6.25] (217) and a *Corynebacterium* sp. [MIC 32 μg/ml] (218) have been reported as causes of endocarditis.

Other Pathogens

Leuconostoc species (219–224), pediococci (225,226), and lactobacilli (227, 228)—Gram-positive cocci increasingly recognized as potential pathogens (229,230)—are intrinsically resistant to the glycopeptides. *Clostridium tertium*, implicated as a cause of septicemia in patients with hematological malignancy, is resistant to β-lactam antibiotics and metronidazole but is susceptible to vancomycin, trimethoprim-sulfamethoxazole, and ciprofloxicin (231). Vancomycin has reportedly been effective in treatment of *Rhodococcus equi* infection (232,233), *Listeria monocytogenes* bacteremia (234–237), *Bacillus cereus* cutaneous infection (238), and in a case of unidentified scotochromogenic mycobacterial bacteremia (239). *Stomatococcus mucilaginosus* has been recognized as a cause of septicemia in immunosuppressed patients (240) and endocarditis (241); these pathogens are usually susceptible to vancomycin. Clinical failure and superimposed meningitis have been reported in vancomycin-treated patients with *L. monocytogenes* infection (242,243).

B. Specific Clinical Entities

Endocarditis and Bacteremia

Although β-lactam agents are considered the therapy of choice for endocarditis due to *S. aureus* and streptococci, vancomycin has a role in the treatment of patients (1) allergic to β-lactam agents or (2) infected with organisms resistant to the agents of choice.

Vancomycin is effective in the treatment of staphylococcal endocarditis (212,244–248); however, it is not uniformly effective. In an open-label study on the efficacy of 2 week antibiotic therapy for *S. aureus* endocarditis in intravenous drug users, vancomycin plus tobramycin was curative in only 1 of 3 patients compared with 47 of 50 patients receiving nafcillin and tobramycin (249). In a retrospective review of 13 consecutive intravenous drug users with *S. aureus* endocarditis treated with intravenous vancomycin, Small and Chambers (250) noted suboptimal outcomes in 5 patients, suggesting that vancomycin may not be as effective as nafcillin in this population. Levine et al. (251), in a study of 42 consecutive patients with MRSA endocarditis, noted a median duration of bacteremia of 9 days and a median duration of fever of 7 days, suggesting a slow response to vancomycin. In one study of recurrent *S. aureus* bacteremia, all the patients with recurrent bacteremia due to the original *S. aureus* strain had received vancomycin, suggesting that this agent may be associated with relapse in some patients (252). Failure of vancomycin in pediatric patients has also been noted (253). The addition of rifampin may improved the efficacy of vancomycin in some patients with *S. aureus* endocarditis (254); however, rifampin resistance can develop during therapy even when combined with vancomycin (255). Synergy of vancomycin plus rifampin is not always present (256), and such combinations may be antagonistic (157).

Vancomycin is effective in the treatment of enterococcal endocarditis (248,257–259) in patients who are allergic to penicillin or who are infected with ampicillin-resistant enterococcal strains. Vancomycin should be combined with an aminoglycoside (260) in the treatment of enterococcal endocarditis, since vancomycin is bacteriostatic against enterococci and animal model studies suggest that it is ineffective when used alone in the treatment of enterococcal endocarditis (261). An early report (262) suggesting the potential utility of vancomycin alone in the treatment of enterococcal endocardis has been questioned (263).

Prosthetic Valve Endocarditis (PVE)

In a review of four studies involving 262 episodes of prosthetic valve endocarditis, the most common isolates were coagulase-negative staphylococci (32%), non-enterococcal streptococci (19%), and *S. aureus* (10%) (264). The incidence of methicillin resistance among coagulase-negative staphylococci appears to be higher in patients presenting within 1 year of surgery (resistance rates 84–87%) compared with those presenting later (resistance rates 22–30%) (176,265). Karchmer et al. (176) demonstrated improved effectiveness of vancomycin combined with gentamicin and/or rifampin in the treatment of *S. epidermidis* PVE; only 3 of 6 patients treated with vancomycin alone were cured compared with 18 of 20 (90%) of patients treated with vancomycin plus an aminoglycoside and/or rifampin. The addition of rifampin may increase the serum bactericidal titer greater than eightfold over that of vancomycin alone. *S. epidermidis* strains resistant to rifampin may develop in patients receiving both vancomycin and rifampin (266,267). The risk of rifampin resistance may be increased if vancomycin concentrations are marginal at the site of infection (267). In a small randomized prospective trial comparing 20 patients receiving vancomycin plus rifampin for 6 weeks versus 13 patients receiving vancomycin plus rifampin for 6 weeks combined with gentamicin for the first 2 weeks, Karchmer and Archer (268) noted similar response rates (77 versus 85%); however, a higher incidence of rifampin resistance (37%) was noted among organisms isolated from patients not receiving gentamicin. Therefore, the optimal regimen for prosthetic valve endocarditis due to coagulase-negative staphylococci appears to be a combination of vancomycin and rifampin for 6–8 weeks accompanied by gentamicin during the first 2 weeks. Surgery may be an important component of therapy in patients with PVE. In one study (176), 30 of 32 patients with complicated coagulase-negative staphylococcal PVE who were cured required surgical intervention.

Prosthetic valve endocarditis due to methicillin-susceptible *S. aureus* or penicillin-susceptible streptococci should be treated with an appropriate penicillin derivative unless precluded by penicillin allergy. Patients allergic to penicillin may benefit from vancomycin treatment; for enterococcal endocarditis an aminoglycoside should be added.

Patients with PVE due to *Corynebacterium* should receive penicillin plus gentamicin (269) unless the organism is resistant to penicillin. If penicillin resistance is demonstrated, some investigators nonetheless suggest a penicillin-gentamicin combination (269), whereas others suggest vancomycin (270). For PVE due to *Corynebacterium jeikeim* or gentamicin-resistant *Corynebacterium*, vancomycin has been recommended (271).

For patients presenting with PVE requiring empirical therapy pending blood culture results, vancomycin plus gentamicin has been recommended (270). If PVE occurs more than 1 year following valve placement, the empirical addition of an agent active against fastidious Gram-negative bacterial pathogens may be justified (272).

Prosthetic Joint Infection

Gram-positive cocci account for over 60% of prosthetic joint infections. Although earlier reports suggested that *S. aureus* was the predominant organism (273), more recent studies suggest that coagulase-negative staphylococci, particularly *S. epidermidis*, have become the most commonly isolated pathogens, accounting for over 40% of cultures yielding single pathogens (274,275). *S. aureus* still predominates as the pathogen (46–55%) in patients with hematogenously acquired prosthetic joint infection (276,277). Medical treatment is usually insufficient to eradicate infection. Therefore, surgical excision of an infected prosthetic joint with either one-stage or two-stage replacement is usually necessary to eradicate infection. For coagulase-negative staphylococcal or MRSA infection, vancomycin is a logical choice based on its in vitro activity. To reduce the risk of relapse, the duration of treatment should generally exceed 4 weeks (275) to 6 weeks (278).

Central Nervous System

A comprehensive review of vancomycin in the treatment of central nervous system infections has been presented by Gump (279). Although vancomycin does not reach high cerebrospinal fluid concentrations following intravenous administration, occasional reports suggest that intravenous vancomycin, without intraventricular or intrathecal administration, may be effective in the treatment of staphylococcal meningitis (59,280), *Flavobacterium meningosepticum* meningitis (281), and resistant *S. pneumoniae* meningitis (205). Because cerebrospinal fluid concentrations of vancomycin following intravenous administration may be marginal, intrathecal or intraventricular routes of administration have been advocated. Dosage recommendations for intrathecal or intraventricular vancomycin have varied from 5 mg (279,282) to 20 mg (283,284) administered daily.

Staphylococci are the predominant cause of cerebrospinal device infections. *S. epidermidis* is implicated in 47–64% and *S. aureus* in 12–29% of meningitis cases complicating cerebrospinal fluid device infections (285). Most *S. epidermidis* strains are resistant to methicillin (286).

Although some cerebrospinal fluid shunt infections respond to antibiotic therapy alone (283,287,288), optimal treatment of infected device infections usually requires device removal. In one series, 6 of 19 patients receiving intraventricular vancomycin required removal of the prosthetic material to effect cure (289). In a review of 50 cases of shunt-associated ventriculitis (284), intraventricular vancomycin, 20 mg/day, in conjunction with systemic antibiotic therapy was associated with an overall cure rate of 66%. In this study, 22 of 24 patients (91.67%) undergoing early shunt removal were cured; in contrast, 44% of patients with shunts left in place required surgical revision for catheter blockage.

Although intraventricular or intrathecal vancomycin is usually administered in combination with intravenous administration, Swayne et al. (290) reported clinical response in four of four patients treated solely with intraventricular vancomycin instilled into the shunt.

Penicillin-allergic patients with enterococcal meningitis may respond to vancomycin combined with an aminoglycoside (291,292) or rifampin (293). Vancomycin has also been reported effective given intravenously and intrathecally in the treatment of meningitis due to *F. meningosepticum* (294) and *L. monocytogenes* (295).

Oral antibiotic therapy has been combined with intravenous and intraventricular vancomycin. In an animal model, the addition of rifampin enhanced the antistaphylococcal bactericidal activity of vancomycin in the cerebrospinal fluid (296). Rifampin may enhance the effectiveness of vancomycin in *S. aureus* (297) and *S. epidermidis* (283) Ommaya reservoir or CSF shunt infections (298–300). Intrashunt vancomycin combined with oral antibiotic therapy (trimethoprim-sulfamethoxazole or rifampin) appeared to be effective in one small series of patients with infected CSF shunts (301).

Osteomyelitis

Although vancomycin inconsistently achieves measurable concentrations in osteomyelitic bone (302), several literature reports (59,61,303–306) suggest that vancomycin is effective in the treatment of acute and chronic (307,308) osteo-myelitis. Clinical failures of vancomycin have been reported despite adequate serum vancomycin concentrations (309). Local application of vancomycin is described in Section II.E, Intraventricular and Intrathecal.

Empirical Therapy of Immunosuppressed Patients

The increasing incidence of Gram-positive coccal bacteremia in neutropenic patients (310) has led to the expanded use of vancomycin in patients with neutropenia and fever (311–314). Vancomycin has been effectively used in combination with other antimicrobial agents (315–318) as initial empirical therapy for patients with neutropenia and fever. It is unclear, however, whether the empirical addition of vancomycin actually reduces mortality in these patients (319,320).

Two conflicting views exist regarding the role of vancomycin as a component of initial empirical therapy of febrile neutropenic patients. The first view contends that vancomycin should be used empirically in these patients since the incidence of Gram-positive bacterial infection is so high in this setting. The second view suggests that vancomycin should be added only if a Gram-positive infection is documented or suspected, since delayed therapy usually results in no added morbidity or mortality.

Evidence Supporting Initial Empirical Vancomycin Use. Several studies suggest that vancomycin should be included in the initial empirical antibiotic regimen for febrile neutropenic patients. Shenep et al. (321) prospectively compared a regimen containing vancomycin, ticarcillin, and amikacin with a combination of ticarcillin-clavulanate and amikacin as initial empirical therapy in 101 febrile, neutropenic children with cancer. The vancomycin-containing regimen was associated with a lower frequency of breakthrough bacteremia (1 of 53 versus 9 of 48, $p = 0.006$) and in increased response rate (85 versus 62%, $p = 0.010$). In a randomized study of 60 adult neutropenic leukemic patients with fever, Karp et al. (322) noted that patients receiving vancomycin experienced fewer days of fever (median 9 versus 14 days) than those not receiving vancomycin. In a randomized study involving granulocytopenic cancer patients, Kramer et al. (323) reported a lower death rate ($p < 0.005$) in patients receiving a carbenicillin-vancomycin regimen (2 of 37 patients) than in patients receiving a cephalothin-gentamicin-carbenicillin regimen (10 of 37 patients).

Evidence Against Initial Empirical Vancomycin Use. Conversely, several studies suggest that vancomycin need not be added to initial empirical antibiotic therapy in febrile neutropenic patients. Viscoli et al. (324) compared ceftazidime plus amikacin versus ceftazidime plus vancomycin in 220 episodes of neutropenic fever in children with cancer. Response rates were similar between groups (66 versus 77%), but adverse events were increased in the vancomycin treatment group (4 versus 35%). Rubin et al. (325), after conducting a retrospective study of 550 episodes of fever in neutropenic patients, suggested that vancomycin was not necessary as empirical therapy even in patients with central lines (326). A large multicenter trial (327) randomized 747 febrile granu-locytopenic patients to either vancomycin, ceftazidime, and amikacin or ceftazidime and amikacin. The vancomycin-containing regimen was not associated with shorter duration of fever or lower mortality, although it was associated with higher overall successful outcome rate (76 versus 63%, $p < 0.001$). However, the vancomycin-containing regimen was associated with a higher rate of nephrotoxicity (6 versus 2%, $p = 0.02$) and hepatocellular dysfunction (22 versus 13.5%, $p = 0.003$). Similar results were noted in a smaller randomized study comparing ceftazidime versus ceftazidime plus vancomycin as initial empirical treatment in 127 febrile neutropenic patients (328). Additional studies (329–332) have noted no increased mortality in febrile granulocytopenic patients who failed to receive vancomycin until the diagnosis of Gram-positive infection was confirmed by positive culture.

Although no consensus exists, vancomycin should be considered a component of empirical antibiotic therapy for febrile neutropenic patients in the following circumstances: (1) if the prevalence of MRSA is high at an institution (333,334), (2) if a patient with a central line has evidence of exit site infection, tunnel infection, or impaired catheter patency, (3) if a patient has failed to respond to empirical therapy for Gram-negative bacterial infections (335), or (4) in institutions where most bloodstream infections are caused by Gram-positive organisms (327,336). Further studies may be needed to define the role of initial empirical vancomycin therapy in specific subgroups, such as bone marrow transplant patients.

Pulmonary Infections
Vancomycin is effective in the treatment of pneumonia and empyema due to methicillin-resistant *S. aureus* (337,338). Intrapleural vancomycin has also been reported to be effective in treatment of empyema (339). Topical and inhalational uses of vancomycin are discussed in Section II.E, Topical Oropharynx and Respiratory Tract.

Vascular Access in Hemodialysis Patients
S. aureus and coagulase-negative staphylococci are the most common pathogens infecting arteriovenous fistulas and prosthetic grafts (340,341). Several studies (82,83,342) have documented the effectiveness of 1 g vancomycin intravenously every 7 days for 4–6 weeks in the treatment of Gram-positive infections occurring in hemodialysis patients.

Catheter-Associated Infection
Coagulase-negative staphylococci, particularly *S. epidermidis*, are the most common organisms infecting intravenous catheters, including central hyperalimenta-

tion catheters (343), subclavian Uldall catheters (344), and Hickman catheters (345), accounting for 50–75% of infections (346).

The optimal treatment of infection involving intravascular catheters is controversial. Immunocompetent patients without symptoms of sepsis may simply require catheter removal and close clinical and microbiological observation. Patients with symptoms suggestive of sepsis, persistent bacteremia, or associated phlebitis should receive antibiotic treatment after catheter removal (347). Therapy of line-associated coagulase-negative catheter sepsis usually requires catheter removal; however, some reports suggest that cure may be achieved with antibiotic therapy alone (313,345,348). Catheter retention may be carefully considered if there is no evidence of subcutaneous tunnel infection or septic phlebitis, and improvement occurs within 48 h of antibiotic initiation. In one report, vancomycin, 2.5 mg/0.5 ml instilled into noninfused catheters, appeared to be more effective in eradicating intraluminal coagulase-negative staphylococcal colonization than intermittent intravenous (IV) infusion of 450 mg every 8 h (349). Local instillation of 2 ml vancomycin, 150 mg/ml for 1 h, with subsequent aspiration did not appear to affect serum vancomycin kinetics in one study (350). Local instillation of 5 ml vancomycin (40 μg/ml) for 2–7 days eradicated *S. epidermidis* from an infected epidural catheter in one report (351). Although not proven by controlled clinical trials, vancomycin appears to a reasonable antibiotic choice for treatment of central line infection due to susceptible coagulase-negative Gram-positive cocci.

Vascular Device-Related Infection

Although earlier studies suggested that *S. aureus* was the most common pathogen implicated in vascular graft infection, accounting for 43% of infections in one review (352), *S. epidermidis* appears to be increasing in importance, accounting for more than 60% of cases in one recent series (353). *S. epidermidis* may be difficult to isolate from resected prosthetic material; however, recovery may be enhanced by sonic disruption of biofilm from the resected prosthetic material before culture (354). Methicillin resistance is common in coagulase-negative staphylococci isolated following vascular graft surgery (355). Treatment usually requires use of both antibiotic therapy and vascular graft excision; however, peripheral graft infection may respond to antibiotics and debridement without the need for excision (356). Vancomycin has been used effectively in the treatment of hemodialysis grafts (357).

Gastrointestinal

Parenteral vancomycin has been used in the setting of necrotizing enterocolitis and peritonitis. In a retrospective study of necrotizing enterocolitis, Scheifele et al. (358) noted a lower risk of culture-positive peritonitis and a lower mortality rate in 44 infants treated with cefotaxime and vancomycin compared with an historical comparison group of infants treated with ampicillin and gentamicin. In one randomized study comparing vancomycin and gentamicin with vancomycin and aztreonam among septic neonates, necrotizing enterocolitis occurred in 14.6% of 41 neonates randomized to vancomycin and gentamicin but in none of 40 neonates randomized to vancomycin and aztreonam (359). The role of vancomycin, if any, in the treatment or prevention of neonatal necrotizing enterocolitis is speculative.

Parenteral vancomycin is effective in the treatment of dialysis-associated peritonitis due to coagulase-negative staphylococci (360–365). Although oral vancomycin or metronidazole is effective in the treatment of pseudomembranous colitis due to *Clostridium difficile*, the intravenous route may not be effective (366). The role of orally administered vancomycin in the treatment of gastro-intestinal disorders is discussed in the section Enteric Infection (p. 341).

Comparison of Vancomycin and Teicoplanin

In Vitro Comparisons. Although vancomycin and teicoplanin have a similar in vitro spectrum, some differences exist. Teicoplanin has a lower MIC_{90} than vancomycin against *S. aureus* in some reports (367); however, teicoplanin activity may be affected by high inocula (368). In vitro selection of resistance among *S. aureus* (369) and *S. epidermidis* (370) strains occurred more frequently during exposure to teicoplanin than to vancomycin. Coagulase-negative staphylo-cocci, particularly *S. haemolyticus*, are more frequently resistant to teicoplanin than to vancomycin (136,371–374). Resistance of *S. aureus* to vancomycin is rare among clinical isolates; however, there are several reports of teicoplanin-resistant *S. aureus* infection (375–377). Teicoplanin appears to have greater in vitro activity than vancomycin against *C. difficile* (378), *Bacteroides melaninogenicus* (379), and enterococci (380), including those isolates with low-level vancomycin resis-tance (135).

Randomized Trials. Several small randomized studies have compared the efficacy and safety of vancomycin and teicoplanin. Cony-Makhoul et al. (381) compared vancomycin and teicoplanin as second-line empirical treatment in neutropenic patients who failed to respond clinically to 48 h of ceftazidime therapy. Fever resolved within 48 h in 21 of 35 patients (60%) randomized to vancomycin compared with 13 of 24 patients (54%) randomized to teicoplanin. Van der Auwera et al. (382) compared teicoplanin and vancomycin in the treatment of 74 immunocompromised hosts, the majority of whom had line-associated bacteremia and skin and soft tissue infections due to various Gram-positive pathogens. Clinical cure or improvement was documented in 74.3% of 35 vancomycin recipients and 75.0% of 36 teicoplanin recipients. Smith et al. (383) compared vancomycin and teicoplanin in an open-label prospective trial involving 72 cases of suspected or proven Hickman catheter-associated infection. The response rate of vancomycin-treated patients (80%) was similar to that of the teicoplanin-treated patients (69%); however, adverse events occurred in 25% of vancomycin recipients compared with 8% of teicoplanin recipients. Gilbert et al. (384) compared teicoplanin, 6 mg/kg/day, and vancomycin, 15 mg/kg every 12 h, in the treatment of 40 patients with Gram-positive infections. All 9 patients with indwelling line infections were cured; however, in the 12 patients with *S. aureus* endocarditis, teicoplanin therapy failed in 6 of 8 patients compared with failure in 1 of 4 vancomycin recipients. Van Laethem et al. (385) compared vancomycin and teicoplanin in an open-label randomized study involving 21 patients. The median duration of therapy was 15 days for vancomycin and 21 days for teicoplanin. Cure or improvement were noted in all 9 vancomycin recipients and in 11 of 12 teicoplanin recipients. Transient renal impairment was noted in 2 patients within each treatment group. Kureishi et al. (386) compared vancomycin versus teicoplanin, each in combination with piperacillin and tobramycin, as initial empirical treatment of febrile neutropenic patients. Although there was no difference in clinical outcomes, nephrotoxicity occurred more frequently in vancomycin recipients, especially those receiving cyclosporine. In contrast, another study (387) failed to detect an increased incidence of cyclosporine-associated nephrotoxicity in bone marrow transplant patients receiving concurrent vancomycin. Gelfand et al. (388) found teicoplanin and vancomycin to have equivalent efficacy and safety in the treatment of 44 patients with Gram-positive bacteremia. De Lalla et al. noted equivalent efficacy of oral vancomycin and teicoplanin in the treatment of *C. difficile*-associated colitis (389). Two small comparative studies (390,391) found teicoplanin and vancomycin to be equally effective in the treatment of peritonitis in CAPD patients.

Adverse Events.   Although several comparative studies (381–386) have suggested that teicoplanin and vancomycin are equally efficacious at a teicoplanin dose of 6 mg/kg/day, the appropriate teicoplanin dosage for more severe infection may be higher (384,392,393). Although vancomycin appeared to have a higher incidence of toxicity than teicoplanin in some studies (383,386; see also Ref. 599), the incidence of teicoplanin toxicity at doses greater than 6 mg/kg are undefined. In one study (392), 5 of 18 patients (28%) receiving teicoplanin doses of 12 mg/kg/day or greater developed drug fever and rash requiring study discontinuation. In a pharmacokinetic study (394) of teicoplanin administered in doses of 15, 20, and 25 mg/kg/day, 5 of 15 volunteers (33%) experienced fever, chills, or skin reactions. Although the incidence of "red man" syndrome is high among vancomycin recipients and rare than among teicoplanin recipients (see Ref. 599), reactions resembling red man syndrome have been reported in teicoplanin recipients (395,396).

Larger comparative trials and extensive clinical experience are needed to assess the relative safety and efficacy of vancomycin and teicoplanin.

C. Prophylaxis

Neurosurgery

Since coagulase-negative staphylococci and *S. aureus* are the predominant etiological agents causing postneurosurgical infection, the use of vancomycin as surgical prophylaxis has been explored in several studies. In two large uncontrolled studies enrolling 1732 and 878 patients, a regimen of intravenous vancomycin and tobramycin combined with topical streptomycin irrigation was associated with infection rates of 0.0–0.9%, respectively (397,398). Comparative studies of vancomycin prophylaxis in neurosurgery, however, have not provided clear evidence of efficacy.

Evidence Supporting Prophylactic Vancomycin.   Several small studies have suggested that vancomycin may have a role as a prophylactic agent in neurosurgery. Quartey et al., after noting a 7% infection rate in clean neurosurgical procedures, noted only 4 infections in 495 cases (0.8%) following institution of vancomycin plus gentamicin prophylaxis (399). Geraghty et al. (783), in a randomized unblinded trial of vancomycin plus gentamicin versus control agents as neurosurgical prophylaxis, noted infection in 1 of 203 patients (0.5%) receiving vancomycin plus gentamicin and 7 of 199 patients (3.5%) receiving other agents ($p < 0.05$). Shapiro et al. (400), using a randomized double-blind trial design, noted postneurosurgical infection in 2 of 71 patients (2.8%) receiving prophylactic intravenous vancomycin plus gentamicin compared with 9 of 77 placebo recipients (11.7%). In a small randomized trial of intravenous vancomycin therapy as prophylaxis for hydrocephalic shunt surgery, 2 of 78 vancomycin recipients developed infection compared to 5 of 80 patients in a control group consisting of either no antibiotic or other comparator agents (401). Bloomstedt and Kytta (402) compared single-dose vancomycin prophylaxis with no prophylaxis in a randomized study of patients undergoing craniotomy; 3 of 169 patients (1.8%) receiving vancomycin developed infections compared with 14 of 191 patients (7.3%) receiving no prophylaxis, a statistically significant reduction.

Evidence Against Vancomycin Prophylaxis.   Several reports, however, have cast doubt on the utility of vancomycin-containing regimes in reducing postneurosurgical infections. Two trials of vancomycin prophylaxis for neurosurgical shunt implant surgery failed to demonstrate a reduction in infection rates among patients receiving vancomycin and were accompanied by a high frequency of adverse events (403,404). In the setting of cerebrospinal fluid shunt placement, low CSF concentrations of vancomycin (<0.6–0.8 µg/ml) are achieved following intravenous administration 15 mg/kg every 8 h for 3 doses (405). These low CSF concentrations following IV prophylaxis are unrelated to the ventricular fluid volume (406). Intrashunt vancomycin administration may provide advantages of high intraventricular concentrations and avoidance of the systemic effects of intravenous administration; however, in a retrospective review of the role of intrashunt vancomycin in cerebrospinal shunt surgery, Younger et al. (407) reported infections in 15 of 103 vancomycin recipients (14.6%) compared with 18 of 127 patients (14.1%) who did not receive vancomycin prophylaxis.

The decision to use vancomycin as neurosurgical prophylaxis should be based upon an assessment of risks and benefits for the individual patient.

Unfortunately, neither the risks nor the benefits of vancomycin prophylaxis in this setting have been adequately defined.

Orthopedic Implant Surgery

Although Bernakis (408) has recommended vancomycin prophylaxis in orthopedic surgery when a significant risk of methicillin-resistant staphylococcal infections is present, no randomized controlled trials of vancomycin prophylaxis for prosthetic implant surgery have been performed. Colwell et al. (409) noted mean vancomycin concentrations of 14.7, 3.78, and 5.5 μg/ml in samples of joint fluid, cortical bone, and cancellous bone, respectively, 30 minutes following a 1 g vancomycin infusion administered over 60 minutes to 6 patients undergoing joint arthroplasty surgery. None of 201 consecutive patients undergoing elective joint arthroplasty developed infectious complications following single-dose prophylaxis with 1 g vancomycin and 80 mg gentamicin preoperatively (410). The small risk of hematogenous seeding to preexisting prosthesis during dental procedures has prompted some authors to advocate prophylactic antibiotics before dental or abdominal surgery (411); however, such measures may not be cost effective for most patients (412).

Vascular Devices

Studies exploring the efficacy of vancomycin as prophylaxis against infection in patients with central lines have provided contradictory results. In a prospective randomized study, heparin-vancomycin flushes (10 units heparin/ml; 25 μg/ml of vancomycin) reduced the incidence of Gram-positive bacteremia in children with oncological or hematological disorders requiring tunneled venous catheters; six infections occurred in the 24 patients receiving heparin flushes alone compared with none in 21 patients receiving vancomycin-heparin flushes (413). Randomized studies have failed to demonstrate the efficacy of prophylactic vancomycin in the prevention of catheter infection in patients undergoing hyperalimentation catheter insertion (414) and in cancer patients undergoing chronic catheter placement (415,416). An additional study failed to demonstrate a reduction in the frequency of infection following vancomycin prophylaxis for placement of indwelling central lines in patients with the acquired immunodeficiency syndrome (417). Prophylactic vancomycin (5 mg/kg twice a day) appeared to reduce the incidence of coagulase-negative staphylococcal septicemia among a group of very low birth weight infants in a randomized placebo-controlled study (418); however, the potential benefit of any intervention must be balanced against the risk of enhancing vancomycin resistance (419).

Vancomycin has also been used to prevent infection of hemodialysis shunts. Vancomycin appeared to reduce the frequency of polytetrafluoroethylene graft infections in pediatric chronic hemodialysis patients (420). Morris and Bilinsky (421) administered 1 g vancomycin every 14 days to 25 patients with external arteriovenous shunts. During 20 patient-years experience, five infections occurred, all of which were due to Gram-negative bacilli.

Cardiovascular Surgery

The role of vancomycin in cardiovascular surgical prophylaxis is undefined. Kaiser (422) suggested that prophylactic vancomycin may be reasonable in institutions with a high prevalence of postoperative methicillin-resistant *S. aureus* infections. An outbreak of methicillin-resistant *S. epidermidis* wound infections, involving 11.9% of patients undergoing coronary artery bypass surgery at one institution, stopped following a change from cephalosporine to vancomycin prophylaxis (423). The pharmacokinetics of vancomycin may be altered in patients undergoing cardiac surgery. Following administration of 15 mg/kg of vancomycin to 13 males undergoing coronary bypass surgery, Austin et al. (424) noted serum concentrations of 26.4 μg/ml upon institution of bypass surgery, 15.4 μg/ml 30 minutes into surgery, 8.8 μg/ml at the end of surgery, and 1.6 μg/ml 12 h postdose. These findings suggest that an initial prophylactic vancomycin dose of 15 mg/kg may provide subtherapeutic concentrations 12 h following cardiopulmonary bypass. Farber et al. (425) recommended that adequate serum concentrations could be maintained if vancomycin is administered at a dose of 15 mg/kg body weight preoperatively followed by 10 mg/kg after initiation of bypass and 10 mg/kg every 8 h postoperately for 48 h. Klamerus et al. (426) reported an initial 4.0 μg/ml (16.8%) decrease in vancomycin concentrations upon initiation of cardiopulmonary bypass and a 2.3 μg/ml rebound increase when the aorta was unclamped. These changes may not be significant enough to justify modification of dosing recommendations made by Farber et al. (425). A prospective study (427) of topically applied vancomycin paste as prophylaxis for median sternotomy infection demonstrated efficacy: 1 of 223 patients (0.45%) receiving vancomycin topically became infected compared with 7 in the 193 control patients ($p = 0.02$). In a double-blind clinical study comparing penicillin G with vancomycin in 113 patients undergoing open-heart surgery, none of 61 vancomycin recipients developed wound infection compared with 5 of 52 patients receiving penicillin (428). In a double-blind placebo-controlled study of vancomycin versus placebo in arterial prosthetic reconstructive surgery (429), vancomycin reduced the incidence of postoperative wound infection: 1 of 62 patients (1.6%) receiving vancomycin developed a wound infection compared with 14 of 66 patients (21.2%) in the placebo group ($p < 0.0008$).

Endocarditis Prophylaxis

The American Heart Association (430) recommends that vancomycin be administered as a single dose of 1 g intravenously infused over 1 h in penicillin-allergic patients in two circumstances: (1) in "high-risk" patients (those with prosthetic cardiac valves, prior bacterial endocarditis, or surgically constructed systemic-pulmonary shunts or conduits) undergoing dental or oral procedures and (2) in combination with gentamicin for patients at risk undergoing genitourinary and/or gastrointestinal procedures. These recommendations are similar to those advanced by other expert groups (431–433). Vancomycin is not uniformly effective in preventing infection (434,435).

Immunosuppressed Patients

Studies addressing the efficacy of vancomycin in preventing Gram-positive infections in immunosuppressed patients have provided conflicting results. In a prospective randomized study (436) vancomycin was effective in a population of bone marrow transplant patients: 11 of 30 controls developed infection compared with none of 30 vancomycin recipients. In contrast, no difference in the incidence of Gram-positive bacteremias was found in another randomized study of 60 consecutive bone marrow transplant patients treated with prophylactic vancomycin beginning 5 days before to 1 day after transplantation (437). Differing dosing regimens may account for these divergent results. An uncontrolled study of 22 bone marrow transplant patients receiving prophylactic intravenous vancomycin demonstrated no episodes of Gram-positive coccal bacteremia, although 1 patient developed septicemia with a vancomycin-resistant *Capnocytophaga ochraceus* (438).

Rolando et al. (439) compared aztreonam and vancomycin to gentamicin and piperacillin as prophylaxis against infection in patients with fulminant hepatic failure. No difference in infectious mortality was noted between the two treatment groups.

D. Oral Vancomycin

Pseudomembranous Colitis (PMC)

Efficacy of Vancomycin. The efficacy of oral vancomycin in the treatment of pseudomembranous colitis has been well documented (9,440–445) and is not extensively reviewed here. Most patients improve within 24–48 h after beginning treatment. Symptoms generally resolve in 5 days (444). Oral solution or semisolid matrix capsules provide similar fecal, serum, and urine concentrations (8). A dosage of 125 mg four times per day is as effective as 500 mg four times per day (446). Intravenous vancomycin may not be effective and has actually been associated with the development of pseudomembranous colitis (see Refs. 712–714).

Alternative Agents. Randomized studies suggest that oral metronidazole (447) or bacitracin (448,449) has clinical efficacy similar to that of vancomycin; however, such small studies may not have been sufficiently large to detect a difference in efficacy between treatment groups (450). In two studies comparing oral vancomycin and bacitracin (448,449), the latter treatment was less effective than vancomycin in stool *C. difficile* eradication. Oral fusidic acid, 0.5 g daily (451), and oral teicoplanin (452) appeared to be effective treatment in uncontrolled studies.

Relapse. Relapse of PMC is not uncommon (453,454), generally occurring 4–21 days following completion of therapy (455) in 0% (9) to 53% (456) of patients initially responding to vancomycin. Vancomycin and metronidazole have similar relapse rates (447). In a large study, Bartlett (442) noted at least one relapse in 46 of 189 patients (24%) and a subsequent relapse in an additional 21 patients (11%). The wide range of relapse rates reported in the literature may be attributed to differences in the definition of relapse, study design, length and intensity of follow-up, patient population, and risk of reinfection with a different *C. difficile* isolate. Although some studies have suggested that relapse is related to a failure to eradicate *C. difficile* from the stool (455,456), eradication of *C. difficile* or elimination of toxin does not assure cure. Young et al. (457) noted clinical relapse in 10 of 35 patients (28.6%) who had eradication of *C. difficile* by culture and cytotoxin assay at the end of treatment. Relapse may be more common in the aged and those recovering from abdominal surgery (457). A longer duration of vancomycin therapy does not appear to reduce the incidence of relapse (442,455). Relapse has been attributed to residual survival of *C. difficile* spores or vegetative forms, residual change in bowel flora following vancomycin therapy, or reinfection by a new *C. difficile* strain (458).

Treatment of relapsing pseudomembranous colitis includes (1) retreatment with oral vancomycin (455), (2) changing to an alternative therapy (459), (3) tapering vancomycin dosage (460), and (4) combining vancomycin with colestipol (461), cholestyramine (462), *Saccharomyces boulardii* (463), or rifampin (464). The rationale of using combination therapy for relapses is unclear, since colestipol as a single agent is not effective (465), rifampin is not always synergistic with vancomycin against *C. difficile* (466), and cholestyramine may bind vancomycin (467). Reconstitution of gut flora using bacteriotherapy (468), oral administration of nontoxigenic *C. difficile* (469), or oral administration of lactobacillus GG (470) appears to be effective in uncontrolled studies. Some authors suggest that vancomycin may be administered rectally in patients with pseudomembranous colitis who cannot tolerate oral administration (see Rectal Administration, p. 342).

Prophylaxis for Pseudomembranous Colitis. In a prospective study, oral vancomycin appeared to prevent acquisition of *C. difficile* among a group of leukemic patients (471). In that report, administration of oral vancomycin to *C. difficile* fecal carriers, in conjunction with environmental decontamination measures, was followed by a decrease in the percentage of patients with positive cultures from 16.6 to 3.6%. It is difficult to attribute the beneficial effect to vancomycin alone. In another study, an outbreak of necrotizing enterocolitis associated with the isolation of *C. difficile* in the stool was terminated upon the institution of oral vancomycin therapy in cases and contacts; however, it is uncertain whether these cases represented pseudomembranous colitis (see Ref. 477).

The efficacy of vancomycin or metronidazole in the eradication of *C. difficile* among 30 asymptomatic hospitalized carriers was assessed in a randomized, placebo-controlled trial (472). Although vancomycin was effective in eradicating carriage in 9 of 10 fecal excretors, such eradication was only transient; 8 of the 9 successfully treated patients began to excrete *C. difficile* again. Many of the vancomycin-treated patients who relapsed were actually reinfected with new strains. Metronidazole was infrequently detectable in stool following oral administration, and *C. difficile* was not eradicated among carriers during or immediately after treatment. Therefore, vancomycin cannot be recommended as prophylaxis for asymptomatic carriers of *C. difficile*.

Staphylococcal Enterocolitis

Whether staphylococcal enterocolitis exists as a distinct clinical entity remains enigmatic. Wallace et al. (473) reported prompt recovery following administration of oral vancomycin, 500 mg every 6 h, in 7 staphylococcal enterocolitis patients. Kahn and Hall (474) reported no mortality among 45 *S. aureus* enterocolitis patients treated with vancomycin compared with 7% in 54 patients treated with other agents.

Enteric Infection

In a nonrandomized trial, Ng et al. (475) noted 1 episode of necrotizing enterocolitis in 84 very low birth weight babies receiving oral vancomycin for 48 h before first feeding compared with 17 episodes occurring in 121 infants not receiving such therapy. Routine use of vancomycin to prevent necrotizing enterocolitis in high-risk neonatal units should be tempered by the knowledge that some neonates receiving oral vancomycin may have measurable serum vancomycin concentrations (476). An outbreak of necrotizing enterocolitis was terminated upon the institution of oral vancomycin therapy in cases and contacts (477); however, many of the cases were associated with *C. difficile* in the stool, raising concern that these cases may have been examples of pseudomembranous colitis. One case report suggested that oral vancomycin may be effective in necrotizing enterocolitis occurring in an adult with leukemia (478). Vancomycin has also been suggested as adjunctive therapy in the treatment of ulcerative colitis (479–481).

Portosystemic Encephalopathy

In one crossover study of 12 patients suffering lactulose-refractory portal systemic encephalopathy, 2 g oral vancomycin daily was associated with clinical and laboratory improvement (482).

Prophylaxis in Neutropenic Patients

Oral vancomycin has been used as prophylaxis in granulocytopenic patients receiving chemotherapy for hematological malignancy (483); however, its utility when added to other agents is unclear (484,485). In two studies, a regimen containing vancomycin and polymyxin proved to less effective and less well tolerated than oral fluoroquinolone agents (486,487). A study (488) comparing vancomycin plus gentamicin and cotrimoxazole revealed the superiority of vancomycin plus gentamicin in decreasing the incidence of infection in neutropenic patients; however, 40% of patients receiving the vancomycin were intolerant to treatment. Another randomized trial (489) comparing the effectiveness of oral gentamicin-colistin-nystatin with or without oral vancomycin failed to demonstrate a difference between treatment groups. Gluckman et al. (490), using a randomized prospective study design in bone marrow transplant recipients, found ofloxacin-amoxicillin prophylaxis to be associated with fewer days of fever (9.2 ± 7.1 days) than vancomycin-tobramycin-colistin prophylaxis (13.7 ± 6.8 days). Classen et al. (491), however, noted an increased incidence of *Streptococcus mitis* sepsis in a population of bone marrow transplant patients following a change in antimicrobial prophylaxis from oral vancomycin-polymyxin-tobramycin to oral norfloxicin.

E. Other Routes of Administration

Intraperitoneal

Several studies have suggested that intraperitoneal administration of vancomycin alone (492,493) or in combination with other agents (360,494–499) is effective treatment of peritonitis in chronic ambulatory peritoneal dialysis patients. Two methods of administration are commonly used: a continuous method using a 30 mg/L concentration of vancomycin in each dialysis bag and an intermittent method using 30 mg/kg doses at weekly intervals (500). These regimens are equally effective (501,502), although the intermittent method may be more convenient. Several studies have suggested that intraperitoneal vancomycin is as effective as quinolones in the treatment of peritonitis in CAPD patients (503–506). Long-term, intraperitoneal vancomycin, 1 g weekly for 13–21 months, was reported to be effective prophylaxis in 3 patients with a history of recurrent peritonitis associated with chronic ambulatory peritoneal dialysis (507). The efficacy, safety, and potential for resistance during prolonged vancomycin prophylaxis remain undefined.

Rectal Administration

Several reports (508–511) have suggested that rectal administration of vancomycin may be useful in pseudomembranous colitis patients who cannot tolerate oral vancomycin. Goodpasture et al. (512) successfully treated a patient with ileus using 1 g vancomycin in 1000 ml diluent as a retention enema every 6 h. For pseudomembranous colitis patients who cannot tolerate oral vancomycin, Silva

(513) has recommended rectal administration of vancomycin 500 mg in 1 L saline solution every 8 h. The efficacy and safety of rectally administered vancomycin have not been objectively assessed in clinical trials, and such use should be considered investigational.

Ocular Use

Coagulase-negative staphylococci have been implicated as ocular pathogens with increasing frequency (514,515). The limited ocular penetration of vancomycin following intravenous use suggests a role for local therapy in the presence of ocular infection; however, the efficacy and safety of topical, intraocular, and subconjunctival routes of administration have not been assessed in well-controlled clinical trials.

Topical Use. Various vancomycin preparations have been used topically to treat keratitis and blepharoconjunctivitis. Several reports (516–519) have described the use of 50 mg/ml of topical vancomycin as treatment for keratitis, blepharitis, and blepharoconjunctivitis with variable results. In one report (517), the concentration of vancomycin required reduction to 5 mg/ml as a result of ocular irritation experienced at the higher vancomycin concentration. An additional patient (520) was successfully treated with 31 mg/ml of vancomycin in methylcellulose tears. Although vancomycin has been found to be stable in various artificial tear preparations (521), sterile water, and sterile saline (517) for at least 1 week, caution must be taken to assure the sterility, stability, physiological pH, osmolality, and viscosity of extemporaneously made formulations.

Vancomycin has also been used as a component in corneal storage medium as a method of preventing bacterial contamination during the storage of corneal grafts (522,523).

Subconjunctival Injection. Subconjunctival administration of 25 mg vancomycin has been advocated as a method for attaining therapeutic vancomycin concentrations in the vitreous fluids; however, such administration does not add significantly to the vancomycin concentrations attained following 1 mg intravitreal administration, and subconjunctival administration may be irritating (see Ref. 527).

Intraocular Injection. Since systemic antibiotics do not generally reach therapeutic concentrations in the vitreous humor, intravitreal injection of antibiotics has often been advocated in the treatment of endophthalmitis (524,525). The U.S. National Eye Institute is conducting a study to establish the role of vitrectomy and intravenous antibiotic therapy in the treatment of endophthalmitis. The antibiotic regimen used in this trial includes the intravitreal injection of 1 mg vancomycin and 0.4 mg amikacin sulfate in conjunction with the subconjunctival injection of 25 mg vancomycin and 100 mg ceftazidime in all treatment groups (526). Barza (527) recommends that the volume of intravitreal injections should not exceed 0.1 ml. Although the safety of intravitreal vancomycin has not been documented in clinical studies, animal studies using electroretinographic (528,

529) or histological (530) end points have failed to detect retinal toxicity following the intravitreal injection of 1 mg vancomycin. Toxicity has been documented, however, in rabbit eyes subjected to repeated injections of vancomycin in combination with an aminoglycoside (531). Vancomycin has also been used in ocular irrigation solutions following cataract surgery (532).

Less Common Intraocular Pathogens. Posttraumatic endophthalmitis is frequently associated with isolation of *Bacillus* species (533), which are usually susceptible in vitro to vancomycin (534). Hemady et al. (535) suggest that parenteral and intravitreal gentamicin in combination with either vancomycin or clindamycin are reasonable as prophylaxis against *Bacillus* species infection following penetrating ocular trauma. *Propionibacterium acnes*, a pathogen usually susceptible to vancomycin in vitro, is increasingly recognized as a cause of chronic endophthalmitis following extracapsular cataract extraction (536). Zambrano et al. (537), in a review of nine cases of *P. acnes* endophthalmitis, suggested intravitreal 1 mg vancomycin as initial treatment in mild cases.

Topical Oropharynx and Respiratory Tract

Neither intravenous (538) nor topical vancomycin (539,540) is effective in eradicating nasal *S. aureus* carriage. The clinical role of locally applied vancomycin in the treatment of oropharyngeal or dental disease has been the subject of several studies (541–546). Vancomycin oral paste was more effective than oral chlorhexidine mouthwashes in reducing the numbers of oral streptococci (547, 548). In one study, topical vancomycin appeared to decrease the incidence of dental caries in schoolchildren (549). The role of topical vancomycin in oral diseases, although undefined, is likely to be minimal.

Aerosolized vancomycin successfully eradicated airway colonization by MRSA in one pediatric patient (550), although such eradication may be transient (551). In one study, vancomycin aerosol (120 mg vancomycin in 1 ml sterile water, preceded by albuterol inhalation, administered every 6 h by face mask in room air) was used in conjunction with nasal drops (using the same vancomycin solution) successfully to eliminate MRSA carriage (552). This regimen was associated with a vancomycin serum concentration of 2.2 µg/ml 30 minutes following the treatment. A randomized study (553) found a decreased frequency of oropharyngeal or tracheal colonization in intubated neonates receiving instillation of 4 drops of a 5% vancomycin solution every 8 h. One placebo-controlled study (554) suggested that oropharyngeal decontamination with a regimen containing polymyxin B, neomycin sulfate, and vancomycin may be effective in reducing the incidence of ventilator-associated pneumonia.

Intraventricular and Intrathecal

Since vancomycin may not achieve adequate cerebrospinal fluid concentrations following intravenous administration (see Cerebrospinal Fluid Concentrations, p.

311), several studies have explored the intraventricular or intrathecal route of administration.

Pharmacokinetics of Intraventricular Vancomycin. The pharmacokinetics of vancomycin following intrathecal or intraventricular administration is variable, and therefore dosages require individualization. Recommended doses have ranged between 5 and 20 mg daily (279,290); however, these doses have resulted in highly variable cerebrospinal fluid vancomycin concentrations. Gaur et al. (555) noted mean peak and 24 h trough CSF vancomycin concentrations of 26.2 μg/ml (4.9–48.8 μg/ml) and 7.8 μg/ml (2.0–15.4 μg/ml), respectively, following intraventricular administration of 1–6 mg to a series of 11 children. Reesor et al. (556) noted the CSF vancomycin half-life to range from 9.3 to 20.5 h following intraventricular administration of 20 mg vancomycin (557). McGee et al. (558) reported ventricular fluid vancomycin concentrations ranging from 4.5 to 58.6 μg/ml in children receiving intraventricular doses from 1 to 45 mg daily. Hirsch et al. (559) noted first-order pharmacokinetics, a calculated half-life of 3.52 h, and an apparent volume of distribution of 60 ml following intraventricular vancomycin administration. Because of the variable pharmacokinetic behavior of vancomycin following intraventricular administration, measurement of cerebrospinal fluid concentrations is a desirable guide to dosage adjustment.

Monitoring of cerebrospinal fluid concentrations following vancomycin administration may be necessary to avoid drug accumulation following multiple doses (560). Congeni et al. (561) noted peak CSF vancomycin concentrations of 810 μg/ml following daily doses of 20 mg. Arrayo and Quindlen (562) noted accumulation of vancomycin following intraventricular administration of 10 mg daily for nine doses; the CSF concentration rose from 76 μg/ml after the first dose to 606 μg/ml after the ninth dose. Pau et al. (563) noted a 1 h postdose concentration as high as 144 μg/ml following administration of 5 mg intraventricular vancomycin every other day in a neonate. Following inadvertent intraventricular administration of 45 mg vancomycin to a hydrocephalic infant, a peak concentration of 800 μg/ml and a cerebrospinal fluid half-life of 77.7 h were noted (563). Bayston et al. (284) noted trough CSF vancomycin concentrations ranging from 5 to 236 μg/ml (median 26.8 μg/ml) following the intraventricular administration of 20 mg vancomycin to 25 patients with CSF shunts.

Clinical Use. Seufert et al. (564) recommend that drugs for intrathecal or intraventricular use be reconstituted with preservative-free sterile water for injection or 0.9% sodium chloride for injection and that the solution be filtered with a 0.5 μ filter needle to remove undissolved drug. Gump (279) also suggested that solutions for intraventricular administration be adjusted to physiological pH, although most clinical reports do not indicate that a pH adjustment was made. Desirable target CSF vancomycin concentrations have not been established but may differ from those recommended for serum. Bayston (565) found that clinical cure of CSF shunt infections was associated with relatively high CSF vancomycin concentrations of 100–300 µg/ml. McGee et al. (558), however, noted a mean ventricular fluid vancomycin concentration of 9.7 µg/ml among meningitis patients whose cerebrospinal fluid was sterilized.

Adverse Events. Adverse events directly attributed to intraventricular vancomycin are infrequently reported despite inadvertent administration of high doses (563). CSF pleocytosis (285), CSF eosinophilia (566), and severe headache (287) attributed to the local effect of vancomycin have been reported. Bayston (565) noted no adverse events attributable to intraventricular vancomycin administration even at CSF vancomycin concentrations as high as 280 µg/ml. Cogenti et al. (561) noted no toxicity in an infant who had peak CSF vancomycin concentrations of 810 µg/ml.

Bone and Joint

Because erratic bone concentrations of vancomycin have been noted in osteomyelitic bone following intravenous administration (302), other methods of administration have been studied. Direct infusion of antibiotics into the site of osteomyelitic bone using an infusion pump has been proposed for the purpose of achieving higher local antibiotic concentrations; in one report, however, vancomycin lost 38% of its activity over a 4 week period in an implantable drug pump (567).

Local application of vancomycin in polymethyl methacrylate-impregnated beads may provide higher local vancomycin concentrations. Antibiotic activity of vancomycin may persist for several days in vitro following incorporation into acrylic bone cement (568–571). Gerhart et al. (572) measured wound concentrations following implantation of vancomycin-impregnated bone cement in rats. At 14 days following implantation of vancomycin-impregnated polymethyl methacrylate and polypropylene fumarate-methyl methacrylate bone cement, wound vancomycin concentrations reached 4.8 ± 1.8 and 120 ± 44 µg/ml, respectively. Scott et al. (573) implanted vancomycin-impregnated beads into two patients with chronic osteomyelitis. Wound drainage concentrations were 58 µg/ml 24 h following implantation in one patient and 38 µg/ml 6 days following implantation in another patient. The clinical usefulness of local vancomycin therapy in osteomyelitis treatment remains to be defined.

III. VANCOMYCIN SAFETY

A. Infusion-Related Events

Red Man Syndrome

Clinical Spectrum. Patients receiving vancomycin may develop symptoms during or soon after completion of the infusion. Such symptoms, variously called red man syndrome (574), red neck syndrome (575), red person syndrome (576), and glycopeptide-induced anaphylactoid reaction (577), all appear to be mediated, at least in part, through the release of histamine. This reaction has been comprehensively reviewed by Polk (577). The manifestations of the red man syndrome are diverse but generally include some combination of findings, including flushing, hypotension, wheezing, dyspnea, urticaria, pruritus, angioedema, cutaneous erythema, and macular or maculopapular rash over the neck, chest, and upper extremities. Additional symptoms considered part of the syndrome include back pain (578,579), chest pain (580), seizures (581), and cardiovascular depression, including bradycardia, tachycardia, and cardiac arrest. Although symptoms may appear rapidly and resolve within minutes to hours, longer lasting reactions have been reported (582). The red neck syndrome has generally been reported following rapid infusions; however, it may occur during slow infusions (404,583–585) and following intraperitoneal administration (586). One report, which described the occurrence of red man syndrome following oral vancomycin administration, did not note similar symptoms following intravenous administration (587). Although the red man syndrome frequently occurs during or soon after infusion, the onset may be delayed (588). The red man syndrome has been associated with the use of generic (588) as well as brand name vancomycin preparations. One report (589) suggests that nifedipine may be associated with an increased likelihood of red man syndrome; however, the patient cited in this report received a rapid vancomycin infusion.

Incidence. The incidence of rash, as illustrated in a sample of studies listed in Table 1, varies greatly. The widely varying incidence rates among these studies is probably due to multiple factors as reviewed by Polk (577), including (1) the rate and concentration of vancomycin infusion, (2) study design, (3) concomitant illnesses (590) (renal insufficiency) (591), (4) race (591), (5) medications (see Ref. 603), and (6) the intensity of surveillance. Prospective pharmacokinetic studies have generally reported higher rates than those estimated by retrospective analysis (577). Symptoms compatible with red man syndrome have also been reported in patients not receiving vancomycin; in one study, 3 of 48 patients receiving ticarcillin-clavulanate, and amikacin had symptoms compatible with red man syndrome (321). Erythromycin was also associated with red neck syndrome in one report (592).

The red man syndrome also occurs in children. Levy et al. (593), in a retrospective review, noted red man syndrome in 11 of 650 children (1.6%) receiving vancomycin. Patients developing red man syndrome did not differ from control children with respect to dose administered or rate of infusion. Odio et al. (404) noted symptoms of red man syndrome in 35% of children receiving vancomycin prophylaxis before neurosurgery. Five children with vancomycin-associated rash subsequently tolerated vancomycin administered at a lower rate. Reactions have also been reported in preterm infants (594).

Table 1 Incidence of Histamine Release Reactions

Study	Year	Incidence No.	%	Study type[a]	Comments
Dangerfield et al. (721)	1959	1/85	1.2	R	59% of patients received drug over 60–90 minutes
Newfield and Roizen (606)	1979	11/76[b]	14.5	R	1 g in 10 ml over 10 minutes
Schaad et al. (63)	1980	4/6	66.6	P	Infusion over >30 minutes
Schaad et al. (63)	1980	0/49	0.0	P	Infusion >30 minutes
Farber and Moellering (720)	1983	3/100	3.0	R	
Odio et al. (404)	1984	7/20	35.0	P	15 mg/kg over 60 minutes
Schifter et al. (779)	1985	8/202[c]	4.0	P	1 g over 20 minutes
Sorrell et al. (155)	1985	4/54	7.4	P	
Healy et al. (110)	1987	9/11	81.8	P	Healthy volunteers, 1 g infused over 60 minutes
Healy et al. (110)	1987	0/11	0.0	P	Health volunteers, 500 mg infused over 60 minutes
Shenep et al. (321)	1988	3/53	5.7	P	>5 mg/ml infused in 1 h
Smith et al. (652)	1989	4/35	11.4	P	1 g over 60 minutes
Comstock et al. (591)	1989	3/10	30.0	P	Renal failure patients, 60 minute infusion
Levy et al. (593)	1990	11/650	1.7	R	Pediatric study
Healy et al. (597)	1990	8/10	80.0	P	1 g infused over 1 h
Healy et al. (597)	1990	3/10	30.0	P	1 g infused over 2 h
Wallace et al. (603)	1991	8/17	47.1	P	1 g over 60 minutes
Levine et al. (251)	1991	1/42	2.4	P	Endocarditis patients
Rybak et al. (590)	1992	0/15	0.0	P	Endocarditis and bacteremia patients

[a]P = prospective; R = retrospective.
[b]Hypotension.
[c]Denominator is number of vancomycin infusions.

Pathophysiology. Histamine release appears to be the major pathogenetic mechanism for these reactions. Levy et al. (595) initially reported elevated plasma histamine concentrations in 2 patients who developed hypotension after vancomycin infusion. Subsequently, Polk et al. (596) compared 1 h infusions of 500 mg every 6 h for 5 doses or 1000 mg every 12 h for 3 doses in 11 volunteers using a crossover design. The red man syndrome developed in 9 of 11 volunteers who received 1000 mg doses and in none of those who received 500 mg doses. Furthermore, plasma histamine concentrations increased in most subjects given the 1000 mg doses but were only slightly increased following administration of the 500 mg dose. The increase in plasma histamine concentrations was correlated with the severity of the reactions. Although the severity of reactions is generally decreased with subsequent doses, patients may experience more severe reactions following the first dose (see Ref. 603). The relationship between rate of vancomycin administration and the incidence of the red man syndrome was further characterized in 10 volunteers by Healy et al. (597), who noted the syndrome in 8 of 10 volunteers receiving 1 g vancomycin over 1 h compared with only 3 of 10 volunteers receiving the same dose over 2 h. The 1 h infusion rate was associated with significantly greater peak plasma histamine concentrations.

In addition to the rate of administration, the concentration of the vancomycin infusion may have an effect on the amount of histamine released. Verberg et al. (598) measured serum histamine concentrations following infusion of 33.3 mg/minute of vancomycin at concentrations of either 20 or 5 μg/ml. After a 30 minute infusion, volunteers receiving the 20 μg/ml infusion had a higher mean plasma histamine concentration (1007 $\pm$ 259 pg/ml) than those receiving the 5 μg/ml infusion (391 $\pm$ 45 pg/ml). Symptoms were more frequently observed in the volunteers receiving the 20 μg/ml vancomycin infusion.

Unlike vancomycin, teicoplanin is not associated with histamine release following infusion. Sahai et al. (599) reported symptoms of red man syndrome in 11 of 12 volunteers receiving vancomycin but in none of 12 volunteers receiving teicoplanin. There have been reports suggesting that symptoms reminiscent of the red man may rarely occur in patients receiving teicoplanin (395,396).

Not all infusion-related events are mediated by histamine release (600). Some rashes with vancomycin may be precipitated by immunological mechanisms (601).

Treatment and Prevention. Antihistamines may prevent or reduce the severity of histamine release reactions. In a double-blind study, Sahai et al. (602) found hydroxyzine 50 mg superior to ranitidine 300 mg in preventing symptoms of red man syndrome following vancomycin infusion. Wallace et al. (603) compared the incidence of histamine release reactions in patients receiving 50 mg diphenhydramine versus placebo before administration of vancomycin. Following the administration of 1 g vancomycin over 60 minutes, 8 of 17 patients (48%) receiving placebo had reactions compared to none of 16 patients receiving diphenhydramine. Individual case reports (583,604) have suggested that diphenhydramine may be effective in reversing vancomycin-induced hypotension.

Identification of patients who might require antihistamine pretreatment is uncertain. Polk et al. were unable to use vancomycin skin tests as a predictor of anaphylactoid reactions (605).

Hypotension

Although hypotension often accompanies rash as a component of the red man syndrome, it may occur without rash. Most reported episodes of hypotension followed the rapid infusion of vancomycin. In a retrospective study, Newfield and Roizen (606) noted hypotension in 11 of 76 patients receiving vancomycin in 10 ml over 10 minutes. None of 100 patients who subsequently received 1 g vancomycin over 30 minutes developed hypotension. Waters and Rosenberg (607) reported a case of hypotension following administration of 1000 mg vancomycin in 250 ml over 30 minutes. Laconture et al. (608) described 2 newborn infants who developed hypotension following administration of 15 mg/kg over 20 minutes. Maternal hypotension following vancomycin has been associated with fetal distress in one case report (609).

The mechanism of hypotension experienced following vancomycin is uncertain but may be related to exaggerated histamine release (577) or an increased sensitivity to the hemodynamic effects of histamine. In an animal study, vancomycin appeared to have a direct myocardial depressant effect (610); however, the relevance of this finding to hypotension in the human is uncertain. Stier et al. (611) measured hemodynamic variables (cardiac index, heart rate, blood pressure, pulmonary venous pressures, and systemic and pulmonary vascular resistance) and histamine concentrations in 16 critically ill patients receiving vancomycin after open-heart surgery. In only 1 patient did plasma histamine concentrations increase over baseline values. No change in any hemodynamic variables were detected during infusion of 1 g vancomycin over 30 minutes.

Hypotension has been reported following the rapid administration of vancomycin during the perioperative period (612–616; see also Refs. 623 and 630). Slight et al. (403) reported that 22.7% of children who received an intravenous dose in the operating room became hypotensive, and another 13.6% had flushing, urticaria, or pruritus for an overall rate of 36.4%. The incidence was decreased by providing the first dose on the ward and reducing the dose from 20 to 15 mg/kg. Odio et al. (404), however, noted hypotension following infusion of 15 mg/kg of vancomycin infused over 1 h in an infant. In a prospective study of vancomycin prophylaxis in 116 patients undergoing cardiac surgery, 26.7% of adults and 20.0% of children had infusion-related events; however, patients received infusions over an abbreviated 30 minute period (617).

Studies exploring the hemodynamic effects of vancomycin infused during the perioperative period have provided conflicting results. Fineberg and Lamantia (618) studied the hemodynamic effects of vancomycin during cardiac surgery in 8 patients. Following the preanesthetic administration of vancomycin, 0.5 mg/kg/minute (35 mg/minute) over 20–30 minutes, no changes in hemodynamic parameters could be documented. Another prospective double-blind study involving 36 patients undergoing orthopedic surgery (619) failed to detect a significant effect of 1 g vancomycin infused over 30–60 minutes upon systolic blood pressure or heart rate. Stable hemodynamic variables were noted in a series of 16 patients given vancomycin within 24 h following cardiovascular surgery (611). In contrast, Romanelli et al. (620) prospectively monitored hemodynamic variables in 20 patients undergoing cardiac surgery randomized to vancomycin 1 g versus placebo preoperatively. Patients receiving vancomycin demonstrated significantly lower systemic vascular resistance, increased heart rate and systolic arterial pressure, and increased norepinephrine use during and after surgery than control patients.

Anaphylactic and Anaphylactoid Reactions

Both anaphylactic (621) and anaphylactoid (622–627) reactions have been described in patients receiving vancomycin with or without concomitant anesthesia. Most reported reactions appeared to be associated with short infusion periods. Anaphylactoid, as opposed to anaphylactic reactions, are not mediated by immunological mechanisms (577).

Cardiac Arrest

Cardiac arrest and bradycardia have been reported in patients receiving vancomycin. Most, but not all cases (631) have occurred following rapid vancomycin administration. Mayhew and Deutsch (628) reported a 2-year-old child who sustained cardiac arrest following central intravenous administration of vancomycin during the perioperative period. Glicklich and Figura (629) reported cardiac arrest following a bolus of 1 g vancomycin administered over 2 minutes. Dajee et al. (630) reported a similar case of cardiac arrest following rapid infusion of 500 mg IV vancomycin over 1 minute. Wade and Mueller (631) reported a case of bradycardia in a patient receiving a slow infusion of vancomycin (500 mg intravenously over 2 h).

Phlebitis

The incidence of phlebitis has varied widely among reports in the medical literature. In an early report, Dangerfield et al. (see Ref. 721) noted the occurrence of thrombophlebitis in 38 of 85 patients (44.7%) receiving intravenous vancomycin. In a retrospective study of 100 vancomycin courses, Farber and Moellering (see Ref. 720) noted a phlebitis incidence of only 13%. Recent prospective studies have documented a high incidence of phlebitis associated with vancomycin infusion. Garrelts et al. (632) noted phlebitis in 32 of 35 peripheral intravenous sites (91.4%) despite infusion of 1 g doses of vancomycin over 1 h through peripheral Teflon catheters. Sorrell and Collignon (633) noted phlebitis in 20 of 54 patients (37.0%). Even chromatographically purified preparations are not free of phlebitis potential; Wang et al. (634) encountered phlebitis in 4 of 14 patients (28.6%) receiving a chromatographically purified preparation. Extravasation of vancomycin, which can result in tissue irritation and injury, is potentially preventable and may respond to local measures (635).

B. Rash and Allergic Reactions

Various cutaneous manifestations, in addition to the cutaneous reactions associated with the red man syndrome have been associated with vancomycin administration; these reactions include maculopapular rash (see Ref. 720 and 777), cutaneous vasculitis (636,637), toxic epidermal necrolysis (638–640), linear IgA bullous dermatosis (641,642), Stevens-Johnson syndrome (643,644), and exfoliative dermatitis (645–647). Patients with renal insufficiency may have prolonged reaction duration as a result of persistent serum vancomycin concentrations (644, 646). Vancomycin allergy may occasionally present as a prolonged fever (648).

Several reports suggest that patients experiencing reactions to vancomycin may be safely given teicoplanin (649–652); however, at least one other report suggests that potential for cross-reactivity (653). One report documented the presence of vancomycin- and teicoplanin-induced IgE-mediated histamine release from basophils isolated from a patient with type 1 allergy to vancomycin (654). Two patients who developed rash and pruritus while receiving vancomycin have been successfully "desensitized" to vancomycin (655,656). Although one of these cases was described as having had an allergic reaction to vancomycin, one cannot exclude nonimmunologically mediated reactions in either patient.

C. Ototoxicity

Incidence

The incidence of ototoxicity associated with vancomycin varies considerably among studies. The studies, some of which are listed in Table 2, use varying methods of detection (clinical, audiometric, or evoked responses) and differing methods of case finding (prospective or retrospective). Sorrell and Collignon (633) noted audiometric evidence of ototoxicity in 1 of 11 vancomycin recipients, although none of 34 patients had clinical evidence of hearing impairment. Mellor et al. (see Ref. 682) noted transient tinnitus and dizziness in 2 of 34 prospectively monitored vancomycin recipients, all but 6 of whom were receiving concomitant aminoglycoside therapy.

Table 2 Incidence of Vancomycin-Associated Ototoxicity in Several Clinical Studies

Study	Year	n/d[a]	%	Type[b]	Comments
Dutton and Elmes (140)	1959	5/9	55.6	R	5 patients on erythromycin, 3 on streptomycin
Morris and Bilinsky (421)	1971	0/25	0.0	R	
Hook and Johnson (212)	1978	3/10	30.0	P	Series of audiometric studies
Sorrell et al. (155)	1982	1/19	5.3	R	1 patient had transient tinnitus
Cafferkey et al. (154)	1982	0/21	0.0	R	
Farber and Moellering (720)	1983	1/100	1.0	R	1 patient had transient tinnitus
Sorrell and Collignon (633)	1985	1/11	9.1	P	Series of audiometric studies finding
Mellor et al. (682)	1985	2/34	5.9	P	Transient tinnitus and dizziness

[a]Number of cases/number of vancomycin courses.
[b]Case ascertainment: R = retrospective; P = prospective.

Evidence from Animal Studies

Animal toxicity studies suggest that vancomycin by itself has a low potential for ototoxicity. Tange et al. (657) were unable to detect functional or histological evidence of cochlear damage after administration of vancomycin, 80 mg/kg/day for 14 days, in gerbils. Similarly, Brummett (658) was unable to detect abnormalities in guinea pig cochlear morphology and electrophysiology after administration of 1000 mg/kg of vancomycin and ethacrynic acid. Lutz et al. were unable to detect significant cochlear hair cell loss in guinea pigs receiving vancomycin in doses up to 300 mg/kg for 11–17 days (659). Vancomycin, however, appears to enhance the ototoxicity of gentamicin in a guinea pig (660) and a rat (661) model.

Association Between Vancomycin and Ototoxicity

Although ototoxicity has been described in patients receiving vancomycin, recent reviews (662,663) suggest that such toxicity (hearing loss, vertigo, tinnitus, and abnormality of audiogram or evoked response) may be less common than previously believed. Brummett and Fox (662) suggested that vancomycin-associated ototoxicity may occur predominantly in patients also receiving other ototoxic agents, such as aminoglycosides. Table 3 presents a listing of cases derived from the review of Brummett and Fox (662) and Bailie and Neal (663). Among the 33 reports of vancomycin-associated ototoxicity listed, concomitant or preceding ototoxic antibiotics were also administered in 17 patients (51.5%). Of the 33 cases reviewed 14 were reversible. Although many of the patients with vancomycin-associated ototoxicity have had renal impairment, some reports suggest that ototoxicity may also occur in patients with normal renal function (664,665).

Serum Levels and Ototoxicity

As indicated in Table 3, no consistent relationship is apparent between high serum vancomycin concentrations and ototoxicity; however, transiently elevated vancomycin concentrations occurring during infusion may not be detected if serum concentrations are measured following completion of the infusion (666). Hekster et al. (see Ref. 773) reported normal brain stem auditory evoked potentials in a patient who had an overdose with maximum vancomycin concentrations of 121 µg/ml.

Detection of Ototoxicity

More sensitive methods for the detection of ototoxicity, such as audiology or evoked potentials, have been advocated as a method for early detection of vancomycin ototoxicity; however, the effectiveness of such measures in preventing toxicity are unproven. In one study of 44 patients receiving prolonged vancomycin or aminoglycoside therapy, the combination of routine twice weekly audiological threshold monitoring, close surveillance of renal function, and frequent review of serum concentrations did not detect any patient with occult ototoxicity (667). In a study of burn patients undergoing periodic auditory brain stem response studies, Hall et al. (668) noted a higher mean vancomycin dosage in patients demonstrating auditory impairment than a group of 7 patients without auditory deficit; however, confounding factors preclude assignment of causality. Combination vancomycin-gentamicin was found to be associated with increased brain stem auditory evoked response failure in low-birth-weight neonates (669). In a prospective trial of ultrahigh-frequency audiometric screening performed in 13 CAPD patients receiving 15 courses of vancomycin therapy, only 1 patient developed asymptomatic 15 dB decrements at two frequencies, indicating a low incidence of such abnormalities in this patient population (670). In a study of 126 neonates, Sharma et al. (671) did not detect significant abnormalities by auditory brain stem evoked responses in 15 patients with elevated peak or trough vancomycin concentrations. Abnormalities detected by audiogram may not be specific; pure-tone auditory threshold changes of at least 15 dB may occur in normal subjects (672). The role of routine monitoring of auditory function in patients receiving vancomycin is of unproven benefit; however, periodic monitoring may be justifiable in patients receiving prolonged courses of vancomycin, in patients receiving concomitant ototoxic drugs, or in patients with previously impaired auditory or renal function.

D. Nephrotoxicity

Manifestations

Nephrotoxicity associated with vancomycin use has been reviewed by Appel et al. (673) and Bailie and Neal (663). Clinical manifestations of nephrotoxicity may include azotemia as well as changes in urinary sediment (hematuria, proteinuria, and casts) (674). Acute interstitial nephritis, associated with rash, eosinophilia, eosinophiliuria, and acute onset of renal impairment, has been described (675–677). In one case, a renal biopsy revealed a granulomatous process (678).

Incidence

Although early clinical reports suggested a relatively high frequency of nephrotoxicity associated with vancomycin (679,680), the incidence of nephrotoxicity in more recent clinical series has been estimated to be in the range of 1.5–15% (see Table 4) when studied prospectively (633,681–685) or retrospectively (686; see also Ref. 720). The higher incidence of nephrotoxicity noted in earlier studies has been attributed to impurities present in earlier vancomycin preparations (see Ref. 720).

Effect of Concomitant Aminoglycoside

Animal studies (687–690) suggest that vancomycin-aminoglycoside combinations are associated with a higher potential for nephrotoxicity. Some clinical studies suggest that the incidence of nephrotoxicity is much higher in patients receiving vancomycin-aminoglycoside combinations than in patients receiving vancomycin alone (633,681,682,685,691–697; see also Ref. 720); however, other studies suggest that synergistic or additive toxicity does not exist (683,686, 698–700).

Factors Associated with Nephrotoxicity

In addition to the effect of aminoglycosides, the incidence of vancomycin-associated nephrotoxicity may be increased among patients with prolonged treatment (681,691,694), elevated peak serum concentrations (694), trough concentrations over 10 μg/ml (681,686,691,694), advanced age (694), neutropenia (694), liver disease (694), peritonitis (694), concurrent amphotericin B (694) or furosemide (685), elevated baseline serum creatinine (686), and preexisting renal impairment (685). Even short courses of vancomycin have been associated with serum creatinine elevations. Gundmundsson and Jensen (see Ref. 782) noted that 14 of 101 patients receiving vancomycin surgical prophylaxis developed creatinine

Table 3 Reports of Ototoxicity Associated with Vancomycin

Investigators	Year	Symptoms and signs	Other ototoxic medications	Renal impairment	Serum levels	Reversible
Geraci et al. (245)	1958	Tinnitus and deafness	None	Yes	3 h = 80 μg/ml 6 h = 95 μg/ml	No
Geraci et al. (245)	1958	Profound deafness	None	Unknown	Unknown	Unknown
Dutton and Elmes (140)	1959	Hearing loss	Erythromycin, streptomycin	Yes	Unknown	Unknown
Dutton and Elmes (140)	1959	Hearing loss	Erythromycin, streptomycin	Yes	Unknown	Unknown
Dutton and Elmes (140)	1959	Hearing loss	Erythromycin, streptomycin	Yes	Unknown	Unknown
Dutton and Elmes (140)	1959	Hearing loss	Erythromycin	Yes	Unknown	Unknown
Dutton and Elmes (140)	1959	Hearing loss	Erythromycin	Yes	Unknown	Unknown
Kirby et al. (780)	1959	Tinnitus	Unknown	Unknown	Unknown	Yes
Kirby et al. (780)	1959	Tinnitus	Unknown	Unknown	Unknown	Yes
Dangerfield et al. (721)	1960	Tinnitus[a]	None	Yes	Unknown	Yes
Dangerfield et al. (721)	1960	Hearing loss	None	Yes	Unknown	Unknown
Dangerfield et al. (721)	1960	Hearing loss	Kanamycin	No	Unknown	Unknown
Kirby et al. (142,780)	1960	Hearing loss	Neomycin	Unknown	Unknown	No
Weisbren et al. (680)	1960	Hearing loss[a]	Erythromycin	Yes	Unknown	Yes
Woodley and Hull (141)	1961	Hearing loss 2 months after vancomycin	None	No	Unknown	Unknown
Louria et al. (139)	1961	Hearing loss	Streptomycin, novobiocin	No	Unknown	Unknown
Morris and Bilinsky (421)	1971	Hearing loss	Unknown	Yes	Unknown	Yes
Hook and Johnson (212)	1978	Slight decrease in hearing[a] above 4000 Hz	Streptomycin	Yes	P = 25 μg/ml	No

Reference	Year	Finding	Concurrent drug		Level	
Hook and Johnson (212)	1978	Slight decrease in hearing above 4000 Hz	Streptomycin	Yes	P = 25 µg/ml	No
Hook and Johnson (212)	1978	Slight decrease in hearing above 4000 Hz	None	Yes	P = 25 µg/ml	Yes
Guerit et al. (781)	1981	Altered auditory evoked potential	Unknown	Unknown	Unknown	Unknown
Traber and Levine (664)	1981	Tinnitus, hearing loss[a]	None	No	P = 49.2 µg/ml, T = 32.0 µg/ml	Yes
Sorrell et al. (155)	1982	Tinnitus	Unknown	Unknown	Unknown	Yes
Ahmad et al. (90)	1982	Dizziness, tinnitus	No	Yes	P = 37 µg/ml	Yes
Farber and Moellering (720)	1983	Vertigo[a]	Unknown	Unknown	Unknown	Yes
Mellor et al. (665)	1984	Tinnitus, vertigo[a]	Gentamicin	No	P < 25 µg/ml, T = 17.3 µg/ml	Yes
Mellor et al. (665)	1984	Tinnitus, deafness, vertigo[a]	Amikacin	No	Unknown	Yes
Mellor (665)	1984	Hearing loss	Gentamicin	No	Unknown	Unknown
Sorrell and Collignon (633)	1985	Audiometric change unilateral	Gentamicin	Unknown	P = 51.3 µg/ml, T = 17.9 µg/ml	Yes
Mellor et al. (682)	1985	Tinnitus, dizziness[a]	None	No	P = 25 µg/ml, T = 19 µg/ml	Yes
Mellor et al. (682)	1985	Tinnitus, dizziness[a]	Amikacin	No	Unknown	Yes
Mellor et al. (682)	1985	No symptoms, audiometric changes	None	No	Unknown	Unknown
Mellor et al. (682)	1985	No symptoms, audiometric changes	Gentamicin	No	Unknown	Unknown
Hall et al. (666)	1985	Middle latency evoked recording	Overdose	Yes	Unknown	Unknown

[a]Transient.

P = peak: T = trough.

Table 4 Incidence of Renal Insufficiency

Study	Year	Criteria for nephrotoxicity[1]	Group[a]	Incidence vancomycin alone		Incidence vancomycin and aminoglycoside		Type[c]
				n/d[b]	%	n/d[b]	%	
Schaad et al. (59)	1980	NS[d]	P	0/55	0.0	ND	ND	P
Farber and Moellering (720)	1981	>0.5 mg/dl	U	3/60	5.0	12/34	35.3	R
Glew et al. (692)	1983	>0.7 mg/dl	U	1/14	7.1	5/19	26.3	R
Sorrell and Collignon (633)	1985	50% increased creatinine	U	0/25	0.0	4/29	14.3[e]	P
Mellor et al. (665)	1985	0.5 mg/dl	U	1/12	8.3	2/27	7.4	P
Dean et al. (696)	1985	≥0.5 mg/dl	P	2/19	10.5	2/9	22.2[f]	R
Cimino et al. (686)	1987	0.5 mg/dl	O	6/41	14.6	6/40	14.6[g]	R
Swinney and Rudd (699)	1987	0.3 mg/dl	P	ND	ND	0/10	0.0	R
Nahata (700)	1987	≥0.2 mg/dl	P	ND	ND	0/90	0.0	P
Cohen et al. (683)	1988	≥0.5 mg/dl	U	1/66	1.5	8/108	7.4[h]	P
Shenep et al. (321)	1988	≥0.3 mg/dl	P	ND	ND	0/53	0.0	P
Downs et al. (685)	1989	44 μmol/L	E	7/54	13.0	4/12	33.0	P
Gudmundsson and Jensen (782)	1989	40 μmol/L	A'	14/101	14.0	ND	ND	P

Eng et al. (684)	1989	≥0.3 mg/dl	A	4/23	17.4	ND	ND	P
Jaresko et al. (695)	1989	≥0.5 mg/dl	M	ND	ND	18/48	37.5[1]	R
Rybak et al. (681)	1990	0.5 mg/dl	U	8/163	4.8	14/63	22.2[k]	P
Pauly et al. (694)	1990	≥0.5 mg/dl	A	ND	ND	28/105	26.7	R
Kureishi et al. (386)	1991	>1.1 mg/dl[P]	A	ND	ND	10/25	4.0[l]	P
Calandra (327)	1991	>45 μmol/L	P	ND	ND	24/383	6.3[m]	P

[a]Patient groups: U = undefined; P = pediatrics; E = elderly; A = adults; M = adults and children; O = oncology patients.
[b]Number of patients with nephrotoxicity/number of patients at risk.
[c]Type of study: P = prospective; R = retrospective.
[d]Not stated.
[e]Of patients receiving aminoglycosides without vancomycin 14% demonstrated nephrotoxicity.
[f]Less than 1% of patients receiving aminoglycosides without vancomycin demonstrated nephrotoxicity.
[g]Of 148 patients receiving aminoglycosides without vancomycin 18% demonstrated nephrotoxicity.
[h]Of 64 patients receiving aminoglycosides without vancomycin 6% demonstrated nephrotoxicity.
[i]Refers to the increment in serum creatinine concentration.
[j]Eleven patients received aminoglycosides for less than 2 days.
[j]Of 131 patients receiving aminoglycosides without vancomycin 11% had nephrotoxicity.
[k]Of 103 patients receiving aminoglycosides without vancomycin 11% demonstrated nephrotoxicity.
[l]Many patients also receiving cyclosporine and/or amphotericin B.
[m]Of 370 patients receiving aminoglycosides without vancomycin 2% had nephrotoxicity.
[P]Any creatinine greater than 1.1 mg/dl for males or greater than 1.0 for females.

increases in excess of 40 mmol/L within the first 5 days following surgery, compared with only 2 of 99 patients receiving placebo. In 10 of the 14 cases the creatinine elevations were transient.

E. Sterile Peritonitis

Sterile peritonitis has been reported following the intraperitoneal administration of vancomycin. Some reports (701–706) but not others (707,708) have suggested that the risk of sterile peritonitis may differ among vancomycin products of various manufacturers. Piriano et al. (701) reported sterile peritonitis after a change from Lilly vancomycin to vancomycin manufactured by Lederle. In that report, none of 15 patients previously receiving intraperitoneal Vancocin (Lilly) developed sterile peritonitis compared with 6 of 7 patients receiving intraperitoneal Vancoled (Lederle). The two brands of vancomycin were of comparable pH (5.1) in dialysis fluid but may have had different degrees of chromatographic purity. Two reports (707,708), however, failed to note the occurrence of peritonitis following the use of Vancoled. Munro (702) described a similar occurrence of sterile peritonitis in 5 patients following a change from Vancocin (Lilly) to a generic vancomycin (Bull). Charney and Gouge (703) described 3 cases of sterile peritonitis following a change from Vancocin (Lilly) to generic vancomycin (Abbott), potentially related to differences in formulation. Newland et al. (705) reported 3 sterile peritonitis cases following intraperitoneal administration of Vancoled, although the same product was used previously without events in 12 patients. Lot-to-lot variation was suggested as a possible explanation for these findings (709). Dubot et al. (706) described 5 cases of peritonitis following intraperitoneal administration of a generic vancomycin; no further episodes occurred after a change to a "purified" form of vancomycin. The U.S. Food and Drug Administration, after a reviewing 51 spontaneous adverse reports of sterile peritonitis following intraperitoneal vancomycin use, issued a letter to dialysis units noting disproportionate numbers of reports following use of products manufactured by some vancomycin manufacturers (710). In addition to the question of purity, Johnson (711) has suggested that the intraperitoneal administration of large does may be associated with peritonitis in some patients.

F. Gastrointestinal Events

Gastrointestinal symptoms, such as nausea, vomiting, and abdominal pain, have been reported in patients receiving vancomycin. Hepatic enzyme and bilirubin elevations (682) have been attributed to vancomycin administration. Shenep et al. (321) noted twofold increases in serum alanine aminotransferase (ALT) in 11 of 53 patients (20.8%) randomized to vancomycin-ticarcillin-amikacin compared with 3 of 48 patients (6.3%) randomized to ticarcillin-clavulanate-amikacin ($p = 0.04$). Similarly, Calandra et al. (327) noted twofold elevations in serum ALT and aspartate aminotransferase (AST) in 85 of 383 patients (22%) randomized to vancomycin-ceftazidime-amikacin compared with 50 of 370 patients (13.5%) randomized to ceftazidime-amikacin ($p = 0.003$). *C. difficile* colitis has been noted following the intravenous administration of vancomycin alone (712,713) or in combination with other agents (714,715).

G. Hematological Toxicity

Erythrocytes

Vancomycin may induce erythrocyte aggregation in vitro at concentrations in excess of 3.0 mg/ml (716). Although these concentrations are higher than is usually achieved in serum, spontaneous aggregation of erythrocytes has been noted in blood samples drawn from intravenous tubing previously used for vancomycin infusion (717). Such aggregation reactions may cause confusion in the evaluation of suspected transfusion reactions (718). Hemolysis has also been described in a patient receiving vancomycin for treatment of *S. aureus* septicemia (719).

Leukocytes

Leukopenia may develop in 2–5% of patients receiving vancomycin (720–722). However, in one study (722), 7 of 42 patients (17%) receiving vancomycin over 14 days developed neutropenia. Morris and Ward (723), in a retrospective study of 50 vancomycin courses, noted the development of neutropenia $|<1000 \times 10(6)|$/L in 4 patients (8%) after a median duration of 22 days of therapy. Reports (724–727) suggest that these reactions are reversible and generally do not become evident until after 14 days of vancomycin administration. Neftel et al. (722) failed to find an association between vancomycin serum concentrations and the incidence of neutropenia. The duration of neutropenia may be prolonged in patients with renal failure (728), presumably because of slow vancomycin elimination. Occasional patients (729,730) may develop agranulocytosis (neutrophil count less than 500/mm^3). Rash (731,732) and fever (733) occasionally accompany the leukopenia. The mechanism for leukopenia is unknown, although antineutrophil antibodies (734–736) and positive lymphocyte transformation studies (737) suggest a possible immunological mechanism. Vancomycin does not appear to directly inhibit human bone marrow colony formation in vitro (738). Patients receiving concomitant vancomycin and azidothymidine may be at an increased risk of developing neutropenia (739).

Platelets and Coagulation

Thrombocytopenia (740,741) and refractoriness to platelet transfusions (742) have been reported in association with vancomycin administration. The thrombocytopenia may be mediated through an immunological mechanism, since vancomycin-dependent antibodies (742) have been detected. Vancomycin appears less likely than cefamandole (743) or cefazolin to prolong prothrombin time in postoperative patients receiving warfarin (744). Heparin, at the high concentrations that may be reached in intravenous lines, has been reported markedly to reduce vancomycin activity, with resulting clinical failure (745). However, dilute solutions of vancomycin and heparin retain antibacterial and anticoagulant activities (746).

H. Miscellaneous Reactions

Vancomycin has been associated with several additional events, including mononeuritis multiplex (747), increased lacrimation (748), and prolonged neuromuscular blockade in combination with vecuronium (749). Conley et al. (750) described a case of severe rigors that developed after a dose of a vancomycin preparation from one manufacturer (Vancoled) but not following administration of a vancomycin preparation from another manufacturer (Vancocin). Caglayan et al. described a patient with hypertension, transient blindness, and generalized seizure associated with vancomycin infusion (751).

Physical incompatibilities between vancomycin and ticarcillin (752) or ceftriaxone (753) have been described. Vancomycin is compatible with methotrexate (754) and with parenteral nutrition solutions (755–757) stored for up to 8 days under refrigeration. The stability of vancomycin in total parenteral nutrition solutions raises the possibility that vancomycin may be used as prophylaxis or treatment of hyperalimentation line infections (758). Concurrent administration of ceftazidime does not affect the pharmacokinetics of vancomycin (759). One report suggests that vancomycin may be safely administered to patients with acute porphyria (760).

I. Safety in Pregnancy

Although vancomycin should be used with caution in pregnant patients (761,762), no study has demonstrated abnormalities in the fetus or newborn after vancomycin administration during gestation. Individual case reports showing no harmful effects (763,764) cannot totally exclude fetal risk, however. Transplacental passage of vancomycin has been confirmed (765). Reyes et al. (766) compared auditory brain stem response and renal functional tests in 10 infants born to mothers who received vancomycin during the second or third trimester of pregnancy and 20 infants born to 20 control mothers. Only 1 infant exposed to vancomycin in utero demonstrated an abnormal auditory brain stem response attributed to a conductive hearing loss. No abnormalities in serum creatinines were noted in the newborn infants exposed to vancomycin, although serum creatinine measurements may not reflect renal function in newborn infants (767).

The pharmacokinetics of vancomycin may be altered during pregnancy. In one report (768), a pregnant patient required a daily dose of 57 mg/kg to achieve adequate peak vancomycin concentrations. This increased dose has been related to the increased creatinine clearance and volume of distribution that commonly occur during pregnancy. An increased volume of distribution and increased plasma clearance were confirmed in a pregnant woman receiving vancomycin during the second trimester (765). Fetal distress due to vancomycin-induced maternal hypotension has also been described (609).

J. Overdosage and Excessive Serum Concentrations

Inadvertent vancomycin overdose has been described in several literature reports. Walczyk et al. (769) described a patient who received 56 g vancomycin over a 10 day period, with subsequent renal insufficiency and serum concentrations reaching 284 µg/ml. In this case, continuous arteriovenous hemofiltration proved useful in decreasing the serum vancomycin concentration. Burkhart et al. (770,771) noted serum concentrations of 427 µg/ml following administration of 720 mg/day for 2 days in a 47-day-old neonate. Although a 1.5 volume exchange transfusion over 3.25 h initiated 32 h after the last dose failed measurably to decrease serum vancomycin concentrations, use of 1 g/kg of activated charcoal enterally every 4 h was accompanied by a decrease in vancomycin half-life from 35 to 12 h. Davis et al. (772), however, were unable to detect any effect of orally administered activated charcoal on the pharmacokinetics of intravenously administered vancomycin in healthy volunteers. Hekster et al. (773) reported a 3.5-year-old CAPD patient who received six loading doses of 185 mg vancomycin, resulting in plasma concentrations of 121 µg/ml. In this patient, an increase in the apparent CAPD flow by shortening the dwell time appeared to increase vancomycin clearance. Schaad et al. (59) described a patient who received two vancomycin doses in error, with resulting serum concentrations of 92.5 µg/ml. No evidence of ototoxicity was detected by brain stem auditory evoked potentials in either of these patients (59,773).

High vancomycin concentrations have also been reported in a patient who received an appropriate vancomycin dose. In one report (774), a patient with IgA myeloma was noted to have extremely high vancomycin concentrations (464 µg/ml), presumably because of abnormal serum protein binding.

K. Safety of Oral Vancomycin

Although enteral vancomycin is poorly absorbed, some patients with renal insufficiency (775) and colitis (10,776) may achieve measurable serum concentrations: in one case serum concentrations of 34 µg/ml were documented following oral vancomycin therapy (11). In addition to adverse events that may result from oral absorption (777), changes in oropharyngeal and bowel flora may occur. Cholestyramine, an anion-exchange resin occasionally used in the treatment of *C. difficile*-associated colitis, binds to both vancomycin and teicoplanin (778). Symptoms compatible with red man syndrome following oral administration of vancomycin have been described; however, this patient did not have similar symptoms following intravenous administration (587).

IV. THE FUTURE

After 35 years of use, vancomycin remains an important component of the clinical antibiotic armamentarium against Gram-positive coccal infections. Few antibiotics, after more than three decades of use, continue to provide the unique properties of expanding clinical use and a low incidence of bacterial resistance. The need for vancomycin or agents with a similar spectrum will continue to grow. Agents with (1) efficacy superior to vancomycin, (2) in vitro activity against vancomycin- and teicoplanin-resistant enterococci and staphylococci, and (3) an improved safety profile are needed. Until such an agent becomes available and clinical experience confirms its utility, vancomycin will continue to be a valuable agent for the treatment of Gram-positive coccal infections.

REFERENCES

1. Cooper G.L., Given D.B. (1986). Vancomycin: A Comprehensive Review of 30 Years Clinical Experience. Park Row Publishers.
2. Levin J.F. (1987). Vancomycin: A review. Med. Clin. North Am. 71(6):1135–1145.
3. Ingerman M.J., Santoro J. (1989). Vancomycin: A new old agent. Infect. Dis. Clin. North Am. 3(3):1989.
4. Hermans P.E., Wilhelm M.P. (1987). Vancomycin. Mayo Clin. Proc. 62:901–905.
5. Matzke G.R., Zhanel G.G., Guay D.R.P. (1986). Clinical pharmacokinetics of vancomycin. Clin. Pharmacokinet. 11:257–282.
6. Moellering R.C. (1984). Pharmacokinetics of vancomycin. J. Antimicrob. Chemother. 14(Suppl.):43.
7. Griffith R.S., Peck F.B. (1955). Vancomycin, a new antibiotic. III. Preliminary clinical and laboratory studies. Antibiot. Annu. 1955–1956:619–622.
8. Lucas R.A., Bowtle W.J., Ryder R. (1987). Disposition of vancomycin in healthy volunteers from oral solution and semi-solid matrix capsules. J. Clin. Pharm. Ther. 12:27–31.
9. Tedesco F., Markham R., Gurwith M., Christie D., Bartlett J.G. (1978). Oral vancomycin for antibiotic-associated pseudomembranous colitis. Lancet 2(8083):226–228.
10. Matzke G.R., Halstenson C.E., Olson P.L., Collins A.J., Abraham P.A. (1987). Systemic absorption of oral vancomycin in patients with renal insufficiency and antibiotic-associated colitis. Am. J. Kidney Dis. 9(5):422–425.
11. Thompson C.M. Jr., Long S.S., Gilligan P.H., Prebis J.W. (1983). Absorption of oral vancomycin—possible associated toxicity. Int. J. Pediatr. Nephrol. 4(1):1–4.
12. Dudley M.N., Quintiliani R., Nightingale C.H., Gontraz N. (1984). Absorption of vancomycin. Ann. Intern. Med. 101:144.
13. Halstenson C.E., Collins A.J., Olson P.L., Matzke G.R. (1985). Systemic absorption of oral vancomycin in hemodialysis patients with antibiotic associated colitis. Kidney Int. 27:163.
14. Vancocin HCL package product label (1991). Eli Lilly and Company, Indianapolis, Indiana.

15. Rubin J. (1990). Vancomycin absorption from the peritoneal cavity during dialysis-related peritonitis. Peritoneal Dial. Int. 10(4):283–285.
16. Rogge M.C., Johnson C.A., Zimmerman S.W., Welling P.G. (1985). Vancomycin disposition during continuous ambulatory peritoneal dialysis (CAPD): A pharmacokinetic analysis of peritoneal drug transport. Antimicrob. Agents Chemother. 27: 578–582.
17. Bastani B., Spyker D.A., Westervelt F.B. Jr. (1988). Peritoneal absorption of vancomycin during and after resolution of peritonitis in continuous ambulatory peritoneal dialysis (CAPD) patients. Peritoneal Dialysis Int. 8:135–136.
18. Neal D., Bailie G.R. (1990). Clearance from dialysate and equilibration of intraperitoneal vancomycin in continuous ambulatory peritoneal dialysis. Clin. Pharmacokinet. 18(6):485–490.
19. MacIlwaine W.A., et al. (1974). Penetration of anti-staphylococcal antibiotics into the human eye. Am. J. Ophthalmol. 77:589.
20. Nolan C.M., Flanigan W.J., Rastogi S.P., Brewer T.E. (1980). Vancomycin penetration into CSF during treatment of patients receiving hemodialysis. South Med. J. 73(10):1333–1334.
21. Redfield D.C., et al. (1980). Cerebrospinal fluid penetration of vancomycin in bacterial meningitis. In Nelson J.D., Grassi C. (Eds.), Proceedings of the 11th International Congress on Chemotherapy and the 19th Interscience Conference on Antimicrobial Agents and Chemotherapy, Vol. 1, pp. 630–640.
22. Barois A., Estournet B., Moranne J.B., Piliot J., Chabenat C., Bataille J. (1986). Infections ventriculaires a staphylocoque. Traitement par la vancomycine en perfusion veineuse continue. Presse Med. 15(36):1805–1808.
23. Levy R.M., Gutin P.H., Baskin D.S., Pons V.G. (1986). Vancomycin penetration of a brain abscess: Case report and review of the literature. Neurosurgery 18(5):632–636.
24. Lindholm D.D., Murray J.S. (1966). Persistence of vancomycin in the blood during renal failure and its treatment by hemodialysis. N. Engl. J. Med. 274:1047–1051.
25. Albrecht L.M., Rybak M.J., Warbasse L.H., Edwards D.J. (1991). Vancomycin protein binding in patients with infections caused by *Staphylococcus aureus*. DICP Ann. Pharmacother. 25:713–715.
26. Tan C.C., Lee H.S., Ti T.Y., Lee E.J.C. (1990). Pharmacokinetics of intravenous vancomycin in patients with end-stage renal failure. Ther. Drug Monit. 12:29–34.
27. Golper T.A., Noonan H.M., Elzinga L., Gilbert D., Brummett R., Anderson J.L., Bennett W.M. (1988). Vancomycin pharmacokinetics, renal handling and nonrenal clearances in normal human subjects. Clin. Pharmacol. Ther. 43:565–570.
28. Macias W.L., Mueller B.A., Scarim S.K. (1991). Vancomycin pharmacokinetics in acute renal failure: Preservation of nonrenal clearance. Clin. Pharmacol. Ther. 50: 688–694.
29. Geraci J.E., Heilman F.R., Nichols D.R., Wellman W.E., Ross G.T. (1956). Some laboratory and clinical experiences with a new antibiotic, vancomycin. Mayo Clin. Proc. 31:564–582.
30. Brown N., Ho D.H.W., Fong K.L., Bogerd L., Maksymiuk A., Bolivar R., Fainstein V., Bodey G.P. (1983). Effects of hepatic function on vancomycin clinical pharmacology. Antimicrob. Agents Chemother. 23(4):603.

31. Cocchetto D.M., Tschanz C., Bjornsson T.D. (1983). Decreased rate of creatinine production in patients with hepatic disease: Implications for estimation of creatinine clearance. Ther. Drug Monit. 5:161–168.
32. Pachorek R.E., Wood F. (1991). Vancomycin half-life in a patient with hepatic and renal dysfunction. Clin. Pharm. 10:297–300.
33. Pryka R.D., Rodvold K.A., Erdman S.M. (1991). An updated comparison of drug dosing methods. Part IV. Vancomycin. Clin. Pharmacokinet. 20(6):463–476.
34. Rotschafer J.C., Crossley K., Zaske D.E., Mead K., Sawchuk R.J., Solem L.D. (1982). Pharmacokinetics of vancomycin: Observations in 28 patients and dosage recommendations. Antimicrob. Agents Chemother. 22(3):391–394.
35. Nielsen H.E., Hansen H.E., Korsager B., Skov P.E. (1975). Renal excretion of vancomycin in kidney disease. Acta Med. Scand. 197:261–264.
36. Moellering R.C., Krogstad D.H., Greenblatt D.J. (1981). Vancomycin therapy in patients with impaired renal function: A nomogram for dosage. Ann. Intern. Med. 94:343–346.
37. Brown D.L., Mauro L.S. (1988). Vancomycin dosing chart for use in patients with renal impairment. Am. J. Kidney Dis. 11(1):15–19.
38. Lake K.D., Peterson C.D. (1985). A simplified dosing method for initiating vancomycin therapy. Pharmacotherapy 5(6):340–344.
39. Matzke G.R., McGory R.W., Halstenson C.E., Keane W.F. (1984). Pharmacokinetics of vancomycin in patients with various degrees of renal function. Antimicrob. Agents Chemother. 25:433–437.
40. Rodvold K.A., Blum R.A., Fischer J.H., Zokufa H.Z., Rotschafer J.C., Crossley K.B., Riff L.J. (1988). Vancomycin pharmacokinetics in patients with various degrees of renal function. Antimicrob. Agents Chemother. 32(6):848–852.
41. Garaud J., Regnier B., Inglebert F., Faurisson F., Bauchet J., Vachon F. (1984). Vancomycin pharmacokinetics in critically ill patients. J. Antimicrob. Chemother. 14(Suppl. Z).
42. Soto J., Sacristan J.A., Alsar M.J. (1991). Necessity of a loading dose when using vancomycin in critical ill patients. J. Antimicrob. Chemother. 27:875.
43. Reed R.L., Wu A.H., Miller-Crotchett P., Crotchett J., Fischer R.P. (1989). Pharmacokinetic monitoring of nephrotoxic antibiotics in surgical intensive care patients. J. Trauma 29(11):1462–1470.
44. Alazia M., Roux T., Lacarelle B., Ducoureau J.P., Durand A., Francois G. (1991). Bayesian prediction of vancomycin serum concentrations in critically ill patients. Interscience Conference on Antimicrobial Agents and Chemotherapy, Abstract 1296.
45. Brater D.C., Bawdon R.E., Anderson S.A., Purdue G.F. (1986). Vancomycin elimination in patients with burn injury. Clin. Pharmacol. Ther. 39:631–634.
46. Garrelts J.C., Peterie J.D. (1988). Altered vancomycin dose vs. serum concentration relationship in burn patients. Clin. Pharmacol. Ther. 44(1):9–13.
47. Rybak M.J., Albrecht L.M., Berman J.R., Warbasse L.H., Svensson C.K. (1990). Vancomycin pharmacokinetics in burn patients and intravenous drug abusers. Antimicrob. Agents Chemother. 34(5):792–795.
48. Musa D.M., Bauly D.J. (1987). Evaluation of a new vancomycin dosing method. Pharmacotherapy 7(3):69–72.

49. Ackerman B.H. (1989). Evaluation of three methods for determining initial vancomycin doses. DICP Ann. Pharmacother. 23:123–127.
50. Lake K.D., Peterson C.D. (1988). Evaluation of a method for initiating vancomycin therapy: Experience in 205 patients. Pharmacotherapy 8:284–286.
51. Zokufa H.Z., Rodvold K.A., Blum R.A., Riff L.J., Fischer J.H., Crossley K.B., Rotschafer J.C. (1989). Simulation of vancomycin peak and trough concentrations using five dosing methods in 37 patients. Pharmacotherapy 9:10–16.
52. Rybak M.J., Boike S.C. (1986). Monitoring vancomycin therapy. Drug Intell. Clin. Pharm. 20:757–761.
53. Birk J.K., Chandler M.H. (1990). Using clinical data to determine vancomycin dosing parameters. Ther. Drug Monit. 12(2):206–209.
54. Garrelts J.C., Godley P.J., Horton M.W., Karboski J.A. (1987). Accuracy of Bayesian, Sawchuck-Zaske, and nomogram dosing methods for vancomycin. Clin. Pharm. 6:795–799.
55. Krogstad D.J., Moellering R.C., Greenblatt D.J. (1980). Single-dose kinetics of intravenous vancomycin. J. Clin. Pharmacol. 20(4):197–201.
56. Ackerman B.H., Olsen K.M., Padilla C.B. (1990). Errors in assuming a one-compartment model for vancomycin (letter). Ther. Drug Monit. 12(3):304–305.
57. Rodvold K.A., Pryka R.D., Garrison M., Rotschafer J.C. (1989). Evaluation of a two compartment Bayesian forecasting program for predicting vancomycin concentrations. Ther. Drug Monit. 11:269–275.
58. Hurst A.K., Yoshinaga M.A., Mitani G.H., Foo K.A., Jelliffe R.W., Harrison E.C. (1990). Application of a Bayesian method to monitor and adjust vancomycin dosage regimens. Antimicrob. Agents Chemother. 34(6):1165–1171.
59. Schaad U.B., Nelson J.D., McCracken H. (1981). Pharmacology and efficacy of vancomycin for staphylococcal infections in children. Rev. Infect. Dis. 3(Suppl): S282–287.
60. Papp C.M., Nahata M.C. (1990). Clinical pharmacokinetics of antibacterial drugs in neonates. Clin. Pharmacokinet. 19(4):280–318.
61. Kaplan E.L. (1984). Vancomycin in infants and children: A review of pharmacology and indications for therapy and prophylaxis. J. Antimicrob. Chemother. 14(Suppl D):59–66.
62. Schaible D.H., Rocci M.L., Alpert G.A., Campos J.M., Paul M.H., Polin R.A., Plotkin S.A. (1986). Vancomycin pharmacokinetics in infants: Relationships to indices of maturation. Pediatr. Infect. Dis. 5:304–308.
63. Schaad U.B., McCracken G.H., Nelson J.D. (1980). Clinical pharmacology and efficacy of vancomycin in pediatric patients. J. Pediatr. 96(1):119–126.
64. Alpert G., Campos J.M., Harris M.C., Preblud S.R., Plotkin A. (1984). Vancomycin dosage in paeidatics reconsidered. Am. J. Dis. Child. 138:20–22.
65. Mercier J.C., Bingen E., Lambert-Zechowsky N., Proux M.C., Salord F., Beaufils F. (1984). Assessment of serum vancomycin levels in newborns and children with severe staphylococcal infections. 24th Interscience Conference on Antimicrob. Agents Chemother. Abstract 494.
66. Gross J.R., Kaplan S.L., Kramer W.G., Mason E.O. (1985). Vancomycin pharmacokinetics in premature infants. Paediatr. Pharmacol. 5:17–22.
67. Naqvi S.H., Reichley R.M., Keenan W.J., Fortune K.P. (1986). Re-evaluation of

vancomycin pharmacokinetics in premature infants using high pressure liquid chromatography (HPLC). Clin. Res. 32:804A.

68.   Reed M.D., Kleigman R.M., Weiner J.S., Huang M., Yamashita T.S., Blumer J.L. (1987). The clinical pharmacology of vancomycin in seriously ill preterm infants. Pediatr. Res. 22:360.

69.   James A., Koren G., Milliken J., Soldin S., Prober C. (1987). Vancomycin pharmacokinetics and dose recommendations for preterm infants. Antimicrob. Agents Chemother. 31:52–54.

70.   Koren G., James A. (1987). Vancomycin dosing in preterm infants: Prospective verification of new recommendations. J. Pediatr. 110(5):797–798.

71.   Kildoo C.W., Lin L., Gabriel M.H., Folli H.L., Modanlou H. (1990). Vancomycin pharmacokinetics in infants: Relationship to postconceptional age and serum creatinine. Dev. Pharmacol. Ther. 14:77–83.

72.   Gabriel M.H., Kildoo C.W., Cennrich J.L., Modanlou H.D., Collins S.R. (1991). Prospective evaluation of a vancomycin dosage guideline for neonates. Clin. Pharm. 10:129–132.

73.   Lisby-Sutch S.M., Nahata M.C. (1988). Dosage guidelines for the use of vancomycin based on its pharmacokinetics in infants. Eur. J. Clin. Pharmacol. 35: 637–642.

74.   Leonard M.B., Koren G., Stevenson D.K., Prober C.G. (1989). Vancomycin pharmacokinetics in very low birth weight neonates. Pediatr. Infect. Dis. J. 8: 282–286.

75.   Hoif E.B., Swigart S.A., Leuschen M.P., Willett L.D., Bolam D.L., Goodrich P.D., Bussey M.E., Nelson R.J. (1990). Vancomycin pharmacokinetics in infants undergoing extracorporeal membrane oxygenation. Clin. Pharm. 9:711–715.

76.   Cutler N.R., Narang P.K., Lesko L.J., Nimos M., Power M. (1984). Vancomycin disposition: The importance of age. Clin. Pharmacol. Ther. 36:803–810.

77.   Vance-Bryan K., Guay D., Gilliland S., Rodvold K., Rotschafer J. (1991). Characterization of vancomycin (V) pharmacokinetic (PK) parameters in young (Y) and elderly (E) patients with normal renal function using a Baysian [sic] forecasting technique. Pharmacotherapy 11(3):274 (No. 67).

78.   Blouin R.A., Bauer L.A., Miller D.D., Record K.E., Griffen W.O. (1982). Vancomycin pharmacokinetics in normal and morbidly obese subjects. Antimicrob. Agents Chemother. 21(4):575–580.

79.   Rotschafer J., Bance-Bryan K., Gilliland S., Rodvold K., Guay D. (1991). Effect of obesity on vancomycin (V) pharmacokinetic (PK) parameters as determined using a Bayesian forecasting technique. Pharmacotherapy 11(3):274 (No. 65).

80.   Cunha B.A., Quintiliani R., Deglin J.M., Izard M.W., Nightingale C.H. (1981). Pharmacokinetics of vancomycin in anuria. Rev. Infect. Dis. 3(Suppl Nov.–Dec.): S269–S272.

81.   Lam F.Y., Lindner A., Plorde J., Blair A., Cutler R.E. (1981). Pharmacokinetics of vancomycin in chronic renal failure. Kidney Int. 19:152.

82.   Eykyn S., Phillips I., Evans J. (1970). Vancomycin for staphylococcal shunt site infections in patients on regular haemodialysis. Br. Med. J. 3:80–84.

83.   Bierman M.H., Needham-Walker C.A., Hammeke M., Egan J.D. (1980). Vanco-

mycin therapy for serious staphylococcal infections in chronic hemodialysis patients. J. Dial. 4:179–184.

84. Bastani B., Spyker D.A., Minoch A., Cummings R., Westervelt F.B. (1988). In vivo comparison of three different hemodialysis membranes for vancomycin clearance: Cuprophan, cellulose acetate, and polyacrylonitrile. Dial. Transplant. 17(10): 527–543.

85. Barth R.H., DeVincenzo N., Zara A.C., Berlyne G.M. (1990). Vancomycin pharmacokinetics in high-flux hemodialysis. Clin. Res. 38:A339.

86. Quale J.M., OHalloran J.J., DeVincenzo N., Barth R.H. (1992). Removal of vancomycin by high-flux hemodialysis membranes. Antimicrob. Agents Chemother. 36:1424–1426.

87. Torras J., Cao C., Rivas M.C., Cano M., Fernandez E., Montoliu J. (1991). Pharmacokinetics of vancomycin in patients undergoing hemodialysis with polyacrylonitrile. Clin. Nephrol. 36(1):35–41.

88. Lanese D.M., Alfrey P.S., Molitoris B.A. (1989). Markedly increased clearance of vancomycin during hemodialysis using polysulfone dialyzers. Kidney Int. 35:1409–1412.

89. Bohler J., Reetze-Bonorden P., Keller E., Kramer A., Schollmeyer P.J. (1992). Rebound of plasma vancomycin levels after haemodialysis with highly permeable membranes. Eur. J. Clin. Pharmacol. 42:635–640.

90. Ahmad R., Raichura N., Kilbane V. (1982). Vancomycin: A reappraisal. Br. Med. J. 284:1953–1954.

91. Matzke G.R., O'Connell M.B., Collins A.J., Keshaviah P.R. (1986). Disposition of vancomycin during hemofiltration. Clin. Pharmacol. Ther. 40:425–430.

92. Golper T.A. (1985). Continuous arteriovenous hemofiltration in acute renal failure. Am. J. Kidney Dis. 6(6):373–380.

93. Rawer, P., Seim, K.E. (1989). Elimination of vancomycin during hemofiltration. Eur. J. Clin. Microbiol. Infect. Dis. 8:529–531.

94. Dupuis R.E., Matzke G.R., Maddux R.W., O'Neil M.G. (1989). Vancomycin disposition during continuous arteriovenous hemofiltration. Clin. Pharm. 8(5):371–374.

95. Golper T.A., Pulliam J., Bennett W.M. (1985). Removal of therapeutic drugs by continuous arteriovenous hemofiltration. Arch. Intern. Med. 145:1651–1652.

96. Bellomo R., Ernest D., Parkin G., Boyce N. (1990). Clearance of vancomycin during continuous arteriovenous hemodiafiltration. Crit. Care Med. 18(2):181–183.

97. Kox W.J., Davies S.P. (1992). Continuous hemodialysis in the critically ill. Care Crit. Ill 8:8–11.

98. Talbert R.L., Brockway B.A., Ludden T.M. (1983). Vancomycin (V) clearance during chronic ambulatory peritoneal dialysis. Drug Intell. Clin. Pharm. 17:442–443.

99. Harford A.M., Sica D.A., Tartaglione T., Polk R.E., Dalton H.P., et al. (1986). Vancomycin pharmacokinetics in continuous ambulatory peritoneal dialysis patients with peritonitis. Nephron 43:217–222.

100. Whitby M., Edwards R., Aston E., Finch R.G. (1987). Pharmacokinetics of single dose intravenous vancomycin in CAPD patients. J. Antimicrob. Chemother. 19: 351–357.

101. Bunke C.M., Aronoff G.R., Brier M.E., Sloan R.S., Luft F.C. (1983). Vancomycin

kinetics during continuous ambulatory peritoneal dialysis. Clin. Pharm. Ther. 34: 631–637.

102. Blevins R.D., Halstenson C.E., Salem N.G., Matzke G.R. (1984). Pharmacokinetics of vancomycin in patients undergoing continuous ambulatory peritoneal dialysis. Antimicrob. Agents Chemother. 25:603–609.

103. Magera B.E., Arroyo J.C., Rosansky S.J., Postic B. (1983). Vancomycin pharmacokinetics in patients with peritonitis on peritoneal dialysis. Antimicrob. Agents Chemother. 23:710–714.

104. Glew R.H., Pavuk R.A., Shuster A., Alfred H.G. (1982). Vancomycin pharmacokinetics in patients undergoing chronic intermittent peritoneal dialysis. Int. J. Clin. Pharmacol. Ther. Toxicol. 20:559–563.

105. Edwards D.J., Pancorbo S. (1987). Routine monitoring of serum vancomycin concentrations: Waiting for proof of its value. Clin. Pharm. 6:652–653.

106. Rodvold K.A., Zokufa H., Rotschafer J.C. (1987). Routine monitoring of serum vancomycin concentrations: Can waiting be justified? Clin. Pharm. 6:655–658.

107. Blaser J., Statzler A., Luthy R. (1987). Optimal scheduling of vancomycin: Should high peak concentrations be avoided? Abstract 438. Interscience Conference on Antimicrobial Agents and Chemotherapy, 1987.

108. Ebert S., Leggett J., Bogelman B. (1987). In vivo cidal activity and pharmacokinetic parameters (PKP's) for vancomycin (VAN) against methicillin-susceptible and methicillin-resistant (MRSA) S. aureus. Abstract 439. Interscience Conference on Antimicrobial Agents and Chemotherapy, 1987.

109. Fitzsimmons W.E., Postelnick M.J., Tortorice P.V. (1988). Survey of vancomycin monitoring guidelines in Illinois hospitals. Drug Intell. Clin. Pharm. 22:598–600.

110. Healy D.P., Polk R.E., Garson M.L., Rock D.T., Comstock T.J. (1987). Comparison of steady-state pharmacokinetics of two dosage regimens of vancomycin in normal volunteers. Antimicrob. Agents Chemother. 31(3):393–397.

111. Walker C.A., Kopp B. (1978). Sensitive bioassay for vancomycin. Antimicrob. Agents Chemother. 13(1):30–33.

112. Rotschafer J., Mead K., Chern M., Crossley K. (1980). Evaluation of a radioimmunoassay for measurements of serum vancomycin concentration. Curr. Chemother. Infect. Dis. 1:521–523.

113. Fong K.L., Ho D.W., Bogerd L., Pan T., Broan N.S., Gentry L., Bodey G.P. (1981). Sensitive radioimmunoassay for vancomycin. Antimicrob. Agents Chemother. 19(1):139–143.

114. Schwenzer K.S., Wang C.J., Anhalt J.P. (1983). Automated fluorescence polarization immunoassay for monitoring vancomycin. Ther. Drug Monit. 5:341–345.

115. Filburn B.H., Shull V.H., Tempera Y.M., Dick J.D. (1983). Evaluation of an automated fluorescence polarization immunoassay for vancomycin. Antimicrob. Agents Chemother. 24(3):216–220.

116. Uhl J.R., Ahnalt J.P. (1979). High performance liquid chromatographic assay of vancomycin in serum. Ther. Drug Monit. 1:75–83.

117. Hoagland R.J., Sherwin J.E., Phillips J.M. (1984). Vancomycin: A rapid HPLC assay for a potent antibiotic. J. Anal. Toxicol. 8:75–77.

118. Sztaricskai F., Borda J., Puskas M.M., Bognar R. (1983). High performance liquid

chromatography (HPLC) of antibiotics of vancomycin type. J. Antibiot. 36(12): 1691–1698.

119. Hosotsubo H. (1989). Rapid and specific method for the determination of vancomycin in plasma by high-performance liquid chromatography on an aminopropyl column. J. Chromatogr. 487:421–427.

120. Greene S.V., Abdalla T., Morgan S.L. (1987). High-performance liquid chromatographic analysis of vancomycin in plasma, bone, arterial appendage tissue and pericardial fluid. J. Chromatogr. 417:121–128.

121. Bauchet J., Pussard E., Garaud (1987). Determination of vancomycin in serum and tissues by column liquid chromatography using solid-phase extraction. J. Chromatogr. 414:472–476.

122. Anne L., Hammad N., Chang C.C., et al. (1988). Development of Syva EMIT assay for the measurement of vancomycin in serum (abstract). Clin. Chem. 34:1256.

123. Crossley K.B., Rotschafer J.C., Chern M.M., Mead K.E., Zaske D.E. (1980). Comparison of a radioimmunoassay and a microbiological assay for measurement of serum vancomycin concentrations. Antimicrob. Agents Chemother. 17(4):654–657.

124. Pohold D.J., Saravolatz L.D., Somerville M.M. (1984). Comparison of fluorescence polarization immunoassay and bioassay of vancomycin. Antimicrob. Agents Chemother. 20(2):159–161.

125. Ackerman B.H., Berg H.G., Strate R.G., Rotschafer J.C. (1983). Comparison of radioimmunoassay and fluorescent polarization immunoassay for quantitative determination of vancomycin concentrations in serum. J. Clin. Microbiol. 18:994–995.

126. Ristuccia P.A., Ristuccia A.M., Bidanset J.H., Cunha B.A. (1984). Comparison of bioassay, high-performance liquid chromatography, and fluorescence polarization immunoassay for quantitative determination of vancomycin in serum. Ther. Drug Monit. 6:238–242.

127. Pfaller M.A., Krogstad D.J., Branich G.G., Murray P.R. (1984). Laboratory evaluation of five assay methods for vancomycin: Bioassay, high-pressure liquid chromatography, fluorescence polarization immunoassay, radioimmunoassay, and fluorescence immunoassay. J. Clin. Microbiol. 20(3):311–316.

128. Morse G.D., Nairn D.K., Bertine J.S., Walshe J.J. (1987). Overestimation of vancomycin concentrations utilizing fluorescence polarization immunoassay in patients on peritoneal dialysis. Ther. Drug Monit. 9:212–215.

129. Hu M.W., Anne L., Forni T., Gottwald K. (1990). Measurement of vancomycin in renally impaired patient samples using a new high-performance liquid chromatography method with vitamin B_{12} internal standard: Comparison of high-performance liquid chromatography, EMIT, and flurorescence polarization immunoassay methods. Ther. Drug Monit. 12:562–569.

130. Yeo K., Traverse W., Horowitz G.L. (1989). Clinical performance of the EMIT vancomycin assay. Clin. Chem. 35(7):1504–1507.

131. White L.O., Edwards R., Hold H.A., Lovering A.M., Finch R.G., Reeves D.S. (1988). The in-vitro degradation at 37 degrees C of vancomycin in serum, CAPD fluid and phosphate-buffered saline. J. Antimicrob. Chemother. 22:739–745.

132. Anne L., Hu M., Chan K., Colin L., Gottwald K. (1989). Potential problem with fluorescence polarization immunoassay cross-reactivity to vancomycin degradation

product CDP-1: Its detection in serum of renally impaired patients. Ther. Drug Monit. 11:585–591.

133. Hortin G.L., Landt M., Wilhite T., Smith C.H. (1991). Cross-reactivity of a vancomycin degradation product in three immunoassays of vancomycin. Clin. Chem. 37:2009.

134. Abbdelhady E., Mortensen J.E. (1991). The bactericidal activity of ampicillin, daptomycin, and vancomycin against ampicillin-resistant *Enterococcus faecium.* Diagn. Microbiol. Infect. Dis. 14:141–145.

135. Johnson A.P., Uttley A.H., Woodford N., George R.C. (1990). Resistance to vancomycin and teicoplanin: An emerging clinical problem. Clin. Microbiol. Rev. 3(3):280–291.

136. Vedel G., Leruez M., Lemann F., Hraoui E., Ratovohery D. (1990). Prevalence of *Staphylococcus aureus* and coagulase-negative staphylococci with decreased sensitivity to glycopeptides as assessed by determination of MIC's. Eur. J. Clin. Microbiol. Infect. Dis. 9(11):820–822.

137. Geraci J.E., Heilman F.R., Nichols D.R. (1957). Some laboratory and clinical experiences with a new antibiotic, vancomycin. Antibiot. Annu. 1956–1957:90–117.

138. Ehrenkranz N.J. (1959). The clinical evaluation of vancomycin in treatment of multi-antibiotic refractory staphylococcal infections. Antibiot. Annu. 1958–1959. New York Medical Encyclopedia, Inc., pp. 587–594.

139. Louria D.B., Kaminski T., Buchman J. (1961). Vancomycin in severe staphylococcal infections. Arch. Intern. Med. 107:225–240.

140. Dutton A.A.C., Elmes P.C. (1959). Vancomycin: Report on treatment of patients with severe staphylococcal infections. Br. Med. J. 1:1141–1149.

141. Woodley D.W., Hull W.H. (1961). The treatment of severe staphylococcal infections with vancomycin. Ann. Intern. Med. 55:235–249.

142. Kirby W.M.M., Perry D.M., Bauer A.W. (1960). Treatment of staphylococcal septicemia with vancomycin. Report of thirty-three cases. N. Engl. J. Med. 262: 49–55.

143. Chambers H.F. (1988). Methicillin-resistant staphylococci. Clin. Microbiol. Rev. 1: 173–186.

144. Milatovic D. (1986). Vancomycin for treatment of infections with methicillin-resistant *Staphylococcus aureus*: Are there alternatives? Eur. J. Clin. Microbiol. 5(6):689–692.

145. Markowitz N., Poholod D.L., Saravolatz L.D., et al. (1983). In vitro susceptibility patterns of methicillin-resistant and susceptible *Staphylococcus* strains in a population of parenteral drug abusers from 1972–1981. Antimicrob. Agents Chemother. 20:450–457.

146. Acar J.F., et al. (1970). Methicillin-resistant staphylococcemia: Bacteriological failure of treatment with cephalosporins. Antimicrob. Agents Chemother. 10:280.

147. Markowitz N., Quinn E.L., Saravolatz L.E. (1992). Trimethoprim-sulfamethoxazole compared with vancomycin for the treatment of *Staphylococcus aureus* infection. Ann. Intern. Med. 117:390–398.

148. Benner E.J., Morthland V. (1967). Methicillin-resistant *Staphylococcus aureus.* Antimicrobial susceptibility. N. Engl. J. Med. 277:673–680.

149. Coppens L., et al. (1983). Therapy of staphylococcal infections with cefamandole or vancomycin alone or with a combination of cefamandole and tobramycin. Antimicrob. Agents Chemother. 23:36.

150. Levine D.P., Cushing R.D., Jui J., et al. (1982). Community-acquired methicillin-resistant *Staphylococcus aureus* endocarditis in the Detroit medical center. Ann. Intern. Med. 97:330.

151. Craven D.E., Kollisch N.R., Hsieh C.R., Connollly M.G., McCabe W.R. (1983). Vancomycin treatment of bacteremia caused by oxacillin-resistant *Staphylococcus aureus*: Comparison with beta-lactam antibiotic treatment of bacteremia caused by oxacillin-sensitive *Staphylococcus aureus*. J. Infect. Dis. 147(1):137–143.

152. Myers J.P., Linnemann C.C. Jr. (1982). Bacteremia due to methicillin-resistant *Staphylococcus aureus*. J. Infect. Dis. 145:532.

153. Klastersky J., Coppens L., Van der Auwera P., Meunier-Carpentier F. (1983). Vancomycin therapy of oxacillin-resistant *Staphylococcus aureus* infections. J. Antimicrob. Chemother. 11(4):361–367.

154. Cafferkey M.T., Hone R., Keane C.T. (1985). Antimicrobial chemotherapy of septicemia due to methicillin-resistant *Staphylococcus aureus*. Antimicrob. Agents Chemother. 28(6):819–823.

155. Sorrell T.C., Packham D.R., Shanker S., et al. (1982). Vancomycin therapy for methicillin-resistant *Staphylococcus aureus*. Ann. Intern. Med. 97:344.

156. Gopal V., Bisno A.L., Silverblatt F.J. (1976). Failure of vancomycin treatment in *Staphylococcus aureus* endocarditis. JAMA 236:1604–1606.

157. Watanakunakorn. C (1982). Treatment of infections due to methicillin-resistant *Staphylococcus aureus*. Ann. Intern. Med. 97:376–378.

158. Archer G.L. (1984). *Staphylococcus epidermidis*: The organism, its disease, and treatment. Curr. Top. Infect. Dis. 5:25–48.

159. Morrison A.J., Freer C.V., Searcy M.A., et al. (1986). Nosocomial bloodstream infections: Secular trends in a statewide surveillance program in Virginia. Infect. Control 7:550–553.

160. Martin M.A., Pfaller M.A., Wenzel R.P. (1989). Coagulase-negative staphylococcal bacteremia: Mortality and hospital stay. Ann. Intern. Med. 110:9–16.

161. Freeman J., Platt R., Sidebottom D.G., et al. (1987). Coagulase-negative staphylococcal bacteremia in the changing neonatal intensive care unit population. JAMA 258:2548–2552.

162. Kernodle D.S., Barg N.L., Kaiser A.B. (1988). Low-level colonization of hospitalized patients with methicillin-resistant coagulase-negative staphylococci and emergence of the organisms during surgical antimicrobial prophylaxis. Antimicrob. Agents Chemother. 32:202–208.

163. Munson D.P., Thompson T.R., Johnson D.E., et al. (1982). Coagulase-negative staphylococcal septicemia: Experience in a newborn intensive care unit. J. Pediatr. 101:602–604.

164. Baumgart S., Hall S.E., Campos J.M., Polin R.A. (1983). Sepsis with coagulase-negative staphylococci in critically ill newborns. Am. J. Dis. Child. 137:461–463.

165. Starr S.E. (1985). Antimicrobial therapy of bacterial sepsis in the newborn infant. J. Pediatr. 106:1043–1048.

166. Archer G.L., Armstrong B.C. (1983). Alteration of staphylococcal flora in cardiac surgery patients receiving antibiotic prophylaxis. J. Infect. Dis. 147:642.

167. Laverdiere M., Peterson P.K., Verhoef J., et al. (1978). In vitro activity of cephalosporins against methicillin-resistant, coagulase negative staphylococci. J. Infect. Dis. 137:245–250.

168. Davenport D.S., Massanari R.M., Pfaller M.A., et al. (1986). Usefulness of a test for slime production as a marker for clinically significant infections with coagulase-negative staphylococci. J. Infect. Dis. 153:332–339.

169. Farber B.F., Kaplan M.H., Clogston A.G. (1990). *Staphylococcus epidermidis* extracted slime inhibits the antimicrobial action of glycopeptide antibiotics. J. Infect. Dis. 161:37–40.

170. Schwalbe R.S., Stapleton J.T., Gilligan P.H. (1987). Emergence of vancomycin resistance in coagulase-negative staphylococci. N. Engl. J. Med. 16(15):927–931.

171. Veach L.A., Pfaller M.A., Barrett M., Koontz F.P., Wenzel R.P. (1990). Vancomycin resistance in *Staphylococcus haemolyticus* causing colonization and bloodstream infection. J. Clin. Microbiol. 28(9):2064–2068.

172. Sanyal D., Johnson A.P., George R.C., Cookson B.D., Williams A.J. (1991). Peritonitis due to vancomycin-resistant *Staphylococcus epidermidis* (letter). Lancet 337(8732):54.

173. Schwalbe R.S., Ritz W.J., Verma P.R., Barranco E.A., Gilligan P.H. (1990). Selection for vancomycin resistance in clinical isolates of *Staphylococcus haemolyticus*. J. Infect. Dis. 161(1):45–51.

174. Lyon B.R., Skurray R. (1987). Antimicrobial resistance of *Staphylococcus aureus*: Genetic basis. Microbiol. Rev. 51:8–10.

175. Yuk J.H., Dignani C., Harris R.L., Bradshaw J.W., Williams T.W. (1991). Minocycline as an alternative antistaphylococcal agent. Rev. Infect. Dis. 13:1023–1024.

176. Karchmer A.W., Archer G.L., Dismukes W.E. (1983). *Staphylococcus epidermidis* causing prosthetic valve endocarditis: Microbiologic and clinical observations as guides to therapy. Ann. Intern. Med. 98:447–455.

177. Ein M.E., Smith H.J., Aruffo J.E., et al. (1979). Susceptibility and synergy studies of methicillin-resistant *Staphylococcus epidermidis*. Antimicrob. Agents Chemother. 16:665–669.

178. Lowy F.D., Chang D.S., Lash P.R. (1983). Synergy of combinations of vancomycin, gentamicin, and rifampin against methicillin-resistant, coagulase-negative staphylococci. Antimicrob. Agents Chemother. 23:932–934.

179. Watanakunakorn C. (1984). Mode of action and in-vitro activity of vancomycin. J. Antimicrob. Chemother. 14:7–18.

180. Rupar D.G., Fisher M.C., Fletcher H., Mortensen J. (1989). Emergence of isolates resistance to ampicillin. Am. J. Dis. Child. 143:1033–1037.

181. Watanakunakorn C., Bakie C. (1973). Synergism of vancomycin-gentamicin and vancomycin-streptomycin against enterococci. Antimicrob. Agents Chemother. 4:120.

182. Zerfos M.J., Terpenning M.S., Schaberg D.R., et al. (1987). High level aminoglycoside-resistant enterococci. Arch. Intern. Med. 147:1591–1594.

183. Uttley A.H., George R.C., Naidoo J., Woodford N., Johnson A.P., Collins C.H.,

Morrison D., Gilfillan A.J., Fitch L.E., Heptostall J. (1989). High-level vancomycin-resistant enterococci causing hospital infections. Epidemiol. Infect. 103(1): 173–181.

184. Uttley A.H., Collins C.H., Naidoo J., George R.C. (1988). Vancomycin-resistant enterococci (letter). Lancet 1(8575–8576):57–58.

185. Bingen E., Lambert-Zechovsky N., Mariani-Kurkdjian P., Cezard J.P., Navarro J. (1989). Bacteremia caused by a vancomycin-resistant enterococcus. Pediatr. Infect. Dis. J. 8(7):475–476.

186. Eliopoulos G.M., Wennersten C., Zighelboim-Daum S., et al. (1988). High-level resistance to gentamicin in clinical isolates of *Streptococcus (Enterococcus) faecium* Antimicrob. Agents Chemother. 32:1528–1532.

187. Courvalain P. (1990). Resistance of enterococci to glycopeptides (minireview). Antimicrob. Agents Chemother. 34(12):2291–2296.

188. Handwerger S., Perman D.C., Altarac D., McAuliffe V. (1992). Concomitant high-level vancomycin and penicillin resistance in clinical isolates of enterococci. Clin. Infect. Dis. 14:655–661.

189. Livornese L.L., Dias S., Samel C., Romanowski B., Taylor S., May P., et al. (1992). Hospital-acquired infection with vancomycin-resistant *Enterococcus faecium* transmitted by electronic thermometers. Ann. Intern. Med. 117:112–116.

190. Kaplan A.H., Gilligan P.H., Facklam R.R. (1988). Recovery of resistant enterococci during vancomycin prophylaxis. J. Clin. Microbiol. 26(6):1216–1218.

191. Karanfil L.V., Murphy M., Josephson A., Gaynes R., Mandel L., Hill B.C., Swenson J.M. (1992). A cluster of vancomycin-resistant *Enterococcus faecium* in an intensive care unit [see Comments]. Infect. Control Hosp. Epidemiol. 13(4): 195–200.

192. Peetermans W.E., Sebens F.W., Guiot H.F.L. (1991). Vancomycin-resistant *Enterococcus faecalis* in a bone-marrow transplant recipient. Scand. J. Infect. Dis. 23: 105–109.

193. Guiot H.F., Peetermans W.E., Sebens F.W. (1991). Isolation of vancomycin-resistant enterococci in haematologic patients. Eur. J. Clin. Microbiol. Infect. Dis. 10(1): 32–34.

194. Green M., Barbadora K., Michaels M. (1991). Recovery of vancomycin-resistant Gram-positive cocci from pediatric liver transplant recipients. J. Clin. Microbiol. 29:2503–2506.

195. Swenson J.M., Hill B.C., Thornsberry C. (1989). Problems with the disk diffusion test for detection of vancomycin resistance in enterococci. J. Clin. Microbiol. 27(9): 2140–2142.

196. (1990). Erratum. J. Clin. Microbiol. 28(2):403.

197. Sahm D.F., Olsen L. (1990). In vitro detection of enterococcal vancomycin resistance. Antimicrob. Agents Chemother. 34(9):1846–1848.

198. Shlaes D., Etter L., Gutmann (1991). Synergistic killing of vancomycin-resistant enterococci of classes A, B, and C by combinations of vancomycin, penicillin, and gentamicin. Antimicrob. Agents Chemother. 35:776–779.

199. Leclercq R., Bingen E., Su Q.H., Lambert-Zechovski N., Courvalin P., Duval J. (1991). Effects of combinations of β-lactams, daptomycin, gentamicin, and glyco-

peptides against glycopeptide-resistant enterococci. Antimicrob. Agents Chemother. 35:92–98.

200. Fraimow H.S., Venuti E. (1992). Inconsistent bactericidal activity of triple-combination therapy with vancomycin, ampicillin, and gentamicin against vancomycin-resistant, highly ampicillin-resistant *Enterococcus faecium*. 36:1563–1566.

201. Cercenado E., Eliopoulos G.M., Wennersten C.B., Moellering R.C. (1992). Absence of synergistic activity between ampicillin and vancomycin against highly vancomycin-resistant enterococci. Antimicrob. Agents Chemother. 36:2201–2203.

202. Caron F., Carbon C., Guttmann L. (1991). Triple combination penicillin-vancomycin-gentamicin for experimental endocarditis caused by a moderately penicillin and highly glycopeptide resistant isolate of *Enterococcus faecialis*. J. Infect. Dis. 164:888–893.

203. Nobel W.C., Virani Z., Cree R.G. (1992). Co-transfer of vancomycin and other resistance genes from *Enterococcus faecium* NCTC 12201 to *Staphylococcus aureus*. FEMS Microbiol. Lett. 72(2):195–198.

204. Shlaes D.M., Marino J., Jacobs M.R. (1984). Infection caused by vancomycin-resistant *Streptococcus sanguis* II. Antimicrob. Agents Chemother. 25(4):527–528.

205. Buzon L.M., Guerrero A., Romero J., Santamaria J.M., Bouza E. (1984). Penicillin-resistant *Streptococcus pneumoniae* meningitis successfully treated with vancomycin (letter). Eur. J. Clin. Microbiol. 3(5):442–443.

206. Laurens E., Poirier T., Viaud J.Y., Bedos J.P., Bannier B., Lorre G. (1992). Pneumopathie nosocomiale a *Streptococcus pneumoniae* resistant a la penicilline au cours d'une pneumocystose grave chez un sideen, guerie par l'assocation vancomycin-amikacine. Med. Mal. Infect. 22:410–412.

207. Dronda, F., Montilla, P., Moreno, S., Martinex-Luengas (1989). Endocarditis por *Streptococcus pneumoniae* resistente a penicilina. Rev. Clin. Esp. 185(7):383–384.

208. Viladrich P.F., Gudiol F., Linares J., Pallares R., Sabate I., Rufi G., Ariza J. (1991). Evaluation of vancomycin for therapy of adult pneumococcal meningitis. Antimicrobial Agents Chemother. 35(12):2467–2472.

209. Bradley J.S., Connor J.D. (1991). Ceftriaxone failure in meningitis caused by *Streptococcus pneumoniae* with reduced susceptibility to beta-lactam antibiotics. Pediatr. Infect. Dis. J. 10:871–873.

210. McNaughton R.D., Villanueva R.R., Donnelly R., Freedman J., Nawrot R. (1988). Cavitating pneumonia caused by *Corynebacterium* group JK. J. Clin. Microbiol. 10: 2216–2217.

211. Cavallo J.D., De Revel T., Baudet J.M., Nedellec G., Schill H., Buisson Y., Auzanneau G. (1991). Les infections a *Corynebacterium jeikeium* a propos de quatre formes septicemiques en milieu hematologique. Med. Mal. Infect. 21:719–724.

212. Hook E.W., Johnson W.D. (1978). Vancomycin therapy of bacterial endocarditis. Am. J. Med. 65:411–415.

213. VanScoy R.E., Cohen S.N., Geraci J.E., Washington J.A. (1977). Coryneform bacterial endocarditis. Mayo Clin. Proc. 52:216–223.

214. Cauda R., Tamburrini E., Ventura G., Ortona L. (1987). Effective vancomycin therapy for *Corynebacterium pseudodiphtheriticum* endocarditis (letter). South. Med. J. 80(12):1598.

215. Soriano F., Aguado J.M., Ponte C., Fernandez-Robias R., Rodriguez-Tudela (1990). Urinary tract infection caused by *Corynebacterium* group D_2: Report of 82 cases and review. J. Rev. Infect. Dis. 12(4):1019–1033.

216. Ena J., Berenguer J., Pelaez T., Bouza E. (1991). Endocarditis caused by *Corynebacterium* group D_2. J. Infect. 22:95–96.

217. Garavelli P.L., Poggio A. (1990). Endocardite su valvola protesica da difteroide insensibile alla vancomicina. Minerva Med. 81(Suppl. 12):111–112.

218. Barnass S., Holland K., Tabaqchali S. (1991). Vancomycin-resistant *Corynebacterium* species causing prosthetic valve endocarditis successfully treated with imipenem and ciprofloxacin. J. Infect. 22:161–169.

219. Handwerger S., Horowitz H., Coburn K., Kolokathis A., Wormser G.P. (1990). Infection due to *Leuconostoc* species: Six cases and review. Rev. Infect. Dis. 12(4):602–610.

220. Horowitz H.W., Handwerger S., Van Horn K.G., Wormser G.P. (1987). *Leuconostoc*, an emerging vancomycin-resistant pathogen (letter). Lancet 2(8571):1329–1330.

221. Buu-Joi A., Branger C., Acar J.F. (1985). Vancomycin-resistant streptococci or *Leuconostoc* sp. Antimicrob. Agents Chemother. 28(3):458–460.

222. Rubin L.G., Vellozzi E., Shapiro J., Isenberg H.D. (1988). Infection with vancomycin-resistant "streptococci" due to *Leuconostoc* species (letter). J. Infect. Dis. 157(1):216.

223. Dyas A., Chauhan N. (1988). Vancomycin-resistant *Leuconostoc* (letter). Lancet 1(8580):306.

224. Hardy S., Ruoff K.L., Catlin E.A., Ignacio Santos J. (1988). Catheter-associated infection with a vancomycin-resistant gram-positive coccus of the *Leuconostoc* sp. Pediatr. Infect. Dis. J. 7(7):519–520.

225. Reibel W.J., Washington J.A. (1990). Clinical and microbiologic characteristics of pediococci. J. Clin. Microbiol. 28(6):1348–1355.

226. Mastro T.D., Spika J.S., Lozano P., Appel J., Facklam R.R. (1990). Vancomycin-resistant *Pediococcus acidilactici*: Nine cases of bacteremia. J. Infect. Dis. 161(5):956–960.

227. Holliman R.E., Bone G.P. (1988). Vancomycin resistance of clinical isolates of lactobacilli. J. Infect. 16(3):279–283.

228. Banter C.E., Relloso S., Castell R., Smayevsky J., Bianchini H.M. (1991). Abscess caused by vancomycin-resistant *Lactobacillus confusus*. J. Clin. Microbiol. 29:2063–2064.

229. Green M., Wadowsky R.M., Barbadora K. (1990). Recovery of vancomycin-resistant Gram-positive cocci from children. J. Clin. Microbiol. 28(3):484–488.

230. Ruoff K.L., Kuritzkes D.R., Wolfson J.S., Ferraro M.J. (1988). Vancomycin-resistant Gram-positive bacteria isolated from human sources. J. Clin. Microbiol. 26(10):2064–2068.

231. Speirs G., Warren R.E., Rampling A. (1988). *Clostridium tertium* septicemia in patients with neutropenia. J. Infect. Dis. 158(6):1336–1340.

232. Rouquet R.M., Clave D., Massip P., Moatti N., Leophonte P. (1991). Imipenem/

vancomycin for *Rhodococcus equi* pulmonary infection in HIV-positive patient (letter). Lancet 337(8737):375.

233. Van Etta L.L., Filice G.A., Ferguson R.M., Gerding D.N. (1983). *Corynebacterium equi*: A review of 12 cases of human infections. Rev. Infect. Dis. 5:1012–1018.
234. Gallagher P.G., Amedia C.A., Watanakunakorn C. (1986). *Listeria monocytogenes* endocarditis in a patient on chronic hemodialysis, successfully treated with vancomycin-gentamicin. Infection 14:125–128.
235. Mossey R.T., Sonheimer J. (1985). Listerosis in patients with long-term hemodialysis and transfusional iron overload. Am. J. Med. 79:397–400.
236. Blatt S.P., Zajac R.A. (1991). Treatment of *Listeria* bacteremia with vancomycin (letter). Rev. Infect. Dis. 13(1):181–182.
237. Zeitlin J., Carvounis C.P., Murphy R.G., Tortora G.T. (1982). Graft infection and bacteremia with *Listeria monocytogenes* in a patient receiving hemodialysis. Arch. Intern. Med. 142:2191–2192.
238. Henrickson K.J., Shenep J.L., Flynn P.M., Ching-Hon P. (1989). Primary cutaneous *Bacillus cereus* infection in neutropenic children. Lancet 1(8638):601–603.
239. Jadeja L., Bolivar R., Wallace R.J., Silcox V.A., Bodey G.P. (1983). Bacteremia caused by a previously unidentified species of rapidly growing *Mycobacterium* successfully treated with vancomycin. Ann. Intern. Med. 99(4):475–477.
240. McWhinney P.H.M., Kibbler C.C., Gillespie S.H., Patel S., Morrison D., Hoffbrand A.V., Prentice H.G. (1992). *Stomatococcus mucilaginosus*: An emerging pathogen in neutropenic patients. Clin. Infect. Dis. 14:641–646.
241. Pinsky R.L., Piscitelli V., Patterson J.E. (1989). Endocarditis caused by relatively penicillin-resistant *Stomatococcus mucilaginosus*. J. Clin. Microbiol. 27(1): 215–216.
242. Baldassarre J.S., Ingerman M.J., Nansteel J., Santoro J. (1991). Development of *Listeria* meningitis during vancomycin therapy: A case report. J. Infect. Dis. 164: 221–222.
243. Dryden M.S., Jones N.F., Phillips I. (1991). Vancomycin therapy failure in *Listeria monocytogenes* peritonitis in a patient on continuous ambulatory peritoneal dialysis. J. Infect. Dis. 164:1239.
244. Caputo G.M., Archer G.L., Calderwood S.B., DiNubile M.J., Karchmer A.W. (1987). Native valve endocarditis due to coagulase-negative staphylococci. Clinical and microbiologic features. Am. J. Med. 83(4):619–625.
245. Geraci J.E., Heilman R.R., Nichols D.R., Wellman W.E. (1958). Antibiotic therapy of bacterial endocarditis. VII. Vancomycin for acute micrococcal endocarditis. Mayo Clin. Proc. 33:172–181.
246. Newsom S.W.B. (1984). The treatment of endocarditis by vancomycin. J. Antimicrob. Chemother. 14(Suppl. D):79–84.
247. Cook F.V., Coddington C.C., Wadland W.C., Farrar W.E. (1978). Treatment of bacterial endocardis with vancomycin. Am. J. Med. Sci. 276(2):153–158.
248. Geraci J.E., Wilson W.R. (1981). Vancomycin therapy for infective endocarditis. Rev. Infect. Dis. 3(Suppl.):S250–258.
249. Chambers H.F., Miller R.T., Newman M.D. (1988). Right-sided *Staphylococcus aureus* endocarditis in intravenous drug users. Antimicrob. Agents Chemother. 34(6):1227–1231.

250. Small P.M., Chambers H.F. (1990). Vancomycin for *Staphylococcus aureus* endocarditis in intravenous drug abusers: Two-week combination therapy. Arch. Intern. Med. 109(8):619–624.

251. Levine D.P., Fromm B.S., Reddy B.R. (1991). Slow response to vancomycin or vancomycin plus rifampin in methicillin-resistant *Staphylococcus aureus* endocarditis. Arch. Intern. Med. 115:674–680.

252. Hartstein A.I., Mulligan M.E., Morthland V.H., Kwok R.Y.Y. (1992). Recurrent *Staphylococcus aureus* bacteremia. J. Clin. Microbiol. 30:670–674.

253. Jackson M.A., Hicks R.A. (1987). Vancomycin failure in staphylococcal endocarditis. Pediatr. Infect. Dis. J. 6(8):750–752.

254. Faville R.J., Zaske D.E., Kaplan E.L., Crossley K., Sabath L.D., Quie P.G. (1978). *Staphylococcus aureus* endocarditis. Combined therapy with vancomycin and rifampin. JAMA 240(18):1963–1965.

255. Simon G.L., Smith R.H., Sande M.A. (1983). Emergence of rifampin-resistant strains of *Staphylococcus aureus* during combination therapy with vancomycin and rifampin: A report of two cases. Rev. Infect. Dis. 5(Suppl. 3):S507–508.

256. Tofte R.W., Solliday J., Rotschafer J., Crossley K.B. (1981). *Staphylococcus aureus* infection of dialysis shunt: Absence of synergy with vancomycin and rifampin. South Med. J. 74(5):612–615.

257. Herzstein J., Ryan J.L., Mangi R.J., Greco T.P., Andriole V.T. (1984). Optimal therapy for enterococcal endocarditis. Am. J. Med. 76(2):186–191.

258. Spiegel C.A., Huycke M. (1989). Endocarditis due to streptomycin-susceptible *Enterococcus faecalis* with high-level gentamicin resistance. Arch. Intern. Med. 149(8):1873–1875.

259. Besnier J.M., Leport C., Bure A., Vilde J.L. (1990). Vancomycin-aminoglycoside combinations in therapy of endocarditis caused by *Enterococcus* species and *Streptococcus bovis*. Eur. J. Clin. Microbiol. Infect. Dis. 9(2):130–133.

260. Westenfenlder G.O., Paterson P.Y., Reisberg B.E., Carlson G.M. (1973). Vancomycin-streptomycin synergism in enterococcal endocarditis. JAMA 223:37–40.

261. Hook E.W. III, Roberts R.B., Sande M.A. (1975). Antimicrobial therapy of experimental enterococcal endocarditis. Antimicrob. Agents Chemother. 8:564.

262. Freiberg C.K., Rosen K.M., Bienstock P.A. (1968). Vancomycin therapy for enterococcal and *Streptococcus viridans* endocarditis. Successful treatment of 6 patients. Arch. Intern. Med. 122:134.

263. Murray B.E. (1991). Antibiotic resistance among enterococci: Current problems and management strategies. In Remington J.S., Swartz M.N. (Eds.), Current Clinical Topics in Infectious Diseases. Blackwell Scientific, Oxford. 1991.

264. Heimberger T.S., Duma R.J. (1989). Infections of prosthetic heart valves and cardiac pacemakers. Infect. Dis. Clin. North Am. 3(2):221–245.

265. Calderwood S.B., Swinshi L.A., Karchmer A.W., et al. (1986). Prosthetic valve endocarditis: Analysis of factors affecting outcome of therapy. J. Thorac. Cardiovasc. Surg. 75:219.

266. Chamovitz B., Bryant R.E., Gilbert D.N., Hartstein A.I. (1985). Prosthetic valve endocarditis caused by *Staphylococcus epidermidis*. Development of rifampin resistance during vancomycin and rifampin therapy. JAMA 253(19):2867–2868.

267. Karchmer A.W., Archer G.L., Dismukes W.E. (1983). Rifampin treatment of

prosthetic valve endocarditis due to *Staphylococcus epidermidis*. Rev. Infect. Dis. 5(Suppl. 3):S543.

268. Karchmer A.W., Archer G.A. (1984). Endocarditis study group. Methicillin-resistant *Staphylococcus epidermidis* prosthetic valve endocarditis: A therapeutic trial. Abstracts of the 24th Interscience Conference on Antimicrobial Agents and Chemotherapy, Washington, D.C., Abstract No. 476.

269. Murry B.E., Karchmer A.W., Moellering R.C. Jr. (1980). Diphtheroid prosthetic valve endocarditis: A study of clinical features and infecting organisms. Am. J. Med. 69:838.

270. Threlkeld M.G., Cobbs C.G. (1990). Infectious disorders of prosthetic valves and intravascular devices. In Mandell G.L., Douglas R.G., Bennett S.E. (Eds.), Principles and Practice of Infectious Disease, 3rd ed. Churchill Livingstone, New York, p. 709.

271. Karchmer A.W. (1986). Prosthetic valve endocarditis: Mechanical valves. In Magilligan D.J., Quinn E.L. (Eds.), Endocarditis: Medical and Surgical Management. Marcel Dekker, New York, p. 241.

272. Karchmer A.W. (1984). Treatment of prosthetic valve endocarditis. In Sande J.A., Root R. (Eds.), Endocarditis. Churchill Livingstone, New York, p. 163.

273. Fitzgerald R.H. Jr., Peterson L.F.A., Washington J.A. II, et al. (1973). Bacterial colonization of wounds and sepsis in total hip arthroplasty. J. Bone Joint Surg. 55: 1242–1250.

274. Fitzgerald R.H., Jones D.R. (1985). Hip implant infection. Treatment with resection arthroplasty and late total hip arthroplasty. Am. J. Med. 78(Suppl. B):225.

275. McDonald D.J., Fitzgerald R.H. Jr. (1987). The two-stage reconstruction of the infected total hip arthroplasty. Orthop. Trans. 11:462.

276. Manderazo E.G., Judson S., Pasternak H. (1988). Late infections of total joint prosthesis. Clin. Orthop. 229:131.

277. Blomgren G. (1981). Hematogenous infection of total joint replacement. An experimental study in the rabbit. Acta Orthrop. Scand. 52(Suppl. 187):1–64.

278. Brause B.D. (1986). Infections associated with prosthetic joints. Clin. Rheum. Dis. 12:523–536.

279. Gump D.W. (1981). Vancomycin for treatment of bacterial meningitis. Rev. Infect. Dis. 3(Suppl.): S289–292.

280. Ehrenkrantz H.J. (1959). The clinical evaluation of vancomycin in treatment of multi-antibiotic refractory staphylococcal infections. In Welch H., Marti-Ibanez F. (Eds.), Antibiotics Annual 1958–1959. Medical Encyclopedia, New York, pp. 587–594.

281. Hawley H.B., Gump D.W. (1973). Vancomycin therapy of bacterial meningitis. Am. J. Dis. Child. 126:261.

282. Cogenti et al. (1979). Kinetics of vancomycin after intraventricular and intravenous administration (abstract). Pediatr. Res. 13:459.

283. Connors J.M. (1982). Cure of Ommaya reservoir associated *Staphylococcus epidermidis* ventriculitis with a simple regimen of vancomycin and rifampin without reservoir removal. Med. Pediatr. Oncol. 10:549–552.

284. Bayston R., Hart C.A., Barnicoat M. (1987). Intraventricular vancomycin in the

treatment of ventriculitis associated with cerebrospinal fluid shunting and drainage. J. Neurol. Neurosurg. Psychiatry 50(11):1419–1423.

285. Kaufman B.A., Tunkel A.R., Pryor J.C., Dacey R.G. Jr. (1990). Meningitis in the neurosurgical patient. Infect. Dis. Clin. North Am. 4(4):677–701.

286. Archer G.L. (1978). Antimicrobial susceptibility and selection of resistance among *Staphylococcus epidermidis* isolates recovered from patients with infections of indwelling foreign devices. Antimicrob. Agents Chemother. 14:353.

287. Sutherland G.E., Palitang E.G., Marr J.J., Luedke S.L. (1981). Sterilization of Ommaya reservoir by instillation of vancomycin. Am. J. Med. 71(6):1068–1070.

288. Ratcheson R.A., Ommaya A.K. (1986). Experience with the subcutaneous cerebrospinal reservoir. N. Engl. J. Med. 279:1025–1028.

289. Obbens E.A.M.T., Leavens M.E., Beal J.W., Ya Y.L. (1985). Ommaya reservoirs in 387 cancer patients: A 15 year experience. Neurology 35:1274–1278.

290. Swayne R., Rampling A., Newsom S.W. (1987). Intraventricular vancomycin for treatment of shunt-associated ventriculitis. J. Antimicrob. Chemother. 19(2): 249–253.

291. Bayer A.S., Seidel J.S., Yoshikawa T.T., et al. (1976). Group D enterococcal meningitis. Arch. Intern. Med. 136:883–887.

292. Barriere S.L., Lutwick L.I., Jacobs R.A., Conte J.E. (1985). Vancomycin treatment for enterococcal meningitis. Arch. Neurol. 42(7):686–688.

293. Ryan J.L., Pachner A., Andriole V.T., Root R.K. (1980). Enterococcal meningitis: Combined vancomycin and rifampin therapy. Am. J. Med. 68(3):449–451.

294. George R.M. et al. (1961). Epidemic meningitis of the newborn caused by flavobacteria. II. Clinical manifestations and treatment. Am. J. Dis. Child. 101:296.

295. Richards S., Lambert C.M., Scott A.C. (1992). Recurrent *Listeria monocytogenes* meningitis treated with intraventricular vancomycin. J. Antimicrob. Chemother. 29: 351–353.

296. Sato K., Lin T.Y., Weintrub L., Olsen K., McCracken G.H. (1985). Bacteriologic efficacy of nafcillin and vancomycin alone or combined with rifampicin or amikacin in experimental meningitis due to methicillin-susceptible or -resistant *Staphylococcus aureus*. Jpn. J. Antibiot. 38(8):2155–2162.

297. Vichyanond P., Olson L.C. (1984). Staphylococcal CNS infections treated with vancomycin and rifampin. Arch. Neurol. 41(6):637–639.

298. Gombert M.E., Landesman S.H., Corrado M.L., Stein S.C., Melvin E.T., Cummings M. (1981). Vancomycin and rifampin therapy for *Staphylococcus epidermidis* meningitis associated with CSF shunts: Report of three cases. J. Neurosurg. 55(4):633–636.

299. Archer G.L., Tenenbaum M.J., Haywood H.B. III (1978). Rifampin therapy of *Staphylococcus epidermidis*. JAMA 240:251.

300. Osborn J.S., Sharp S., Hanson E.J., MacGee E., Brewer R.H. (1986). *Staphylococcus epidermidis* ventriculitis treated with vancomycin and rifampin. Neurosurgery 19(5):824–827.

301. Frame P.T., McLaurin R.L. (1984). Treatment of CSF shunt infections with intrashunt plus oral antibiotic therapy. J. Neurosurg. 60(2):354–360.

302. Graziani A.L., Lawson L.A., Gibson G.A., Steinberg M.A., MacGregor R.T.

(1988). Vancomycin concentrations in infected and noninfected human bone. Antimicrob. Agents Chemother. 32(9):1320–1322.

303. Weeks J.L. et al. (1981). Methicillin-resistant *Staphylococcus aureus* osteomyelitis in a neonate. JAMA 245:1662.

304. Kirby W.M. (1981). Vancomycin therapy in severe staphylococcal infections. Rev. Infect. Dis. 3(Suppl.):236.

305. Cafferkey M.T. et al. (1982). Severe staphylococcal infections treated with vancomycin. J. Antimicrob. Chemother. 9:69.

306. Ish-Horowicz M.R., McIntyre P., Nade S. (1992). Bone and joint infections caused by multiply resistant *Staphylococcus aureus* in a neonatal intensive care unit. Pediatr. Infect. Dis. J. 11(2):82–87.

307. Wagner D.K., Collier B.D., Rytel M.W. (1985). Long-term intravenous antibiotic therapy in chronic osteomyelitis. Arch. Intern. Med. 145:1073–1078.

308. Sheftel T.G., Mader J.T., Pennick J.J., Cierny G. (1984). Methicillin-resistant *Staphylococcus aureus* osteomyelitis. Clin. Orthop. 198:231–239.

309. Tuazon C.U., Decker C.F. (1990). *Staphylococcus aureus* infections of bone, joints, and bursae. Curr. Opin. Infect. Dis. 3:662–665.

310. Cohen J., Worsley A.M., Donnelly J.P., et al. (1984). Septicaemia caused by viridans streptococci in neutropenic patients with leukaemia. Lancet 2(8365):1452–1454.

311. Del Favera A., Menichetti F., Bucaneve G., Minotti V., Pauluzzi S. (1988). J. Antimicrob. Chemother. 21(Suppl. C):157–165.

312. Wade J.C., Schimpff S.C., Newman K.A., Wiernick P.H. (1982). *Staphylococcus epidermidis*: An increasing cause of infection in patients with granulocytopenia. Ann. Intern. Med. 97:503–508.

313. Winston D.J., Dudnick D.J., Chapin M., et al. (1983). Coagulase-negative staphylococcal bacteremia in patients receiving immunosuppressive therapy. Arch. Intern. Med. 143:32–36.

314. Langley J., Gold R. (1988). Sepsis in febrile neutropenic children with cancer. Pediatr. Infect. Dis. J. 7(1):34–37.

315. Jadeja L., Bolivar R., Fainstein V., Keating M., McCredie K., Hay M., Bodey G.P. (1984). Piperacillin plus vancomycin in the therapy of febrile episodes in cancer patients. Antimicrob. Agents Chemother. 26(3):295–299.

316. Smith G.M., Leyland M.J., Farrell I.D., Geddes A.M. (1988). A clinical, microbiological and pharmacokinetic study of ciprofloxacin plus vancomycin as initial therapy of febrile episodes in neutropenic patients. J. Antimicrob. Chemother. 21(5):647–655.

317. Schaison G., Leverger G., Pappo M., Chiche D. (1990). Efficacy of ceftazidime/vancomycin combination as the first line treatment in neutropenic children. Pathol. Biol. (Paris) 38(5):557–560.

318. Kelsey S.M., Shaw E., Newland A.C. (1992). Aztreonam plus vancomycin versus gentamicin plus piperacillin as empirical therapy for the treatment of fever in neutropenic patients: A randomized study. J. Chemother. 4:107–113.

319. Viscoli C., Van der Auwera P., Meunier F. (1988). Gram-positive infections in granulocytopenic patients: An important issue? J. Antimicrob. Chemother. 21 (Suppl. C):149.

320. Viscoli C. (1988). Aspects of infections in children with cancer. Recent Results Cancer Res. 108:71–81.
321. Shenep J.L., Hughes W.T., Roberson P.K., Blankenship K.R., Baker D.K., Meyer W.H. (1988). Vancomycin, ticarcillin, and amikacin compared with ticarcillin-clavulanate and amikacin in the empirical treatment of febrile, neutropenic children with cancer. N. Engl. J. Med. 319(16):1053–1058.
322. Karp J.E., Dick J.D., Angelopulos C., et al. (1986). Empiric use of vancomycin during prolonged treatment-induced granulocytopenia: Randomized, double-blind, placebo-controlled clinical trial in patients with acute leukemia. Am. J. Med. 81: 237–242.
323. Kramer B.S., Ramphal R., Rand K.H. (1986). Randomized comparison between two ceftazidime-containing regimens and cephalothin-gentamicin-carbenicillin in febrile granulocytopenic cancer patients. Antimicrob. Agents Chemother. 30:64–68.
324. Viscoli C., Moroni C., Boni L., Bruzzi P., Comelli A., Dini G., Fabbri A., Secondo V., Terragna A. (1991). Ceftazidime plus amikacin versus ceftazidime plus vancomycin as empiric therapy in febrile neutropenic children with cancer. Rev. Infect. Dis. 13:397–404.
325. Rubin M., Hathorn J.W., Marchal D., Gress J., Steinberg S.M., Pizzo P.A. (1988). Gram-positive infections and the use of vancomycin in 550 episodes of fever and neutropenia. Ann. Intern. Med. 108:30–35.
326. Rubin M., Todeschini G., Marshall D., Gress I., Pizzo P.A. (1987). Does the presence of an indwelling venous catheter affect the type of infections in neutropenic cancer patients? Interscience Conference on Antimicrob. Agents Chemother. 961.
327. Callendra, T., et al. (1991). Vancomycin added to empirical combination antibiotic therapy for fever in granulocytopenic cancer patients. J. Infect. Dis. 163:951–958.
328. Ramphal R., Bolger M., Oblon D.J., Sheretz R.J., Malone J.D., Rand K.H., Gilliom M., Shands J.W. Jr., Kramer B.S. (1992). Vancomycin is not an essential component of the initial empiric treatment regimen for febrile neutropenic patients receiving ceftazidime: A randomized prospective study. Antimicrob. Agents Chemother. 36(5):1062–1067.
329. Pizzo P.A., Hathorn J.W., Hiemenz J., et al. (1986). A randomized trial comparing ceftazidime alone with combination antibiotic therapy in cancer patients with fever and neutropenia. N. Engl. J. Med. 315:552–558.
330. Granowetten L., Wells H., Lang B.J. (1988). Ceftazidime with or without vancomycin vs. cephalothin, carbenicillin and gentamicin as initial therapy of the febrile neutropenic pediatric cancer patient. Pediatr. Infect. Dis. J. 7:165–170.
331. Wade J.C. (1989). Antibiotic therapy for the febrile granulocytopenic cancer patient: Combination therapy vs. monotherapy. J. Infect. Dis. 11(Suppl. 7):S1572–S1581.
332. Rhkonen P. (1991). Imipenem compared with ceftazidime plus vancomycin as initial therapy for fever in neutropenic children with cancer. Pediatr. Infect. Dis. J. 10: 918–923.
333. Pizzo P.A. (1990). Current issues in the antibiotic primary management of the febrile neutropenic cancer patient: A perspective from the National Cancer Institute. J. Hosp. Infect. (Suppl. A):41–48.
334. Hughes W.T., Armstrong D., Bodey G.P., et al. (1990). Guidelines for the use of

antimicrobial agents in neutropenic patients with unexplained fever. J. Infect. Dis. 161:381–396.

335. Koeppler H., Pflueger K.H., Seitz R., Havemann K. (1989). Three-step empiric treatment for severely neutropenic patients with fever: Ceftazidime-vancomycin-amphotericin B. Infection 17(3):142–145.

336. Callendra, T. (1992). Reply. J. Infect. Dis. 165:591.

337. Cafferkey M.T., Abranamson E., Bloom A., Keane C.T. (1988). Pulmonary infection due to methicillin-resistant *Staphylococcus aureus*. Scand. J. Infect. Dis. 20(3):297–301.

338. Fujita K., Murono K., Sakata H., Kaeriyama M. (1992). Methicillin-resistant *Staphylococcus aureus* empyema in children. Acta Paediatr. Jpn. 34(2):151–156.

339. Vercelloni M., Ravini M., Maioli M., Schlacht I., Caggese I., Belloni P.A. (1991). Pleural administration of vancomycin in the treatment of patients with empyema. Curr. Ther. Res. 50(2):211–214.

340. Nsouli K.A., Lazarus M., Schoenbaum S.C., et al. (1979). Bacteremic infection in hemodialysis. Arch. Intern. Med. 139:1255.

341. Dobkin J.F., Miller M.H., Steigbigel N.H. (1978). Septicemia in patients on chronic hemodialysis. Ann. Intern. Med. 88:28.

342. Barcenas C.G., et al. (1976). Staphylococcal sepsis in patients on chronic hemo-dialysis regimens. Arch. Intern. Med. 136:1131.

343. Snydman D.R., Murray S.A., Kornfeld S.J., et al. (1982). Total parenteral nutrition-related infections; prospective epidemiologic study using semiquantitative methods. Am. J. Med. 73:695–699.

344. Shererty R.J., Falk R.J., Huffman K.A., et al. (1983). Infections associated with subclavian Udall catheters. Arch. Intern. Med. 143:52–56.

345. Press O.W., Ramsey P.G., Larson E.B., et al. (1984). Hickman catheter infections in patients with malignancies. Medicine (Baltimore) 63:189–200.

346. Archer G.L. (1990). *Staphylococcus epidermidis* and other coagulase-negative staphylococci. In Mandell G.L., Douglas R.G., Bennett J.E. (Eds.), Principles and Practice of Infectious Diseases. Churchill Livingstone, New York, p. 1513.

347. Maki D.G. (1984). Infections associated with intravascular lines. Curr. Top. Infect. Dis. 1:167–193.

348. Heimenz J., Skelton J., Pizzo P.A. (1987). Perspective on the management of catheter related infections in cancer patients. Pediatr. Infect. Dis. 5:6–11.

349. Gaillard J.L., Merlino R., Pajot N., Goulet O., Fauchere J.L., Ricour C., Veron M. (1990). Conventional and nonconventional modes of vancomycin administration to decontaminate the internal surface of catheters colonized with coagulase-negative staphylococci. J. Perenter. Enteral Nutr. 14(6):593–597.

350. Elian J.C., Frappaz D., Ros A., Gay J.P., Guichard D., Dorche G., et al. (1992). Etude de la cinetique serique de la vancomycine au decours de la technique de "blocage in situ." Arch. Fr. Pediatr. 49:357–360.

351. Hahn M.G., Bettencourt J.A., McCrea W.B. (1992). In vivo sterilization of an infected long-term epidural catheter. Anaesthesiology 76:645–646.

352. Bunt T.J. (1983). Synthetic vascular graft infections. I. Graft infections. Surgery 93:733–736.

353. Bandyk D.F., Berni G.A., Thiele B.L., et al. (1984). Aortofemoral graft infection due to *Staphylococcus epidermidis*. Arch. Surg. 119:102–108.

354. Tollefson D.F., Bandyk D.F., Kaebnick H.W., et al. (1987). Surface biofilm disruption: Enhanced recovery of microorganisms from vascular prostheses. Arch. Surg. 122:38–43.

355. Levy M.F., Schmitt D.D., Edmiston C.E., Bancyk D.F., et al. (1990). Sequential analysis of staphylococcal colonization of body surfaces of patients undergoing vascular surgery. J. Clin. Microbiol. 28(4):664–669.

356. Golan J.F. (1989). Vascular graft infection. Infect. Dis. Clin. North Am. 3(2): 247–258.

357. Gault M.H., Costerton J.W., Paul M.D., et al. (1987). Staphylococcus epidermidis infection of a hemodialysis button-graft complex controlled by vancomycin for 11 months. Nephron 45(2):126–128.

358. Scheifele D.W., Ginter G.L., Olsen E., Fussell S., Pendray M. (1987). Comparison of two antibiotic regimens for neonatal necrotizing enterocolitis. J. Antimicrob. Chemother. 20(3):421–429.

359. Millar M.R., MacKay P., Levene M., Langdale V., Martin C. (1992). Enterobacteriaceae and neonatal necrotising enterocolitis. Arch. Dis. Child. 67:53–56.

360. Millikin S.P., Matzke G.R., Keane W.F. (1991). Antimicrobial treatment of peritonitis associated with continuous ambulatory peritoneal dialysis. Peritoneal Dial. Int. 11:252–260.

361. Gruer L.D., Bartlett R., Ayliffe G.A.J. (1984). Species identification and antibiotic sensitivity of coagulase-negative staphylococci from CAPD peritonitis. J. Antimicrob. Chemother. 13(6):577–583.

362. Fenton P. (1984). Treatment of peritonitis complicating continuous ambulatory peritoneal dialysis. J. Antimicrob. Chemother. 13(5):411–413.

363. Gokal R., Ramos J.M., Francis D.M.A., et al. (1982). Peritonitis in continuous ambulatory peritoneal dialysis. Laboratory and clinical studies. Lancet 2(8312): 1388–1391.

364. Krothapalli R.K., Senekjian H.O., Ayus J.C. (1983). Efficacy of intravenous vancomycin in the treatment of Gram-positive peritonitis in long-term peritoneal dialysis. Am. J. Med. 75(2):345–348.

365. Obermiller L.E., Tzamaloukas A.H., Leymon P., Avasthi P.S. (1985). Intravenous vancomycin as initial treatment for gram-positive peritonitis in patients on chronic peritoneal dialysis. Clin. Nephrol. 24(5):256–260.

366. Oliva S.L., Guglielmo B.J., Jacobs R., Pons V.G. (1989). Failure of intravenous vancomycin and intravenous metronidazole to prevent or treat antibiotic-associated pseudomembranous colitis (letter). J. Infect. Dis. 159(6):1154–1155.

367. Maple P.A., Hamilton-Miller J.M., Brumfitt W. (1989). Comparative in-vitro activity of vancomycin, teicoplanin, ramoplanin (formerly A16686) paldimycin, DuP 721 and DuP 105 against methicillin and gentamicin resistant *Staphylococcus aureus*. J. Antimicrob. Chemother. 23(4):517–525.

368. Greenwood D., Bidgood K., Turner M. (1987). A comparison of the responses of staphylococci and streptococci to teicoplanin and vancomycin. J. Antimicrob. Chemother. 20(2):155–164.

369. Watanakunakorn C. (1990). In-vitro selection of resistance of *Staphylococcus aureus* to teicoplanin and vancomycin. J. Antimicrob. Chemother. 25(1):69–72.
370. Watanakunakorn C. (1988). In-vitro induction of resistance in coagulase-negative staphylococci to vancomycin and teicoplanin. J. Antimicrob. Chemother. 22(3): 321–324.
371. Goldstein F.W., Coutrot A., Seiffer A., Acar J.F. (1990). Percentages and distributions of teicoplanin-and vancomycin- resistant strains among coagulase-negative staphylococci. Antimicrob. Agents Chemother. 34(5):899–900.
372. Maugein J., Pellegrin J.L., Brossard G., Fourche J., Leng B., Reiffers J. (1990). In vitro activities of vancomycin and teicoplanin against coagulase-negative staphylococci isolated from neutropenic patients. Antimicrob. Agents Chemother. 35(5): 901–903.
373. Low D.E., McGeer A., Poon R. (1989). Activities of daptomycin and teicoplanin against *Staphylococcus haemolyticus* and *Staphylococcus epidermidis*, including evaluation of susceptibility testing recommendations. Antimicrob. Agents Chemother. 33(4):585–588.
374. Moore E.P., Speller D.C. (1988). In-vitro teicoplanin-resistance in coagulase-negative staphylococci from patients with endocarditis and from a cardiac surgery unit. J. Antimicrob. Chemother. 21(4):417–424.
375. Brunet F., Vedel G., Dreyfus F., Vaxelaire J.F., Giraud T., Schremmer B., Monsallier J.F. (1990). Failure of teicoplanin therapy in two neutropenic patients with staphylococcal septicemia who recovered after administration of vancomycin. Eur. J. Clin. Microbiol. Infect. Dis. 9(2):145–147.
376. Kaatz G.W., Seo S.M., Dorman N.J., Lerner S.A. (1990). Emergence of teicoplanin resistance during therapy of *Staphylococcus aureus* endocarditis. J. Infect. Dis. 162:103–108.
377. Manquat G., Croize J., Stahl J.P., Meyran M., Hirtz P., Micoud M. (1992). Failure of teicoplanin treatment associated with an increase in MIC during therapy of *Staphylococcus aureus* septicaemia. J. Antimicrob. Chemother. 30:791–792.
378. Pantosti A., Luzzi I., Cardines R., Gianfrilli P. (1985). Comparison of the in vitro activities of teicoplanin and vancomycin against *Clostridium difficile* and their interactions with cholestyramine. Antimicrob. Agents Chemother. 28(6):847–848.
379. Robbins M.J., Marais R., Felmingham D., Ridgway G.L., Grunberg R.N. (1987). In vitro activity of vancomycin and teicoplanin against anaerobic bacteria. Drugs Exp. Clin. Res. 13(9):551–554.
380. Ravizzola G., Pirali F., Foresti I., Turano A. (1987). Comparison of the in vitro antibacterial activity of teicoplanin and vancomycin against gram-positive cocci. Drugs Exp. Clin. Res. 13(4):225–229.
381. Cony-Makhoul P., Brossard G., Marit G., Pellegrin J.L., Texier-Maugein J., Reiffers J. (1990). A prospective study comparing vancomycin and teicoplanin as second line empiric therapy for infection in neutropenic patients. Br. J. Haematol 76(Suppl. 2):35–40.
382. Van der Auwera P., Aoun M., Meunier F. (1991). Randomized study of vancomycin versus teicoplanin for the treatment of gram-positive bacterial infections in immunocompromised hosts. Antimicrob. Agents Chemother. 35:451–457.

383. Smith S.R., Cheesbrough J., Spearing R., Davies J.M. (1989). Randomized prospective study comparing vancomycin with teicoplanin in the treatment of infections associated with Hickman catheters. Antimicrob. Agents Chemother. 33(8):193–197.

384. Gilbert D.N., Wood C.A., Kimbrough R.C. (1991). Failure of treatment with teicoplanin at 6 milligrams/kilogram/day in patients with *Staphylococcus aureus* intravascular infection. Infectious Diseases Consortium of Oregon. Antimicrob. Agents Chemother. 35(1):79–87.

385. Van Laethem Y., Hermans P., De Wit S., Goosens H., Clumeck N. (1988). Teicoplanin compared with vancomycin in methicillin-resistant *Staphylococcus aureus* infections: Preliminary results. J. Antimicrob. Chemother 21(Suppl. A):81–87.

386. Kureishi A., Jewesson P.J., Rubinger M., Cole C.D., Reece D., Phillips G.L., Smith J.A., Chow A. (1991). Double-blind comparison of teicoplanin versus vancomycin in febrile neutropenic patients receiving concomitant tobramycin and piperacillin: Effect on cyclosporin A-associated nephrotoxicity. Antimicrob. Agents Chemother. 35(11):2246–2252.

387. Chandrasekar P.H., Cronin S.M. (1991). Nephrotoxicity in bone marrow transplant recipients receiving aminoglycoside plus cyclosporine or aminoglycoside alone. J. Antimicrob. Chemother. 27(6):845–849.

388. Gelfand M.S., Simmons B.P., Threlkeld M.G., Winters D.B., Grogan J.T., Amarshi N., Elsabawy M.A. (1991). Randomized double-blind study of teicoplanin (T) and vancomycin (V) in gram-positive bacteremia (GPB) and endocarditis (GPE). Clin. Res. 39(4):809A.

389. De Lalla F., Nicolin R., Rinaldi E., Scarpellini P., Rigoli R., Manfrin K.V., Tramarin A. (1992). Prospective study of oral teicoplanin versus oral vancomycin for therapy of pseudomembranous colitis and *Clostridium difficile*-associated diarrhea. Antimicrob. Agents Chemother. 36:2192–2196.

390. Bowley J.A., Pickering S.J., Scantlebury A.J., Ackrill P., Jones D.M. (1988). Intraperitoneal teicoplanin in the treatment of peritonitis associated with continuous ambulatory peritoneal dialysis. J. Antimicrob. Chemother. 21(Suppl. A):133–139.

391. Al-Wali W.I., Hamilton-Miller J.M., Foo J., Baillod R., Brumfilt W. (1989). Specific treatment for peritonitis in patients undergoing CAPD: A comparative study of teicoplanin and vancomycin. In Phillips I. (Ed.), Focus on Coagulase Negative Staphylococci. Royal Society of Medicine, London, pp. 119–127.

392. Greenberg R.N. (1990). Treatment of bone, joint, and vascular-access-associated Gram-positive bacterial infections with teicoplanin. Antimicrob. Agents Chemother. 34:2392–2397.

393. Greenberg R.N., Benes C.A. (1990). Time-kill studies with oxacillin, vancomycin, and teicoplanin versus *Staphylococcus aureus* (letter). J. Infect. Dis. 161(5):1036–1037.

394. Favero A.D., Patoia L., Rosina R., Buniva G., Danese A., Bernareggi A., Molini E., Cavenaghi L. (1991). Pharmacokinetics and tolerability of teicoplanin in healthy volunteers after single increasing doses. Antimicrob. Agents Chemother. 35:2551–2557.

395. Dubettier S., Boibieux A., Lagable M.L., Crevon L., Peyramond D., et al. (1991). Red man syndrome with teicoplanin. Rev. Infect. Dis. 13:770.

396. Grek V., Andrien F., Collignon J., Fillet G. (1991). Allergic crossreaction of teicoplanin and vancomycin. J. Antimicrob. Chemother. 28:476–477.
397. Haines S.J., Goodman M.L. (1982). Antibiotic prophylaxis of postoperative neurosurgical wound infection. J. Neurosurg. 56:103–105.
398. Malis L.I. (1979). Prevention of neurosurgical infection by intraoperative antibiotics. Neurosurgery 5:339–343.
399. Quartey G.R.C., Polyzoidis K. (1981). Intraoperative antibiotic prophylaxis in neurosurgery: A clinical study. Neurosurgery 8(6):669–671.
400. Shapiro M., Wald U., Simchen E., et al. (1986). Randomized clinical trial of intraoperative antimicrobial prophylaxis of infection after neurosurgical procedures. J. Hosp. Infect. 8:283–295.
401. Bayston R., Bannister C., Boston V., Burman R., Burns B., Cooke F., Cooke R., Cudmore R., Fitzgerald R., Goldberg C., et al. (1990). A prospective randomised controlled trial of antimicrobial prophylaxis in hydrocephalus shunt surgery. Z. Kinderchir. 45(Suppl. 1):5–7.
402. Bloomstedt G.C., Kytta J. (1988). Results of a randomized trial of vancomycin prophylaxis in craniotomy. J. Neurosurg. 69(2):216–220.
403. Slight P.H., Gundling K., Plotkin S.A., Schut L., Bruce D., Sutton L. (1985). A trial of vancomycin for prophylaxis of infections after neurosurgical shunts (letter). N. Engl. J. Med. 312(14):921.
404. Odio C., Mohs E., Sklar F.H., Nelson J.D., McCracken G.H. (1984). Adverse reactions to vancomycin used as prophylaxis for CSF shunt procedures. Am. J. Dis. Child. 138:17–19.
405. Fan-Havard P., Nahata M.C., Bartkowski M.H., Barson W.J., Kosnik E.J. (1990). Pharmacokinetics and cerebrospinal fluid (CSF) concentrations of vancomycin in pediatric patients undergoing CSF shunt placement. Chemotherapy 36(2):103–108.
406. LeRoux P., Howard M.A., Winn H.R. (1990). Vancomycin pharmacokinetics in hydrocephalic shunt prophylaxis and relationship to ventricular volume. Surg. Neurol. 34(6):366–372.
407. Younger J.J., Simmons J.C., Barrett F.F. (1987). Failure of single-dose intraventricular vancomycin for cerebrospinal fluid shunt surgery prophylaxis. Pediatr. Infect. Dis. J. 6(2):212–213.
408. Burnakis T.G. (1984). Surgical antimicrobial prophylaxis: Principles and guidelines. Pharmacotherapy 4:248–271.
409. Colwell C.W., Frey C., Morris B., Freedman S.D. (1988). Vancomycin distribution in patients undergoing total hip replacement. Thirty-fourth Annual Meeting. Orthopedic Research Society, Atlanta, Georgia.
410. Ritter M.A., Barzilauskas C.D., Faris P.M., Keating E.M. (1989). Vancomycin prophylaxis and elective total joint arthroplasty. Orthopedics 12(10):1333–1336.
411. Cioffi G.A., Terezhalmy G.T., Taybos G.J. (1988). Total joint replacement: A consideration for antimicrobial prophylaxis. Oral Surg. 66:124.
412. Gillespie W.J. (1990). Infection in total joint replacement. Infect. Dis. Clin. North Am. 4(3):465–484.
413. Schwartz C., Henrickson K.J., Roghmann K., Powell K. (1990). Prevention of bacteremia attributed to luminal colonization of tunneled central venous catheters with vancomycin-susceptible organisms. J. Clin. Oncol. 8(9):1591–1597.

414. McKee R., Dunsmuir R., Whitby M., Garden O.J. (1985). Does antibiotic prophylaxis at the time of catheter insertion reduce the incidence of catheter-related sepsis in intravenous nutrition. J. Hosp. Infect. 6(4):419–425.

415. Ransom N.R., Oppenheim B.A., Jackson A., Kamthan A.G., Scarffe J.H. (1990). Double-blind placebo controlled study of vancomycin prophylaxis for central venous catheter insertion in cancer patients. J. Hosp. Infect. 15(1):95–102.

416. Kumar M.K., Bruce J., Stevens R.F., Shanbhogue L.K.R., Jones P.H.M. (1989). The role of vancomycin in immunosuppressed patients requiring central venous access. Med. Pedatr. Oncol. 17(4):329.

417. Battan R., Raviglione M., D'Amore T., Pablos-Mendez A., Raggi P., Hedni R. (1991). Vancomycin prophylaxis for infections related to long term central venous catheters in AIDS patients: A pilot study. AIDS Patient Care 5(3):120–124.

418. Moller J.C., Nachtrodt G., Richter A., Tegtmeyer F.K. (1992). Prophylactic vancomycin to prevent staphylococcal septicaemia in very-low-birth-weight infants. Lancet 340:424.

419. Nadel S. (1992). Prevention of staphylococcal septicaemia in very-low-birthweight infants. Lancet 340:728.

420. Fivush B.A., Bock G.H., Guzzetta P.C., Salcedo J.R., Ruley E.J. (1985). Vancomycin prevents polytetrafluoroethylene graft infections in pediatric patients receiving chronic hemodialysis. Am. J. Kidney Dis. 5(2):120–123.

421. Morris A.J., Bilinsky R.T. (1971). Prevention of staphylococcal shunt infections by continuous vancomycin prophylaxis. Am. J. Med. Sci. 262(2):87–92.

422. Kaiser A.B. (1986). Antimicrobial prophylaxis in surgery. N. Engl. J. Med. 315(18):1129–1137.

423. Jauregui E., Montesinos E., Van Bergen R., Allman J., Matzke D. (1983). Wound infections, sternal osteomyelitis and mediastinitis due to methicillin-resistant staphylococcus epidermidis in coronary bypass patients. 16th World Congress of the International Society for Cardiovascular Surgery, Rio de Janeiro, Brazil. J. Cardiovasc. Surgery 24:445–446.

424. Austin T.W., Leake J., Coles J.C., Goldbach M.M. (1981). Vancomycin blood levels during cardiac bypass surgery. Can. J. Surg. 24(4):423–425.

425. Farber B.F., Karchmer A.W., Buckley M.J., Moellering R.C. (1983). Vancomycin prophylaxis in cardiac operations: Determination of an optimal dosage regimen. J. Thorac. Cardiovasc. Surg. 85(6):933–935.

426. Klamerus K.J., Rodvold K.A., Silverman N.A., et al. (1988). Effect of cardiopulmonary bypass on vancomycin and netilmicin disposition. Antimicrob. Agents Chemother. 32(5):631–635.

427. Vander Salm T.J., Okike O.N., Pasque M.K., Pezzella A.T., Lew R., Traina V. (1989). Reduction of sternal infection by application of topical vancomycin. J. Thorac. Cardiovasc. Surg. 98(4):618–622.

428. Joyce F.S., Szczepanski K.P. (1986). A double-blind comparative study of prophylactic antibiotic therapy in open heart surgery: Penicillin G versus vancomycin. Thorac. Cardiovasc. Surg. 34(2):100–103.

429. Jensen L.J., Aagaard M.T., Schifter S. (1985). Prophylactic vancomycin versus placebo in arterial prosthetic reconstructions. Thorac. Cardiovasc. Surg. 33(5):300–303.

430. Dajani, A.S. et al. (1990). Prevention of bacterial endocarditis: Recommendations by the American Heart Association. JAMA 264:2919–2922.

431. British Society of Antimicrobial Chemotherapy (1990). Antibiotic prophylaxis of infective endocarditis: Recommendations from the Endocarditis Working Party. Lancet 1:88–89.

432. Cars O., Nord C.E., Nordbring F. (1988). Antibiotikaprofylax mot endokardit. Lakartidningen 85:1046–1047.

433. Schweizerische Arbeitsgruppe Fur Endokarditisprophylaxe (1984). Prophylaxe der bakteriellen Endokarditis. Schweiz. Med. Wochenschr 114:1246–1252.

434. Eng R.H., Smith S.M., Goldstein E.J., Miyassaki K.T., Quah S.E., Buccini F. (1986). Failure of vancomycin prophylaxis and treatment for *Actinobacillus actinomycetemcomitans* endocarditis. Antimicrob. Agents Chemother. 29(4):699–700.

435. Prior R.B., Spagna V.A., Perkins R.L. (1979). Endocarditis due to strain of *Cardiobacterium hominis* resistant to erythromycin and vancomycin. Chest 75(1):85–86.

436. Attal M., Schlaifer D., Rubie H., Huguet F., Charlet J.P., Bloom E., Lemozy J., Massip P., Pris J., Laurent G. (1991). Prevention of Gram-positive infections after bone marrow transplantation by systemic vancomycin: A prospective, randomized trial. J. Clin. Oncol. 9(5):865–870.

437. Maraninchi D., Hartman O., Benhamou E., Viens P., Blaise D., Vassal G. (1988). Prophylaxis of Gram positive infections after bone marrow transplantation: A randomized study of prophylactic I.V. vancomycin. Interim analysis in sixty patients. Pathol. Biol. 36(7):915–919.

438. Rubie H., Attal M., Lemozy J., Massip P., Huguet F., Pris J. (1988). Initial antimicrobial prophylaxis after allogenic bone marrow transplantation. Pilot study with systemic vancomycin. Pathol. Biol. 36(7):912–914.

439. Rolando N., Wade J.J., Fagan E., Philpott-Howard J., Casewell M.W., Williams R. (1992). An open, comparative trial of aztreonam with vancomycin and gentamicin with piperacillin in patients with fulminant hepatic failure. J. Antimicrob. Chemother. 30:215–220.

440. Silva J., Batts D.H., Fekety R., Plouffe J.F., Rifkin G.D., Baird I. (1981). Treatment of *Clostridium difficile* colitis and diarrhea with vancomycin. Am. J. Med. 71(5): 815–822.

441. Fekety R., Silva J., Buggy B., Deery H.G. (1984). Treatment of antibiotic-associated colitis with vancomycin. J. Antimicrob. Chemother. 14(Suppl. D):97–102.

442. Bartlett J.G. (1984). Treatment of antibiotic-associated pseudomembranous colitis. Rev. Infect. Dis. 6S:235–241.

443. Batts D.H., Martin D., Holmes R., Silva J., Fekety F.R. (1980). Treatment of antibiotic-associated *Clostridium difficile* diarrhea with oral vancomycin. J. Pediatr. 97(1):151–153.

444. Keighley R.R., Burdon D.W., Arabi Y., Williams J.A., Thompson H., Youngs D., Johnson M., Bentley S., George R.H., Mogg G.A. (1978). Randomised controlled trial of vancomycin for pseudomembranous colitis and postoperative diarrhoea. BMJ 16;2(6153)1667–1669.

445. Fekety R., Silva J., Armstrong J., Allo M., Browne R., Ebright J., Lusk R., Rifkin G., Toshniwal R. (1981). Treatment of antibiotic-associated enterocolitis with vancomycin. Rev. Infect. Dis. 3(Suppl.):S273–281.

446. Fekety R., Silva J., Kauffman C., Buggy B., Deery H.G. (1989). Treatment of antibiotic-associated *Clostridium difficile* colitis with oral vancomycin: comparison of two dosage regimens. Am. J. Med. 86(1):15–19.

447. Teasley D.G., Gerding D.N., Olson M.M., Peterson L.R., Gebhard R.L., Schwartz M.J., Lee J.T. (1983). Prospective randomised trial of metronidazole versus vancomycin for *Clostridium difficile*-associated diarrhoea and colitis. Lancet 2(8358): 1043–1046.

448. Dudley M.N., McLaughlin J.C., Carrington G., Frick J., Nightingale C.H., Quintilliani R. (1986). Oral bacitracin vs vancomycin therapy for *Clostridium difficile*-induced diarrhea. A randomized double-blind trial. Arch. Intern. Med. 146(6): 1101–1104.

449. Young G.P., Ward P.B., Bayley N., Gordon D., Higgins G., Trapani J.A., McDonald M.I., Labrooy J., Hecker R. (1985). Antibiotic-associated colitis due to *Clostridium difficile*: Double-blind comparison of vancomycin with bacitracin. Gastroenterology 89(5):1038–1045.

450. Gordon R.S. (1983). Metronidazole or vancomycin for *Clostridium difficile* associated diarrhoea (letter). Lancet 2(8364):1417.

451. Cronberg S., Castor B., Thoren A. (1984). Fusidic acid for the treatment of antibiotic-associated colitis induced by *Clostridium difficile*. Infection 12(4): 276–279.

452. De Lala F., Privitera G., Rinaldi E., Ortisi G., Santoro D., Rizzardini G. (1989). Treatment of *Clostridium difficile*-associated disease with teicoplanin. Antimicrob. Agents Chemother. 33(7):1125–1127.

453. Rampling A., Warren R.E., Sykes H.V. (1980). Relapse of *Clostridium* colitis after vancomycin therapy. J. Antimicrob. Chemother. 6(4):551–552.

454. George W.L., Volpicelli N.A., Stiner D.B., Richman D.D., Liechty E.J., Mok H.Y., Rolfe R.D., Finegold S.M. (1979). Relapse of pseudomembranous colitis after vancomycin therapy. N. Engl. J. Med. 301(8):414–415.

455. Bartlett J.G., Tedesco F.J., Shull S., Lowe B., Tewen C. (1980). Symptomatic relapse after oral vancomycin therapy of antibiotic-associated pseudomembranous colitis. Gastroenterology 78:431–434.

456. Walters B.A., Roberts R., Stafford R., Seneviratne E. (1983). Relapse of antibiotic associated colitis: Endogenous persistence of *Clostridium difficile* during vancomycin therapy. Gut 24(3):206–212.

457. Young G.P., Bayley N., Ward P., St. John J.B., McDonald M.I. (1986). Antibiotic-associated colitis caused by *Clostridium difficile*: Relapse and risk factors. Med. J. Aust. 144:303–306.

458. Johnson S., Adelmann A., Clabots C.R., Peterson L., Gerding D. (1989). Recurrences of *Clostridium difficile* diarrhea not caused by the original infecting organism. J. Infect. Dis. 159(2):340–343.

459. Oldenburger D., Miller J.A. (1980). Treatment of pseudomembranous colitis with oral metronidazole after relapse following vancomycin. Am. J. Gastroenterol. 74: 359–360.

460. Tedesco F.J., Gordon D., Fortson W.C. (1985). Approach to patients with multiple relapses of antibiotic-associated pseudomembranous colitis. Am. J. Gastroenterol. 80(11):867–868.

461. Tedesco F.J. (1982). Treatment of recurrent antibiotic-associated pseudomembranous colitis. Am. J. Gastroenterol. 77(4):220–221.

462. Pruksananonda P., Powell K.R. (1989). Multiple relapses of *Clostridium difficile*-associated diarrhea responding to an extended course of cholestyramine. Pediatr. Infect. Dis. J. 8:175–178.

463. Surawicz C.M., McFarland L.V., Elmer G., Chinn J. (1989). Treatment of recurrent *Clostridium difficile* colitis with vancomycin and *Saccharomyces boulardii*. Am. J. Gastroenterol. 84(10):1285–1287.

464. Buggy B.P., Fekety R., Silva J. (1987). Therapy of relapsing *Clostridium difficile*-associated diarrhea and colitis with the combination of vancomycin and rifampin. J. Clin. Gastroenterol. 9(2):155–159.

465. Moog A.G., Arabi Y., Youngs D., Johnson M., Bentley S., Burdon W., Keighley M.R.B. (1980). Therapeutic trials of antibiotic associated colitis. Scand. J. Infect. Dis. (Suppl.) 22:41–45.

466. Bacon A.E., McGrath S., Fekety R., Holloway W.L.J. (1991). In vitro synergy studies with *Clostridium difficile*. Antimicrob. Agents Chemother. 35:582–583.

467. Taylor N.S., Barlett J.G. (1980). Binding of *Clostridium difficile* cytotoxin and vancomycin by anion-exchange resins. J. Infect. Dis. 140:92–97.

468. Tvede M., Rask-Madsen J. (1989). Bacteriotherapy for chronic relapsing *Clostridium difficile* diarrhoea in six patients. Lancet 1:1156–1160.

469. Weal D., Borriello S.P., Barclay F., Welch A., Piper M., Bonnycastle M. (1987). Treatment of relapsing *Clostridium difficile* diarrhoea by administration of a nontoxigenic strain. Eur. J. Clin. Microbiol. 6(1):51–53.

470. Gorbach S.L., Chang T., Goldin B. (1987). Successful treatment of relapsing *Clostridium difficile* colitis with *Lactobacillus* GG. Lancet 2:1519.

471. Delmee M., Vandercam B., Avesani V., Michaux J.L. (1987). Epidemiology and prevention of *Clostridium difficile* infections in a leukemia unit. Eur. J. Clin. Microbiol. 6(6):623–627.

472. Johnson S., Homann S.R., Bettin K.M., Quick J.N., Clabots C.R., Peterson L.R., Gerding D.N. (1992). Treatment of asymptomatic *Clostridium difficile* carriers (fecal excretors) with vancomycin or metronidazole. A randomized, placebo-controlled trial. Ann. Intern. Med. 117(8):297–302.

473. Wallace J.F., Smith R.H., Petersdorf R.G. (1965). Oral administration of vancomycin in the treatment of staphylococcal enterocolitis. N. Engl. J. Med. 272:1014–1015.

474. Kahn M.Y., Hall W.H. (1966). Staphylococcal enterocolitis—treatment with oral vancomycin. Ann. Intern. Med. 65:1–8.

475. Ng P.C., Dear P.R., Thomas D.F. (1988). Oral vancomycin in prevention of necrotizing enterocolitis. Arch. Dis. Child. 63(11):1390–1393.

476. Tulloh R., Brownlee K., Dear P. (1986). Vancomycin and necrotising enterocolitis (letter). Arch. Dis. Child. 61(7):719–720.

477. Han V.K.M., Sayed H., Chance G.W., et al. (1983). An outbreak of *Clostridium difficile* necrotizing enterocolitis: A case for oral vancomycin therapy? Pediatrics 71(6):935–941.

478. Freeman H.J., Rabeneck L., Owen D. (1981). Survival after necrotizing enterocolitis of leukemia treated with oral vancomycin. Gastroenterology 81(4):791–794.

479. Pinder I.F., Hamilton I., Dickinson R.J., et al. (1982). Vancomycin as adjunctive therapy in acute exacerbations of idiopathic colitis. Gut 23(5):f18.
480. Gardner M.E. (1981). Use of vancomycin in treating ulcerative colitis (letter). Am. J. Hosp. Pharm. 38(4):471–473.
481. Dickinson R.J., O'Connor H.J., Pinder I., et al. (1985). Double blind controlled trial of oral vancomycin as adjunctive treatment in acute exacerbations of idiopathic colitis. Gut 26(12):1380–1384.
482. Tarao K., Ikeda T., Hayashi K., Sakurai A., Okada T., Ito T., Karube H., Nomoto T., Mizuno T., Shindo K. (1990). Successful use of vancomycin hydrochloride in the treatment of lactulose resistant chronic hepatic encephalopathy. Gut 31(6):702–706.
483. Bodey G.P., Keating M.J., McCredie K.B., Elting L., Rosenbaum B., Freireich E.J. (1985). Prospective randomized trial of antibiotic prophylaxis in acute leukemia. Am. J. Med. 79:497–514.
484. Henry S.A. (1984). Chemoprophylaxis of bacterial infections in granulocytopenic patients. Am. J. Med. 76:645–651.
485. Bender J.F., Schimpff S.C., Young V.M., Fortner C.L., Love L.J., Brouillet M.D., Wiernik P.H. (1979). A comparative trial of tobramycin vs gentamicin in combination with vancomycin and nystatin for alimentary tract suppression in leukemic patients. Eur. J. Cancer 15(Suppl.):35–44.
486. Winston D.J., Ho W.G., Bruckner D.A., Gale R.P., Champlin R.E. (1990). Ofloxacin versus vancomycin/polymyxin for prevention of infections in granulocytopenic patients. Am. J. Med. 88(1):36–42.
487. Winston D.J., Ho W.G., Nakao S.L., Gale R.P., Champlin R.E. (1986). Norfloxicin versus vancomycin/polymyxin for prevention of infections in granulocytopenic patients. Am. J. Med. 80(5):884–890.
488. Malarme M., Meunier-Carpentier F., Klastersky J. (1981). Vancomycin plus gentamicin and cotrimoxazole for prevention of infections in neutropenic cancer patients (a comparative, placebo-controlled pilot study). Eur. J. Cancer Clin. Oncol. 17(12):1315–1322.
489. Guyotat D., Plonton C., Fiere D. (1985). A randomized trial of oral vancomycin in neutropenic patients. Prog. Clin. Biol. Res. 181:263–265.
490. Gluckman E., Roudet C., Hirsch I., Devergie A., Bourdeau H., Arlet C., Perol Y. (1991). Prophylaxis of bacterial infections after bone marrow transplantation. A randomized prospective study comparing oral broad-spectrum nonabsorbable antibiotics (vancomycin-tobramycin-colistin) to absorbable antibiotics (ofloxicin-amoxicillin). Chemotherapy 37(Suppl. 1):33–38.
491. Classen D.C., Burke J.P., Ford C.D., Dvershed S., Aloia M.R., Wilfahrt J.K., Elliott J.A. (1990). Streptococcus mitis sepsis in bone marrow transplant patients receiving oral antimicrobial prophylaxis. Am. J. Med. 89(4):441–446.
492. Flanigan M.J., Lim V.S. (1991). Initial treatment of dialysis associated peritonitis: A controlled trial of vancomycin versus cefazolin. Peritoneal Dialysis Int. 11:31–37.
493. Bastani B., Freer K., Read D., Bailey S., Sherman R.A., Davis M., Engels D., Westervelt F.B. (1987). Treatment of Gram-positive peritonitis with two intraperitoneal doses of vancomycin in continuous ambulatory peritoneal dialysis patients. Nephron 45(4):283–385.

494. Brauner L., Kahlmeter G., Lindholm T., Simonsen O. (1985). Vancomycin and netilmicin as first line treatment of peritonitis in CAPD patients. J. Antimicrob. Chemother. 15(6):751–758.
495. Beaman M., Solaro L., McGonigle R.J.S., Michael J., Adu D. (1989). Vancomycin and ceftazidime in the treatment of CAPD peritonitis. Nephron 51(1):51–55.
496. Maskill R., Crump H. (1985). Vancomycin in the treatment of CAPD peritonitis. J. Antimicrob. Chemother. 15(5):647–648.
497. Gruer L.D., Turney J.H., Curley J., Michael J., Adu D. (1985). Vancomycin and tobramycin in the treatment of CAPD peritonitis. Nephron 41(3):279–282.
498. Gray H.H., Goulding S., Eykyn S.J. (1985). Intraperitoneal vancomycin and ceftazidime in the treatment of CAPD peritonitis. Clin. Nephrol. 23(2):81–84.
499. Were A.J., Marsden A., Tooth A., Ramsden R., Mistry C.D., Gokal R. (1992). Netilmycin and vancomycin in the treatment of peritonitis in CAPD patients. Clin. Nephrol. 37:209–213.
500. Morse G.D., Nairn D.K., Walshe J.J. (1987). Once weekly intraperitoneal therapy for Gram-positive peritonitis. Am. J. Kidney Dis. 10(4):300–305.
501. Boyce N.W., Wood C., Thompson N.M., Kerr P., Atkins R.C. (1988). Intra-peritoneal (IP) vancomycin therapy for CAPD peritonitis—a prospective, randomized comparison of intermittent v continuous therapy. Am. J. Kidney Dis. 12(4): 304–306.
502. Bailie G.R., Morton R., Ganguli L., et al. (1987). Intravenous or intraperitoneal vancomycin for the treatment of continuous ambulatory peritoneal dialysis associated Gram-positive peritonitis. Nephron 46(3):316–318.
503. Friedland J.S., Iveson T.J., Fraise A.P., Winearls C.G., Selkon J.B., Oliver D.O. (1990). Comparison between intraperitoneal ciprofloxicin and intraperitoneal vancomycin and gentamicin in the treatment of peritonitis associated with continuous ambulatory peritoneal dialysis (CAPD). J. Antimicrob. Chemother. 26(Suppl. F): 77–81.
504. Bennett-Jones D.N., Russell G.I., Barrett A. (1990). A comparison between oral ciprofloxicin and intra-peritoneal vancomycin and gentamicin in the treatment of CAPD peritonitis. J. Antimicrob. Chemother. 26(Suppl. F):73–76.
505. Tapson J.S., Orr K.E., George J.C., Stansfield E., Bint A.J., Ward M.K. (1990). A comparison between oral ciprofloxicin and intraperitoneal vancomycin and netilmicin in CAPD peritonitis. J. Antimicrob. Chemother. 26(Suppl. F):63–71.
506. Cheng I.K.P., Chan C., Wong W.T. (1991). A randomised prospective comparison of oral ofloxicin and intraperitoneal vancomycin plus aztreonam in the treatment of bacterial peritonitis complicating continuous ambulatory peritoneal dialysis (CAPD). Peritoneal Dialysis Int. 11:27–30.
507. Lam T.Y., Vas S.I., Oreopoulos D.G. (1991). Long-term, intraperitoneal vancomycin in the prevention of recurrent peritonitis during CAPD: Preliminary results. Peritoneal Dialysis Int. 11:281–282.
508. Bolton R.P., Thomas D.F.M. (1986). Pseudomembranous colitis in children and adults. Br. J. Hosp. Med. 1986:37–42.
509. Griebie M., Adams G.L. (1985). *Clostridium difficile* colitis following head and neck surgery. Arch. Otolaryngol. 111:550–553.

510. Osler T., et al. (1986). Cefazolin-induced pseudomembranous colitis resulting in perforation of the sigmoid colon. Dis. Colon Rectum 29(2):140–143.

511. Lee J.T., Olson M. (1986). Clostridium difficile colitis/diarrhea. Arch. Otolaryngol. Head Neck Surg 112:335.

512. Goodpasture H.C., Donan P.J., Jacobs E.R., Meredith W.T. (1986). Pseudomembranous colitis and antibiotics. Kans. Med. 1986:133–146.

513. Silva J. Jr. (1989). Update on pseudomembranous colitis (medical staff conference). West. J. Med. 151:644–648.

514. Davis J.L., Koidou-Tsiligianni A., Pflugfelder S.C., Miller D., Flynn H.W., Forster R.K. (1988). Coagulase-negative staphylococcal endophythalmitis. Ophthalmology 95:1404–1410.

515. Weber D.J., Hoffman K.L., Thoft R.A., Baker A.S. (1986). Endophthalmitis following intraocular lens implantation: Report of 30 cases and review of the literature. Rev. Infect. Dis. 8(1):12–20.

516. Goodman D.F., Gottsch J.D. (1988). Methicillin-resistant Staphylococcus epidermidis keratitis treated with vancomycin. Arch. Ophthalmol. 106:1570–1571.

517. Fleischer A.B., Hoover D.L., Khan J.A., Parisi J.T., Burns R.P. (1986). Topical vancomycin formulation for methicillin-resistant Staphylococcus epidermidis blepharoconjunctivitis. Am. J. Ophthalmol. 101(3):283–287.

518. Zabel R.W., Mintsioulis G., MacDonald I., Tuft S. (1988). Infectious crystalline keratopathy. Can. J. Ophthalmol. 23:311–314.

519. Kahn J.A., Hoover D.L., Ide C.H. (1984). Methicillin-resistant Staphylococcus epidermidis blepharitis. Am. J. Ophthalmol. 98:562–565.

520. Ross J., Abate M.A. (1990). Topical vancomycin for the treatment of Staphylococcus epidermidis and methicillin-resistant Staphylococcus aureus conjunctitis. DICP 24(11):1050–1053.

521. Osborne E., Baum J.L., Ernst C., Koch P. (1976). The stability of ten antibiotics in artificial tear solutions. Am. J. Ophthalmol. 82:775–780.

522. Garcia-Ferrer F.J., Pepose J.S., Murray P.R., Glaser S.R., Lass J.H., Green W.R. (1991). Antimicrobial efficacy and corneal endothelial toxicity of Dexsol corneal storage medium supplemented with vancomycin. Ophthalmology 98:863.

523. Steinemann T.L. (1992). Vancomycin-enriched corneal storage medium. Am. J. Ophthalmol. 113:555–560.

524. Baum J., Peyman G.A., Barza M. (1982). Intravitreal administration of antibiotics in the treatment of endophthalmitis. III. Consensus. Surv. Ophthalmol. 26:204–206.

525. Mandelbaum S., Forster R.K. (1987). Postoperative endophthalmitis. Int. Opthalmol. Clin. 27(2):95–106.

526. Doft B.H. (1991). The endophthalmitis vitrectomy study. Arch. Ophthalmol. 109:487–488.

527. Barza M. (1989). Antibacterial agents in the treatment of ocular infections. Infect. Dis. Clin. North Am. 3(3):533–551.

528. Kawasaki K., Mochizuki K., Toisaki M., Yamashita Y., et al. (1990). Electroretinographical changes due to antimicrobials. Lens Eye Toxicity Res. 7(3,4):693–704.

529. Mochizuki K., Torisaki M., Wakabayashi K. (1991). Effects of vancomycin and ofloxacin on rabbit ERG in vivo. Jpn. J. Ophthalmol. 35:435–445.

530. Pflugfelder S.C., Hernandez E., Fliesler S.J., Alvarez J., et al. (1987). Intravitreal vancomycin: Retinal toxicity, clearance, and interaction with gentamicin. Arch. Ophthalmol. 105(6):831–837.

531. Oum B.S., D'Amico D.J., Kwak H.W., Wong K.W. (1992). Intravitreal antibiotic therapy with vancomycin and aminoglycoside: Examination of the retinal toxicity of repetitive injections after vitreous and lens surgery. Graefes Arch. Clin. Exp. Ophthalmol. 230(1):56–61.

532. Gills J.P. (1987). Antibiotics in irrigating solutions (letter). J. Cataract Refract. Surg. 13(3):344.

533. Davey R.T., Tauber W.B. (1987). Posttraumatic endophthalmitis: The emerging role of *Bacillus cereus* infection. Rev. Infect. Dis. 9:110–123.

534. Weber D.J., Ratula W.A. (1988). Bacillus species. Infect. Control. Hosp. Epidemiol. 9:368–373.

535. Hemady R., Zaltas M., Paton B., Foster C.S., Baker A.S. (1990). Bacillus-induced endophthalmitis: New series of 10 cases and review of the literature. Br. J. Ophthalmol. 74:26–29.

536. Meisler D.M., Mandelbaum S. (1989). *Propionibacterium*-associated endophthalmitis after extracapsular cataract extraction: Review of reported cases. Ophthalmology 96:54–61.

537. Zambrano W., Flynn H.W. Jr., Pflugfelder S.C., Roussel T.J., Culbertson W.W., Holland S., Miller D. (1989). Management options for *Propionibacterium acnes* endophthalmitis. Ophthalmology 96:1100–1105.

538. Yu V.L., Goetz A., Wagener M., Smith P.B., Rihs J.D., Hanchett J., Zuravleff J.J. (1986). *Staphylococcus aureus* nasal carriage and infection in patients on hemodialysis. Efficacy of antibiotic prophylaxis. N. Engl. J. Med. 10;315(2):91–96.

539. Williams J.D., Waltho C.A., Ayliffe G.A., Lowbury E.J. (1967). Trials of five antibacterial creams in the control of nasal carriage of *Staphylococcus aureus*. Lancet 2(512):390–392.

540. Bryan C.S., Wilson R.S., Meade P., Sill L.G. (1980). Topical antibiotic ointments for staphylococcal nasal carriers: Survey of current practices and comparison of bacitracin and vancomycin ointments. Infect. Control 1(3):153–156.

541. Bartlett R.C., Howell R.M. (1973). Topical vancomycin as a deterrent to bacteremias following dental procedures. Oral Surg. Oral Med. Oral Pathol. 35(6): 780–788.

542. Collins J.F. (1968). Utilization of topical vancomycin dental ointment in the treatment of recurrent herpes, aphthous, and traumatic lesions of the oral mucous membranes. J. Oral Med. 23(3):99–103.

543. Collins J.F., Hood H.M. (1967). Topical antibiotic treatment of acute necrotizing ulcerative gingivitis. J. Oral Med. 22(2):59–64.

544. Kaslick R.S., Tuckman M.A., Chasens A.I. (1973). Effect of topical vancomycin on plaque and chronic gingival inflammation. J. Periodontol. 44(6):366–368.

545. Mitchell D.F., Baker B.R. (1968). Topical antibiotic control of necrotizing gingivitis. J. Periodontol. 39(2):81–82.

546. Emslie R.D., Ashley F.P. (1971). Topical treatment of acute ulcerative gingivitis: A comparison of vancomycin with penicillin and metronidazole. Parodontologie 25(1):3–8.

547. Borthen Svinhufvud L., Heimdahl A., Nord C.E. (1988). Effect of topical administration of vancomycin versus chlorhexicdine on alpha-hemolytic streptococci in the oral cavity. Oral Surg. Oral Med. Oral Pathol. 66(3):304–309.

548. Jordan H.V., DePaola P.F. (1977). Effect of prolonged topical application of vancomycin on human oral *Streptococcus mutans* populations. Arch. Oral Biol. 22(3): 193–199.

549. DePaola P.F., Jordan H.V., Soparkar P.M. (1977). Inhibition of dental caries in school children by topically applied vancomycin. Arch. Oral Biol. 22(3):187–191.

550. Weathers L., Riggs D., Santeiro M., Weibley R.E. (1990). Aerosolized vancomycin for treatment of airway colonization by methicillin-resistant *Staphylococcus aureus*. Pediatr. Infect. Dis. J. 9(3):220–221.

551. Nakano A. (1992). Effect of inhalation treatment with vancomycin hydrochloride on MRSA. Chemotherapy (Jpn.) 40(4):481–490.

552. Gradon J.D., Wu E.H., Lutwick L.I. (1992). Aerosolized vancomycin therapy facilitating nursing home placement. Ann. Pharmacother. 26:209–210.

553. Marchand S., Poisson D., Borderon J.C., Gold F., Chantepie A., Saliba E., Laugier J. (1990). Randomized study of vancomycin pharyngeal instillation as a prophylaxis of bronchopulmonary infection in intubated neonates. Biol. Neonate 58(5):241–246.

554. Pugin J., Auckenthaler R., Lew D.P., Suter P.M. (1991). Oropharyngeal decontamination decreases incidence of ventilator-associated pneumonia: A randomized, placebo-controlled, double-blind clinical trial. JAMA 265:2704–2710.

555. Gauer S.M., Lisby S.R., Deeter R.G., Kesarwala H.H., Frenkel R.D. (1989). Pharmacokinetics of vancomycin (vanc) after intraventricular (inv) administration in CSF shunt infection (si). Clin. Pharmacol. Ther. Abstract PPA-4. 45(2):122.

556. Reesor C., Chow A.W., Kureishi A., Jewesson P.J. (1988). Kinetics of intraventricular vancomycin in infections of cerebrospinal fluid shunts (letter). J. Infect. Dis. 158(5):1142–1143.

557. Kay E.A., Sanders P.J., Slesenger J.P., Bailie G.R. (1988). Disposition of intrathecal vancomycin. Drug Intell. Clin. Pharm. 22(3):267–268.

558. McGee S.M., Kaplan S.L., Mason E.O. (1990). Ventricular fluid concentrations of vancomycin in children after intravenous and intraventricular administration. Pediatr. Infect. Dis. J. 9(2):138–139.

559. Hirsch B.E., Amodio M., Einzig A.L., Halevy R., Soeiro R. (1991). Instillation of vancomycin into a cerebrospinal fluid reservoir to clear infection: pharmacokinetic considerations. J. Infect. Dis. 163(1):197–200.

560. Golledge C.L., McKenzie T. (1988). Monitoring vancomycin concentrations in CSF after intraventricular administration (letter). J. Antimicrob. Chemother. 21(2): 262–263.

561. Congeni B.L., Tan J., Salstrom S.D. (1979). Kinetics of vancomycin after intraventricular and intravenous administration. Pediatr. Res. 13:459–463.

562. Arrayo J.C., Quindlen E.A. (1983). Accumulation of vancomycin after intraventricular infusions. South Med. J. 76(12):1554–1555.

563. Pau A.K., Smego R.A., Fisher M.A. (1986). Intraventricular vancomycin: Observations of tolerance and pharmacokinetics in two infants with ventricular shunt infections. Pediatr. Infect. Dis. 5(1):93–96.

564. Seufert R.J., Blackman P., Hudson W. (1990). Recommended procedures for the

preparation of intrathecal dosing in the hospital pharmacy. Hosp. Pharm. 25: 849–851.

565.   Bayston R. (1988). CSF vancomycin concentrations (letter). J. Antimicrob. Chemother. 22(2):265–266.

566.   Grabb P.A., Albright A.L. (1992). Intraventricular vancomycin-induced cerebrospinal fluid eosinophilia: Report of two patients. Neurosurgery 30:630–635.

567.   Greenberg R.N., Saeed A.M.K., Kennedy D.J., McMillian R. (1987). Instability of vancomycin in Infusaid drug pump model 100. Antimicrob. Agents Chemother. 31(4):610–611.

568.   Beeching N.J., et al. (1986). Comparative in-vitro activity of antibiotics incorporated in acylic bone cement. J. Antimicrob. Chemother. 17:173–184.

569.   Kuechle D.K., Landon G.C., Musher D.M., Noble P.C. (1991). Elution of vancomycin, daptomycin, and amikacin from acrylic bone cement. Clin. Orthop. 264: 302–308.

570.   Adams K., Couch L., Cierny G., Calhoun J., Mader J.T. (1992). In vitro and in vivo evaluation of antibiotic diffusion from antibiotic-impregnated polymethylmethacrylate beads. Clin. Orthop. May(278):244–252.

571.   Lawson K.J., Marks K.E., Brems J., et al. (1990). Vancomycin vs. tobramycin elution from polymethymethacrylate: An in vitro study. Orthopedics 13:521–524.

572.   Gerhart T.N., Roux R.D., Horowitz G., Miller R.L., Hanff P., Hayes W.C. (1988). Antibiotic release from an experimental biodegradable bone cement. J. Orthop. Res. 6(4):585–592.

573.   Scott D.M., Rotschafer J.C., Behrens F. (1988). Use of vancomycin and tobramycin polymethylmethacrylate impregnated heads in the management of chronic osteomyelitis. Drug Intell. Clin. Pharm. 22(6):480–483.

574.   Garrelts J.C., Peterie J.D. (1985). Vancomycin and the "red man's syndrome." N. Engl. J. Med. 312:245.

575.   Ackerman B.H., Bradsher R.W. (1985). Vancomycin and red necks. Ann. Intern. Med. 102:723–724.

576.   Royston K.V., Hoy J. (1986). Man/woman achieves whole red personhood through vancomycin (letter). JAMA 255(18):2445.

577.   Polk R.E. (1992). Anaphylactoid reactions to glycopeptide antibiotics. J. Antimicrob. Chemother. 27(Suppl. B):17–29.

578.   Gatterer G. (1984). Spasmodic low back pain in a patient receiving intravenous vancomycin during continuous ambulatory peritoneal dialysis. Clin. Pharm. 3: 87–89.

579.   Cohen L.G., Souney P.F., Taylor S.J. (1988). Paresthesia and back pain in a patient receiving vancomycin during hemodialysis. Drug Intell. Clin. Pharm. 22(10): 784–785.

580.   Miralles R., Pedro-Botet J., Rubies-Prat J. (1990). Chest pain—another adverse reaction to vancomycin. Chest 97:1504.

581.   Bailie G.R., Yu R., Morton R., Waldek S. (1985). Vancomycin, red neck syndrome, and fits (letter). Lancet 2(8449):279–280.

582.   Mellor J. (1983). Vancomycin and chronic hypotension (letter). J. R. Soc. Med. 76(1):83.

583. Arroyo J.C., Rosansky S.J., Rosenzweig P.N. (1986). Red neck syndrome with slow infusion of vancomycin (letter). Am. J. Kidney Dis. 7(6):511.

584. Davis R.L., Smith A.L., Koup J.R. (1986). The "red man's syndrome" and slow infusion of vancomycin (letter). Ann. Intern. Med. 104(2):285–286.

585. Pau A.K., Khakoo R. (1985). "Red-neck syndrome" with slow infusion of vancomycin (letter). N. Engl. J. Med. 313(12):756–757.

586. Bailie G.R., Kowalsky S.F., Eisele G. (1990). Red-neck syndrome associated with intraperitoneal vancomycin (letter). Clin. Pharm. 9(9):671–672.

587. Killian A.D., Sahai J.V., Memish Z.A. (1991). Red man syndrome after oral vancomycin (letter). Ann. Intern. Med. 115(5):410–411.

588. Muoghalu B.U., Lattimer G.L. (1988). Delayed red neck syndrome with generic vancomycin (letter). Drug Intell. Clin. Pharm. 22(2):173.

589. Daly B.M., Sharkey I. (1986). Nifedipine and vancomycin-associated red man syndrome. Drug Intell. Clin. Pharm. 20(12):986.

590. Rybak M.J., Bailey E.M., Warbasse L.H. (1992). Absence of "red man syndrome" in patients being treated with vancomycin or high-dose teicoplanin. Antimicrob. Agents Chemother. 36:1204–1207.

591. Comstock T.J., Sica D.A., Fichtl R.E., Fakhry I., Davis J. (1989). Vancomycin-induced histamine release in patients with end-stage renal disease (ESRD). Clin. Pharmacol. Ther. 45(2):181.

592. Estrada V., Algarra J., Vargas E., Jimenez-De-Diego (1992). Red neck syndrome induced by erythromycin. Rev. Clin. Esp. 190:100–101.

593. Levy M., Koren G., Dupuis L., Read S.E. (1990). Vancomycin-induced red man syndrome. Pediatrics 84(4):572–580.

594. Barton L.L. (1988). Vancomycin reactions in preterm infants (letter). J. Pediatr. 112(2):334.

595. Levy J.H., Kettlekamp N., Goertz P., Hermens J., Hirshman C.A. (1987). Histamine release by vancomycin: A mechanism for hypotension in man. Anesthesiology 67(1):122–125.

596. Polk R.E., Healy D.P., Schwartz L.B., Rock D.T., Garson M.L., Roller K. (1988). Vancomycin and the red-man syndrome: Pharmacodynamics of histamine release. J. Infect. Dis. 157(3):502–507.

597. Healy D.P., Sahai J.V., Fuller S.H., Polk R.E. (1990). Vancomycin-induced histamine release and "red man syndrome": Comparison of 1- and 2-hour infusions. Antimicrob. Agents Chemother. 34(4):550–554.

598. Verberg K.M., Bowsher R.R., Israel K.S., Black H.R., Henry D.P. (1985). Histamine release by vancomycin in humans. (Abstract 4890). Fed Proc. 44:1632.

599. Sahai J., Healy D.P., Shelton M.J., Miller J.S., Ruberg S.J., Polk R. (1990). Comparison of vancomycin- and teicoplanin-induced histamine release and "red man syndrome." Antimicrob. Agents Chemother. 34(5):765–769.

600. Sahai J.V., Polk R.E., Schwartz L.B., Healy D.P., Westin E.H. (1988). Severe reaction to vancomycin not mediated by histamine release and documented by rechallenge (letter). J. Infect. Dis. 158(6):1413–1414.

601. Rimailho A., Riou B., Teboul J.L., Richard C., Auzepy P. (1984). Erythrodermie immuno-allergique a la vancomycine. Presse Med. 13(9):567.

602. Sahai J., Healy D.P., Garris R., Berry A., Polk R.E. (1989). Influence of anti-
      histamine pretreatment on vancomycin-induced red-man syndrome. J. Infect. Dis.
      160(5):867–881.
603. Wallace M.R., Mascola J.R., Oldfield E.C. (1991). Red man syndrome: Incidence,
      etiology and prophylaxis. J. Infect. Dis. 164:1180–1185.
604. Lyon G.D., Bruce D.L. (1988). Diphenhydramine reversal of vancomycin-induced
      hypotension. Anesth. Analg. 67(11):1109–1110.
605. Polk R., Isreal D., Wang J., Stotka J., Miller J. (1991). Vancomycin (V) skin test as a
      predictor of anaphylactoid reactions. Abstract 1300. Interscience Conference of
      Antimicrobial Agents and Chemotherapy.
606. Newfield P., Roizen M.F. (1979). Hazards of rapid administration of vancomycin.
      Ann. Intern. Med. 91:581.
607. Waters B.G., Rosenberg M. (1981). Vancomycin-induced hypotension. Oral Surg.
      52:239–240.
608. Laconture P.G., Epstein M.F., Mitchell A.A. (1987). Vancomycin-associated shock
      and rash in newborn infants. J. Pediatr. 111(4):615–616.
609. Hill L.M. (1985). Fetal distress secondary to vancomycin-induced maternal hypo-
      tension. Am. J. Obstet. Gynecol.153(1):74–75.
610. Cohen L.S., Wechler A.S., Mitchell J.H., Glick G. (1970). Depression of cardiac
      function by streptomycin and other antimicrobial agents. Am. J. Cardiol. 26:
      505–511.
611. Stier G.R., McGory R.W., Spotnitz W.D., Schwenzer K.J. (1990). Hemodynamic
      effects of rapid vancomycin infusion in critically ill patients. Anesth. Analg.
      71(4):394–399.
612. Best C.J., Ewart M., Somner E. (1989). Perioperative complications following the
      use of vancomycin in children: A report of two cases. Br. J. Anaesth. 62(5):
      576–577.
613. Southorn P.A., Plevak D.J., Wright A.J., Wilson W.R. (1986). Adverse effects of
      vancomycin administration in the perioperative period. Mayo Clin. Proc. 61(9):
      721–724.
614. Himmet P., Hiller L., Miller J., Oren R. (1984). Profound hypotension from rapid
      vancomycin administration during cardiac operation. J. Thorac. Cardiovasc. Surg.
      87(1):145–146.
615. Tetzlaff J.E., Spitzer L.E. (1987). Hemodynamic parameters during a perioperative
      vancomycin reaction. J. Am. Assoc. Nurse Anaesthetists 55(2):126–128.
616. Forestner J.E., Wright J.A. (1984). Vancomycin-induced hypotension in a child with
      congenital heart disease. Clin. Pediatr. (Phila.) 23(7):416.
617. Valero R., Gomar C., Fita G., Bonzalez M., Pacheco M., Mulet J., Nalda M.A.
      (1991). Adverse reactions to vancomycin prophylaxis in cardiac surgery. J. Cardio-
      thorac. Vasc. Anesth. 5:574–576.
618. Feinberg B.I., LaMantia K.R. (1985). Hemodynamic effects of vancomycin during
      cardiac surgery. Unpublished abstract.
619. Von Koenel W.E., Bloomfield E.L., Wilde A.H., Hicks S.R., Nethers B.A. (1989).
      Vancomycin does not potentiate hypotension under anesthesia. Anaesthesiology
      71(3A):Abstract A307.

620. Romanelli V.A., Howie M.B., Zvara D.A., McSweeney T.D., Myerowitz P.D. (1989). Double-blind evaluation of vancomycin administration during cardiac surgery. Anaesthesiology 71(3A):Abstract A82.

621. Lognon P., Dabout C., Laugier J., Santini J.J. (1987). [Anaphylactic shock caused by vancomycin (letter).] Choc anaphylactique a la vancomycine (French). Presse Med 16(14):682.

622. Symons N.L., Hobbes A.F., Leaver H.K. (1985). Anaphylactoid reactions to vancomycin during anaesthesia: Two clinical reports. Can. Anaesth Soc. J. 32(2): 178–181.

623. Miller R., Tausk H.C. (1977). Anaphylactoid reaction to vancomycin during anaesthesia: A case report. Anesth. Analg. 56:870–872.

624. Crumpton M.W., Thornton J.B., Mackall L.L. (1983). Anaphylactoid reaction to vancomycin during general anesthesia in a child patient. Pediatr. Dent. 5(4): 276–279.

625. Roelofse J.A., Joubert J.J. (1989). Anaphylactoid reaction to vancomycin: Report of a case. J. Oral Maxillofac. Surg. 47(1):69–71.

626. Rothenburg M.J. (1959). Anaphylactoid reaction to vancomycin. JAMA 171:123–124.

627. Levy J.H. (1989). Anaphylactic/anaphylactoid reactions during cardiac surgery. J. Clin. Anesth. 1(6):426–430.

628. Mayhew J.F., Deutsch S. (1985). Cardiac arrest following administration of vancomycin. Can. Anaesth. Soc. J. 32(1):65–66.

629. Glicklich D., Figura I. (1984). Vancomycin and cardiac arrest (letter). Ann. Intern. Med. 101(6):880.

630. Dajee H., Laks H., Miller J., Oren R. (1984). Profound hypotension from rapid vancomycin administration during cardiac operation. J. Thor. Cardiovasc. Surg. 87: 145–146.

631. Wade T.P., Mueller G.L. (1986). Vancomycin and the "red-neck syndrome." Arch. Surg. 121:859–860.

632. Garrelts J.C., Smith D.F., Ast D., LaRocca J., Peterie J. (1988). Phlebitis associated with vancomycin therapy. Clin. Pharm. 7:720–721.

633. Sorrell T.C., Collignon P.J. (1985). A prospective study of adverse reactions associated with vancomycin therapy. J. Antimicrob. Chemother. 16:235–241.

634. Wang L., Liu C., Wang F., Fung C., Chiu Z., Cheng D. (1988). Chromatographically purified vancomycin: Therapy of serious infections caused by Staphylococcus aureus and other Gram positive bacteria. Clin. Ther. 10(5):1988.

635. Smith R. (1985). Extravasation of intravenous fluids. Br. J. Parenter. Ther. 6(2): 30–35.

636. Markman M., Lim H.W., Bluestein H.G. (1986). Vancomycin-induced vasculitis. South Med. J. 79:382–383.

637. Rawlinson W.D., George C.R.P. (1987). Vancomycin-induced vasculitis. Med. J. Aust. 147:470–471.

638. Hannah B.A., Kimmel P.L., Dosa S., Turner M.L. (1990). Vancomycin-induced toxic epidermal neurolysis. South Med. J. 83:720–722.

639. Vidal C., Quintela A.G., Fuente R. (1992). Toxic epidermal neurolysis due to vancomycin. Ann. Allergy 68:345–347.

640. McDonald B.J., Singer J.W., Bianco J.A. (1992). Toxic epidermal neurolysis possibly lined to aztreonam in bone marrow transplant patients. Ann. Pharmacother. 26:34–35.

641. Baden L.A., Apovian C., Imber M.J., Dover J.S. (1988). Vancomycin-induced linear IgA bullous dermatosis. Arch. Dermatol. 124:1186–1188.

642. Carpenter S., Berg D., Sidhu-Malik N., Hall R.P., Rico M.J. (1992). Vancomycin-associated linear IgA dermatosis. J. Am. Acad. Dermatol. 26:45–48.

643. Matthews S.J., Kephart P.A., Cersosimo R.J. (1987). Vancomycin-induced interstitial nephritis and Stephens-Johnson Syndrome. Twenty-second Annual ASHP Midyear Clinical Meeting (abstract):85D.

644. Packer J., Olshan A.R., Schwartz A.B. (1987). Prolonged allergic reaction to vancomycin in end-stage renal disease. Dial. Transplant. 16:86–88.

645. Gutfeld M.B., Reddy P.V., Morse G.D. (1988). Vancomycin-associated exfoliative dermatitis during continuous ambulatory peritoneal dialysis. Drug Intell. Clin. Pharm. 22:881–882.

646. Forrence E.A., Goldman M.P. (1990). Vancomycin-associated exfoliative dermatitis. DICP 24(4):369–371.

647. Neal D., Morton R., Bailie G.R., Waldek S. (1988). Exfoliative reaction to vancomycin. Br. Med. J. 296(6615).

648. Clayman M.D., Capaldo R.A. (1989). Vancomycin allergy presenting as fever of unknown origin. Arch. Intern. Med. 149:1425–1426.

649. Schlemmer B., Falkman H., Boudadja A., Jacob L., Le Gall J.R. (1988). Teicoplanin for patients allergic to vancomycin. N. Engl. J. Med. 318:1127–1128.

650. Wood G., Whitby M. (1989). Teicoplanin in patients who are allergic to vancomycin (letter). Med. J. Aust. 150:468.

651. Van Laethem Y., Goossens H., Cran S., Butzler J.P., Clumeck N. (1984). Teichomycin (Te) in severe methicillin resistant (Me/R) gram positive septicemia. Abstract presented at the 24th Interscience Conference on Antimicrobial Agents and Chemotherapy.

652. Smith S.R., Cheesbrough J.S., Makris M., Davies J.M. (1989). Teicoplanin administration in patients experiencing reactions to vancomycin. J. Antimicrob. Chemother. 23(5):810–812.

653. McElrath J.J., Goldberg D., Neu H.C. (1986). Allergic cross-reactivity of teicoplanin and vancomycin (letter). Lancet 1(8471):47.

654. Knudsen J.D., Pedersen M. (1992). IgE-mediated reaction to vancomycin and teicoplanin after treatment with vancomycin. Scand. J. Infect. Dis. 24:395–396.

655. Lerner A., Dwyer J.M. (1984). Desensitization to vancomycin (letter). Ann. Intern. Med. 100(1):157.

656. Lin R.Y. (1990). Desensitization in the management of vancomycin hypersensitivity. Arch. Intern. Med. 150(10):2197–2198.

657. Tange R.A., Kieviet H.L., von Marle J., Bagger-Sjoback D., Ring W. (1989). An experimental study of vancomycin-induced cochlear damage. Arch. Otorhinolaryngol. 246(2):67–70.

658. Brummett R.E. (1981). Effects of antibiotic diuretic interactions in the guinea pig model of ototoxicity. Rev. Infect. Dis. 3(Suppl.):S216–S223.

659. Lutz H., Lenarz T., Weidauer H., Federspil P., Hoth S. (1991). Ototoxicity of vancomycin: An experimental study in guinea pigs. J. Otorhinolaryngol. 53:273–278.
660. Brummett R.E., Fox K.E., Jacobs F., Kempton J.B., Stokes Z., Richmond A.B. (1990). Augmented gentamicin ototoxicity induced by vancomycin in guinea pigs. Arch. Otolaryngol. Head Neck Surg. 116(1):61–64.
661. Fauconneau B., De Lemos E., Pariat C., Bouquet S., Courtois P., Piriou A. (1992). Chrononephrotoxicity in rat of a vancomycin and gentamicin combination. Pharmacol. Toxicol. 71:31–36.
662. Brummett R.E., Fox K.E. (1989). Vancomycin- and erythromycin-induced hearing loss in humans. Antimicrob. Agents Chemother. 33(6):791–796.
663. Bailie G.R., Neal D. (1988). Vancomycin ototoxicity and nephrotoxicity. A review. Med. Toxicol. 3:376–386.
664. Traber P.G., Levine D.P. (1981). Vancomycin ototoxicity in patient with normal renal function. Ann. Intern. Med. 95(4):458–460.
665. Mellor J.A., Kingdom J., Cafferkey M., Keane C. (1984). Vancomycin ototoxicity in patients with normal renal function. Br. J. Audiol. 18(3):179–180.
666. Lackner T.E. (1984). Relationship of vancomycin concentrations and ototoxicity (letter). Arch. Intern. Med. 144(2):419.
667. Meyerhoff W.L., Maale G.E., Yellin W., Roland P.S. (1989). Audiologic threshold monitoring of patients receiving ototoxic drugs. Preliminary report. Ann. Otol. Rhinol. Laryngol. 98(12 Pt I):950–954.
668. Hall J.W., Herndon D.N., Gary L.B., Winkler J.B. (1986). Auditory brainstem response in young burn-wound patients treated with ototoxic drugs. Int. J. Pediatr. Otorhinolaryngol. 12(2):187–203.
669. Zuerlein T.J., Daily D.K., Kilbride H.W. (1991). Combination vancomycin-gentamicin (V-G) is associated with increased BSAER failure in <800 gram birthweight infants. Clin. Res. 39:721A.
670. Van der Hulst R.J., Boeschoten E.W., Nielsen F.W., Struijk D.G., Dreschler W.D., Tange R.A. (1991). Ototoxicity monitoring with ultra-high frequency audiometry in peritoneal dialysis patients treated with vancomycin or gentamicin. ORL. J. Oto-rhinolaryngol. 53:19–22.
671. Scharma S., Pramanik A.K., Diaz-Blanco J., Manno J.L., Otto W. (1988). Elevated serum vancomycin levels and ototoxicity in high-risk neonates (abstract 1644). Pediatr. Res. 23(4,part 2):476A.
672. Brummett R.E., Morrison R.B. (1990). The incidence of aminoglycoside antibiotic-induced hearing loss. Arch. Otolaryngol Head Neck Surg. 116:406–410.
673. Appel G.B., Given D.B., Levine L.R., Cooper G.L. (1986). Vancomycin and the kidney. Am. J. Kidney Dis. 8(2):75–80.
674. Geraci J.E. (1977). Vancomycin. Mayo Clin. Proc. 52:631–634.
675. Bergman M.M., Glew R.H., Ebert T.H. (1988). Acute interstitial nephritis associated with vancomycin therapy. Arch. Intern. Med. 148 (10):2139–2140.
676. Eisenberg E.S., Robbins N., Lenci M. (1981). Vancomycin and interstitial nephritis (letter). Ann. Intern. Med. 95(5):658.
677. Ratner S.J. (1988). Vancomycin-induced nephritis (letter). Am. J. Med. 84:541–542.

678. Codding C.E., Ramseyer L., Allon M., Pitha J., Rodriguez R. (1989). Tubulointerstitisal nephritis due to vancomycin. Am. J. Kidney Dis. 14(6):512–515.

679. Alexander M.R. (1974). A review of vancomycin after 15 years use. Drug Intell. Clin. Pharm. 8:520.

680. Waisbren B.A., et al. (1960). The comparative toxicity and clinical effectiveness of vancomycin, ristocetin, and kanamycin. In Antibiotics Annual 1960. Medical Encyclopedia. p. 497.

681. Rybak M.J., Albrecht L.M., Boike S.C., Chandresekar P.H. (1990). Nephrotoxicity of vancomycin, alone and with an aminoglycoside. J. Antimicrob. Chemother. 25: 679–687.

682. Mellor J.A., Kingdom J., Cafferkey M., Keane C.T. (1985). Vancomycin toxicity: A prospective study. J. Antimicrob. Chemother. 15:773–780.

683. Cohen L.G., Figge H.L., Simpson C., Rovers J., Souney P.F. (1988). Analysis of the frequency of nephrotoxicity in patients receiving vancomycin alone and in combination with aminoglycosides. ACCP Abstracts No. 34, p. 123.

684. Eng R.H.K., Wynn L., Smith S.M., Tecson-Tumang F. (1989). Effect of intravenous vancomycin on renal function. Chemotherapy 35:320–325.

685. Downs N.J., Neihart R.E., Dolezal J.M., Hodges G.R. (1989). Mild nephrotoxicity associated with vancomycin use. Arch. Intern. Med. 149:1777–1781.

686. Cimino M.A., Rotstein C., Slaughter R.L., Emrich L.J. (1987). Relationship of serum antibiotic concentrations to nephrotoxicity in cancer patients receiving concurrent aminoglycoside and vancomycin therapy. Am. J. Med. 83:1091–1097.

687. Wood C.A., Kohlhepp S.J., Kohnen P.W., Houghton D.C., Gilbert D.N. (1986). Vancomycin enhancement of experimental tobramycin nephrotoxicity. Antimicrob. Agents Chemother. 30(1):20–24.

688. Marre R., Schulz E., Hedtke D., Sack K. (1985). Influence of fosfomycin and tobramycin on vancomycin-induced nephrotoxicity. Infection 13(4):190–192.

689. Ngeleka M., Beauchamp D., Tardif D., Auclair P., Gourde P., Bergeron M.G. (1990). Endotoxin increases the nephrotoxic potential of gentamicin and vancomycin plus gentamicin. J. Infect. Dis. 161:721–727.

690. Beauchamp D., Pellerin M., Gourde P., Pettigrew M., Bergeron M.G. (1990). Effects of daptomycin and vancomycin on tobramycin nephrotoxicity in rats. Antimicrob. Agents Chemother. 34(1):139–147.

691. Rybak M.J., Boike S.C. (1983). Additive toxicity in patients receiving vancomycin and aminoglycosides. Clin. Pharm. 2(6):508.

692. Glew R.H., Parnk R.A., Hennick K.A. (1983). Vancomycin pharmacokinetics and toxicity. International Congress of Chemotherapy, Vienna (abstract).

693. Odio C., McCracken G.H., Nelson J.D. (1984). Nephrotoxicity associated with vancomycin-aminoglycoside therapy in four children. J. Pediatr. 105(3):491–493.

694. Pauly D.J., Musa D.M., Lestico B.S., Lindstrom M.J., Hetsko C.M. (1990). Risk of nephrotoxicity with combination vancomycin-aminoglycoside antibiotic therapy. Pharmacotherapy 10(6):378–382.

695. Jaresko G.S., Boucher B.A., Dole E.J., Tolly E.A., Fabian T.C. (1989). Risk of renal dysfunction in critically ill trauma patients receiving aminoglycosides. Clin. Pharm. 8:43–48.

696. Dean R.P., Wagner D.J., Tolpin M.E. (1985). Vancomycin/aminoglycoside nephrotoxicity (letter). J. Pediatr. 106:861–862.
697. Rybak M.J., Frankowski J.J., Edwards D.J., Albrecht L.M. (1987). White blood cell count and vancomycin (letter). Am. J. Med. 81(6):1114.
698. Goren M.P., Baker D.K., Shenep J.L. (1989). Vancomycin does not enhance amikacin-induced tubular nephrotoxicity in children. Pediatr. Infect. Dis. J. 8(5): 278–282.
699. Swinney V., Rudd C.C. (1987). Nephrotoxicity of vancomycin-gentamicin therapy in pediatric patients (letter). J. Pediatr. 110(3):497–498.
700. Nahata M.C. (1987). Lack of nephrotoxicity in pediatric patients receiving concurrent vancomycin and aminoglycoside therapy. Chemotherapy 33(4):302–304.
701. Piriano B., Bernardini J., Johnson J., Sorkin M. (1987). Chemical peritonitis due to intraperitoneal vancomycin (Vancoled). Peritoneal Dialysis Bull. 7:156–159.
702. Munro B. (1989). Vancomycin induced chemical peritonitis. Renal Educ. 9:10–12.
703. Charney D.I., Gouge S.F. (1991). Chemical peritonitis secondary to intraperitoneal vancomycin. Am. J. Kidney Dis. 17:76–79.
704. FDA letter Issued to Dialysis Centers, May 1990.
705. Newland L., Friedlander M., Tessman M. (1990). More experience with Vancoled induced chemical peritonitis. Peritoneal Dialysis Int. 10:182.
706. Dubot P., Cabanne J.F., Bidault C., Ramirez A. (1991). "Peritonites" aseptiques: A propos de cinq observations. Sem. Hop. Paris. 67(15):534.
707. Abel S.R. (1989). Lack of chemical peritonitis after intraperitoneal use of two brands of vancomycin hydrochloride. Clin. Pharm. 91–93.
708. Johnson C.A., Zimmerman S.W., Engeseth S., O'Brien M. (1989). Intraperitoneal Vancoled does not cause chemical peritonitis. Peritoneal Dialysis Int. 9(Suppl. 1): Abstract 86.
709. Smith T.A., Bailie G.R., Eisele G. (1991). Chemical peritonitis associated with intraperitoneal vancomycin. DICP Ann. Pharmacother. 25:602–603.
710. Freiman J.P., Graham D.J., Reed T.G., McGoodwin E.B. (1992). Chemical peritonitis following the intraperitoneal administration of vancomycin. Peritoneal Dialysis Int. 12:57–60.
711. Johnson C.A. (1991). Intraperitoneal vancomycin administration (editorial). Peritoneal Dialysis Int. 11:9–11.
712. Hecht J.R., Olinger E.J. (1989). Clostridium difficile colitis secondary to intravenous vancomycin. Dig. Dis. Sci. 34(1):148–149.
713. Miller S.N., Ringler R.P. (1987). Vancomycin induced pseudomembranous colitis (letter). J. Clin. Gastroenterol. 9(1):114–115.
714. Bingley P.J., Harding G.M. (1987). Clostridium difficile colitis following treatment with metronidazole and vancomycin. Postgrad. Med. J. 63:993–994.
715. Arning M., Gehrt A., Wolf M., Aul C., Chlebowski H., Hadding U., Schneider W. (1992). Lateler Verlauf einer pseudomembranosen Enterokolitis unter parenteraler Gabe von Vancomycin und Imipenem. Dtsch. Med. Wochenschr. 117:91–95.
716. Williams L., Domen R.E. (1989). Vancomycin-induced red cell aggregation. Transfusion 29:23–26.
717. Gilbert D.M., Domen R.E. (1989). Case report: ABO discrepancy due to vanco-

mycin complicating a transfusion reaction investigation. Immunohematology 5(4): 119–120.

718. Williams L., Domen R. (1987). Effect of vancomycin on red blood cell serologic testing. Abstract. Blood 70:116a.

719. Markowitz N., Saravolatz L.D. (1987). Use of trimethoprim-sulfamethoxazole in a glucose-6-phosphate dehydrogenase-deficient population. Rev. Infect. Dis. 9(Suppl. 2):218–229.

720. Farber B.F., Moellering R.C. (1983). Retrospective study of the toxicity of preparations of vancomycin from 1974–1981. Antimicrob. Agents Chemother. 23:138–141.

721. Dangerfield H.G., Hewitt W.I., Monzon O.T., Kudinoff Z., Blackman B., Finegold S.M. (1960). Clinical use of vancomycin. In Antimicrobial Agents Annual. Plenum Press, New York, pp. 428–437.

722. Neftel K., Blaser J., Koelliker F. (1988). Vancomycin induced neutropenia: Impact of duration of therapy and blood level monitoring. Abstract 569. Interscience Conference on Antimicrobial Agents and Chemotherapy.

723. Morris A., Ward C. (1991). High incidence of vancomycin-associated leucopenia and neutropenia in a cardiothoracic surgical unit. J. Infect. 22:217–223.

724. Koo K.B., Bachand R.I., Chow A.W. (1986). Vancomycin-induced neutropenia. Drug Intell. Clin. Pharm. 20:780–782.

725. Henry K., Steinberg I., Crossley K.B. (1986). Vancomycin-induced neutropenia during treatment of osteomyelitis in an outpatient. Drug Intell. Clin. Pharm. 20: 783–785.

726. West B.C. (1981). Vancomycin-induced neutropenia. South Med. J. 74(10):1255–1256.

727. Comer J.B., Goodwin R.A., Calamari L.A., Catrini V.J. (1992). Vancomycin-induced granulocytopenia in a home-care patient. Ann. Pharmacother. 26:563–564.

728. Millsteen S., Welik R., Heyman M.R. (1987). Case report: Prolonged vancomycin-associated neutropenia in a chronic hemodialysis patient. Am. J. Med. Sci. 294: 110–113.

729. Adrouny A., Meguerdichian S., Koo C.H., Gadallah M., Rasgon S., Idroos M., Oppenheimer E., Glowalla M. (1986). Agranulocytosis related to vancomycin therapy. Am. J. Med. 81:1059–1061.

730. Mordenti J., Ries C., Brooks G.F., Unadkat N., Tseng A. (1986). Vancomycin-induced neutropenia complicating bone marrow recovery in a patient with leukemia. Am. J. Med. 80:333–335.

731. Strinas R., Studlo J., Venezio F.R., O'Keefe J.P. (1982). Vancomycin-induced neutropenia (letter). J. Infect. Dis. 146:575.

732. Farwell A.P., Kendall L.G., Vakil R.D., Glew R.H. (1984). Delayed appearance of vancomycin-induced neutropenia in a patient with chronic renal failure. South Med. J. 77:664–665.

733. Kaufman C.A., Severance P.J., Silva J., Huard T.K. (1982). Neutropenia associated with vancomycin therapy. South. Med. J. 75:1131–1133.

734. Weitzman S.A., Stossel T.P., Desmond M. (1978). Drug induced immunological neutropenia. Lancet 1:1068–1072.

735. Roveix B., Leport C., Sirinelli A., Bourdarias J.P., Vilde J.L. (1986). Neutropenia induced by vancomycin: Immuno-allergic mechanism. Presse Med. 15:1732.
736. Domen R.E., Horowitz S. (1990). Vancomycin-induced neutropenia associated with anti-granulocyte antibodies. Immunohematology 6(2):41–51.
737. Mackett R.L., Quay D.R.P. (1985). Vancomycin-induced neutropenia. Can. Med. Assoc. J. 132:39–40.
738. De Bock R., Van Bockstaele D., Snoeck H., Lardon F., Peetermans M. (1992). The effect of vancomycin and teicoplanin on normal human bone marrow progenitor cells. J. Antimicrob. Chemother. 30:559–560.
739. Kitchen L.W., Clark R.A., Hanna B.H., Pollock B., Valainis G.T. (1990). Vancomycin and neutropenia in AZT-treated AIDS patients with staphylococcal infections. J. Acquir. Immune Defic. Syndr. 3:925–931.
740. Walker R.W., Heaton A. (1985). Thrombocytopenia due to vancomycin. Lancet 1:932.
741. Zenon G.J., Cadle R.M., Hamill R.J. (1991). Vancomycin-induced thrombocytopenia. Arch. Intern. Med. 151:995–996.
742. Christie D.J., van Buren N., Lennon S.S., Putnam J.L. (1990). Vancomycin-dependent antibodies associated with thrombocytopenia and refractoriness to platelet transfusion in patients with leukemia. Blood 75:518–523.
743. Angeran D.M., Dias V.C., Arom K.V., Northrup W.F., Kersten T.E., Lindsay W.G., Nicoloff D.M. (1984). The influence of prophylactic antibiotics on the warfarin anticoagulation response in the postoperative prosthetic cardiac valve patient: Cefamandole versus vancomycin. Ann. Surg. 199:107–111.
744. Angaran D.M., Dias V.C., Arom K.V., Northrum W.F., Kersten T.G., Lindsay W.G., Nicoloff D.M. (1984). The comparative influence of prophylactic antibiotics on prothrombin response to warfarin in the postoperative prosthetic cardiac valve patient. Cefamandole, cefazolin, vancomycin. Ann. Surg. 206(2):155–161.
745. Barg N.L., Supena R.B., Fekety R. (1986). Persistent staphylococcal bacteremia in an intravenous drug abuser. Antimicrob. Agents Chemother. 29(2):209–211.
746. Henrickson K.J., Powell K.R., Schwartz K.L. (1988). A dilute solution of vancomycin and heparin retains antibacterial and anticoagulant activities. J. Infect. Dis. 157(3):600–601.
747. Leibowitz G., Golan D., Jeshurun D., Brezis M. (1990). Mononeuritis multiplex associated with prolonged vancomycin treatment (letter). Br. Med. J. 1344.
748. Temperley D., Casey E., Connolly R., Fitzsimon S., Mulvihull E., Feely J. (1987). Vancomycin-associated lacrimation (letter). Lancet 2(8571):1337.
749. Huang K.C., Heise A., Shrader A.K., Tsueda K. (1990). Vancomycin enhances the neuromuscular blockade of vecuronium. Anesth. Analg. 71(2):194–196.
750. Conley N.S., Weiner R.S., Hiemenz J.W. (1991). Rigors with vancomycin. Ann. Intern. Med. 115:330.
751. Caglayan S., Ozdogru E., Aksit S., Kansoy S., Senturk H. (1992). Vancomycin-induced hypertension with transient blindness and generalized seizure. Acta Paediatr. Jpn. 34:90–91.
752. Teresi M., Allison J. (1985). Interaction between vancomycin and ticarcillin. Am. J. Hosp. Pharm. 42:2420–2421.

753. Pritts D., Hancock D. (1991). Incompatibility of ceftriaxone with vancomycin. Am. J. Hosp. Pharm. 48(1):77.

754. Seay R., Bostrom B. (1990). Apparent compatibility of methotrexate and vancomycin. Am. J. Hosp. Pharm. 47:2658–2659.

755. Fox A.S., Boyer K.M., Sweeney H.M. (1988). Antibiotic stability in a pediatric parenteral alimentation solution. J. Pediatr. 112(5):813–817.

756. Shilling C.G., Watson D.M., McCoy H.G., Uden D.L. (1989). Stability and delivery of vancomycin hydrochloride when admixed in a total parenteral nutrition solution. J. Parenter. Enteral Nutr. 13:63–64.

757. Nahata M.C. (1989). Stability of vancomycin hydrochloride in total parenteral nutrient solutions. Am. J. Hosp. Pharm. 46:2055–2057.

758. Yao J.D., Arkin C.F., Karchmer A.W. (1992). Vancomycin stability in heparin and total parenteral nutrition solutions: Novel approach to therapy of central venous catheter-related infections. J. Parenter. Enter. Nutr. 16(3):268–274.

759. Boeckh M., Lode H., Borner K., Hoffken G., Wagner J., Koeppe P. (1988). Pharmacokinetics and serum bactericidal activity of vancomycin alone and in combination with ceftazidime in healthy volunteers. Antimicrob. Agents Chemother. 32(1):92–95.

760. Taft R., Sohnius U., Meissner P.N. (1990). Safety of vancomycin in acute porphyria. Br. J. Clin. Pharmacol. 29(2):273–275.

761. Knothe H., Dette G.A. (1985). Antibiotics in pregnancy: Toxicity and teratogenicity. Med. Verlag 13:49–51.

762. Sirrat G.M., Beard R.W. (1973). Drugs to be avoided or given with caution in the second and third trimesters of pregnancy. Prescribers J. 13:135–140.

763. MacCulloch D. (1981). Vancomycin in pregnancy. N. Z. Med. J. 93:93–94.

764. Payne D.G., Fishburne J.I., Rufty A.J., Johnston F.R. (1982). Bacterial endocarditis in pregnancy. Obstet. Gynecol. 60:247–250.

765. Bourget P., Fernendez H., Delouis C., Ribou F. (1991). Transplacental passage of vancomycin during the second trimester of pregnancy. Obstet. Gynecol. 78: 908–911.

766. Reyes M., Ostrea E.M., Cabinian A.E., Schmitt C., Rintelmann W. (1989). Vancomycin during pregnancy: Does it cause hearing loss or nephrotoxicity in the infant. Am. J. Obstet. Gynecol. 161(4):977–981.

767. Gouyon J.B., Petion A.M. (1990). Toxicity of vancomycin given during pregnancy (letter; comment). Am. J. Obstet. Gynecol. 163(4 Pt. 1):1375–1376.

768. Salzman C., Weingold A.B., Simon G.L. (1987). Increased dose requirements of vancomycin in a pregnant patient with endocarditis (letter). J. Infect. Dis. 156(2):409.

769. Walczyk M.H., Hill D., Arai A., Wolfson M. (1988). Acute renal failure owing to inadvertent vancomycin overdose. Vancomycin removal by continuous arteriovenous hemofiltration. Ann. Clin. Lab. Sci. 18(6):440–443.

770. Burkhart K., Metcalf S., Shurnas E., O'Meara O., Brent J., et al. (1990). Exchange transfusion and multi-dose activated charcoal following vancomycin overdose. Vet. Hum. Toxicol. 32:353.

771. Burkhart K.K., Metcalf S., Shurnas E., O'Meara O., Brent J., Kulig K., Rumack

B.H. (1992). Exchange transfusion and multidose activated charcoal following vancomycin overdose. Clin. Toxicol. 30(2):285–294.

772. Davis R.L., Roon R., Koup J.R., Smith A.L. (1987). Effect of oral administered activated charcoal on vancomycin clearance. Antimicrob. Agents Chemother. 31(5): 720–722.

773. Hekster Y.A., Vree T.B., Weemaes C.M., Rotteveel J.J. (1986). Toxicologic and pharmacokinetic evaluation of a case of vancomycin intoxication during continuous ambulatory peritoneal dialysis. Pharm. Weekbl. |Sci.| 8(6):293–297.

774. Cantu T.G., Dick J.D., Elliott D.E., Humphrey R.L., Kornhauser D.M. (1990). Protein binding of vancomycin in a patient with immunoglobulin A myeloma. Antimicrob. Agents Chemother. 34(7):1459–1461.

775. Bryan C.S., White W.L. (1978). Safety of oral vancomycin in functionally anephric patients. Antimicrob. Agents Chemother. 14(4):634–635.

776. Spitzer P.G., Eliopoulos G.M. (1984). Systemic absorption of enteral vancomycin in a patient with pseudomembranous colitis. Ann. Intern. Med. 100:533.

777. McCullough J.M., Dielman D.G., Peery D. (1991). Oral vancomycin-induced rash: Case report and review of the literature. DICP Ann. Pharmacother. 25:1326–1328.

778. Pantosti A., Luzzi I., Cardines R., Gianfrilli P. (1985). Comparison of the in vitro activities of teicoplanin and vancomycin against *Clostridium difficile* and their interactions with cholestyramine. Antimicrob. Agents Chemother. 28(6):847–848.

779. Schifter S., Aagaard M.T., Jensen L.J. (1985). Adverse reactions to vancomycin. Lancet 2:499.

780. Kirby W.M., Perry D.M., Lane J.L. (1959). Present status of vancomycin therapy of staphylococcal and streptococcal infections. Antibiotics Annual 1958–1959. New York Medical Encyclopedia, Inc. pp. 587–594.

781. Guerit J.M., Mahiew P., Houben-Giurgea S., Herbay S. (1981). The influence of ototoxic drugs on brainstem auditory evoked potentials in man. Arch. Otorhinolaryngol. 233:189–199.

782. Gundmundsson G.H., Jensen I.J. (1989). Vancomycin and nephrotoxicity. Lancet 1:625.

783. Geraghty J., Feely M. (1984). Antibiotic prophylaxis in neurosurgery: A randomized controlled trial. J. Neurosurg. 60:724–726.

Epilogue

RAMAKRISHNAN NAGARAJAN

Lilly Research Laboratories, Eli Lilly and Company, Indianapolis, Indiana

The preceding nine chapters attempt to cover in one volume the present state of knowledge of vancomycin and related glycopeptide antibiotics. Because of the intense interest in this subject, several significant publications—especially in the areas of resistance development to these antibiotics, SAR, dimerization of these antibiotics in aqueous solution, and the relevance of these phenomena to biological activity and total synthesis—have appeared since the manuscript was submitted for this book. These publications are listed below and cover the period through the end of 1993.

1. Walsh C.T. (1993). Vancomycin resistance: decoding the molecular logic. Science 261:308–309.
2. Cohen M.L. (1992). Epidemiology of drug resistance: Implications for the post-antimicrobial era. Science 257:1050–1055.
3. Neu H.C. (1992). The crisis in antibiotic-resistance. Science 257:1064–1073.
4. Arthur M., Courvalin P. (1993). Genetics and mechanisms of glycopeptide resistance in entercocci. Antimicrob. Agents Chemother. 37:1563–1571.
5. Reynolds P.E. (1992). Modified peptidoglycan precursors produced by glycopeptide-resistant entercocci. FEMS Microbiol. Lett. 94:195–200.
6. Polk R.E., Israel D., Wang J., Venitz J., Miller J., Stotka J. (1993). Vancomycin skin tests and prediction of "red man syndrome" in healthy volunteers. Antimicrob. Agents Chemother 37:2139–2143.
7. Nagarajan R. (1993). Structure–activity relationships of vancomycin-type glycopeptide antibiotics. J. Antibiot. 46:1181–1195.
8. Gerhard U., Mackay J., Maplestone R.A., Williams D.H. (1993). The role of the sugar and chlorine substituents in the dimerization of vancomycin antibiotics. J. Am. Chem. Soc. 115:232–237.
9. Batta G., Sztaricskai F., Kover K.E., Rudel C., Berdnikova T.H. (1991). An nmr-study of eremomycin and its derivatives full H-1 and C-13 assignment, motional

behavior, dimerization and complexation with Ac-D-Ala-D-Ala. J. Antibiot. 44: 1208–1221.

10. Holroyd S.E., Groves P., Searle M.S., Gerhard U., Williams D.H. (1993). Rational design and binding of modified cell-wall peptides to vancomycin-group antibiotics: Factorizing free-energy contributions to binding. Tetrahedron. 49:9171–9182.

11. Rao A.V. (1993). Studies directed towards the total synthesis of vancomycin and related antibiotics. In Krohn F. (Ed.), Antibiotics and antiviral compounds: Chemical synthesis and modification. VCH Publications, pp. 263–277.

12. Evans D.A., Dinsmore C.J., Evrard D.A., DeVries K.M. (1993). Oxidative coupling of arylglycine-containing peptides. A biomimetic approach to the synthesis of the macrocylcic actinoidinic-containing vancomycin subunit. J. Am. Chem. Soc. 115: 6426–6427.

13. Shonekan D., Mildvan D., Handwerger S. (1992). Comparative in vitro activities of teicoplanin, daptomycin, ramoplanin, vancomycin and PD127,391 against blood isolates of Gram-positive cocci. Antimicrob. Agents Chemother. 36:1570–1572.

14. Cavalleri B., Parenti F. (1992). Glycopeptides (dalbahepides). Kirk-Othmer Encyclopedia of Chemical Technology, 4th ed., John Wiley & Sons. pp. 995–1018.

Index

T - #0156 - 101024 - C0 - 229/152/24 [26] - CB - 9780824791933 - Gloss Lamination